Fish Evolution and Systematics: Evidence from Spermatozoa.

Fish Evolution and Systematics: Evidence from Spermatozoa.

With a survey of lophophorate, echinoderm and protochordate sperm and an account of gamete cryopreservation.

Barrie G.M. Jamieson

Department of Zoology
University of Queensland
Brisbane, Australia

With contributions by L. K.-P. Leung

The right of the University of Cambridge to print and sell all manner of books was granted by Henry VIII in 1534. The University has printed and published continuously since 1584.

CAMBRIDGE UNIVERSITY PRESS

Cambridge

New York Port Chester Melbourne Sydney

Published by the Press Syndicate of the University of Cambridge
The Pitt Building, Trumpington Street, Cambridge CB2 1RP
40 West 20th Street, New York, NY 10011–4011, USA
10 Stamford Road, Oakleigh, Melbourne 3166, Australia

First published 1991

Printed in Great Britain at the University Press, Cambridge

British Library cataloguing in publication data available

Library of Congress cataloguing in publication data available

ISBN 0 521 41304 4 hardback

Dedicated to Björn Afzelius

Contents

Foreword

The relationship between the ultrastructure of spermatozoa and the evolution and phylogeny of fish taxa has received relatively little attention from systematists, despite their being much research on both fish spermatozoa and on fish systematics. Professor Barrie G.M. Jamieson, well-known for his work on oligochaetes and his book on insect spermatozoa, has provided us with a masterful synthesis of the literature and his own findings on fish spermatozoa with the voluminous recent work on the evolutionary history of fishes. He starts with invertebrate phyla related to the chordates and goes through the lower chordates and early fishes to the line leading to amphibians and to the highest teleosts. This book gives readers a review of fish systematics based on the classical evidence of gross morphology in a cladistic framework and a critical integration of this with information on the degree to which spermatozoa support or conflict with the various hypotheses of relationship. Of special interest to readers involved with aquaculture and genetic studies, Professor Jamieson is joined by Mr. Luke K.-P. Leung, who also contributes to some phylogenetic chapters, to give a comprehensive review of the principles of biological cryopreservation and on the live preservation of fish gametes.

Systematists seek new characters to support or refute current hypotheses of relationship. Although spermatozoa have been used in the phylogenetic studies of many invertebrate groups, ichthyologists have seldom used this penetrating character complex (in contrast to fishes, the ultrastructure of sperm is an integral part of the classification of some invertebrate groups). My introduction to this potential came in 1968 at a meeting of the American Society of Ichthyologists and Herpetologists. It was also at this meeting that some ichthyologists discussed the cladistic (phylogenetic) approach to systematics. Thankfully, we now have a synthesis of the two fields of fish spermatology and fish systematics. Professor Jamieson largely restricts his work to studies using ultrastructure, with little reference to the less reliable observations of the light microscope and is truly at the forefront in spermiocladistics.

The reader is treated to the wealth of diversity that exists in fish spermatozoa. Professor Jamieson identifies some 70 apomorphic (derived) character states in spermatozoan ultrastructure relative to a putative cephalochordate ancestor. For example, fish sperm may lack a flagellum or have up to two flagella, the acrosome may be present or absent; some other differences involve such structures as the nucleus, mitochondria, and the centrioles. Particularly impressive are the number of apomorphies, revealed largely by Drs. X. and C. Mattei, showing the monophyly of the elopomorphs, comprising the elopiforms, notacanthiforms, and anguilliforms. Many basic problems in fish systematics are discussed in view of these apomorphic characteristics. The reader is, for example, shown the extent to which fish sperm may help us with arguments of relationship involving sharks, *Latimeria*, lungfish, and tetrapods and involving esocids, ostariophysans, and salmonids. Professor Jamieson makes several original contributions to our classification such as formally recognising the order Esociformes. He explores many challenging ideas, such as the possibility that ancestral agnathans were internally fertilizing, and that the so-called "primitive" sperm is a secondary simplification. He identifies areas of conflict between spermiocladistics and classical systematics. The identification of groups for which we lack information on sperm ultrastructure is also helpful, especially in cases where such knowledge might be useful in resolving problems of relationship using classical characters or where incomplete information teases the curiosity about some exciting new hypothesis of relationship.

We are increasingly seeking to understand the functional morphology of the characters we use in systematics, and the differences that Professor Jamieson documents draws our attention to problems in trying to understand the selective advantage for the observed diversity (for instance, why should mormyroid sperm be aflagellate in contrast to that of notopteroids, and why should apogonids be biflagellate but other known percoids be uniflagellate?)

Ichthyologists over the last 25 years have made many advances in our understanding of the evolutionary history of fishes. We have recently developed a better appreciation for the contributions that can be made to fish systematics from larval characters. Now we have no excuse for not employing spermatozoa. With a high degree of specialization in biologists, it is highly welcomed to have a synthesis of disciplines that has a strong potential for helpful interaction but where few individuals share a working knowledge of the two areas.

Professor Jamieson is to be thanked for providing us with a detailed synthesis and critical evaluation of the literature interfacing fish sperm ultrastructure and fish systematics and, with Mr. Leung, for providing much new information. Ichthyologists will wish to interpret their findings in light of this information. In addition, spermatologists, in having a guide to their work and a guide to fish systematics, will be able to focus their research on basic questions and relate their work to phylogenetic studies. No doubt, as a result of this book, there will be an increased effort on comparative work that will allow us to determine the extent to which the ultrastructure of fish spermatozoa can be used as an independent suite of systematic characters.

Department of Zoology, University of Alberta, Edmonton, Alberta. November 1990

Joseph S. Nelson

Preface

For brevity, the title of this book is 'Fish Evolution and Systematics: Evidence from Spermatozoa' but its subject is, in fact, the spermatozoa of the protochordates (hemichordates, urochordates and cephalochordates) in addition to the artificial (paraphyletic) group we term fish. Some comparison with the spermatozoa of the echinoderms and Lophophorata (Ectoprocta, Brachiopoda and Phoronida) is also made and two chapters are included on cryopreservation of gametes. The primary aim of the work has been to make available in a single volume a critical resumé of all research reports on the ultrastructure of fish sperm, the literature of which is large and scattered. Treatment of cryopreservation is limited to fish gametes. New data of the author are added and contributions by Mr. Luke K.-P. Leung, chiefly from his Honours thesis, University of Queenland (1987) are included.

To keep the size of the volume within acceptable limits and because light microscope observations have often proved incorrect on ultrastructural examination, only ultrastructure and only that of the mature spermatozoon is considered, with few exceptions. Phylogenetic studies are critically discussed but this work does not purport to review the vast literature on fish phylogeny, rather works particularly relevant to the findings from spermatology are included.

The use of spermatozoal ultrastructure for taxonomy and phylogeny is now widely accepted as it has proved effective in resolving hitherto intractable problems of taxonomic placement and relationship. No better endorsement of its validity could be given than the recent ratification from molecular biology by Abele *et al.* (1989) of placement of the Pentastomida in the Crustacea from consideration of spermatozoal ultrastructure by Wingstrand (1972). The terms **spermiocladistics** and **spermiotaxonomy** coined (Jamieson, 1987a) for phylogenetic and taxonomic applications of sperm ultrastructure are gaining increasing currency. Modern phylogenetic analysis or cladistics is based on the principles of phylogenetic systematics developed by Hennig (1966) which is predicated on the requirement that relationship between taxa be recognized on the basis of shared *advanced* (i.e. derived) characters. An advanced character is termed an **apomorphy** and an apomorphy shared between two or more taxa is a **synapomorphy**. Each taxonomic or phylogenetic entity is often definable by at least one distinctive apomorphy, an **autapomorphy**. A corollary of determination of relationship from synapomorphies is that relationship is not recognized on the basis of shared unmodified (i.e. primitive) characters. A primitive character is termed a **plesiomorphy** and when shared with another entity is termed a **symplesiomorphy**.

Examples of works which formally employed, and acknowledged their debt to, the methodology established by Willi Hennig are increasingly numerous and some are mentioned here. The first was that of Schmidt and Zissler (1979) for anthozoan coelenterates, a particularly interesting example as it showed that even 'primitive' sperm (aquasperm in the author's terminology) have apomorphies distinctive of their various groups, a fact to which Afzelius (1979) had independently drawn our attention and which Jamieson and Rouse (1989) have shown for polychaete sperm. The author (Jamieson, 1983), working on oligochaete sperm, related sperm structure to a holomorphological phylogram, later (Jamieson, 1984) used numerical, computer procedures to model Hennig's methods, and recently (Jamieson, 1987b; Jamieson *et al.* 1987) employed the excellent program of Swofford (1984) for phylogenetic analysis using parsimony. This Wagner program attains one of the chief aims of Hennigian phylogeny in minimizing homoplasy (homoiology or parallelism). The term **symparamorphy** has been introduced (Jamieson, 1984, main references), as a theoretical concept at least, for a homoplasy which is considered to be underlain by a common genetical basis. It is envisaged that a symparamorphic state, such as possession of a flagellar rootlet which appears homoplasic, may be turned on or off by the action of a regulator gene or genes on a DNA sequence shared by members of a clade (in the sense of a ramus) but not always expressed.

A further example of use of computer techniques for spermiocladistics has been that of Harding *et al.* (1986) for the sperm of marsupials. Use of sperm ultrastructure reached its ultimate expression, appropriately in the school of Ax who has done so much to establish the value of Hennigian theory, with the definition and naming of a taxonomic group (and a major one), in an otherwise somatic analysis, on the basis of a spermatozoal apomorphy: recognition of the new taxon Trepaxonemata for platyhelminthes from the Polycladida through Cestoda on the autapomorphy of a single central element in the spermatozoal axoneme by Ehlers (1984), a work enriched by the work on platyhelminth sperm ultrastructure of Hendelberg (see Hendelberg, 1986). Similarly, in the present work a new name is given to the Ascidiacea-Thaliacea assemblage, in the tunicates, on the basis of shared sperm ultrastructure. Numerous valuable phylogenetic papers have been published by Justine and Mattei on monogeneans and by Swiderski on cestodes.

Spermiocladistics, with preparation of formal cladograms, has been painstakingly utilized by Wirth (1984), for a number of groups, of which his analysis of chelicerate phylogeny is particularly impressive; Alberti's meticulous studies of chelicerate sperm have laid the foundation for this and further cladistic analyses. In the Crustacea, Grygier (1982) has made a spermiocladistic analysis of flagellated, maxillopod sperm; Pochon-Masson and colleagues, Hinsch and others too numerous to mention have informally shown the value of sperm ultrastructure for phylogeny; and Wingstrand (1978) has illustrated the phylogeny of sperm ultrastructure in the Branchiopoda, though without reference to Hennigian methods. Franzén (1977a) has related sperm structure to a conventional phylogeny of the Metazoa and has pioneered and continues to examine the use of sperm in phylogeny in a long and exemplary series of works. The author (Jamieson, 1985) has made a prelimnary spermiocladistic study of echinoderms. The taxonomy of gastropod molluscs is undergoing confirmation and emendment as a result of the far-reaching investigations by Healy of sperm ultrastructure though he has yet to apply formal cladistics. In the Nematoda, Anya (1976) and later, Baccetti and colleagues (1983), at Siena, have attempted spermatozoal phylogenies. Recently Holland (e.g. 1989) has applied sperm ultrastructure to the phylogeny of the Urochordata. The most prodigious investigations of sperm ultrastructure have been those of Baccetti and his team, chiefly on Uniramia, studies which they have frequently directed to development of phylogenies, again, however, not in an Hennigian framework. The author (Jamieson, 1987a) reviewed these and other insectan studies and analyzed the congruence of spermatozoal characters, in Hennigian terms, to a pre-exisisting, chiefly somatic cladograms. The value of holomorphological cladistics, sampling the entire anatomy of the organism, was upheld but it was shown that spermatozoa offered an 'independent arbiter' capable in many cases of resolving contentious taxonomic and phylogenetic problems.

The value of sperm structure for phylogeny and taxonomy is particularly apparent in groups which have complex sperm offering many synapomorphies and in which radiation to give numerous taxa from one or more complex ground plans has occurred, with an apparent minimum of parallelism (homoplasy), as in the Mollusca and Crustacea and in euclitellate annelids. In fish acquisition of internal fertilization has apparently occurred independently several times with much homoplasic evolution of similar sperm characters and the resulting sperm may tell little of the phylogenetic affinities of their bearers. Nevertheless, sperm ultrastructure has contributed strikingly to resolution of phylognetic and systematic problems in some fish groups, the most noteworthy being the demonstration of anguilliform-elopid relatioships (Mattei and Mattei, 1974). Leaving aside phylogenetic relationships, sperm ultrastructure and its diversity in fish is an absorbing topic. It has also illuminated work in this an other laboratories on cryopreservation of sperm.

The present work relates sperm ultrastructure in fish to pre-existing, holomorphological phylogenies of which those of Lauder and Liem (1983), based on the firm foundations laid by Greenwood and his colleagues, have particularly been utilized. Strict parsimony analysis has been restricted here to a prelimnary analysis of mixed spermatological and somatic data for Gnathostomata as a whole, with particular focus on *Latimeria*. Owing to the great amount of homoplasy in fish sperm, especialy in relation to internal fertilization, the relative simplicity of the most common type of teleostean sperm, and, above all, the gaps in our knowledge of sperm ultrastructure of many fish taxa, it has not yet been found practicable to present a parsimony analysis of the various orders and their interrelationships. Nevertheless, some groups, such as the Elopomorpha, could be effectively subjected to parsimony spermiocladistics. In these cases "intuitive"study of their sperm has been sufficient to indicate the broad relationships in the present work.

This book could not have been written without the generous co-operation of many fellow research-workers who have donated reprints and illustrations. They are too numerous to all be mentioned individually but where their work has been cited they are fully acknowledged in the text.

Colleagues whose micrographs or line drawings are reproduced are: L.C. Abbott, B.A. Afzelius, P.E. Ahlberg, K. Ahlfors, A. Alldredge, D.T. Anderson, J. Anthony, J.W. Atz, B. Baccetti, R.S. Barnes, N.J. Berrill, R. Billard, D. Bogoraze, L. Boissin, S. Brusle, F. Buckland, P.Burighel, R.D. Butler, M.T. Casas, G.M. Cherr, J.-L. Chevalier, S.H. Chuang, W.H. Clark, R.A. Cloney, B.B.Collette, A.L. and L.H. Colwin, P. F. Cotelli, M.A. R. Dallai, L.F. De Beaufort, B. van Deurs, R. Fenaux, M. Ferraguti, W.L. Fink, P.R. Flood, E. Follenius, M. Folliot, P.L. Forey, A. Franzén, L.R. Fuiman, M. Fukumoto, D.M. Gardiner, B.H. Gibbons, I.R. Gibbons, G. Gorsky, P.-P. Grassé, P.H. Greenwood, H.J. Grier, Guha, J.M. Healy, W. Herdman, L. Holland, D.S. Jordan, Å. Jespersen, P.R. Jones, P. Kott, F. Lafargue, U. Lastein, G.V. Lauder, S.M. Lester, L. K.-P. Leung, K.F. Liem, G.U. Lindberg, F. Magri, G.B.Martinucci, K. Matsubara, X. and C. Mattei, G.E. McGowen, S. Mito, J. Millot, S. Munoz-Guerra,J.S. Nelson, L. Nicander, N. Nicholson, J.R. Norman, G. Northcutt, N.V. Parin, T.S. Parsons,G.R. Poirier, L. Pratsch,B.J. Pusey, C.S. Rand, C. Reizer, M. Renieri, A.S. Romer, F. Rosati, G.W. Rouse, F.W.E. Rowe, B. Schaeffer, K. Selman, T. Sensenbaugh, B. Seret, Y. Siau, C.L. Smith,D.G. Smith, H.P.Stanley, H. Stein, G. Sterba, V. Storch, J.A. Subirana, P.R. Todd, O. Tuzet, R.B. Wallace, M. Weber, J.H. Wickstead, E.O. Wiley, K.H.Woodwick, R.M. Woolacott, R. Yanagimachi, J.Z. Young, B.R. Zirkin.

The following publishers gave permission for reproduction of the illustrations. These are listed here with the titles of journals or books from which they originated: Academic Press, Inc: Baccetti, *Comparative Spermatology*, 5.2, 5.3, 6.5, 7.4, 8.7, 10.13, 11.12, 11.16, 12.16, 14.2, 14.13, 15.3, 15.5, 15.12, 15.22, 15.24, 16.4, 16.10, 16.11, 16.18, 16.19, 17.12; Giese and Pearse, *Reproduction of Marine Invertebrates II (Ectoprocta and Lesser Deuterostomes*), 4.3; *Cryobiology*, 19.2, 20.3; *Journal of Fish Biology*, 16.7; *Journal of Ultrastructure Research/ Journal of Ultrastructure and Molecular Structure Research*, 1.4, 3.11, 3.12, 6.8, 6.9, 6.10, 7.4, 7.6, 7.7, 9.4, 9.11, 10.3, 10.4, 10.5, 10.11, 11.15, 12.14, 13.8, 15.11, 15.20, 15.21, 16.15, 16.16; *Journal of the Linnean Society of London (Zoology)/Zoological Journal of the Linnean Society*, 8.1, 8.4, 10.7. Académie des Sciences, Paris: *Comptes Rendus Hebdomadaires des Séances de l'Académie des Sciences. Paris*, 10.16, 17.7. Alan R. Liss. Inc: *Gamete Research* 12.7, 13.11, 17.15, 17.16; *Journal of Morphology*, 3.4; American Society of Ichthyologists and Herpetologists: 10.8, 12.5, 13.2, 17.9. American Association for the Advancement of Science: *Science*, 9.7. American Society of Ichthyologists:*Copeia*, 12.8. American Society of Zoologists: *American Zoologist*, 17.4. Australia New Guinea Fishes Association: 13.10. Australian Coral Reef Society: 3.5. Balaban Publishers: *International Journal of Invertebrate Reproduction and Development*, 3.12. Brill, Leiden: Weber and De Beaufort. *The Fishes of the Indo-Australian Archipelago*, 17.14. Bulletin Biologique: *Bulletin de Biologie de France et de Belgique*, 3.12. Cambridge University Press: 1.1. Clarendon Press, Oxford: 2.1, 2.2, 2.7, 4.1, 4.2, 6.7. Company of Biologists Ltd.: *Journal of Cell Science*, 3.8, 3.9. Duncker und Humblot: *Zeitschrift für Angewandte Zoologie*, 12.3, 13.6. Editions Scientifiques Elsevier: *Biologie Cellulaire/Biology of the Cell*, 14.8;14.10. Editrice Compositori Bologna: *Journal of Submicroscopic Cytology*, 1.2, 11.6, 20.1. Electron Microscope Society of America: 16.5. Elsevier Science Publishers: *Biochimica et Biophysica Acta*, 19.9. Henry Holt and Company: 7.2, 10.6, 11.8, 11.14, 12.2, 12.13, 13.4, 13.5, 15.6, 15.10, 16.6, 17.5. International Standing Committee on Physiology and Pathology of Animal Reproduction: 6.6. Institut Fondamental d'Afrique Noire, Université Cheikh Anta Diop de Dakar: 9.9, 9.10.Institut National de la Recherche Agronomique: *Annales de Biologie Animale, Biochemie, Biophysique*, 11.9. IRL Press Ltd: *Human Reproduction*, 19.6. Japanese Society of Developmental Biologists: 3.12. John Wiley and Sons Ltd.: 1.4, 3.3, 9.2, 9.5, 12.4, 13.3, 14.1, 14.4, 14.7, 14.9, 14.11, 14.12, 15.4, 15.19, 16.2, 16.14, 17.1, 17.10. Masson et Cie., Paris: 9.3, 10.10, 10.15, 11.7, 12.6, 12.15, 15.2, 15.7, 16.3. Museum of Comparative Zoology, Harvard University: 8.5, 11.1, 11.13, 12.1, 12.4, 14.3, 15.1,

16.1, 16.17, 17.1. Pergamon Press: Afzelius *The Functional Anatomy of the Spermatozoon*, 11.2; *Acta Zoologica (Stockholm)*, 2.8, 6.3, 6.4; *Zoologica Scripta*, 1.5, 1.7, 4.4. Phi Sigma Biological Honor Society: *Biologist*, 9.1. Plenum Press: *Journal of in Vitro Fertilization and Embryo Transfer*, 19.7. Ray Society (London): Berrill, *The Tunicata*, 3.15. The Royal Society of London: *Proceedings of the Royal Society of London B*, 7.1, 9.6. Société Français de Microscope Electronique: *Journal de Microscopie et Biologie Cellulaire*, 10.14, 14.5, 14.6. Scientific American Inc: *Scientific American*, 3.16. Springer Verlag: *Cell and Tissue Research*, 3.6, 3.12, 3.17, 11.10, 13.7, 15.23, 17.11; *Marine Biology*, 3.14; *Zeitschrift für Zellforschung*, 2.4, 11.3, 11.4, 11.5; *Zoomorphologie/Zoomorphology*, 2.5, 2.6, 3.19. The Rockefeller University Press: *Journal of Cell Biology*, 2.3. Trustees of the British Museum: 3.1, 8.8, 9.8, 10.2, 11.11. Trustees of the Queensland Museum: 3.7, 3.10. W.B. Saunders Company, Barnes, *Invertebrate Zoology*, 3.13; Romer and Parsons, *The Vertebrate Body*, 6.1, 7.2, 8.6. Wistar Institute Press: *Journal of Experimental Zoology* 8.9; *Journal of Experimental Zoology Supplement* 7.3. Vista Books: 10.12, 12.9, 12.17. Zoological Society of Japan: *Zoological Magazine*, 5.4, 5.5.

Special thanks are due to Drs. X. and C. Mattei without whose many works on fish sperm this book would be greatly diminished. I am grateful, too, to those who read and commented on part or the whole of the typescript: Dr. Patricia Kott who freely gave of her time and valuable advice in greatly improving the chapter on urochordates; and my colleagues Dr. John Healy and Mr. Christopher Tudge. Professor J.D. Pettigrew gave advice and encouragement throughout. Dr. J.C. Cummins gave much practical help and advice in our early cryopreservation work. Mr. C. Marshall kindly allowed access to his unpublished Honours thesis on atheriniform and galaxiid and sperm and his review, here augmented, of literature on the micropyle.

Mrs. Lina Daddow, Zoology Department histologist, has given invaluable technical assistance in my electron microscope work and that of my students. Mr. C. Tudge, research assistant, supported by Australian Research Grants Scheme funding, and Dr. John Healy, University of Queensland Post-doctoral Fellow gave valued technical assistance. The many figures redrawn from the literature were prepared by Christopher Tudge. Other original drawings and all sperm phylograms were prepared by the author. Additional assistance in library work was provided by Mr. John Kennedy who also prepared the photographic plates.

The generous financial support of the Australian Research Council was essential for the production of this book. Work could not have proceeded without the unfailing help of Miss Mary O'Sullivan and the staff of the Biological Sciences Library, University of Queensland. The University of Queensland, and particularly Professors Clifford Hawkins and Gordon Grigg, are thanked for providing facilities and support and for maintaining an environment for scholarly research. The understanding and support of my wife and family in the many weeks which I devoted to my word processor made the tasks of preparation much easier and absolved me from those less academic duties which I neglected. The confidence of my publishers in accepting and seeing through the press this offering is gratefully acknowledged. Any errors of typography or citation are my own.

University of Queensland, Brisbane, May 1990 B.G.M. Jamieson

References cited in the Preface

(Not included in the final bibliography)

Abele, L.G., Kim, W. and Felgenhauer, B.E. (1989). Molecular evidence for inclusion of the phylumPentastomida in the Crustacea. *Molecular Biology and Evolution* **6**, 685-691.

Jamieson, B.G.M. (1984). A phenetic and cladistic study of spermatozoal ultrastructure in the Oligochaeta (Annelida). *Hydrobiologia* **115**, 3-13.

Wingstrand, K.G. (1972). Comparative spermatology of a pentastomid, *Raillietiella hemidactyli* and a branchiuran crustacean, *Argulus foliaceus*, with a discussion of pentastomid relationships. *Kongelige Danske Videnskabernes Selskab Biologiske Skrifter* **19**, 1-72.

Wingstrand, K.G. (1978). Comparative spermatology of the Crustacea Entomostraca I. Subclass Branchiopoda. *Kongelige Danske Videnskabernes Selskab Biologiske Skrifter* **22**, 1-66.

Wirth, U. (1984). Die Struktur der Metazoen-Spermien und ihre Bedeutung für die Phylogenetik. *Verhandlungen des Naturwissenschaftlichen Vereins in Hamburg* **27**, 295-362.

Chapter 1

INTRODUCTION LOPHOPHORATES AND ECHINODERMS

Types of animal sperm

Animal sperm types have been recently defined by the mode of transmission and fertilization with no primary relation to morphology or presumed phylogeny. Primitive sperm *sensu* Franzén have been termed 'aquatic sperm' by Baccetti (e.g. 1979) or 'aquasperm' by Jamieson (1986a, b). Two chief types of aquasperm are recognized: ect-aquasperm and ent-aquasperm. Both are freed into the ambient water but whereas ectaquasperm fertilize externally, entaquasperm are drawn into the female via the

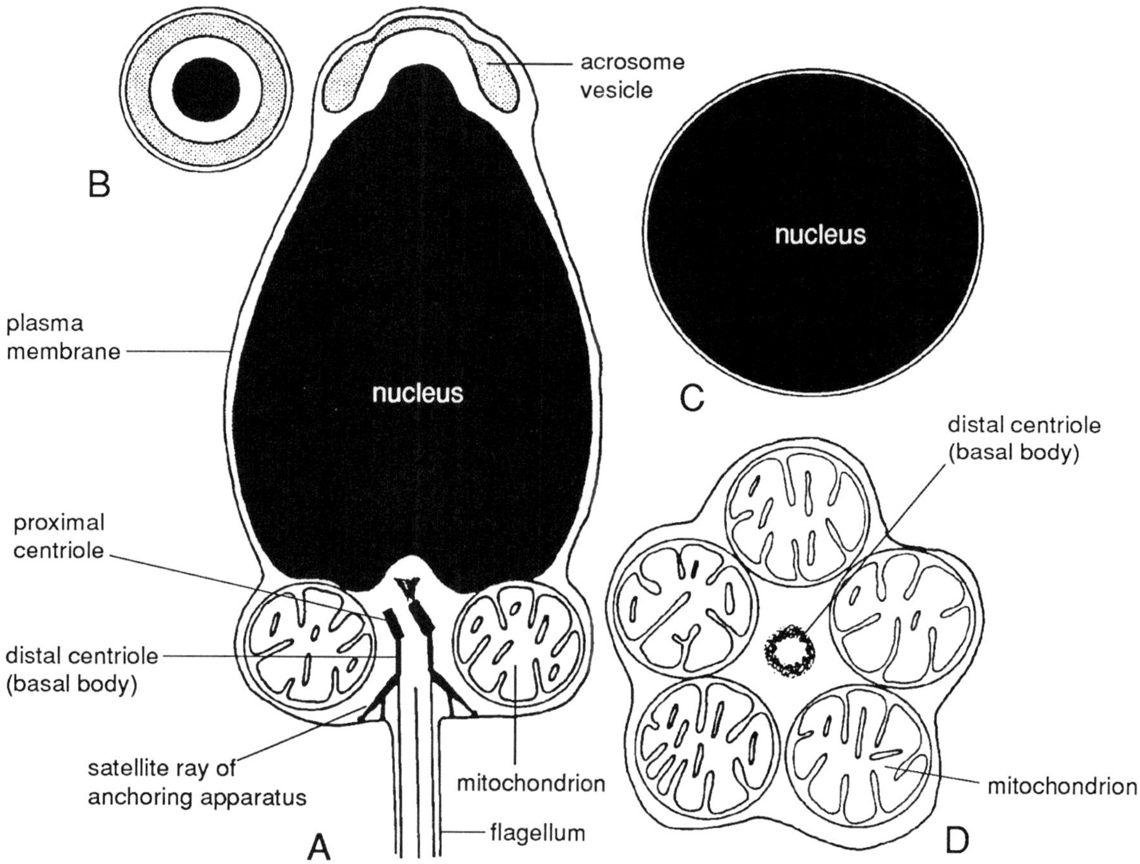

Fig. 1.1. Basic type of ect-aquasperm, approaching the morphology of the hypothetical plesiosperm. Exemplified by the polychaete *Lepidonotus carinulatus*. In the plesiosperm the proximal centriole would be perpendicular to the basal body. A. Longitudinal section of sperm. B. Transverse section (TS) through acrosome and tip of nucleus. C. TS nucleus. D. TS midpiece. After Jamieson B.G.M. (1987). *The Ultrastructure and Phylogeny of Insect Spermatozoa*. Cambridge University Press. Fig. 1.3. Courtesy of G.W. Rouse.

inhalant current, or otherwise enter it, and fertilize internally (Rouse and Jamieson, 1987; Jamieson, 1987). Aquasperm contrast with internally fertilizing sperm which do not enter the water, for which these authors have proposed the term introsperm. These terms define sperm with regard to fertilization biology and complement but do not replace terms based on spermatozoal morphology such as 'modified' and 'aberrant' sperm of Franzén (1977a) or 'eupyrene', 'oligopyrene' and 'apyrene' sperm of Meves (1903). Some widely used terms must, however, be rejected. Thus the terms 'typical' and 'atypical' can no longer usefully be retained for sperm as they have been used not only for alternative sperm types in animals in general, despite the morphological and phylogenetic heterogeneity of each type, but also for intraspecific dimorphism (see Healy and Jamieson, 1981, who introduced the terms eusperm and parasperm for such dimorphism).

Either of the chief sperm types (aquasperm or introsperm) may be acrosomal (possessing an acrosome) or anacrosomal (lacking an acrosome). An acrosome, present in most metazoans, is sporadically, and secondarily, absent in deuterostomes and their lophophorate relatives having been lost in phoronids and in the vast majority of neopterygian fish.

Ect-aquasperm usually correspond with the basic primitive sperm of Retzius (1904), Franzén (1956, 1970) and Afzelius (1972). This type is characterized, in 'anterior-posterior' sequence, by a caplike acrosome (sometimes absent, as in Cnidaria and the possibly secondary aquasperm of teleosts), a subspherical nucleus, a small number of rounded, cristate mitochondria and a free axoneme with the 9+2 arrangement of microtubules. In its most basic form (Fig. 1.1) two centrioles are retained, the distal (posterior) of which forms the basal body of the axoneme, and a satellite- or anchoring-apparatus arises from this centriole. The hypothetical, idealized form, or ground plan, often closely approached in reality, has been termed the plesiosperm (Jamieson, 1987b). The term plesiosperm recognizes that, whether or not aquasperm have evolved in some sections of the Metazoa from more complex sperm, the "primitive" sperm facies may be genuinely plesiomorphic in many metazoan groups. This is not to exclude from further investigation the possibility (a) that the earliest metazoans may have been free-spawning with sperm of an even simpler form than the plesiosperm or (b) that they were internally fertilizing with more or less complex sperm. In the present work the term plesiosperm will be used to denote spermatozoal morphology agreeing with that of the "primitive" sperm with no necessary connotation of its being primitive.

Ect-aquasperm of more complex structure occur in urochordates in which, like their ent-aquasperm, the nucleus is elongate and mitochondria are located laterally to it; these modifications appear to be related to complication of the egg envelopes. In urochordates the sperm of the same morphological type may be drawn into the egg-bearing individuals and bring about internal fertilization, thereby constituting ent-aquasperm and in some species internal fertilization is facultative so that the same sperm morphology exists in ect-aquasperm and ent-aquasperm within the same species. Highly "modified" or complex externally fertilizing sperm occur in the Agnatha (e.g. *Eptatretus*, Myxinidae, Jespersen, 1975, and *Lampetra*, Petromyzontidae, Nicander and Sjödén, 1971) and in a few teleosts (e.g. the elopomorph, *Albula vulpes*, Mattei and Mattei, 1973; the lophiiform *Neoceratias*, Jespersen, 1984) in all of which it appears possible, though unlikely, that external fertilization has been secondarily acquired.

To gain a fuller understanding of the structure and evolution of the sperm of protochordates and fish, brief reference will now be made to the sperm of the oligomerous phyla: lophophorates and lower deuterostomes, including the echinoderms.

SUPERPHYLUM LOPHOPHORATA

The oligomerous phyla are animals with three coelomic cavities in longitudinal sequence, the proto-, meso- and meta- coel, corresponding with three body divisions, the proto-, meso- and metasome. Animals with this basic organization, though it may be obscured in the adult, range from lophophorates and echinoderms to hemichordates. The metacoel forms the main body cavity.

The oligomerous phyla include forms which are undoubted deuterostomes and those which are intermediate in some features between protostomes and deuterostomes. Protostomes are typically animals in which the blastopore becomes the mouth, cleavage of the embryo is spiral and determinate; the mesoderm derives from a single cell, the 4d cell; and a coelom, if present, is schizocoelic, produced by splitting of the mesoderm. Major protostome phyla are the Platyhelminthes, Annelida, Mollusca and Arthropoda. Deuterostomes may be defined as animals in which, basically, the blastopore does not form the mouth, the anus typically arising at its site; cleavage is radial and indeterminate; and the mesoderm and enclosed coelom arise enterocoelically, by budding from the archenteron.

Three oligomerous phyla, Phoronida, Ectoprocta and Brachiopoda, comprising the superphylum Lophophorata of Valentine (1973, 1977) are structurally and developmentally on the border line between protostomes and deuterostomes, hence the extensive debate as to which group they belong or on the validity of recognizing these two groups (Clark, 1964; Jägersten, 1972; Zimmer, 1973; Emig, 1977a,b; Farmer, 1977; Ishikawa, 1977; Løvtrup, 1977; Nielsen, 1977).

Phoronida. Despite the commonly accepted view that the Phoronida are the most plesiomorph lophophorates, they have highly modified sperm which show resemblances to those of the concentricycloid echinoderm *Xyloplax*, to the pterobranch hemichordate *Rhabdopleura* and less obviously to those of some ectoprocts (Fig. 1. 2).

The sperm of *Phoronis pallida* (Franzén and Ahlfors, 1980) (Fig. 1.2) is highly modified and consists of two limbs arranged as a narrow inverted V from the tip of which the acrosome protrudes. One limb is formed by the flagellum, emerging from a centriolar region. The other limb contains the nucleus, the hind-part of the acrosome, a "pseudo-acrosome" and two mitochondria. The anterior part of the elongated, somewhat flattened and concave nucleus is modified as the pseudo-acrosome. The nucleus is kidney-shaped in cross section, with many dense granules in the concavity. The pseudo-acrosome seems to be composed of two components: a bandlike nuclear process and a parallel ribbon-like structure of condensed material. It seems probable to the writer that at least the second band is a centriolar rootlet (see also *Xyloplax*). At the posterior end of the nucleus the two flattened mitochondria are attached as a ribbon-shaped structure. They appear to form by

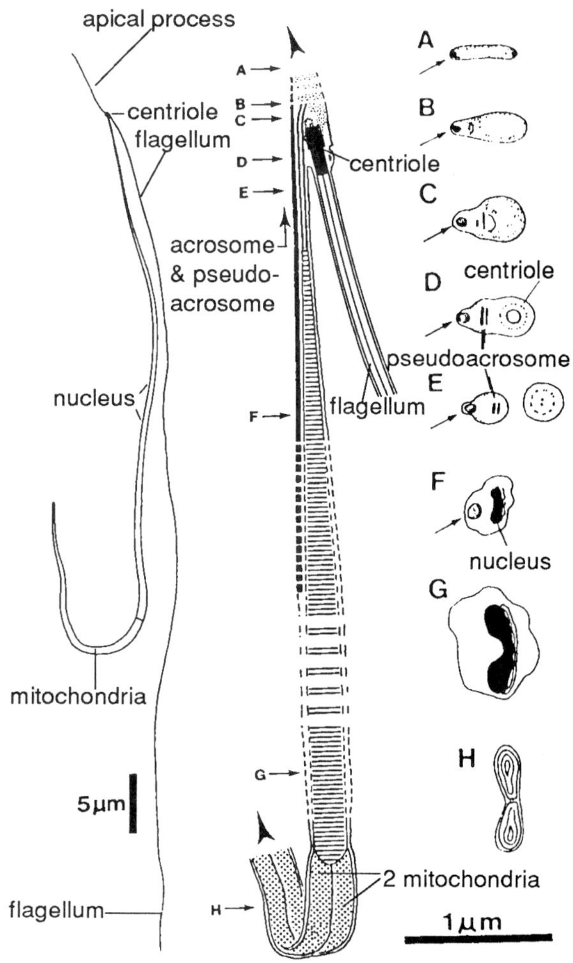

Fig. 1.2. *Phoronis pallida*. A. Diagram of the entire spermatozoon. B. Diagram of spermatozoal components. Arrowheads indicate that the full extent of the acrosomal process and mitochondria are not shown. The sections A-H refer to cross sections shown in Fig. C. C. Cross sections at the levels indicated in A. Arrows indicate the acrosome. After Franzén, A. and Ahflors, K. (1980). *Journal of Submicroscopic Cytology* **12**, 585-597. Figs. 11, 5 and 8.

fusion of several small mitochondria of the spermatid and move posteriorly along the nuclear membrane from their origin near the centriole. A short, approximately 4 µm long, filament-like acrosome projects from the point of insertion of the normal 9+2 flagellum where it is in contact with the pseudo-acrosome. The acrosome continues backwards, alongside the pseudo-acrosome and anterior part of the nucleus, as a tubelike structure (Franzén and

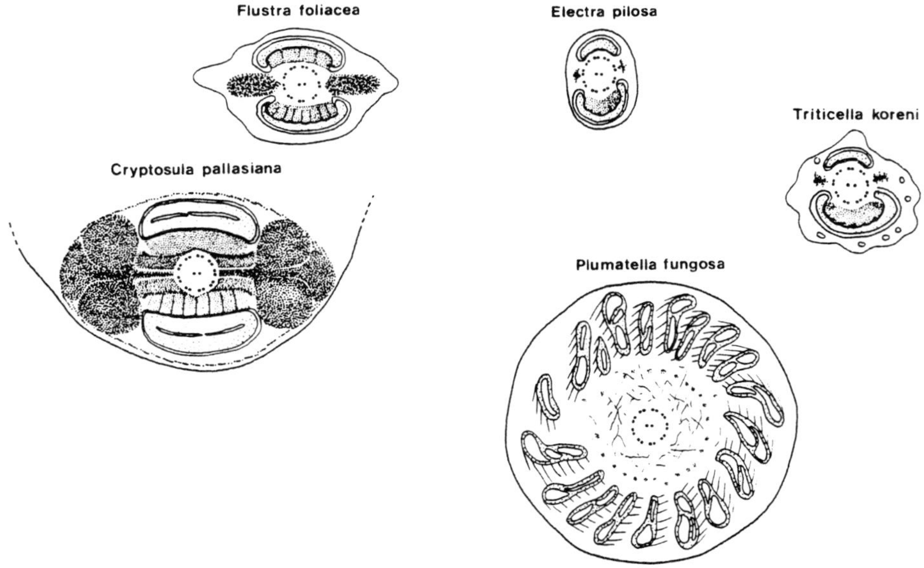

Fig. 1.3. Bryozoan spermatozoa. Drawings of transverse sections of the mitochondrial midpiece regions in the Gymnolaemata, *Flustra foliacea*, *Electra pilosa*, *Cryptosula pallasiana* and *Triticella koreni* and the Phylactolaemata, *Plumatella fungosa*. After Franzén, A. 1981. In *Recent and fossil Bryozoa*. (Eds G.P. Larwood, and C. Nielsen. Olsen and Olsen, Fredensborg. Fig. 2.

Ahlfors, 1980).

It is presumed that the sperm of *P. pallida* are released in spermatophores, as in other species, which somehow reach the eggs (possibly by pseudocopulation) and that this entails sufficient modification of fertilization to account for the great modification of the sperm. In other species fertilized eggs are brooded on the lophophore (Silén, 1952; Franzén and Ahlfors, 1980; Emig, 1977a,b; Barnes, 1980).

Forward dislocation of the centriole and a consequent approximately parallel course of nucleus and axoneme are seen also in pentastomids, branchiuran and other maxillopod crustaceans, in the concentricycloid echinoderm *Xyloplax*, chaetognaths and acanthocephalans but only in *Phoronis* and *Xyloplax* does the flagellum lose the cytoplasmic connection to the nucleus which exists in the spermatid so that the nucleus and the mitochondria lie in a lateral appendage. This dissociation of the axoneme also occurs in the pterobranch *Rhabdopleura* (Fig. 2.8). The centriole is, however, postnuclear in *Rhabdopleura* and the appendage contains only the mitochondrion, while an acrosome is absent. Other hemichordates (the enteropneusts) have "primitive" sperm. Branchiurans and the related pentastomids share with *Phoronis* the filamentous form of the acrosome. The acrosome is also elongate, though otherwise distinctive, in *Xyloplax*. Continuation of the acrosome along the nucleus is restricted to *Phoronis*.

Because the lateral appendage cannot parsimoniously be construed as plesiomorphic for hemichordates (*Rhabdopleura*) or echinoderms (*Xyloplax*) in view of the occurrence of "primitive" sperm in enteropneusts, asteroids, ophiuroids and crinoids, it is concluded that fertilization biology rather than phylogeny is chiefly dictating these spermatozoal similarities. It is nevertheless possible that development of the appendage and especially the close similarities of the sperm of *Xyloplax* and *Phoronis* represent the phenomenon of paramorphy (Jamieson, 1984): recurrence of similar structure, in response here to similar demands of fertilization biology, because of genetic (phylogenetic) relationship. Franzén and Ahlfors (1980) have attempted to relate the elongate acrosome to aspects of fertilization, notably the mode of penetration of the sperm into the egg and the structure of the egg envelope.

Bryozoa (ectoproct Polyzoa). The sperm of the gymnolaemate *Bugula* is virtually of the "primitive"

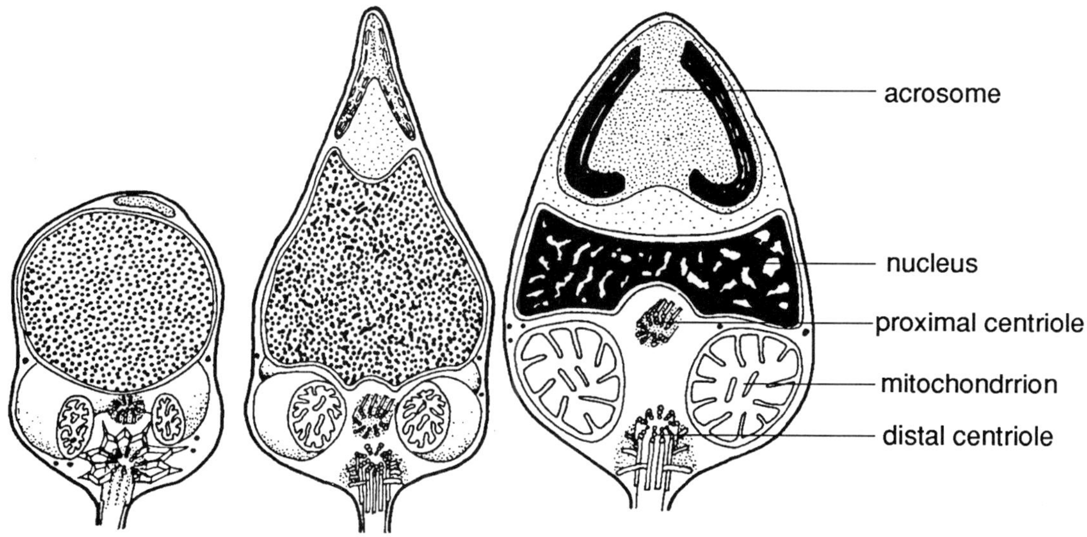

Fig. 1.4. Semidiagrammatic drawing of the spermatozoa of brachiopods. A. The articulate, *Terebratulina caputserpentis*. B and C. The inarticulates, *Crania anomala* and *Lingula anatina*. From Chuang, S.H. (1983). Brachiopoda. In *Reproductive Biology of Invertebrates*. Volume II. Spermatogenesis and Sperm Function. (Eds. K.G. and R.G Adiyodi), pp. 517-530. John Wiley and Sons, Chichester. Fig. 8. A and B after Afzelius, B.A. and Ferraguti, M. 1978. *Journal of Ultrastructure Research* **63**, 308-315.

type (Reger, 1971), excepting its lack of a nuclear membrane and considerable elongation of the nucleus; the absence of an acrosome may well be secondary and it is possible that the oocytes lack outer envelopes requiring penetration by an acrosome. Mitochondria are situated at the base of the flagellum around the basal centriole. From a micrograph is appears that the proximal centriole is at the side of and perpendicular to the basal body. The flagellum is of the 9+2 type.

The sperm of other gymnolaemate bryozoans examined ultrastructurally are advanced. Absence of an acrosome, presence of two composite rod-like mitochondrial derivatives (reaching a length of 50 µm in *Flustra foliacea* and *Triticella korenii*) and of two variably developed granular rods alternating with these; and of a tubelike cell membrane around the axoneme are special features of gymnolaeme bryozoan sperm, including *Triticella*, *Flustra*, *Membranipora* and *Electra* (Zimmer and Woollacott, 1974; Franzén, 1976a, 1977b, 1981, 1983) (Fig. 1.3). The sperm of the Phylactolaemata, examined in *Plumatella*, are complex but of a very different type. The head contains a conical nucleus which is asymmetrically curved and terminates with a small acrosome; a flagellar basal body is situated in a posterior invagination of the nucleus; the elongated midpiece contains a helix of mitochondrial derivatives (Fig. 1.3). The axoneme has the typical 9+2 structure and in the tail region it is surrounded by a spacious tube containing granular material (Franzén, 1981).

In *Bugula*, *Membranipora* and *Electra*, the sperm are shed into the seawater through terminal pores in two or more tentacles and are caught in the feeding tentacles of other individuals. They adhere to the tentacular surface or enter the intertentacular organ. In the fomer case the eggs are fertilized as they leave this organ. In the latter case fertilization occurs in the intertentacular organ. In *Bugula*, eggs are brooded in a special external chamber of the body wall termed an ovicell; the single brooded embryo is nourished by placenta-like connections to the ovicell wall (Silén, 1966; Woollacott and Zimmer, 1972).

The existence of modified sperm in many and probably most bryozoan species is correlated by Franzén (1976a) with such modified fertilization biology. Peculiar undulating movements of the swimming sperm of *Flustra* and *Triticella* are also correlated by Franzén with the absence of radial symmetry of the sperm.

Brachiopoda. We have seen that brachiopods are grouped with ectoproct Bryozoa and phoronids in the superphylum Lophophorata. The molecular configuration of their ribosomal RNA supports placement in the Lophophorata and suggests a relationship with protostomes (Ishikawa, 1977). However, embryological features, such as the tripartite (oligomerous) coelom, the monociliated epidermis of the lophophore and the indeterminate development suggest close affinity with deuterostomes (Zimmer, 1973; Rieger, 1976; Storch and Welsch, 1976). Monophyly of the Brachiopoda (Inarticulata and Articulata) has been questioned (Valentine, 1973, 1977) but appears to be supported by the very full fossil record of the brachiopods (Williams and Rowell, 1965).

From ultrastructural studies of the sperm of the inarticulates *Crania anomala* (Afzelius and Ferraguti, 1978) *Lingula unguis* (Sawada, 1973) and *L. anatina* (Chuang, 1983), which are externally fertilizing, with ect-aquasperm, and the articulate *Terebratulina caputserpentis* (Ferraguti, 1978), which is internally fertilizing, with ent-aquasperm, it can be said that all known brachiopod sperm approximate to the "primitive" model (Fig. 1.4). Chuang (1983) lists several features of the sperm of *Terebratulina* which he considers adaptations to internal fertilization (reduction in the size and differentiation of the acrosome; reduction in the subacrosomal space; changes in the shape and condensation of the nucleus; reduction in number of the mitochondria and fusion as a single annulus; parallel rather than mutually perpendicular centrioles; and increased complexity of the centriolar anchoring apparatus. All of these features or trends occur in ect-aquasperm in other groups and none of them can confidently be attributed to the ent-aquasperm mode of fertilization.

DEUTEROSTOMES

The pterobranch hemichordates, alone of the deuterostomes, retain the lophophore. This is borne on the mesosome of the tripartite, oligomerous body; they thus retain a body form which may be attributed to the earliest deuterostomes. From animals with this form it is reasonable to derive the enteropneusts, echinoderms and ascidians. Further investigation needs, nevertheless, to be directed to the contention of Løvtrup (1977) that echinoderms are further from chordates than are the Mollusca. Phylogenetic analysis of 18S ribosomal RNA base sequences has been equivocal in this respect. One study divorced the echinoderms from chordates, showing them as the sister-group of the Annelida (Field *et al.*, 1988) (molluscs were not considered). Another, probably more reliable as more taxa, including molluscs, were used, showed echindoderms as the sister-group of the chordates (Lake, 1990). To date, most phylogenies derived from molecular biology require confirmation from sampling of larger and discrete portions of the genome.

A rich and varied sessile hemichordate fauna occurred in the lower Ordovician (including graptolites, pterobranchs and acanthastids) and it is presumably from among these groups that the enteropneusts and chordates emerged (references in Clark, 1964). Unity of these with pterobranchs is underlined by the shared presence of gill clefts (pharyngotremy) in extant pterobranchs, albeit a single pair in *Cephalodiscus* and none in *Rhabdopleura* . It is conjectured that enteropneusts have lost the lophophore in acquiring a worm-like burrowing body. The origins of the echinoderms are obscure but because of close similarities between the enteropneust tornaria larva, the bipinnaria larva of asteroids and the auricularia of holothurians, it is deduced that that hemichordates and echinoderms are closely related (Clark, 1964).

Clark argues cogently for derivation of echinoderms from a pterobranch-like ancestor (after Grobben) rather than a dipleurula-like ancestor (as proposed by MacBride) but points out that the dipleurula is essentially the same as a pterobranch with paired lophophores. Similarities of the coelom, and in other regards, between hemichordates and echinoderms are considered by Clark to be too great to be regarded as convergence. The fact that echinoderms are undoubtedly highly modified animals, far removed from any common ancestor between them and hemichordates, implies that the hemichordates (or at least the pterobranchs in these) bear the closest resemblance to that ancestor. From these also arose the enteropneusts. The pterobranchs are thus confirmed as animals with a structure closest to that envisaged for the earliest deuterostomes. Indeed, the pterobranch epistome, used in locomotion, is equivalent to the attachment pit of crinoid and asteroid larvae and the madreporic vesicle of echinoderms is considered to be homologous with the cardio-pericardial vesicle of pterobranchs (Clark, 1964). Clark considers holothurian organization to approach that of unsegmented coelomate deuterostomatous worms whereas that of the other echinoderms does not. Asteroids approach the echinoids much more closely than holothuroids in the functional attributes of the metacoel.

Despite the obscurity of their origins, the relationship between tunicates and the Chordata is supported by overwhelming evidence of the close and fundamental similarities between the ascidian tadpole larva and chordates (Berrill, 1955; Clark, 1964) and by the identical mode of development of the unique pharyngotremy of the 'protochordates' (enteropneusts, tunicates and cephalochordates) despite enormous hypertrophy of the pharynx and multiplication and subdivision of gill slits in tunicates. Presence (though whether by interpolation is highly debatable) of the very larval stage, the ascidian tadpole, which underlines the chordate affinities of the tunicates, nevertheless obscures the origin of tunicates from oligomerous deuterostome ancestors, though a *Cephalodiscus*-like ancestor is envisaged. It is somewhat surprising that the chordates retain in their ontogeny the three coelomic cavities of the oligomerous ancestors before they develop the segmented musculature associated with the internal skeleton.

Echinodermata. The sperm of most species of crinoids, holothuroids, asteroids, and ophiuroids are morphologically similar and allow us to recognize a basic echinoderm sperm type which, though referable to the "primitive" type, has distinctive characteristics, some plesiomorphic but others (embedment of the acrosome in the nucleus, the annular mitochondrion) clearly apomorphic and approaching in magnitude those of modified sperm. It has been termed the basic echinoderm type, or echinosperm (Jamieson, 1985), with no implication that it is necessarily more plesiomorphic than the distinctive sperm (with conical nucleus not enclosing the acrosome) of the echinoids. In allusions to the echinosperm in the original account below, only some of the references to the very large bibliography on echinoderm sperm will be included. The reader is referred to the review by Chia and Bickell (1983) for an extensive bibliography.

In the echinosperm (Fig. 1.5, 1.6A-K) the head is essentially spherical, with a spherical acrosome surrounded, except at its anterior pole, by a wide layer of periacrosomal material, the whole more or less deeply embedded in the anterior end of the nucleus in what is here termed the periacrosomal fossa; the periacrosomal material typically contains the actomere (responsible for production of the acrosmal filament) and this may be located (as in asteroids) in a subacrosomal depression or pit; the nucleus is typically spherical; the single mitochondrion (little if at all demarcated as a midpiece) has the form of a postnuclear annular band surrounding the two centrioles, of which the distal forms the basal body of the long flagellum; the proximal centriole may or may not be contained in a posterior nuclear fossa which is presumably an apomorphy.

In echinosperm the position of the acrosome is indicated externally by a navel-like depression, as shown by SEM for the crinoid *Florometra serratissima*, the ophiuroid *Ophiopholis aculeata*, the asteroids *Luidia foliolata*, *Pycnopodia helianthoides*, *Pisaster ochraceus*, *Crossaster papposus*, *Hippasteria spinosa* and the holothuroids *Psolus chitonoides*, *Eupentacta pseudoquinquisemita*, and *Cucumaria miniata* (Chia *et al.*, 1975). The conclusion that the acrosome membrane is incomplete anteriorly in crinoids and holothuroids (Dan, 1970; Fontaine and Lambert, 1976) has convincingly been attributed by Afzelius (1977), for the crinoid *Antedon petasus*, to fixation artefact. It seems reasonable to conclude that a complete membrane is present but that anterior disruption with some fixatives may, nevertheless, reflect some differentiation of the lost region which is known to be disrupted (e.g. Colwin, Colwin and Summers, 1974) in the normal acrosome reaction.

In **crinoids** the echinosperm is well exemplified by *Antedon petasus*, *Florometra serratissima*, and *Comanthus japonica*. In *Antedon bifida*, however, the acrosome is not significantly recessed into the nucleus [this emergence being conceivably a truly primitive condition for the echinoderms, see below] and the nucleus is elongate barrel-shaped and almost completely penetrated by an anterior fossa or endonuclear canal. In the crinoid *Isometra vivipara* a

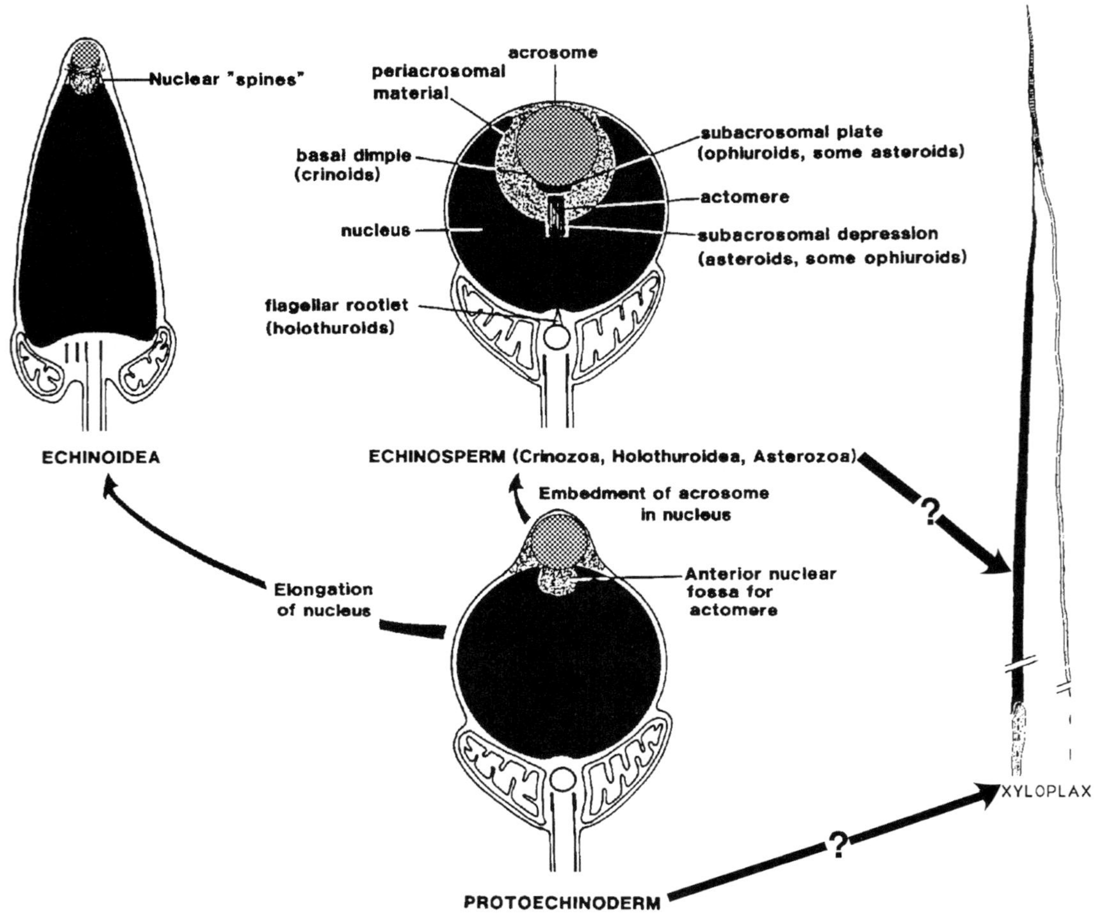

Fig. 1.5. Evolution of echinoderm sperm. Modified from Jamieson, B.G.M. (1985). *Zoologica Scripta* **14**, 123-135. Fig. 26.

remarkable modification has occurred: the acrosome is embedded laterally and pre-equatorially in the nucleus while the mitochondrion is embedded posterolaterally on the opposite side. Some species have a small posterior dimple-like invagination of the acrosome vesicle which is here regarded as an autapomorphy, albeit inconstant, of the Crinoidea. All investigated species are characterized by a wide zone of periacrosomal material lateral to the acrosomal vesicle. As this zone occurs in enteropneusts also it probably should be regarded as a plesiomorph condition for echinoderms. A subacrosomal depression is present, though not always well developed, and in *A. bifida* is probably represented by the deep endonuclear canal.

Holothuroids typically have the echinosperm (Fig. 1.6A-C). Only two exceptions to this, in 11 examined species, are known: *Cucumaria lubrica*, in which the nucleus is elongate, cigar-shaped and the midpiece somewhat modified and *C. pseudocurata*, in which the acrosome is lateral and the bulk of the mitochondrion is on the opposite side near the base (which also displays a prominent striated rootlet and groove) so that the sperm is bilaterally symmetrical. This modification has clearly been acquired independently of the crinoid *Isometra vivipara* (lacking the striated rootlet) and is an example of a paramorphy *sensu* Jamieson (1984), specifically a symparamorphy within the Echinodermata. The holothuroid periacrosomal fossa is always well developed but, unlike the crinoids, asteroids, and ophiuroids, there is no discrete subacrosomal depression. Nevertheless, as in asteroids, a sagittally located actomere has been demonstrated within the periacrosomal material. Possibly, however, a deep, somewhat pinched off continuation of the periacrosomal fossa in *Holothuria atra* should be regarded as a subacrosomal depression; if not, the absence of a depression would appear to be a unifying feature of the Holothuroidea, though whether an apomorphic loss or a plesiomorphy is indeterminable. The rootlet-like extension from the proximal centriole, known for *Leptosynapta clarki*, Atwood, 1974; *Cucumaria lubrica*, Atwood and Chia, 1974; *C. miniata*, Fontaine and Lambert 1976; and *C. pseudocurata*, Atwood 1975; *Holothuria tubulosa*, Pladellorens and Subirana 1975; *H. atra*, Jamieson, 1985b (Fig. 1.6B); *Thyone briareus*, Summers *et al.* 1975) and (there also involving the distal centriole) in *C. pseudocurata*, Atwood 1975, appears to be a synapomorphy and possible autapomorphy for holothuroids. A weakly developed strand similarly extending from the proximal centriole into the centriolar fossa is known elsewhere in the echinoderms only from micrographs of the ophiuroid *Ophiopolis aculeata* (Summers *et al.*, 1975) and the echinoid *Strongylocentrotus droebachiensis* (Afzelius and Murray, 1957)and is possibly homologous with the holothuroid rootlet. If not, the rootlet is diagnostic of holothuroids, though whether monothetically has yet to be established.

Asteroids have classical echinosperm, and are not yet known to depart from this form which has been demonstrated, for instance, in *Asterias amurensis*; *A. forbesi*, *Ctenodiscus crispatus*; *Asterina pectinifera*; *Culcita novaeguineae* (Fig. 1.6D,E), *Nardoa novaecaledoniae* (Fig. 1.6F-H) and in the primitive platyasterid *Luidia clathrata*. A discrete subacrosomal depression seems invariably to be present and probably always houses an actomere.

Ophiuroids, as far as is known, again have exclusively echinosperm, the periacrosomal nuclear fossa varying from wide and shallow (*Ophiopolis aculeata*) to typically deep (*Ophiocoma echinata*) or with intermediate conditions (*Ophiocoma wendti*; *O. variegata*, Fig. 1.6I, J, K; *Gorgonocephalus eucnemis*). A subacrosomal depression is prominent only in *Ophiocoma echinata* but no species possesses a rod-like actomere, though material acting as an actomere in that it produces a tubule or filament at the acrosome reaction is present as a subacrosomal plate in *Oc. echinata*, *Oc. wendti* , *Op. aculeata* and *Op. variegata* (Fig. 1.6I, J). This plate appears to be a distinctive apomorphy of ophiuroids but is approached in some asteroids.

Echinoids. Sperm of sea urchins (Fig. 1.6 L,M), though aquasperm, do not conform to the echinosperm type. The chief departure is that the heads are conical. In addition, although the nucleus is anteriorly recessed, the acrosome, while spherical, is only fractionally, if at all, embedded and may be carried, as in *Echinocardium cordatum*, on a long 'post-acrosomal rod'. The equivalent of the periacrosomal fossa is therefore chiefly subacrosomal and is, functionally at least, the equivalent of the asteroid subacrosomal depression. In *Echinocardium cordatum* this forms a very deep 'nuclear invagination' (endonuclear canal) containing the actomeric postacrosomal rod. Development of a centriolar fossa is variable; it is well developed in *Arbacia punctulata*, weakly developed in *Echinocardium cordatum* and is merely a shallow depression in *Echinarachnius parma*.

Concentricycloidea. The structure of the spermatozoon of *Xyloplax turnerae*, in the only genus of the Class Concentricycloidea, is exceptional for the Echinodermata (Healy *et al.*, 1988; Rowe *et al.*, 1990) (Fig. 1.7).

The *Xyloplax* spermatozoon consists of an elongate, tapered acrosome, segmented internally; a nucleus, finely tapered anteriorly, extremely elongate and rod-shaped posteriorly; a single flagellum attached via a centriolar rootlet to the *anterior* portion of the nucleus; and a single elongate mitochondrion located posterior to the nucleus. Resemblances to several taxa have been noted above but the resemblances to the sperm of *Phoronis* are the most striking. These include attachment of the flagellum anterior to the nucleus with projection of the acrosome apically from the V-shaped structure so formed; the albeit consequential location of the nucleus, followed by the mitochondrial material (one mitochondrion in *Xyloplax*, two in *Phoronis*) in an appendage relative to the flagellum; and the striking and unusual similarity of a rootlet-like structure connecting the centriole with the nucleus. *Xyloplax turnerae* is presumed to deposit fertilized eggs following copulation involving the penial projection of the male. Modification of sperm in *Phoronis* appears to be related to packaging in spermatophores irrespective

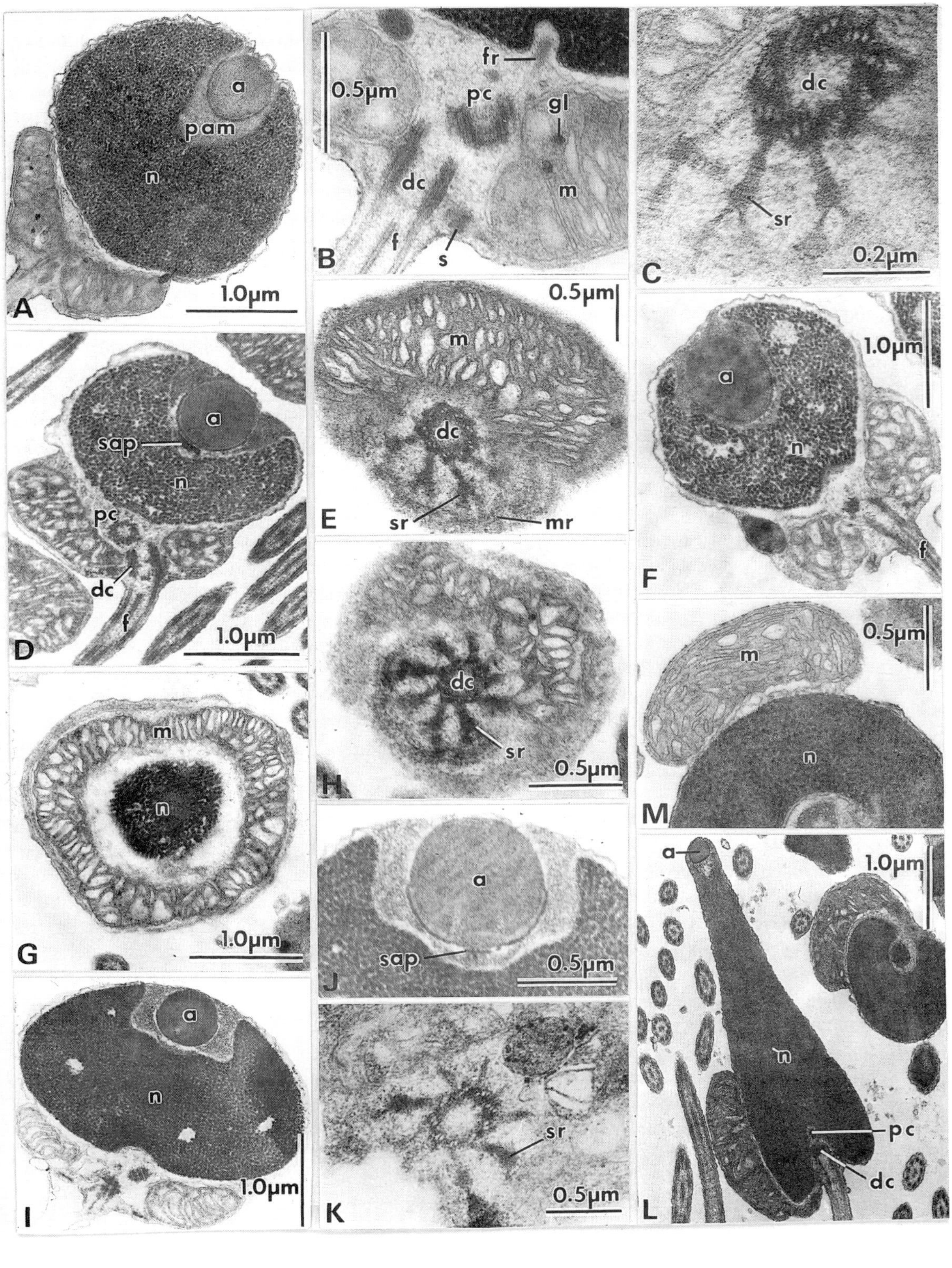

A
a
pam
n
1.0μm
B
fr
0.5μm
pc
gl
dc
m
f
s
C
dc
sr
0.2μm
D
a
sap
n
pc
dc
f
1.0μm
E
0.5μm
m
dc
sr
mr
F
1.0μm
a
n
f
G
m
n
1.0μm
H
dc
sr
0.5μm
M
0.5μm
m
n
I
a
n
1.0μm
J
a
sap
0.5μm
K
sr
0.5μm
L
a
1.0μm
n
pc
dc

of whether fertilization of the eggs occurs in the ambient water or on the tentacles.

Unless we postulate that the the near-plesiosperm morphology of other echinoderm sperm is due to secondary simplification from a complex form resembling that of *Xyloplax* and *Phoronis*, a seemingly unjustifiable proposition, we must conclude that the remarkable similarities between the sperm of these two genera have been separately acquired. It is nevertheless possible, as stated under *Phoronis* above, that these similarities are paramorphic.

From somatic morphology, the Concentricycloidea are believed to be descendants of the Asteroidea, in which only echinosperm are known. One can but concur with Healy *et al.* (1988) that *Xyloplax* spermatozoa graphically demonstrate the potential for modification of sperm structure in the Echinodermata in relation to sperm transfer and internal fertilization, the latter phenomenon restricted in the phylum, so far as is known, to this genus.

Protochordates

Protochordates form a paraphyletic grouping consisting of the subphyla Hemichordata, Tunicata, and Cephalochordata. With the Craniata (Vertebrata) they are generally considered to comprise a monophyletic taxon, the Chordata. Three chapters will now be devoted to protochordates.

Fig. 1.6. (Opposite). Spermatozoa of echinoderms (from Heron Island, Great Barrier Reef). A-C. Holothuroidea, *Holothuria atra*. A. Longitudinal section (LS) of spermatozoon. B. LS through centriolar region, showing rootlet extending from proximal centriole into the centriolar (nuclear) fossa. C. Transverse section (TS) distal centriole, showing triplets and satellite rays. D-H. Asteroidea. D, E. *Culicita novaeguineae*. D. LS spermatozoon. E. TS distal centriole, showing triplets, satellite rays and mitochondrion. F, G, H. *Nardoa novaecaledoniae*. F. LS spermatozoon. G. TS mitochondrion. H. TS distal centriole, showing triplets and satellite rays with encircling ring. I, J, K. Ophiuroidea, *Ophiocoma variegata*. I. LS spermatozoon. J. Acrosome region, showing subacrosomal plate. K. TS distal centriole, showing satellite rays. L, M. Echinoidea sp. L. LS spermatozoon. M. TS nucleus through mitochondrion. a. acrosome vesicle. dc. distal centriole. f. flagellum. fr. flagellar rootlet. m. mitochondrion. mr. marginal ring. n. nucleus. pam. periacrosomal material. pc. proximal centriole. sap. subacrosomal plate. sr. satellite ray. All original.

A discussion of lophophorate-deuterostome evolution in the light of sperm ultrastructure will be deferred to the end of Chapter 4, on the Cephalochordata.

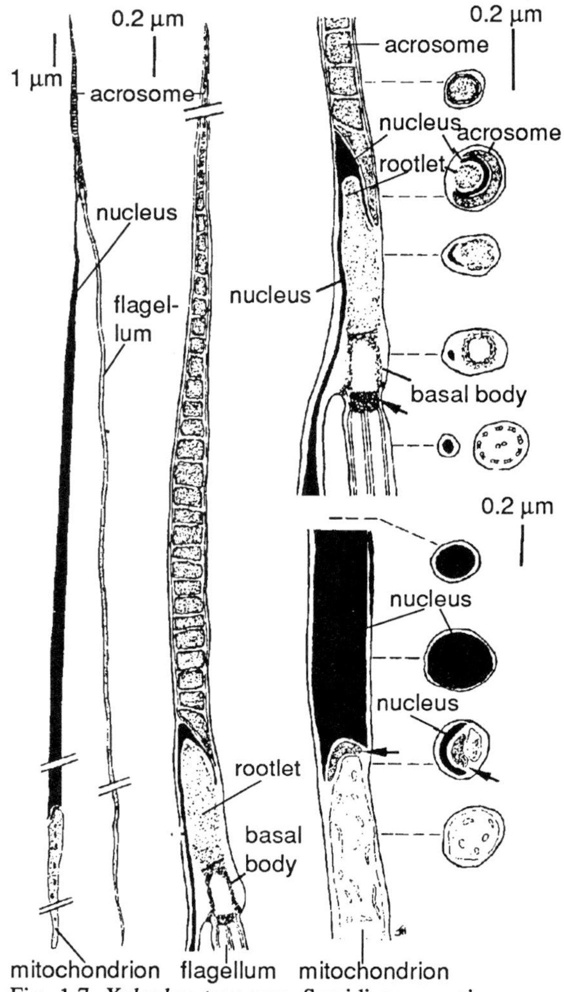

Fig. 1.7. *Xyloplax turnerae*. Semidiagrammatic reconstruction of spermatozoon. A. Whole spermatozoon. B. Acrosome and flagellum attachment zone (at nuclear apex). The acrosome consists of numerous dense cylinders and tapers anteriorly. C. Flagellum attachment zone, showing spatial relationship of acrosome base to anterior tapered anterior extremity of nucleus and flagellum attachment complex (centriolar rootlet, basal body). A dense body (arrow) [basal plate] is the point of origin of the central tubules of the axoneme. After Healy *et al.* (1988). *Zoologica Scripta* **17**, 297-310. Fig. 60-62.

Chapter 2

PHYLUM CHORDATA
SUBPHYLYM HEMICHORDATA

PHYLUM CHORDATA

The chordates are so-named because of the presence of an elastic dorsal rod, the notochord. They can, however, be defined only polythetically, that is to say there is no one character which is present in all members, thus the notochord cannot with certainty be said to be represented in hemichordates. The phylum Chordata is commonly considered to include the subphyla Hemichordata (acorn worms), Urochordata (tunicates or sea-squirts), Cephalochordata (lancelets) and Vertebrata (fish, amphibians, reptiles, birds and mammals). Hemichordates, urochordates and cephalochordates are traditionally termed protochordates. Protochordates are rarely regarded as a formal taxonomic rank and in the absence of the vertebrates do not constitute a monophyletic group. Jefferies (1979, 1981) advocates inclusion of the extinct group Calcichordata in the Chordata.

SUBPHYLUM HEMICHORDATA

Diagnosis. The hemichordates comprise a small group of worm-like marine animals possessing gill clefts. They contain two or possibly three extant classes. The extinct classes Graptolithina (Graptolites) and Acanthastida are also usually included. The third extant class, the Planctosphaeroidea is based on two planktonic specimens which may be larvae of an unknown type of hemichordate (Bullman, 1970) and will not be further considered here. The Enteropneusta (Acorn worms) include *Saccoglossus* and *Balanoglossus* and have a row of paired gill slits. The Pterobranchia include *Cephalodiscus* organized in aggregations and *Rhabdopleura* which forms stolons and is colonial. Pterobranchs differ from enteropneusts in having ciliated feeding tentacles, the lophophore, and in having, in *Cephalodiscus*, only one pair or, in *Rhabdopleura*, no gill slits. There is a widespread view that the "notochord" or stomochord of hemichordates is not homologous with a true notochord and on this basis the subphylum is often excluded from the Chordata and raised to the rank of an independent though unquestionably deuterostomatous phylum. In view of the uniformity of gill skeleton structure with that of other chordates, and general chordate characteristics (below), evolution of gill slits is here treated as a monophyletic event and the subphylum is retained in the Chordata.

CLASS ENTEROPNEUSTA

Diagnosis. Enteropneusts are worm-shaped marine animals mostly in burrows in shallow water but with representation near the Galapagos hydrothermal vent. They vary in length in different species from 2 cm to over 2 m. They have three body divisions conforming with the protosome, mesosome and

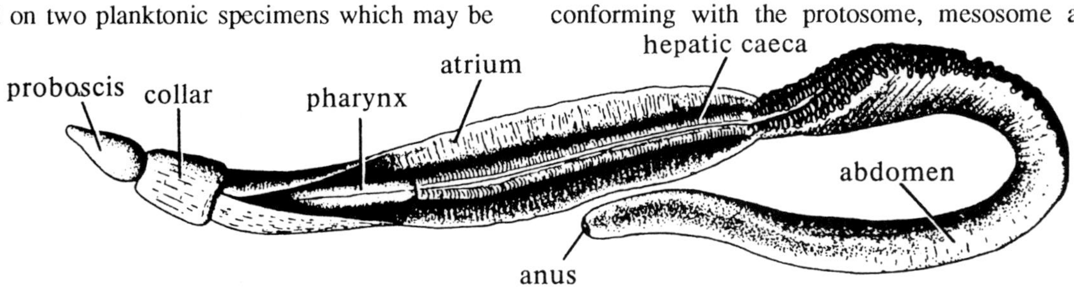

Fig. 2.1. *Balanoglossus*. An enteropneust, removed from its tube. From Young, J.Z. (1981). *The Life of Vertebrates*. Third Edition. Clarendon Press, Oxford. Fig. 3.2. After van der Horst.

metasome of oligomerous phyla and here constituing a proboscis, collar and trunk (Fig. 2.1).

Notable chordate characters are the paired pharyngeal gill slits, situated in the anterior part of the trunk, and the dorsal tubular nerve cord, in the collar. Elsewhere the nervous system resembles that of echinoderms in consisting of a sheet of nerve fbres and cells lying under the epidermis over the entire body. In some species the gill slits open into an atrium formed by lateral folds usually turned upwards to leave a long middorsal opening. In other species each slit opens to a gill pouch. The gill slits, with a skeletal system resembling that in the cephalochordates, are not associated with actual gills and probably serve to filter off excess water from the material, often sand and mud, collected by mucus secreted by the proboscis and conveyed by cilia to the mouth. The mouth lies in a groove between the proboscis and collar. A short structure beneath the dorsal nerve cord at the junction of the probosics and trunk and forming a dorsal diverticulum of the pharynx constitutes the dubious notochord or stomochord and is associated with a skeletal plate (Fig. 2.2). Its vacuolated cells are reminiscent of those of a true notochord.

There is no endostyle but the pharynx possesses a ventral ciliated groove. The anterior part of the intestine bears numerous hepatic caeca. In contrast with vertebrates, the blood is said to flow forwards in the dorsal blood vessel and backwards in the ventral vessel. The dorsal vessel expands into a heartlike sinus, surrounded by a muscular pericardium, differing in its dorsal location from a vertebrate heart. The front of the sinus forms a series of glomeruli which are covered by a region of the proboscis coelom which is specialized to form excretory cells. The entire body surface is ciliated. A postanal region is present in some species during embryonic development. The tornaria larva has many features of echinoderm larvae (Young, 1981; Nelson, 1984).

Sperm literature. Our knowledge of enteropneust sperm was until recently limited to description of the acrosome of *Saccoglossus kowalevskii* and its reaction during fertilization (Colwin and Colwin, 1963, 1967; Colwin *et al.*, 1957), with some additional data on sperm structure by Afzelius (1979), here augmented from light microscope observations in these papers and from scrutiny of micrographs there presented. An exemplary paper by Franzén *et al.*

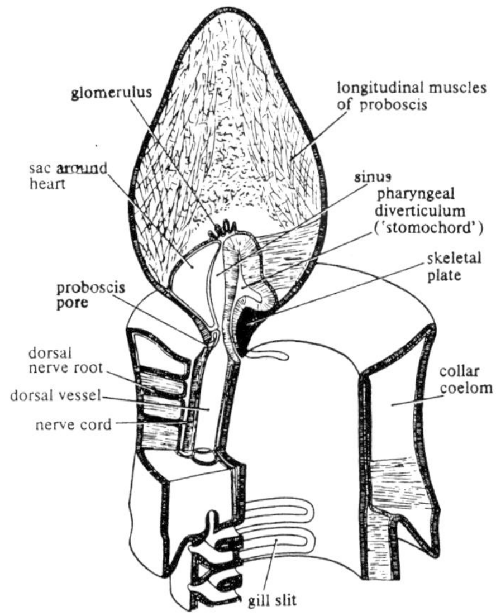

Fig. 2.2. *Balanoglossus*. Diagrammatic section of the anterior end. From Young, J.Z. (1981). *The Life of Vertebrates*. Third Edition. Clarendon Press, Oxford. Fig.3.5. After Spengel.

(1985) for *Saxipendium coronatum*, the "spaghetti worm" of the Galapagos hydrothermal vent, at 2478 m, has greatly augmented knowledge for the group.

***Saccoglossus* sperm.** The spermatozoon is of the "primitive" type, that is a basic ect-aquasperm.

Acrosome. The acrosome (Fig. 2.3) resembles that of echinoderms in the presence of an almost isodiametric acrosome vesicle with well developed periacrosomal material whereas this material, if present, is mainly subacrosomal in most phyla with primitive sperm. Unlike echinoderms (see Jamieson, 1985) in which in all but echinoids and the concentricycloid *Xyloplax* the acrosome is embedded in a deep anterior excavation of the nucleus, the acrosome in *Saccoglossus* only slightly indents the nucleus and a distinct excavation (seen in asteroids and some ophiuroids) for subacrosomal material is not developed.

The external boundary of the acrosome (Fig. 2.3) is the plasma membrane and its inner boundary is the nuclear envelope. Between these lies the membrane bound acrosome vesicle and the periacrosomal material which surround the vesicle except at the apex. Proximally the region intrudes into a shallow depression in the nucleus and distally it protrudes beyond the spherical nuclear outline. The periacrosomal material

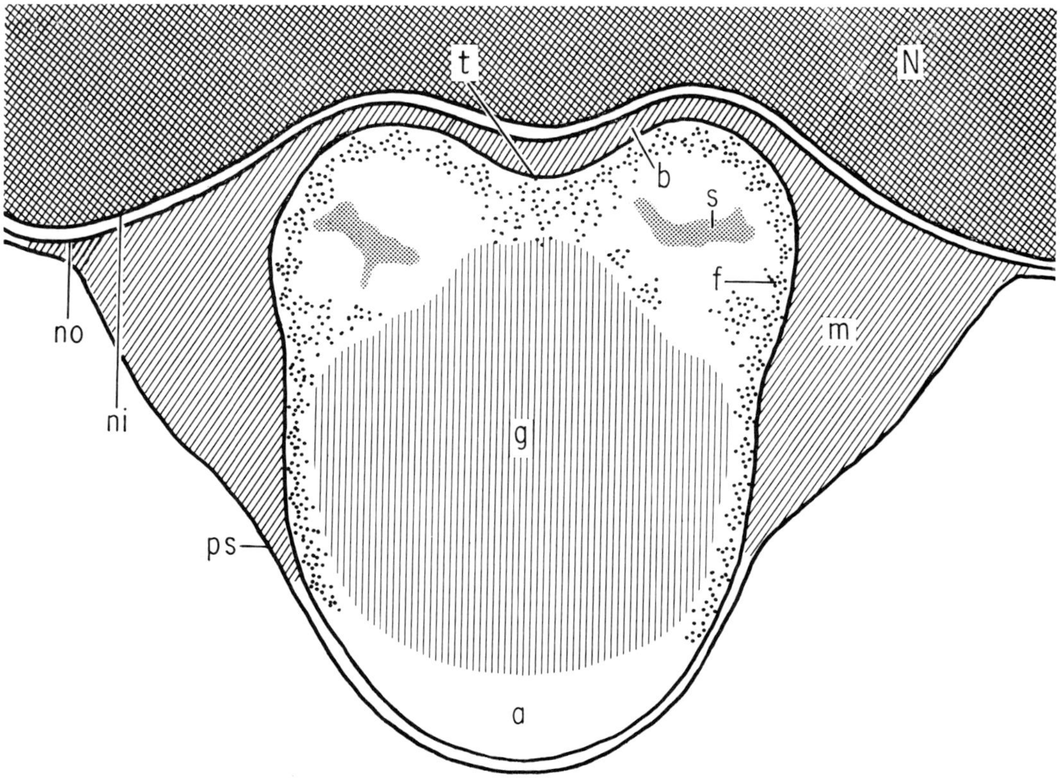

Fig. 2.3. *Saccoglossus kowalevskii.* Diagram of median longitudinal section of unactivated acrosomal region. Except apically, where the acrosomal and sperm plasma membranes are almost contiguous, periacrosomal material forms a sheet (b) between nucleus and acrosome, and elsewhere (m) is confined by sperm plasma membrane (ps), outer membrane of nuclear envelope (no), and acrosomal membrane. Within the acrosome an apical space (a) separates the acrosomal granule (g) from the acrosomal membrane; except apically, this membrane is lined by finely granular material (f). A ring-shaped space containing stellate bodies (s) lies within the base of the acrosome. A single shallow invagination (t) indents the adnuclear end of the acrosome. From Colwin, A.L. and Colwin, L.H. (1963). *Journal of Cell Biology* **19**, 477-500. Fig. 1.

is dense and finely granular, thick around the sides of the vesicle, forming a surrounding annulus, but forming only a thin sheet on the adnuclear side. The acrosome vesicle is bounded by a well developed membrane. The vesicle is nearly filled by a large dense acrosomal granule and is lined, except at the apex, by a thin finely granular layer of material. An apical space separates the acrosomal granule from the unlined apical region of the membrane. Where the lining layer is present it adjoins the granule except

Fig. 2.4. (Opposite). *Saccoglossus kowalevskii.* Diagrams of the acrosome reaction and association of the spermatozoon with egg and blastomere. Small triangles indicate surface of acrosome membrane initially facing into acrosome vesicle but subesquently (after insertion into plasma membrane) facing external medium. A-E. Sperm-egg association: confrontation, sperm activation, and membrane fusion with envelope-enclosed egg. (Subsequent states of sperm-egg association resemble those shown in F-I except that the egg envelope (en) would be present). A, acrosomal region of unactivated sperm cell; except apically, periacrosmal material (p) surrounds the membrane bound acrosome vesicle. B-D. Sperm activation: acrosome membrane, now continuous with sperm plasma membrane (s), forms acrosomal tubule (t) and everts, externalizing acrosome contents; periacrosomal material transforms into ring (r), vesicles (v), adnuclear sheet (a), etc., within the tubule. D. Tubule makes contact with egg plasma membrane (e). E. Early zygote established by sperm-egg membrane fusion. F-I. Sperm-blastomere association; internal sperm organelles, such as nucleus (n), mitochondria (m), and fibrils of flagellar axis (f),

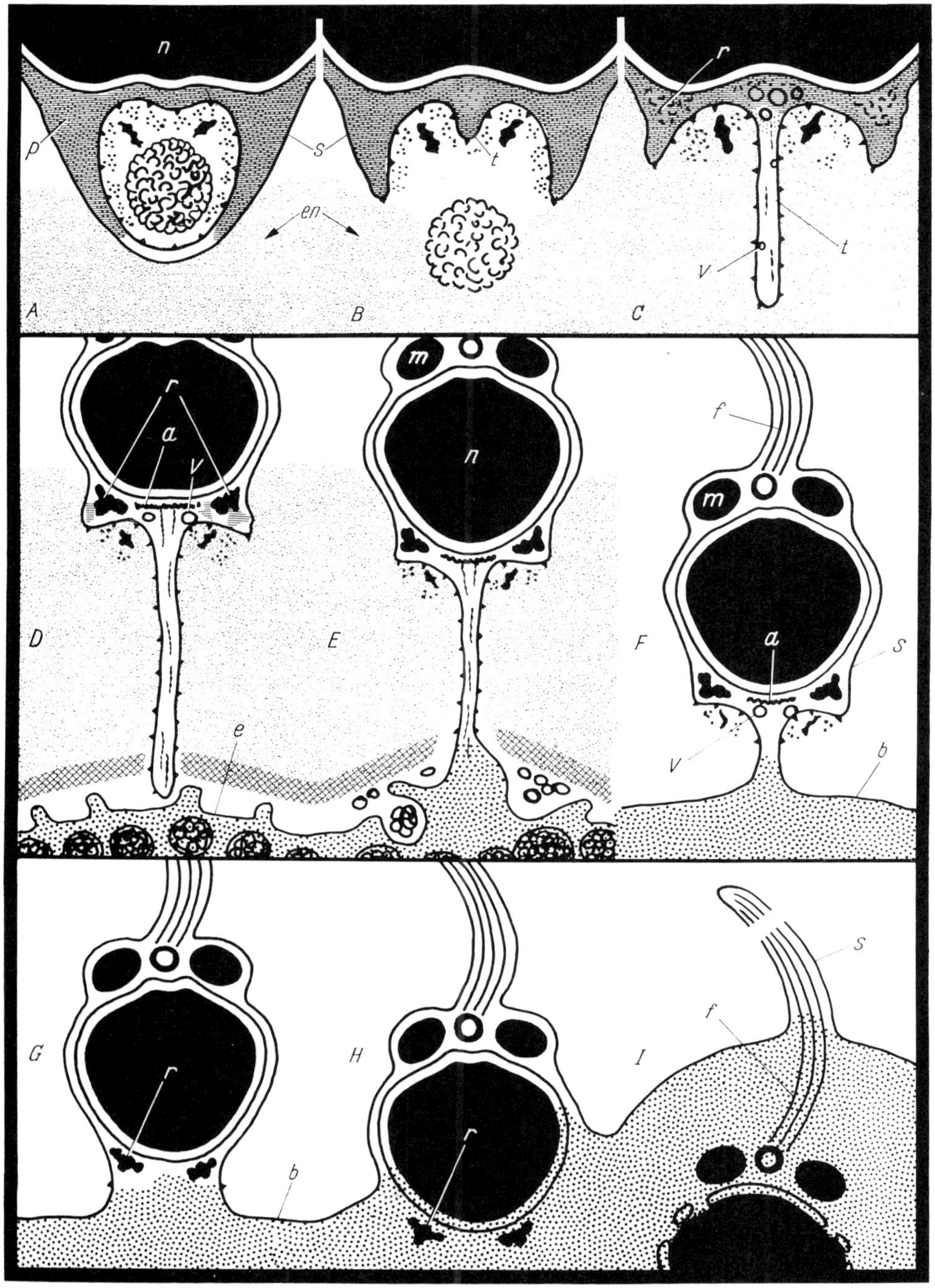

progressively intermingle with cytoplasm of denuded blastomere; blastomere plasma membrane (b) is continuous with sperm plasma membrane. (The stages of sperm activation and membrane fusion which precede this intermingling resemble those shown in A-E, except that the egg envelope would be absent from the denuded blastomere). From Colwin, A.L. and Colwin, L.H. (1967). *Zeitschrift fur Zellforschung* **78**, 208-220. Fig. 1.

that a ring-shaped space separates them around the sides of the adnuclear end of the vesicle; about a dozen irregularly stellate bodies are arranged in a circle within this space. The base of the vesicle is indented by a single shallow axial invagination. The plasma membrane, the acrosomal membrane and the inner and outer nuclear membranes are tripartite, or unit, membranes.

Acrosome reaction. Detailed description of the acrosome reaction is beyond the scope of this work but the elegant analysis by Colwin and Colwin (1967), augmenting the work in the other two cited papers, of the reaction and penetration of the egg is summarized in Fig. 2.4.

Nucleus. The almost spherical nucleus is slightly indented anteriorly by the acrosome and posteriorly in the vicinity of the centrioles.

Mitochondria. Around the centrioles are the mitochondria, constituting with it a somewhat flattened compact midpiece; in longitudinal sections the mitochondria are seen as rounded, internally cristate structures on each side; several discrete structures are described for the midpiece of the reacted sperm and this presumably indicates the existence of a number of separate mitochondria as is characteristic of "primitive" sperm.

Centrioles and flagellum. The centrioles appear to include a proximal centriole perpendicular to the distal centriole which forms the basal body. The flagellum is continuous with and in the same axis as the basal body. Afzelius (1979) illustrates the anchoring apparatus, consisting of a 9-pointed skewed star formed by lamellae radiating from the distal centriole to the plasmalemma. Each of these primary processes branches into two secondary processes which in turn give tertiary processes. The structure of the axoneme has not been described but, from micrographs, is probably of the 9+2 pattern.

***Saxipendium* sperm.** The sperm of *Saxipendium coronatum* (Fig. 2.5, 2.6) is about 29 μm long.

It has a pyramidal head, 3.3 μm in diameter at the base and 3 μm long, including an apical acrosome. A flattened midpiece is closely applied to the base of the head region. A long flagellum emerges from the centre of the midpiece.

Acrosome. The acrosome vesicle is rounded with a slightly flattened or concave adnuclear face and has a thin limiting membrane. The contents of the vesicle may appear to be differentiated into an anterior electron dense part and a posterior more electron lucent part, possibly as a result of an incipient acrosome reaction during processing. Periacrosomal material

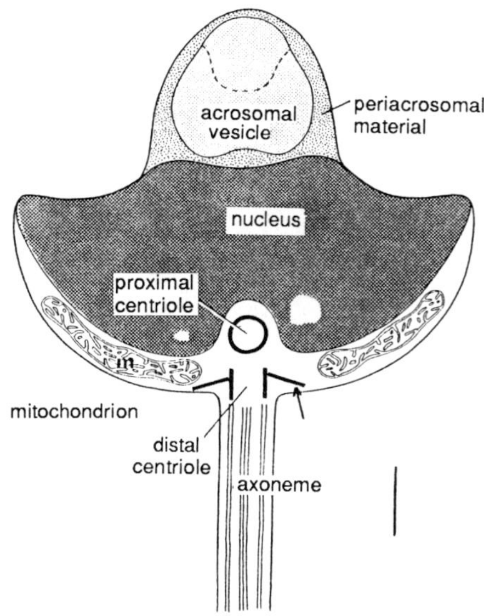

Fig. 2.5. *Saxipendium coronatum.* Diagram of the mature spermatozoon. (Scale bar = 0.5 μm). After Franzén, Å, Woodwick, K.H. and Sensenbaugh, T. (1985). *Zoomorphology* **105**, 302-307. Fig. 12.

surrounds the vesicle but at the apex the acrosome membrane lies close to the sperm plasma membrane.

Nucleus. The anterior surface of the nucleus has a ring-shaped depression around the base of the acrosome and the posterior face is indented by the centriolar fossa. The ring-shaped depression is interrupted by four extended elevations, which by SEM are seen as ridges, radiating from the acrosomal region. The nucleus is electron dense with small scattered areas of low density.

Mitochondria. The midpiece contains the mitochondria, the centriolar region and the anchoring fibre [satellite] apparatus. A number of small mitochondria are present in the late spermatid but it was not clear, owing to difficulties of fixation, whether these are reduced to four or five mitochondria or a single annular mitochondrion though the latter view is favoured. Franzén (1956), by light microscopy, demonstrated a few spherical mitochondria in the enteropneusts *Glossobalanus sarniensis* and *Protoglossus* sp. The posterior part of the midpiece regularly

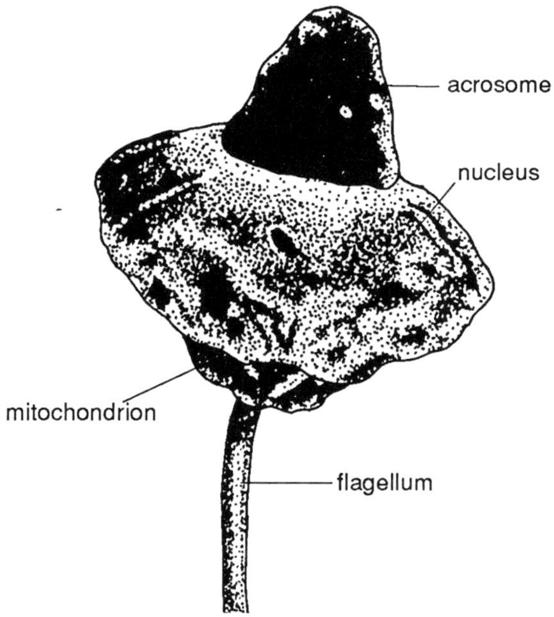

Fig. 2.6. *Saxipendium coronatum.* External appearance of a spermatozoon. Drawn from an SEM micrograph of Franzén, Å, Woodwick, K.H. and Sensenbaugh, T. (1985). *Zoomorphology* **105**, 302-307. Fig. 15.

shows vacuoles which appear to have released their contents and therefore appear by SEM as craters.

Centrioles and flagellum. The flagellum-bearing centriole is said by Franzén *et al.* (1985) to lie in the centriolar fossa but from the illustration (Fig. 2.5) this is clearly the proximal centriole which is at right angles to but in the same axis as the distal centriole (basal body). The posterior part of the distal centriole gives rise to satellite rays; these branch into secondary fibres which join the plasmalemma at electron dense thickenings.

Fertilization biology. Fertilization in *Saccoglossus* is external (Colwin and Colwin, 1963, 1967; Colwin*et al.*, 1957) and this is suspected for *Saxipendium*, the small eggs of which, observed in the female, suggest indirect development, presumably via a tornaria as in other enteropneusts (Franzén *et al.*, 1985).

Summary of enteropneust sperm. This summary is drawn from accounts of the two species examined ultrastructurally. The acrosome is subspheroidal but sufficiently depressed on the adnuclear face to appear dome shaped. It rests in a slight depression of the nucleus and is surrounded by a thick layer of periacrosomal material. The nucleus is electron dense; it is subspheroidal in *Saccoglossus*, though scalloped posteriorly by the mitochondria, but has a hemispherical-bicornuate longitudinal section in *Saxipendium.* There are a few rounded mitochondria; it is uncertain whether these are fused into a single annular mass in *Saxipendium* in which they do not intrude on the nucleus. A proximal centriole is present, perpendicular to the distal centriole. An "anchoring apparatus" of dichotomous satellite rays connects the basal body to the plasma membrane. The flagellum arises centrally from the midpiece.

Sperm phylogeny. The hemichordate (enteropneust) sperm thus has much the appearance which we might attribute to a precursor of the echinoderms, it is an ect-aquasperm distinguished like that of echinoderms (the echinosperm) in having an inflated acrosome enveloped in periacrosomal material whereas the basic aquasperm of the Metazoa (in its hypothetical manifestation termed the plesiosperm) is attributed with a depressed acrosome with at most subacrosomal material. In *Saxipendium* a further resemblance to the echinosperm may be the uncertainly demonstrated annular mitochondrion. It would not seem unreasonable to suggest that in their spermatozoa enteropneusts have retained features of a common ancestral stock of echinoderms and hemichordates whereas echinoderms have developed more apomorphic sperm. However, *Rhabdopleura* (see below) which, as a member of the Pterobranchia, might be expected to have sperm near the ground plan for echinoderm sperm has highly modified sperm.

CLASS PTEROBRANCHIA

Diagnosis. Pterobranchs are colonial or aggregated sedentary marine lophophorates. An external cuticular skeleton is present and they are therefore known as fossils, from the Lower Ordovician to Eocene (Nelson, 1984).

There are two orders, the Cephalodisca and Rhabdopleurida.

Order Cephalodisca

Diagnosis. Zooids free, produced by budding but not forming true colonies though aggregated in a many chambered gelatinous house. Living forms

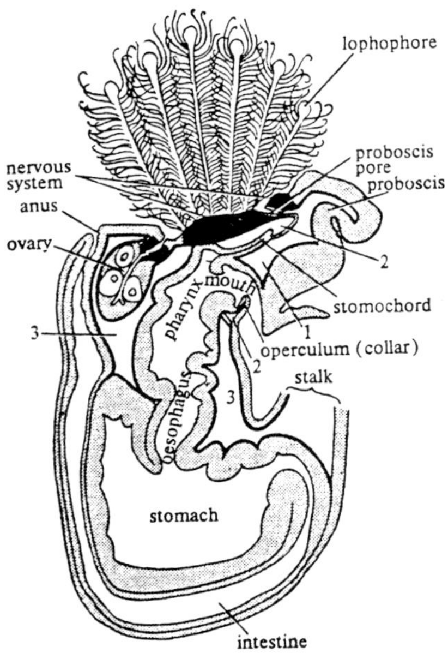

Fig. 2.7. *Cephalodiscus*. Longitudinal median section. 1-3 denote the three coelomic cavities. From Young, J.Z. (1981). *The Life of Vertebrates*. Third Edition. Clarendon Press, Oxford. Fig. 3.12. After Harmer.

with four to nine pairs of arms with tentacles. Extant genera are *Atubaria* (one species, near Japan) and *Cephalodiscus* (mostly Antarctic).

Each zooid of *Cephalodiscus* (Fig. 2.7) has a proboscis, collar, and trunk, each possessing one or more coelomic cavities.

The collar is prolonged into a number of ciliated feeding arms, the lophophore. The single pair of gill slits serves for egress of water drawn in by the tentacles in feeding. The intestine is reflected so that the anus is near the mouth. A thickening of the roof of the pharynx corresponds exactly with the stomochord of enteropneusts and again contains vacuolated cells. The nervous system does not form a hollow tube but the blood system resembles that of *Balanoglossus*. The larva somewhat resembles that of ectoprocts but, though different from echinoderm larvae is derivable from the same plan (Young, 1981).

The sperm of cephalodiscans have not been examined ultrastructurally.

Order Rhabdopleurida

Diagnosis. Zooids attached, forming true colonies. The zooids of the living genus *Rhabdopleura* have two tentacular arms. The coenecium or colony of *Rhabdopleura* is made up of a series of translucent tubes, each with a regular annulated pattern. A stolon running through the bases of all the tubes and interconnecting all the zooids of the colony is the site of asexual budding of new zooids. The zooid, approximately 1 mm long, is divided into three regions. The oral shield (also termed the protosome, cephalic shield, buccal shield or proboscis) is a disc-shaped organ used in forward locomotion and in secretion of the coenecium. It contains an unpaired coelomic cavity, the protocoel. The collar region (mesosome) contains the paired mesocoels which surround the pharynx. This region bears the two tentaculate arms and the oral lamellae (two ventral projections, each with a ciliated groove used to transport particles collected by the arms into the mouth). The trunk sac or metasome encloses the recurved digestive tract, the single gonad and the paired metacoels (Lester, 1988, and references therein). The nervous system appears very primitive, with cell bodies and fibres confined within the epithelial layer. The short-lived larva has cilia not in bands (Young, 1981). Atlantic, Pacific and Antarctic.

Sperm literature. The ultrastructure of the sperm of *Rhabdopleura normani*, a species living in shallow water around Bermuda, has been briefly described by Lester (1988) in an account of gonads and of larval development .

***Rhabdopleura normani*. sperm**. The spermatozoon (Fig. 2.8) is filiform and uniflagellate. No acrosome has been observed. The elongate, conical nucleus is 3.6 µm long and 0.5 µm wide. A midpiece appendix, 0.24 µm wide and at least 4 µm long, containing a mitochondrial filament, joins the base of the "head piece" near the flagellar basal body. It is free distally and parallels the flagellum which is of the 9+2 type (Lester, 1988). The mitochondrion filament is presumably a single mitochondrial derivative as there is a single mitochondrion in the spermatid.

Fertilization biology. The method of sperm release is unknown. Presumably the males are broadcast spawners and the sperm are released through the anterior gonoduct into the surrounding seawater while the zooid projects from the ostium of its tube. The derived condition of the sperm is tentatively related to an unspecified altered method of fertiliza-

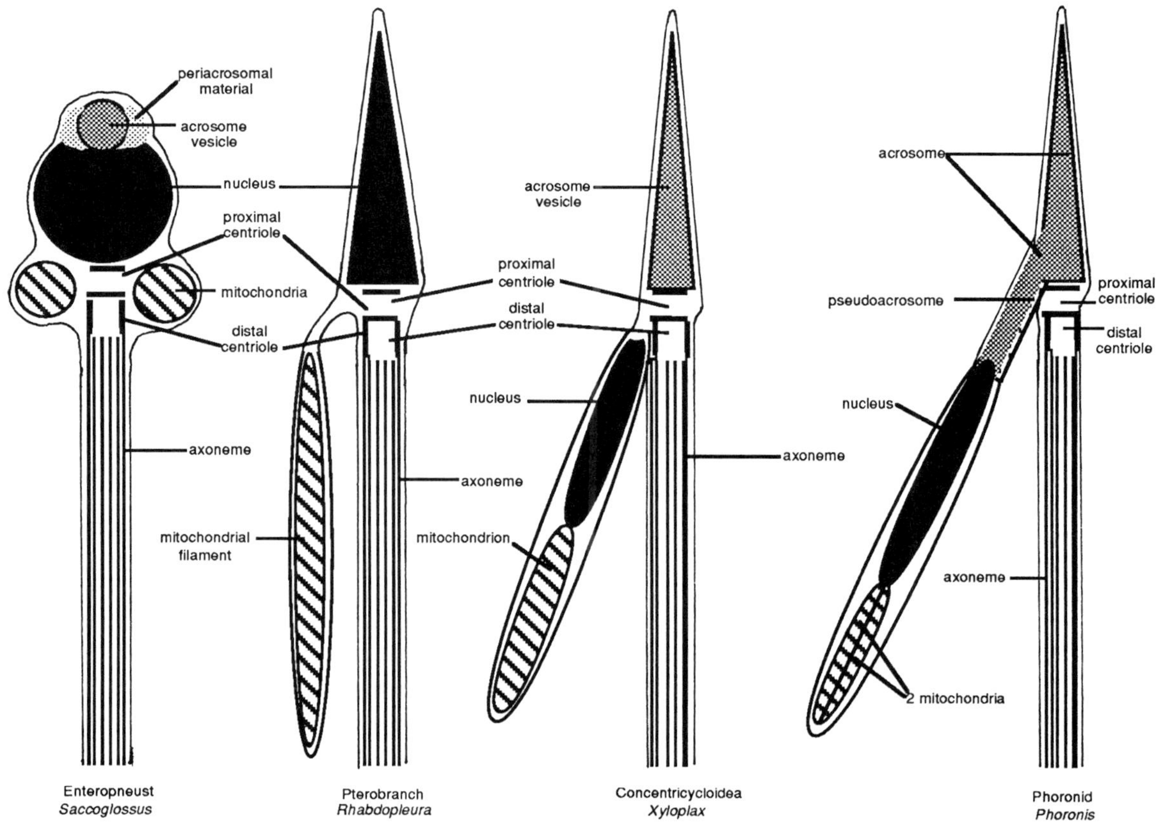

Fig. 2.8. Highly schematic comparison of the spermatozoon of the pterobranch*Rhabdopleura normani* with that of an enteropneust (*Saccoglossus*), a concentricycloid echinoderm (*Xyloplax*) and a phoronid (*Phoronis*). Location of the mitochondrial material in an appendage of the head is shared by *Rhabdopleura*, *Xyloplax* and *Phoronis*, while *Xyloplax* and *Rhabdlopleura* further resemble each other in location of the nucleus in the appendage. These similarities are here seen as symparamorphies, parallel developments by virtue of relationship. The diagram of the spermatozoon of *Rhabdopleura normani* is drawn from the account of Lester, S.M. (1988). *Acta Zoologica* (Stockholm) **69**, 95-109. Original.

tion and brood protection. Fertilization within the brood chamber of *R. normani* is suspected because of the large number of unsegmented eggs observed lying free in the brood chambers. However, Lester does not rule out the possibility of internal fertilization as in many phoronids.

The apparent absence of an acrosome is attributed to the absence of an egg envelope, eliminating the need to enzymatically degrade a path to contact the oolemma.

Chapter 3

SUBPHYLUM UROCHORDATA

Diagnosis. Urochordates (tunicates) are filter feeders with the body covered by a complex tunic secreted by the ectoderm. There is no true coelom. They have an endostyle (homologous with the vertebrate thyroid, and also seen in cephalochordates and larval lampreys), a perforated pharynx, and a dorsal neural ganglion. They demonstrate other characteristic features of chordates in the tadpole larva which has a dorsal hollow nerve cord, and a notochord-like structure in a muscular postanal tail, though the tail differs from that of cephalochordates and vertebrates in lacking segmentation. The tadpole-like form persists in the adult of the Appendicularia (Larvacea).

Tunicates consist of three classes: the sessile Ascidiacea, containing the majority of the 1300 or so known species, and the pelagic Thaliacea and Appendicularia (Larvacea).

CLASS ASCIDIACEA

Diagnosis. Covered with a tunic containing a type of cellulose termed tunicin. Larva free-swimming, tadpole-like (the "ascidian tadpole"), short-lived and non-feeding. Adult sessile, benthic, solitary or colonial; lacking the tail. Marine and worldwide; intertidal to abyssal.

The anatomy of an ascidian (phlebobranch) is shown in Fig. 3.1. Stolidobranchia differ mainly in having a folded pharyngeal wall and gonads on both sides of the body. In aplousobranchs the gut loop is posterior to the pharynx, rather than folded beside it (Fig. 3.2).

Ascidian sperm. These are aquasperm with an exceptional structure which contrasts them with the plesiosperm as a distinct "ascidian type" (Franzén, 1956, 1976b, 1983) (Fig. 3.2), here termed the ascidiosperm. Afzelius (1979) has retained the basic ascidian type within the category of primitive sperm while emphasizing its specializations. Ascidian sperm are always shed into the ambient water, hence designation as aquasperm in the terminology of Jamieson (1986). Whether they fertiize the eggs externally, as

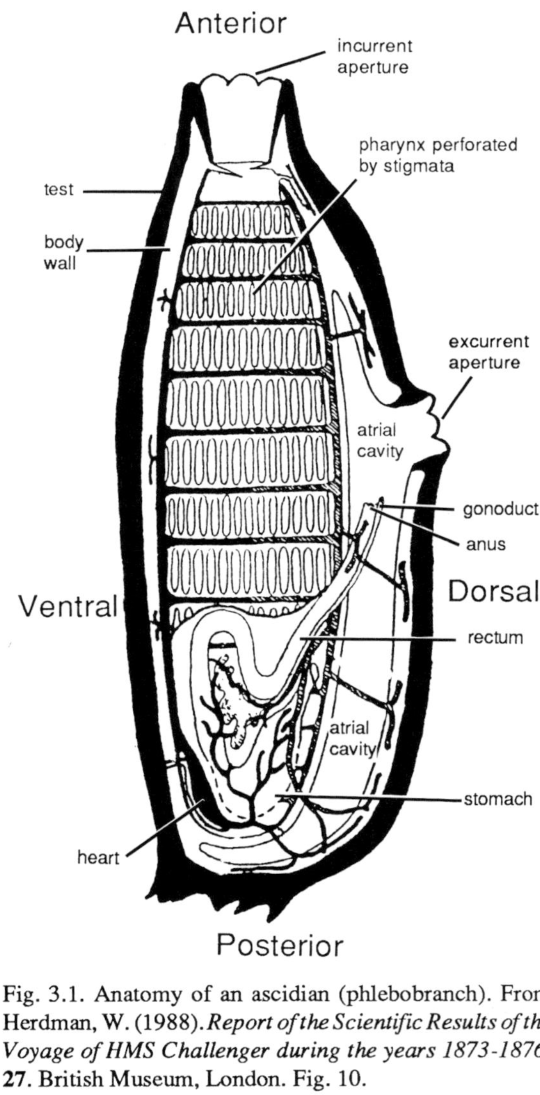

Fig. 3.1. Anatomy of an ascidian (phlebobranch). From Herdman, W. (1988). *Report of the Scientific Results of the Voyage of HMS Challenger during the years 1873-1876*. **27**. British Museum, London. Fig. 10.

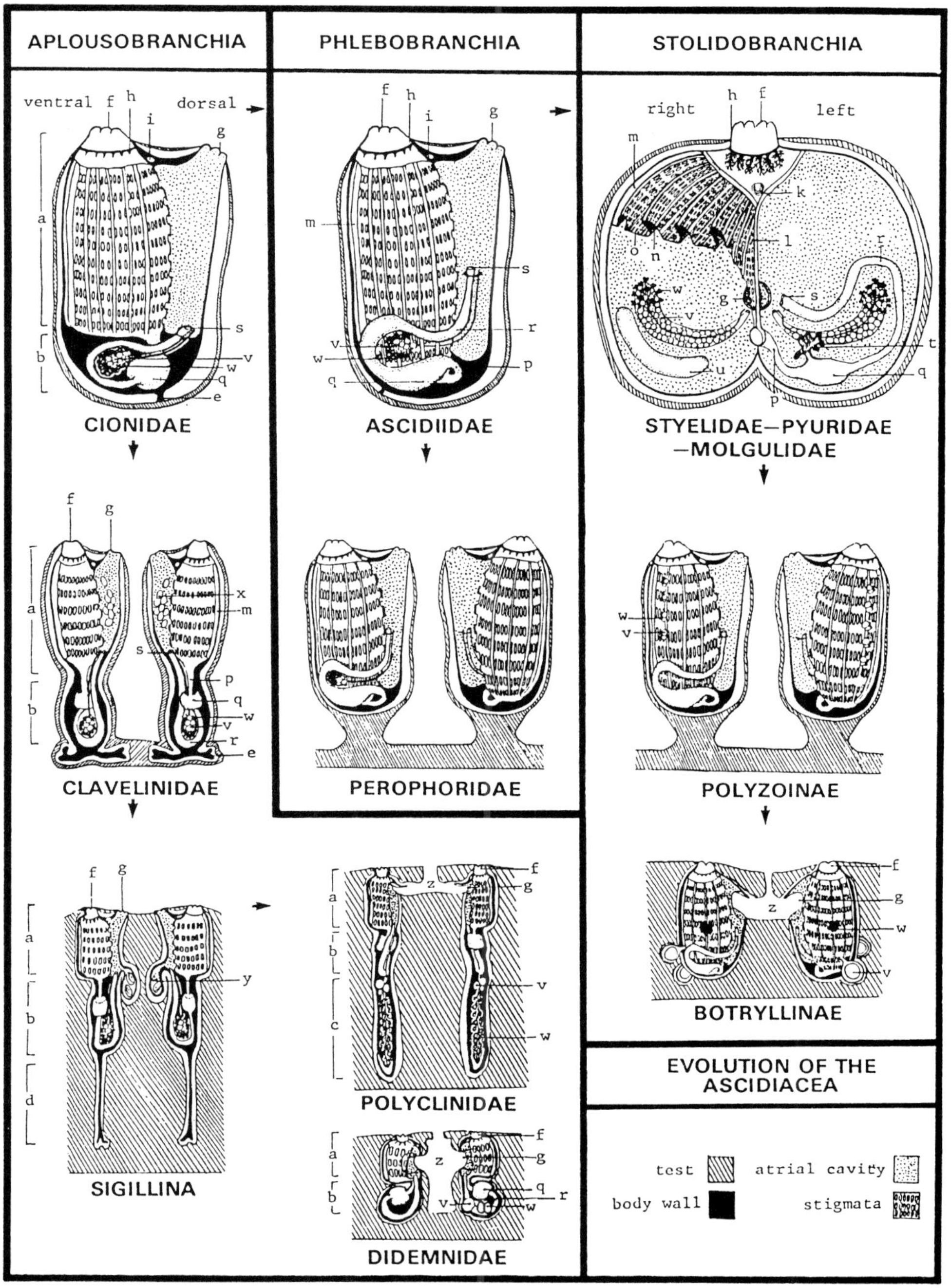

Fig. 3.2. Evolution of the Ascidiacea, showing presumed sequence of development of the types of organization. From Kott (pers. comm). a. Thorax. b. Abdomen. c. Posterior abdomen. e. Blood vessel. f. incurrent (branchial) aperture. g. excurrent (atrial) aperture. h. branchial tentacles. i. cerebral ganglion. k. opening of neural gland. l. Dorsal lamina. m. Endostyle. n. Branchial fold. o. Flat part of branchial sac. p. Oesophagus. q. Stomach. r. Intestine. s. Anus. t. Digestive gland. u. Kidney. v. Ovary. w. Testis follicles. y. Brood pouch. z. Common cloaca.

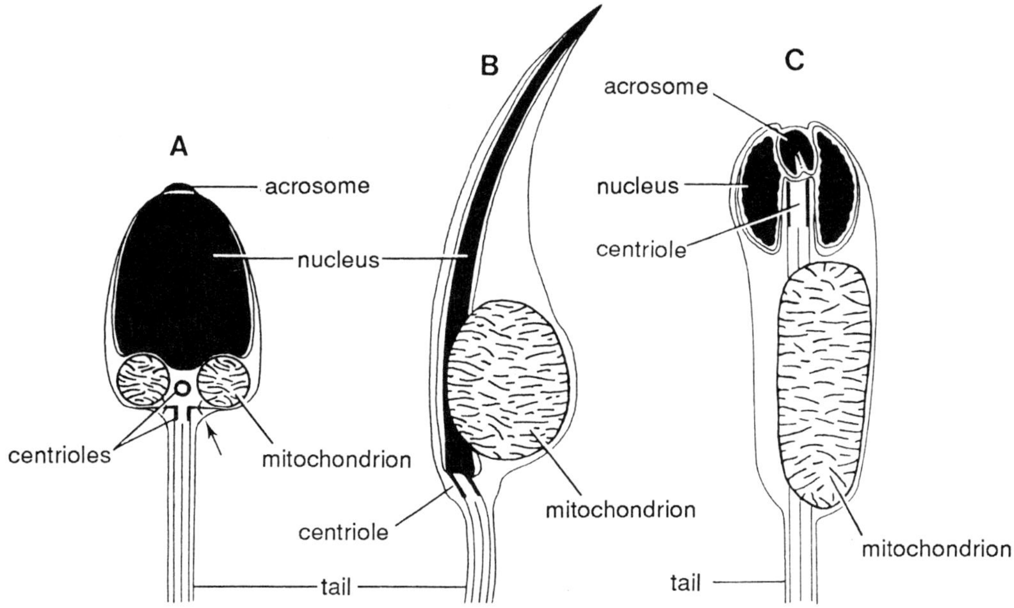

Fig. 3.3. Schematic diagrams of types of spermatozoa. A. the primitive type, here termed the ect-aquasperm. B. The ascidian type, here termed the ascidiosperm. The ancestral ascidiosperm is, however, envisaged in the present work as having a short or only slightly elongated nucleus and both centrioles. C. The larvacean (appendicularian) type. After Franzén, A. (1983). Urochordata. In Adiyodi, K.G. and R.G. (Eds). *Reproductive Biology of Invertebrates*, Volume II: Spermatogenesis and Sperm Function, pp. 621-632. John Wiley and Sons, Chichester. Fig. 8. Arrow. Satellite apparatus.

ect-aquasperm, or enter the body, as ent-aquasperm, in the terminology of Rouse and Jamieson (1987), they are sufficiently altered relative to the plesiosperm morphology in placement of the mitochondrion lateral to the nucleus (an invariable apomorphy of non-appendicularian tunicates) to be termed modified sperm.

Definition of the ascidiosperm. Characteristics of the ascidiosperm (Fig. 3.3B) recognized here are: an elongate, rodlike nucleus, though plesiomorphically, as in *Ciona*, this was little longer, at 3 μm, than in most "primitive" sperm; no acrosome or only a very simple acrosome; a single large mitochondrion, formed by fusion of smaller mitochondria in the spermatid, and located lateral to the nucleus which it often enwraps to give a C-shaped transverse profile; shedding of the mitochondrion at or before penetration of the egg membranes; a single centriole forming the basal body (proximal centriole often represented by a "centriolar adjunct", but plesiomorphically, as in *Ciona intestinalis*, fully developed and coaxial); absence, at maturity, of a pericentriolar anchoring apparatus (a 9-rayed anchoring apparatus is described only for the spermatid of *C. intestinalis* by Cotelli *et al.* (1980), who imply that it persists in the mature sperm); and presence of a flagellum with the 9+2 arrangement of microtubules, usually ending in a terminal filament (endpiece). Also claimed as a general feature of ascidian sperm is organization of the chromatin in filaments and concentric laminae before the final condensation, demonstrated for the stolidobranch *Botryllus schlosseri* and the aplousobranch *Diplosoma listerianum* (Burighel *et al.*, 1985).

An acrosome in the form of one or more minute vesicles, is probably present in all stolidobranchs and occurs in some aplousobranchs and phlebobranchs. Origin of the single vesicle by fusion of two vesicles, in *Pyura haustor* and *Styela plicata* (Fukumoto, 1983), and of three or four vesicles in *Molgula manhattensis* (Fukumoto, 1985) has been claimed. Origin of the contributing vesicles from the Golgi apparatus is presumed but not proven. In contrast, in *Pyura annectens* several "proacrosomal granules"

(70-80 nm in diameter) fragment into smaller vesicles (20-30 nm in diameter) without involving the formation of an acrosome vesicle. These smaller vesicles seem to be incorporated into the apical plasmalemma prior to completion of spermiogenesis. They coexist with a 4 µm long helical, cross striated perforatorium-like structure (Fukumoto, 1984, 1986) (Fig. 3.8). Further details of acrosomal structures are given under each order.

As the large mitochondrion is shed at or before penetration of the egg membranes by the sperm; it is very unlikely that any paternal mitochondrial DNA contributes to the zygote (*C. intestinalis*, Ezell, 1963; *Ascidia nigra*, Ursprung and Schabtach, 1965; *A. malaca*, Villa, 1977). Interesting reviews and investigations of the physiological control of this 'mitochondrial reaction' are given by Lambert (1982) and Lambert and Lambert (1983).

Externally fertilizing ascidian sperm are 50-60 µm long with a head-length from 3 µm (*Ciona intestinalis*) to, rare in the chordates, 90 µm (*Perophora formosana*). Variants of sperm structure can be related to fertilization biology: those members of the orders Phlebobranchia and Stolidobranchia with external fertilization have sperm with relatively short heads and long tails (Lambert, 1982); while sperm from internally fertilizing Aplousobranchia (most species) have proportionately longer heads and shorter tails, the exception being the externally fertilizing *Ciona*, included in the Aplousobranchia by Kott (1969), which has a short nucleus.

Ascidian fertilization. The ascidian egg is enclosed in a complex investment consisting at fullest development, as in *Ciona* (Fig. 3.4), of the thick, noncellular chorion (vitelline coat), with or without external follicle cells and, in the perivitelline space, test cells which are adherent to the egg surface or to the inner surface of the chorion. Sperm have to penetrate the chorion before fusing with the oolemma.

In the self-sterile species *Ciona intestinalis* binding of sperm to the chorion involves "self" and "not self" discrimination (Rosati and De Santis, 1978) believed to be mediated by the interaction of fucosidase on the sperm and fucosyle sites on the vitelline coat (Hoshi, 1984). However, as pointed out by Honneger (1986), in the self-sterile *Halocynthia roretzi*, fertilization depends also on the presence of the follicle cells and in contrast to *Ciona* no difference has been found between autologous or heterologous sperm in the strength and frequency of binding to the vitelline coat (Fuke, 1983). In the self-fertile ascidian *Phallusia mammillata*, *N*-acetylglucosamine and probably sialic acid residues have been visualized with wheat germ agglutinin (WGA) on the vitelline coat. WGA, which binds to the vitelline coat but not to follicle cells and sperm, inhibits fertilization by interfering with sperm-vitelline coat binding (Honneger, 1982, 1986). In addition, a high beta-D-*N*-acetylcosaminidase activity was found in sperm (Hoshi *et al.*, 1983; Hoshi, 1984).

In *Ciona intestinalis* and *Halocynthia roretzi*, there is evidence that the sperm contains proteases and that these are necessary for fertilization (Woollacott, 1977b; Hoshi *et al.*, 1981; Sawada *et al.*, 1982, 1984) but, although one at least is closely similar to mammalian acrosin (Sawada *et al.* 1982) it is not known whether they are located in the acrosome. The fact that in *Ciona intestinalis* the acrosome vesicle appears unchanged after passage of the sperm through the chorion and that the "proacrosomal granules" of *Perophora annectens* become incorporated in the plasmalemma, together with the observation that the "fuzzy extracellular material" [glycocalyx] surrounding the sperm apex is the first to make contact with the egg, have led Fukumoto (1986) to suggest that this glycocalyx is the site of the lysins and that it plays an important role in sperm-chorion interactions at fertilization. The relatively diffuse nature of the acrosome vesicle and its minute size are cited as further indications that it is not a major site of storage of lysins. Rosati *et al.* (1985), however, attribute binding directly to the vesicle. They demonstrated a definite though minute acrosome vesicle in *Ciona intestinalis* by several means, including freeze-fracture, and reported an acrosome reaction which did not diverge significantly from that in other animals. They claim that contact with the chorion [vitelline coat] triggers the fusion of the acrosomal with the sperm plasma membrane; the contents of the acrosomal vesicle flow out and establish close contact with the vitelline coat thus suggesting that the contents are similar to the bindin described for sea urchin sperm; a fibrillar material appears in the subacrosomal space, and the apical region of the nucleus extends, losing its flattened form. The region of the sperm plasma membrane involved in fusion with the acrosome membrane is almost free of particles (as seen by

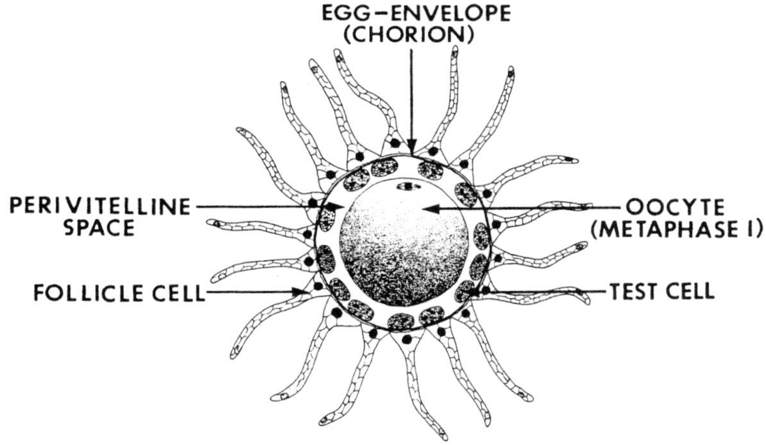

Fig. 3.4. *Ciona intestinalis*. The unfertilized egg and its investments. From Woolacott, R.M. (1977b). *Journal of Morphology* **152**, 77-88. Fig. 1.

freeze-fracture). This particle free area is encircled by two rows of particles: the first is a moniliform circlet of small particles in close mutual contact; the second contains regularly spaced clustered particles 150 Å in diameter. The position of the apical circlet suggests that its particles play a role in receiving and/ or transducing the stimulus derived from the interaction with the glycoprotein(s) of the vitelline coat. Thin filaments seen in negative stained preparations may be the external components of the receptor complex but may be concerned with anchoring the plasma membrane to internal structures. The larger particles, forming the second row, are probably directly related to membrane fusion as in other animals (Rosati *et al.*, 1985). An acrosome-like reaction was also claimed for *Ciona* sperm by De Santis *et al.* (1980, see below).

Overview of ascidian sperm ultrastructure.

To avoid repetition for individual species, ultrastructural investigations will now be reviewed in a comparative account for each order.

Aplousobranchia

Diagnosis. Colonial, rarely (*Ciona, Rhopalaea*) solitary, tunicates with gut loop behind pharynx; pharynx never folded; gonads not paired (Fig. 3.5). *Ciona*, previously placed in the Phlebobranchia is included in the Aplousobranchia in accordance with Kott (1969, 1985). This placement is supported from electron spin resonance and atomic absorption studies which indicate a high level of vanadium (IV) in *Ciona* and in other primitive aplousobranch taxa. Vanadium always is present as vanadium (IV) when

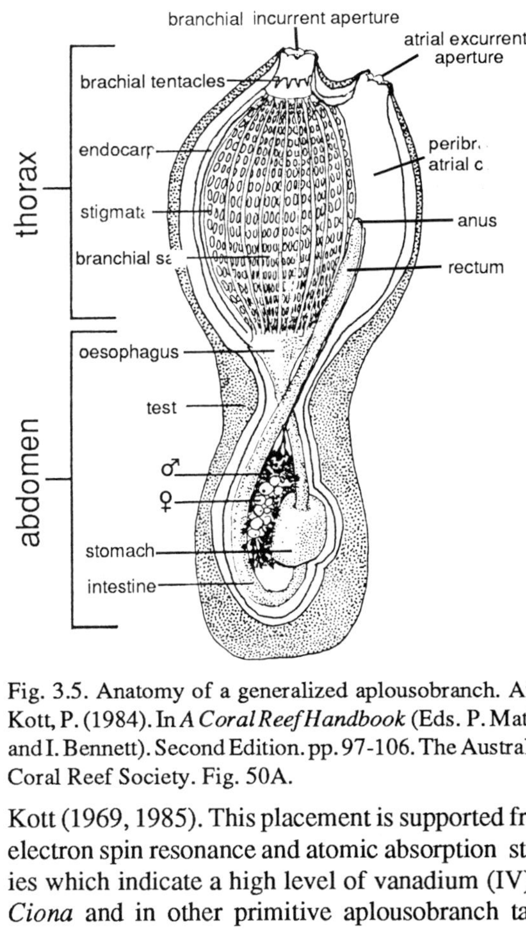

Fig. 3.5. Anatomy of a generalized aplousobranch. After Kott, P. (1984). In *A Coral Reef Handbook* (Eds. P. Mather and I. Bennett). Second Edition. pp. 97-106. The Australian Coral Reef Society. Fig. 50A.

it occurs in aplousobranchs. In Phlebobranchia it is present as vanadium (III). No significant amounts of vanadium are present in Stolidobranchia. Vanadium absorption is considered a plesiomorphic condition for ascidians (Hawkins *et al.*, 1983).

Sperm literature. Cionidae. *Ciona intestinalis*, Ezell, 1963; Georges, 1969; Woollacott and Porter, 1975; Woollacott, 1977a, b; Dale *et al.*, 1978; Rosati and Santis, 1978; Cotelli *et al.*, 1980; De Santis *et al.*, 1980; Rosati *et al.*, 1985; Fukumoto, 1988 (Fig. 3.11). **Clavelinidae.** *Distaplia occidentalis* and *Clavelina huntsmani*, Fukumoto, 1985; *C. oblonga*, Holland, 1989. **Euherdmanidae.** *Aplidium californicum*, Fukumoto, 1985. **Didemnidae.** *Diplosoma listerianum*, Tuzet *et al.*, 1972; Burighel *et al.*, 1982, 1985 (Fig. 3.6, 3.11); *Lissoclinum perforatum* (=*pseudoleptoclinum*), Tuzet *et al.*, 1972); Burighel *et al.*, (1985); *Polysyncraton lacazei* and *Trididemnum cereum*, Tuzet *et al.*, 1974) (Fig. 3.11).

Acrosome. There has been much debate as to whether ascidian sperm have acrosomes. In the Aplousobranchia, denial of the existence of a discrete acrosome by Woollacott and coworkers for the solitary, externally fertilizing *Ciona intestinalis* has not been supported by later work. While not recognizing the existence of a discrete acrosome, they nevertheless recognized an apical reaction, with aryl sulphatase activity (Woollacott and Porter, 1975; Woollacott, 1977a,b) located only in an anterior region of the plasma membrane with surface ornamentation conceivably of Golgi origin; there was evidence for large amounts of proteases in the sperm in the supposed absence of the acrosome; actin, which might form a perforatorium was not demonstrable but actin is not an obligate prerequisite of acrosome reactions; and the sperm undergoes no gross morphological change in ammoniated sea water (see, however, induction of an acrosome reaction by lowered pH by Lambert, 1982), which triggers an acrosome reaction in other invertebrate sperm (Woollacott and Porter, 1975; Woollacott, 1977a,b). It is tentatively hypothesized by these workers that proteases and other acrosome-related materials are intercalated directly into the plasma membrane by the Golgi apparatus during spermiogenesis (Woollacott, 1977b), as is held for *Pyura* and *Styela* by Fukumoto (1983) who, nevertheless, recognizes an acrosome-like structure in these species and in *Ciona*. Actin has been demonstrated anterior to the nucleus in the aplousobranch *Aplidium californicum* (T.S. Schroeder unpublished, cited by Lambert, 1982).

In contrast to the findings of Woollacott and colleagues, an anterior zone of the *Ciona* spermatozoon was thought to be acrosomal by Ezell (1963) and Georges (1969) and a vesicle was demonstrated by Fukumoto (1988). In further confirmation, one or two vesicles lying between a dense plate on the tip of the nucleus and the anterior plasma membrane appeared to originate from small Golgi vesicles, a pattern normal for acrosome development (Cotelli, Santis, Rosati and Monroy, 1980). Furthermore, a small acrosome-like vesicle with weakly PAS-PTA positive contents was reported by De Santis *et al.* (1980) to undergo changes similar to an acrosome reaction upon binding of the spermatozoon to the vitelline coat; these changes consist in the breakdown of the plasma membrane at the tip of the head, in the opening of the 'acrosomal vesicle', in blebbing of the inner 'acrosomal membrane', and in the formation of tubules from the exposed 'inner acrosomal membrane' which make contact with the fibrillar network of the chorion. Rosati *et al.* (1985) further documented an "acrosome reaction" in this species (see fertilization, above).

Dale *et al.* (1978) had previously described, with a micrograph, binding of the *C. intestinalis* sperm to the egg chorion by numerous very thin filaments produced in what they considered to be an acrosome reaction from the outer surface of a similar dome-shaped membrane bound structure *enclosing the anterior tip of the nucleus* [my italics]. It was considered by De Santis *et al.* (1980) that these outer filaments represented only a preliminary gamete binding. The filaments or fibrils were shown to be strongly ruthenium red positive and to bind ferritin-conjugated concanavalin A, as were the plasma membrane at the tip of the sperm head and tufts of fibrils on the outer surface of the chorion and branching fibrils on the inner surface of this layer. Only the outer tufts of the chorion could undergo a binding reaction with the sperm; this indicates that ability of the outer tufts to bind depends on a finer level of molecular organization than is detectable by these methods. These authors hypothesized that while the chorion is the specific and selective fertilization barrier, the follicle cells could function as a non-specific palisade limiting the numbers of sperm reaching the chorion and that the observed sperm-phagcytozing ability of these cells may also be an important component in the overall fertilization process. Bates (1981), who similarly made an SEM study of the distribution of the follicle cells on the *Ciona* egg, found that sperm binding and sperm-oocyte fusion could occur in dechorionated eggs. He nevertheless suggested that for normal eggs the clefts between the follicle cells would allow the sperm to interact directly with the species specific fucosyl sites on the chorion identified by Rosati and De Santis (1980) and suggested that blebs from the cells which partially occluded the clefts might regulate sperm passage.

The reaction phenomena described by de Santis *et al.* (1980) established the existence of a true acrosome and its reaction, though the demonstration of tubules from the inner membrane requires confirmation. Recent work by Fukumoto (1988) has shown unequivocally that the dome-shaped apical vesicle in the sperm of *Ciona intestinalis* has its own bounding membrane, distinct from the overlying, adpressed plasma membrane. It is a flattened vesicle containing moderately electron dense material with an electron dense plate at its centre. Contiguous to this acrosome, an apical substance is located in the anteriormost tip of the sperm head. The acrosome remains intact after sperm-egg binding and there is no observable alteration in the apical plasmalemma. In contrast, spermatozoa which have entered the perivitelline space lack an acrosome. In place of an acrosome, apical processes, considered equivalent to acrosomal processes in other marine invertebrates, are observed at the apex of the spermatozoon. Gamete fusion seems to occur between the egg membrane (oolemma) and some of these apical processes at the tip of the sperm head (Fukumoto, 1988). In his most recent paper, however, Fukomoto (1990) states that in *C. intestinalis* an acrosome reaction occurs by fusion between the acrosomal outer membrane and the plasmalemma involving vesiculation as in mammalian sperm.

In other aplousobranchs, an amorphous anterior structure in the internally fertilizing didemnids *Diplosoma listerianum* and *Lissoclinum perforatum* (=*pseudoleptoclinum*) was claimed to be an acrosome by Tuzet *et al.* (1972). Burighel *et al.* (1982, 1985) report that in these species there are, at the tip of the spermatozoon, some densities and a fuzzy coat, two features which they consider consistent with the hypothesis that ascidians have a poorly developed acrosome in the form of small vesicles and a chorionlysin(s) intercalated into the apical plasmalemma as proposed by Fukumoto (1983); other putatively acrosomal structures (see dense grooves, below) were demonstrated (Fig. 3.6). Fukumoto (1985) reports a small apical vesicle for *Diplosoma macdonaldi*. Anterior coils of the helical head were thought to be acrosomal in the colonial, oviparous species *Polysyncraton lacazei* and *Trididemnum cereum* (Tuzet *et al.*, 1974).

Apical vesicles are reported but not illustrated for *Clavelina huntsmani* (Fukumoto, 1985) and the possibility of the occurrence of some tiny apical vesicles has not been ruled out in a brief reference to *C. oblonga* sperm by Holland (1989).

Perhaps the most compelling proof of the existence of an acrosome, at least in some species, has been the demonstration (by interference light microscopy) that the sperm of the aplousobranch *Aplidium californicum* and *Distaplia occidentalis* (both species with anterior vesicles, Fukumoto, 1985), and the phlebobranch *Ecteinascidia turbinata*, produce a long thin thread from the anterior end when exposed to seawater at pH 9.5 (Lambert, 1982). Confirmation of this observation would be valuable.

Dense groove. In *Diplosoma listerianum* and *Lissoclinum perforatum* (=*pseudoleptoclinum*) (Fig. 3.6), a groove, formed by a narrow invagination of the plasma membrane, runs along the head of the spermatozoon and is in connection with the nuclear membrane through electron dense material; in addition two long, parallel, 20 nm high, ridges run in the bottom of the groove. The groove makes one spiral turn along the length of the head in *L. perforatum* and two in *D. listerianum*; it is less deep but wider in the latter species compared with the former and differs in that the two basal parallel ridges each show two subunits, giving four ridges. In *L. perforatum* the lateral borders of the groove are marked by two couples of narrow laminae with the appearance of two cisternae with a 7 nm wide lumen. In *D. listerianum* the groove ends posteriorly at a right angle at the level of the insertion of the flagellum. On the cytoplasmic side, the groove in both species displays strongly electron dense material which links the nucleus to the plasma membrane; at this juncture the two nuclear membranes are always mutually adherent (Burighel *et al.*, 1982, 1985). These authors observe that the two to four ridges resemble the superficial ornamentation of the head of *Ciona intestinalis* (Georges, 1969; Woollacott, 1977), *Halocynthia roretzi* (Kubo *et al.*, 1978); *Molgula impura* (Villa, 1981); *M. socialis* and *Ascidiella aspersa* (Burighel and Martinucci, unpublished). They suggest that the ridges in *L. perforatum* and *D. listerianum* have a role analogous to that proposed for more diffuse ornamentation in *Ciona* by Woollacott (1977), namely binding of proteases to the plasma membrane which may be involved in fertilization. It is here tentatively proposed that the "helical string" in the sperm head of the stolidobranch *Perophora annectens*, illustrated by Fukumoto (1984b) is the equivalent (homologue?) of the didemnid dense groove.

Nucleus and mitochondrion. The close association of mitochondrion and nucleus, in the absence of a distinct midpiece, which is characteristic of ascidians warrants joint consideration of the two organelles.

In the sperm of the aplousobranch *Ciona intestinalis* (Fig. 3.12) the nucleus is only about 3 µm long and 0.5 µm wide compared with a length of approximately 47 µm for the tail. Ridge-like surface projections, from the plasma membrane, of electron dense material which are 10 nm in height and width extend along the anterior-posterior axis of the head region but are absent from the tail (Woollacott, 1977b).

All but a small anterior region of the *Ciona* nucleus is flanked by the large mitochondrion which partly surrounds it, with a C-shaped transverse profile. Glycogen granules are consistently present in the remaining cytoplasm, especially in the large space between mitochondrion and nucleus, but are absent between the mitochondrion and the plasma membrane (Woollacott, 1977b; Cotelli *et al.*, 1980). The mitochondrion cannot be traced within the egg membranes after fertilization and is presumed to be shed in the "lateral body" at penetration (Ezell, 1963).

Clavelinid sperm are very incompletely known. In *C. oblonga* sperm a single mitochondrion spirals a few times around the nucleus (Holland, 1989).

Didemnids appear more modified than cionids in the marked elongation of the nucleus which they display. Thus the 'head' (the nucleus and mitchondrion and such acrosomal material as it present) has a length and width, respectively, of 12 µm x 1.1 µm in *Lissoclinum perforatum* and 14 µm x 0.8 µm in *Diplosoma listerianum.* In both of these species the nucleus has the form of an inverted club (also described as needle-shaped for *D. listerianum*), with a wide base and long, narrow anterior region. There are a few nuclear pores towards the apex of the head in *Lissoclinum perforatum.*

In *Lissoclinum perforatum* (Fig. 3.6B) the mitochondrion accompanies the nucleus throughout the length of the head. At the opposite side of the head relative to the dense groove it forms an association with the nucleus, embracing it closely though separated by a thin 7 nm space filled with dense material. A "giant cistern", with homogeneous granular contents occupies most of the cytoplasm of the head. It is horseshoe-shaped and embraces the mitochondrion in the anterior two thirds of the head. In the basal third it is reduced to a narrow laminar cistern surrounding the nucleus. The cistern originates from the endoplasmic reticulum and the Golgi complex and it is suggested that it may contain enzymes or other substances to be utilized during internal fertilization (Burighel *et al.*, 1985).

In *D. listerianum* (Fig. 3.6D) the nucleus-mitochondrion association is more extended than in *Lissoclinum perforatum.* In the anterior half, the outer mitochondrial membrane is said to form two very flattened wings embracing almost the entire surface of the nucleus although this is not certainly apparent in micrographs. Dense material again fills the space (8 nm) between mitochondrion and nucleus. The cristae in *D. listerianum* take on a distinctive herring-bone configuration. A remarkable difference from *Lissoclinum perforatum* is that the sperm of *D. listerianum* has long parallel tubules (70 nm in diameter) filling the head cytoplasm; they run spirally and have each moderately dense contents with a central core. They are not continuous along the entire length of the head as there are different numbers (12 to 30) at different levels. It is suggested that the close association and linkage by dense material of the mitochondrion and nucleus may be related to energy requirements for internal fertilization in these two dideminds and in holding the two organelles together (Burighel *et al.*, 1985).

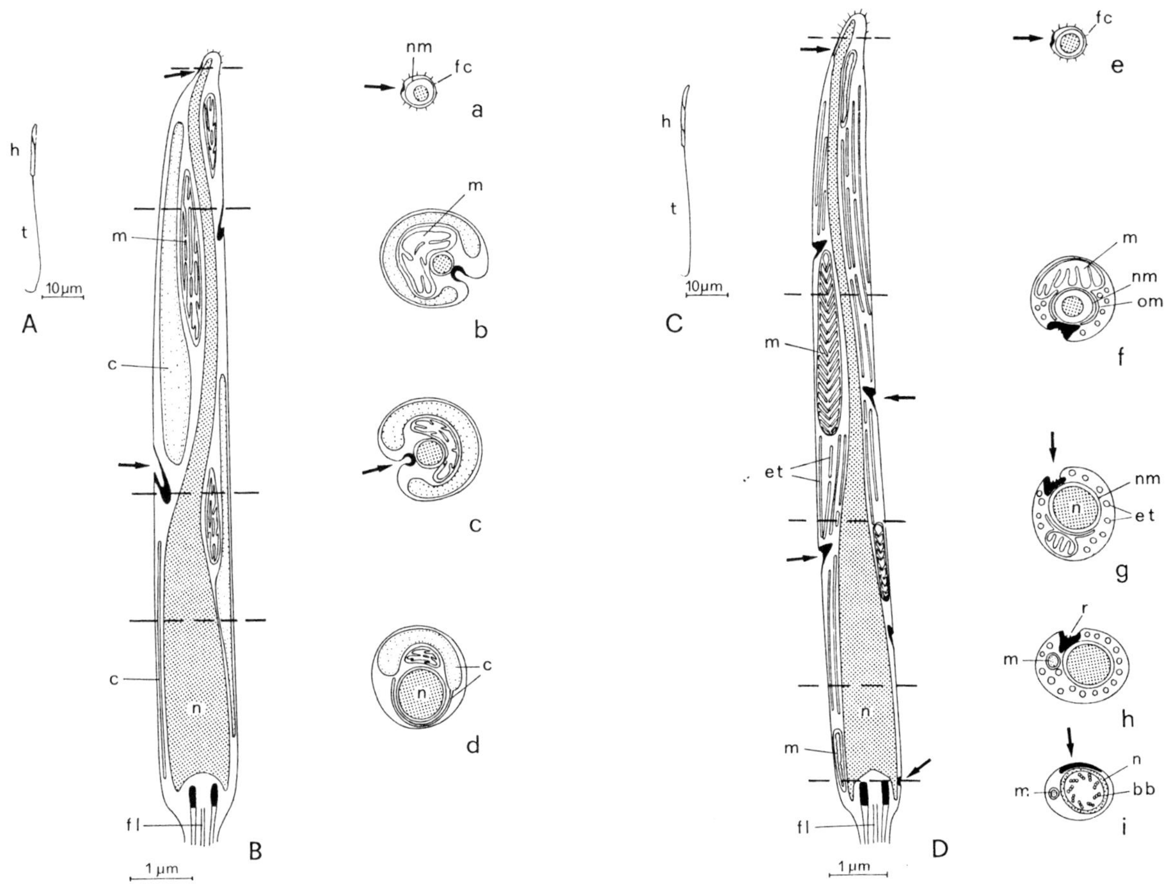

Fig. 3.6. Diagrammatic drawings of the spermatozoa of *Lissoclinim perforatum* (A, B, a-d) and *Diplosoma listerianum* (C, D, e-j). A, C entire spermatozoa. B, D. the fine structure of the head in longitudinal sections. The arrows point to a furrow in the plasmalemma, i.e., the dense groove, bound to the nucleus. The groove runs spirally on the head surface; the long mitochondrion associated with the nucleus, and the endoplasmic derivatives also run spirally, within the head. In *L. perforatum*, the endoplasmic derivatives form a giant cisterna (c) which in the basal third, where the groove is lacking, narrows around the nucleus; in *D. listerianum* they form a series of parallel tubules (et). bb.basal body. c. cisterna. et. endoplasmic tubules. fc. fuzzy coat. fl. flagellum. h. head. m. mitochondrion. n. nucleus. nm. nuclear membrane. om. outer mitochondrial membrane. r. ridges at bottom of groove. t. tail. After Burighel, P., Martinucci, G.B. and Magri, F. (1985). *Cell and Tissue Research* **241**, 513-521. Fig. 30.

In the didemnid *Polysyncraton lacazei*, behind the putatively acrosomal coils there is the long straight nucleus along which the mitochondrion extends as far as the basal body of the flagellum.

In *Trididemnum cereum* (Fig. 3.12) the entire head of the sperm is spiral but the core of the nucleus remains a linear rod. The helical structure, from micrographs, is a strongly protuberant, lamina-like flange tilted, though not strongly, relative to the core. The mitochondrion, said by Tuzet *et al.* (1974) to be absent at maturity has been shown to form a helix wound between the gyres of the nuclear flange and an acrosome is absent (Holland, pers. comm.)

Centriole and flagellum. There is a strong trend in ascidian sperm to loss of the proximal centriole. In *Ciona intestinalis*, consistent with its phylogenetically primitive nature (Fig. 3.2), both centrioles are present and coaxial, though the proximal is displaced

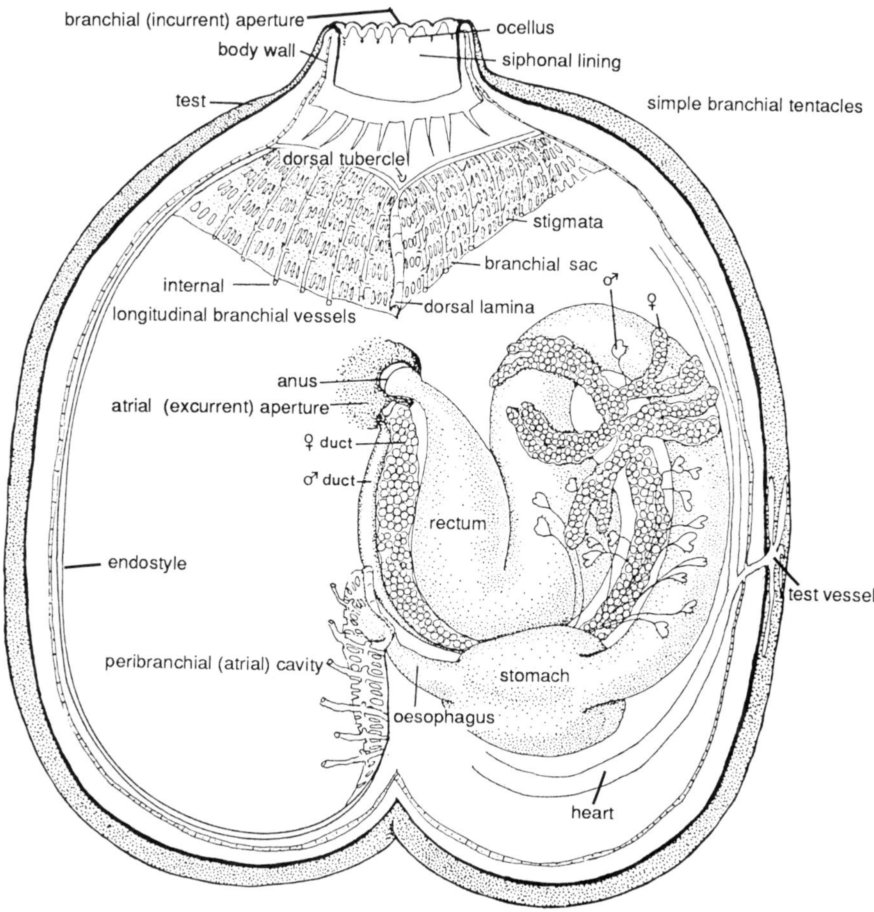

Fig. 3.7. Morphology of a phlebobranch ascidian (diagrammatic). The body is shown opened around the ventral midline (from the branchial aperture). The branchial sac has been largely removed to expose the peribranchial cavity and organs embedded in the parietal body wall. After Kott, P. (1985). *Memoirs of the Queensland Museum* **23**, 1-440. Fig. 1.

laterally (Cotelli *et al.*, 1980). This is substantiated by a micrograph for a presumably mature spermatozoon though other workers do not mention or (Georges, 1969) specifically deny the persistence of the proximal centriole in *Ciona* as in all other investigated species. The distal centriole gives the 9+2 flagellum. A 9-rayed anchoring apparatus is described for the spermatid with an implication that it persists in the mature sperm (Cotelli *et al.*, 1980). The normal electron dense arms are present on the A tubules, and radial links extending to a "sheath" which surrounds the two central singlets (Woollacott, 1977b). As in the sea urchin *Lytechinus pictus*, an endpiece 5-6 μm long contains two central singlets and a variable, small number of A-tubule extensions, lacking dynein arms, of the doublets. This structure suggests that the bending resistance of the endpiece should be approximately an order of magnitude less than that of the 9+2 mainpiece of the axoneme; a computer simulation using this interpretation did not accord with observed swimming but closer agreement was found when it was assumed that bending resistance of the endpiece decreased gradually from the transitional region to the tip; bends are propagated smoothly off the end of the flagellum but if the endpiece is removed in actual or computer-simulated flagella, rapid unbending of bends which have reached the distal end occurs. It is estimated that the elastic bending resistance of an individual microtubule is 0.01×10^9, while that of the 9+2 region is 0.2×10^9 pN nm^2 (Brokaw and Omoto, 1982; Omoto and Brokaw, 1982).

Phlebobranchia

Diagnosis. Usually solitary and rarely (Perophoridae, Plurellidae) colonial tunicates with gut loop embedded in the parietal wall. Pharyngeal wall not folded; gonads enclosed in the gut loop, on one side of the body (Kott, 1985) (Fig. 3.7).

Sperm literature. Perophoridae. *Perophora formosana*, Fukumoto, 1981 (Fig. 3.12); *Perophora annectens*, Fukumoto, 1983, 1985 (Fig. 3.8). **Corellidae.** *Corella parallelogramma*, Franzén, 1976b; 1983; *C. inflata*, Lambert *et al.*, 1981; *C. pacifica*, Fukumoto, 1985. **Ascidiidae.** *Ascidia nigra*, Schabtach and Ursprung, 1965; Ursprung and Schabtach, 1965. *A. callosa* Cloney and Abbott, 1980 (Fig. 3.12); *A. ceratodes*, *A. malaca*, Villa, 1975, 1977; *Phallusia mammillata*, Honneger, 1986.

Acrosome. Although an acrosome appears to have been demonstrated for some phlebobranchs, evidence for its presence in others is equivocal.

In the oviparous perophorid *Perophora formosana* (Fukumoto, 1981) (Fig. 3.12) and *P. annectens* (Fig. 3.8) (Fukumoto, 1983, 1984, 1986) anterior vesicles are at least transiently present but a unique feature is a long cross-striated apical structure behind these which is presumably part of an acrosomal complex. It is an elongate, digitiform, apical structure about 2 μm long present at the anterior end of the sperm and is composed of an electron opaque substance in the form of dense transverse striations, each about 7 nm thick. In the young spermatid, vesicles present anterior to an apical dense plate on the nucleus are reminiscent of the putative acrosome of *Ciona.* The transverse striation of what may here be termed the acrosomoid of *P. formosana* and *P. annectens* is similar to that seen in the polychaete

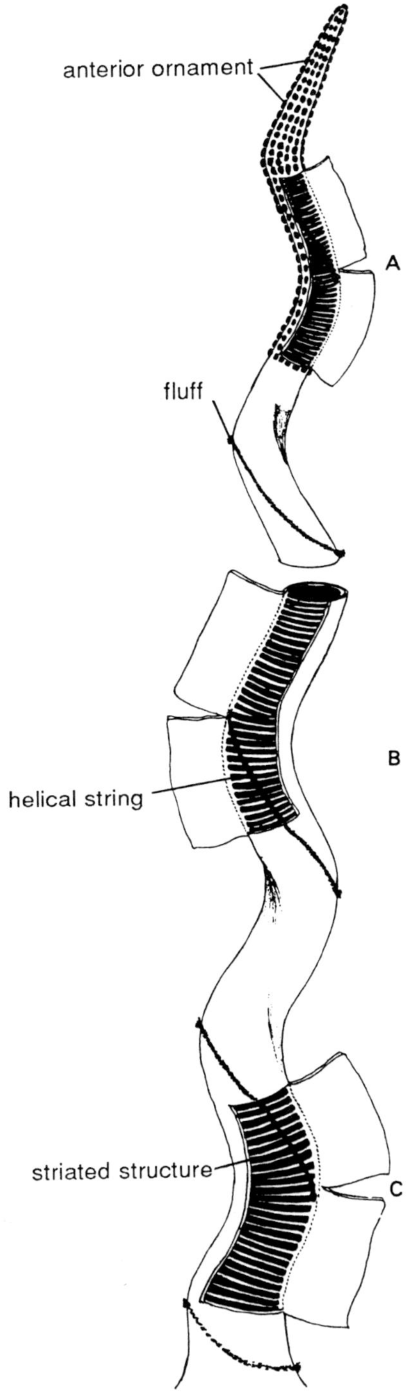

Fig. 3.8. *Perophora annectens.* Schematic illustration of the apical structure. The plasmalemma enclosing the anterior quarter of the apical structure is decorated by the anterior ornaments. An helical string can be seen running within the ridge of the helix. "Fluff" is present on the plasmalemma immediately outside the region corresponding to the helical string. A, B and C are views inside the anterior, middle and posterior regions respectively. After Fukumoto, M. (1984b). *Journal of Cell Science* **66**, 175-187. Fig. 6.

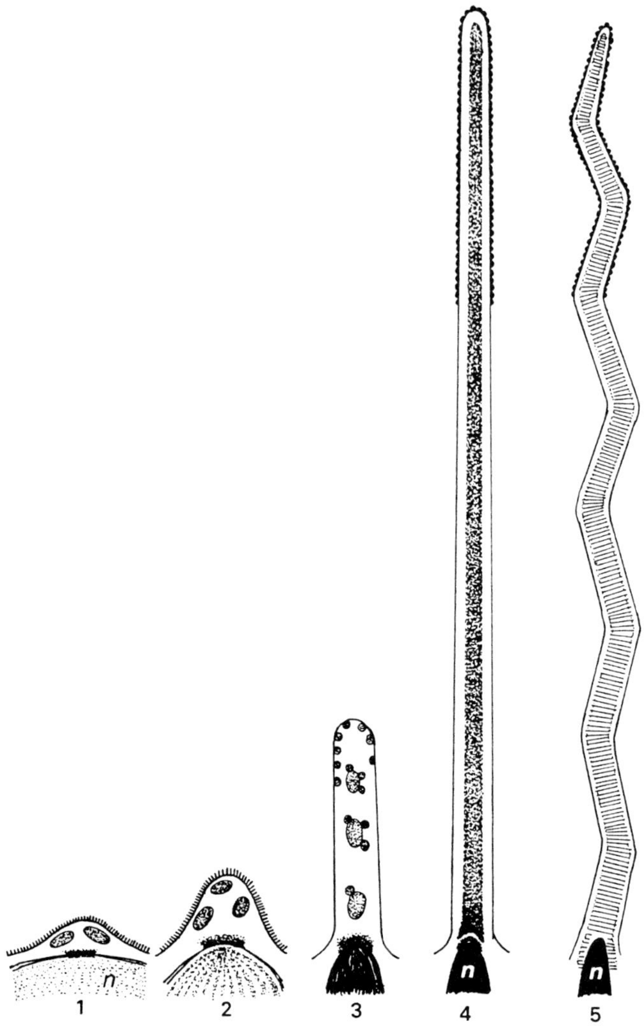

Fig. 3.9. *Perophora annectens*. Schematic illustration showing the differentiation of the apical structure. In younger spermatids, the plasmalemma at the apex expands to form a small blister (stage 1) which is covered with fuzzy extrcellular material. The blister develops further through a conical projection (stage 2) into a finger-like process (stage 3) similarly decorated. This becomes an elongated process with electron dense material in its core (stage 4). Finally, the elongated process is helically coiled to form the apical structure in which electron dense material becomes aggregated periodically to form a striated structure (stage 5). The anterior quarter (approximately 1 µm) of both the elongated process (stage 4) and the apical structure (stage 5) corresponds to the finger-like process at stage 3. Vesicles (presumably proacrosomal vesicles) have been recognized in younger stages (stages 1-3). In the finger-like process at stage 3, these vesicles appear to transform into smaller vesicles which ultimately fuse with the anterior plasmalemma of the finger-like process. After Fukumoto, M. (1984b). *Journal of Cell Science* **66**, 175-187. Fig. 17.

Idanthyrsus (Jamieson and Rouse, 1989) in which it is known to participate in an acrosome reaction, though there unqestionably a component of the acrosome vesicle itself. In *P. annectens* the 50-70 nm vesicles of presumed Golgi origin are incorporated into the plasma membrane but the cross striated structure persists (Fukumoto, 1983, 1984, 1986) (Fig. 3.9).

In *Perophora annectens* the apical structure is even more developed than in *P. formosana*. It is 4 µm long and helical with a 0.3 µm repeat. It is composed of electron dense material which appears to have aggregated periodically to form a cross striated structure (Fig. 3.8) with an helical ridge in which runs longitudinally an "helical string". It has tentatively been proposed above that this string is the equivalent of the dense groove which spirals around the acrosome in the didemnids *Diplosoma listerianum* and *Lissoclinum perforatum* (=*pseudoleptoclinum*) demonstrated by Burighel *et al.* (1985) but this hypothesis has not been incorporated into the tentative phylogeny given in Fig. 3.20.

Each cross-band in *Perophora annectens* is about 7 nm thick. The striated structure lacks a limiting membrane. Extracellular materials, the "anterior ornaments", coat the anterior quarter, approximately 1 µm, of the plasma membrane enclosing the apical structure in fairly regular array (Fukumoto, 1984). Development of the apical structure is illustrated and described in Fig.3.9.

With regard to other phlebobranchs, the dense plate in *Corella parallelogramma* may be an acrosome (Franzén, 1976b). The supposed acrosome in *Ascidia nigra* described by Schabtach and Ursprung (1965) appears in a micrograph to be better developed than in most other ascidians but its acrosomal nature is questionable. A very electron dense line, 100 Å thick, at the tip of the nucleus, was identified as acrosomal. This line was sometimes seen to be continuous further posteriorly with a vesicle enclosed in 80 Å membranes and filled with moderately electron dense material on each side of the nucleus which it presumably encircles. From the posterior rim of the vesicle an electron dense line (sheath) is seen in the diagram extending to the posterior end of the nucleus. The writer has come independently to the conclusion of Cloney and Abbott (1980) that this extensive acrosome is in fact the equivalent of the nuclear envelope, with dense material between the inner and outer membranes which they describe for the sperm of *Ascidia callosa*.

Cloney and Abbott (1980) identify nuclear pores in micrographs by Schabtach and Ursprung (1965) of the so-called acrosome of *A. nigra* and demonstrate these for the nuclear envelope of *A. callosa*. In the sperm of *A. callosa*, between the nuclear envelope and the plasma membrane at the wedge-shaped tip of the head, they identify two vesicles, 45-55 nm in diameter, much as described by Cotelli *et al.* (1980) for *Ciona intestinalis*. The cytoplasm surrounding the vesicles is more electron dense than elsewhere, and the vesicles are regarded by Cloney and Abbott (1980) as a putative acrosome, though no functional data are available.

Villa (1975, 1977) states categorically in a brief description of the sperm of *A. malaca* that an acrosome does not exist though none of her micrographs convincingly endorses this.

In *Phallusia mammillata* freeze-fracture studies after cryofixation reveal that the sperm apex is encircled by three rows of 8-10 nm large particles forming a cap-like structure. In sperm cryofixed and freeze-substituted, up to eight vesicles (50-60 nm in diameter) containing moderately electron dense material were observed apically; in such sections an electron dense membrane structure corresponds to the circlets of particles revealed by freeze-fracture. No dramatic changes occur in sperm which have penetrated the vitelline coat. Some of the vesicles have fused with the sperm membrane and have released their contents, considered probably to be lytic enzymes to digest the vitelline coat. Some vesicles remain intact. Dissolution of the central layer of the vitelline coat occurred without physical contact with the sperm, suggesting that released lysins acted over a considerable distance. Membrane tubules or blebbing, described by De Santis *et al.* (1980) was not observed in this species. Apart from some reduction in vesicle numbers, sperm swimming in the perivitelline space or just about to establish contact with the egg membrane were similar to unreacted sperm. Sperm-egg fusion appeared to occur between the plasma membrane of the postacrosomal region of the sperm head and the egg membrane (Honneger, 1986).

Nucleus and mitochondrion. The sperm of phlebobranchs (*Corella*, *Ascidia*, and especially *Perophora*) are apomorphic with regard to elongation of the nucleus. Investigation of the sperm of the phlebobranch *Ascidia nigra* by Schabtach and Ursprung (1965) gave the first ultrastructural delineation of the now familiar ascidian type of sperm (Fig. 3.3): rodlike nucleus; the mitochondrion wrapped around the nucleus, in this case leaving only a small interruption and extending (Ursprung and Schabtach, 1965) for about 85% of the nuclear length. The base of the nucleus has a shallow sigmoid depression fuctioning as an implantation fossa.

In *Ascidia callosa*, the mitochondrion lies lateral to the posterior three quarters of the long rodlike nucleus which it only partly invests (Cloney and Abbott, 1980). In *A. malaca* the nucleus though described as conical (Villa, 1975, 1977), appears rod-shaped. Villa reports a break down of the plasma- and nuclear- membranes on contact of the sperm with the egg, an observation questioned by Cloney and Abbott (1980). In *Phallusia mammillata* the head, though elongated, is only 5.5 µm long. The mitochondrion is wrapped laterally around about one third of the length of the nucleus (Honneger, 1986).

The sperm of the simple, externally fertilizing correlid *Corella parallelogramma* was used by Franzén (1976b, 1983) to exemplify the basic "ascidian type" of spermatozoon (Fig. 3.3); the nucleus is, nevertheless, elongate with a dense plate or line anteriorly which questionably represents an acrosome; a large mitochondrion lies lateral to but does not embrace the nucleus.

We have seen that *Perophora formosana* has a highly modified sperm (Fukumoto, 1981) (Fig. 3.12). It is approximately 140 µm long and contains an apical structure, an extremely filiform nucleus, about 90 µm long (about 30 times as long as that of the sperm of *Ciona intestinalis*), and a single unusually long mitochondrion, about 80 µm long, which is loosely wound around the nucleus as the "lateral body". The anterior third of the head is helically coiled. Filamentous structures about 10 nm thick are observed exclusively in the mitochondrial matrix. They are arranged parallel to one another along the long axis of the mitochondrion. It is suggested that these contribute to mitochondrial elongation. The sperm of *Perophora annectens* is 70 µm long, 30 µm of which is the head including the 4 µm long apical structure (Fukumoto, 1984).

Shedding of the mitochondrion, typical of ascidians, has been investigated by Koch and Lambert (1986) in *Ascidia ceratodes*. During fertilization the single mitochondrion swells and migrates along the entire length of the tail and is shed completely. This process can be artificially induced by low Na^+ or high pH (9.3) seawater. The outer mitochondrial membrane of unreacted sperm is tightly associated with the nuclear membrane. During translocation of the mitochondrion along the tail, its outer membrane, usually at the leading and trailing tips of the organelle, was attached to sites immediately adjacent to the axoneme. These membrane associations may be the 20-25 nm knobby connections seen in high voltage TEM. Electron dense structures were seen to project from the outer microtubules of the axoneme towards the mitochondrial membrane at many locations (Koch and Lambert, 1986). This process of shedding is also described for *Phallusia mammillata* (Honneger, 1986).

Centriole and flagellum. Parallel to the basal body of the 9+2 axoneme in *Perophora formosana* there is a recognizable vestige of the proximal centriole. The tail is about 50 µm long and ends in a 5 µm long tapered region (endpiece) (Fukumoto, 1981). In *Corella parallelogramma* only one distinct centriole persists, forming the basal body which, perhaps typically for ascidian sperm, is set at a slight angle to the long axis of the sperm (Franzén, 1976b). In a light microscope study it has been shown that in the sperm of the externally fertilizing *Corella willmeriana*, which closely resemble those of *C. parallelogramma* in size, the flagellum occupies a significantly greater part of the sperm than in the internally fertilizing *C. inflata*. In addition, the sperm heads are more uniformly cylindrical in *C. inflata* (Lambert *et al.*, 1981).

The anterior end of the basal body lies in a concavity in the posterior end of the nucleus in *Ascidia callosa* (Cloney and Abbott, 1980) or in an oblique shallow fossa in *A. malaca* (Villa, 1975, 1977); in both species dense material at its side is termed a centriolar adjunct. This is probably a remnant of the proximal centriole (Cloney and Abbott, 1980).

Taxonomic and phylogenetic value. Lambert (1982) considers the Phlebobranchia to be monophyletic as the sperm of colonial and solitary members are considered by him to be relatively similar. He nevertheless recognizes similarities between perophorid and aplousobranch sperm which suggest a special relationship between these entities. Perophorids are here considered highly derived phlebobranchs in, for instance, possession of the apical structure.

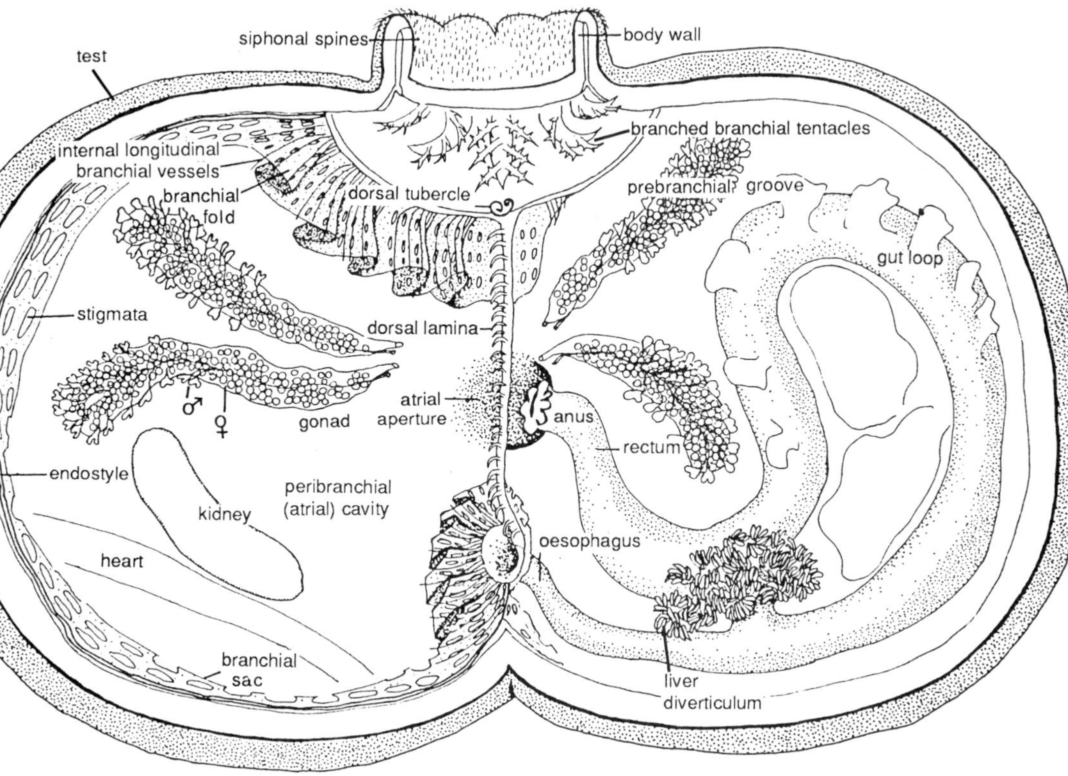

Fig. 3.10. Morphology of a composite stolidobranch ascidian, showing morphological characteristics of Styelidae, *Pyura* and Mogulidae (diagrammatic). The body is shown opened around the ventral midline (from the branchial aperture). The branchial sac has been largely removed to expose the peribranchial cavity and organs embedded in the parietal body wall. After Kott, P. (1985). *Memoirs of the Queensland Museum* **23**, 1-440. Fig. 2.

Stolidobranchia

Diagnosis. Solitary or colonial tunicates with folded pharyngeal wall; gonads in gut loop on both sides of the body; gut loop in parietal body wall beside the pharynx (Kott, 1985) (Fig. 3.10).

Sperm literature. Styelidae. *Metandrocarpa taylori*, Fukumoto, 1985. *Botryllus schlosseri* (Fig. 3.11), *Cnemidocarpa finmarkiensis* (Fig. 3.12), *Styela clava*, Fukumoto, 1986; *S. plicata*, Villa, 1981); Fukumoto, 1983 (Fig. 3.12). **Pyuridae.** *Halocynthia roretzi*, Kubo *et al.*, 1978; *Pyura vittata*, Fukumoto, 1979 (spermatids); *P. haustor*, Fukumoto, 1983 (Fig. 3.12); *Boltenia villosa*, Cloney and Abbott (1980; Fukumoto, 1986 (Fig. 3.12); *Herdmania momus*, Fukumoto, 1986, and original. **Molgulidae.** *Molgula impura*, Villa, 1981; *M. manhattensis*, Fukumoto, 1985, 1986 (Fig. 3.11); *Microcosmus sabatieri*, Villa and Tripepi 1981, 1983. These are solitary species with the exception of the colonial *B. schlossseri* and *Metandrocarpa taylori*.

Acrosome. Villa (1981) found no evidence of an acrosome in the simple, externally fertilizing stolidobranch *Styela plicata*, stating that a small flattened vesicle about 200 Å thick forming a cap over the tip of the nucleus was in fact a section of the nuclear envelope pierced by pores and containing dense material in the lumen. However, Fukumoto (1983) has described an acrosome-like structure [hereafter termed the acrosome for brevity] for this species (Fig. 3.12) and for the pyurid *Pyura haustor* and (Fukumoto, 1985) for *P. vittata*. In *S. plicata* the acrosome is an antero-posteriorly elongated, flattened vesicle, approximately 200 x 100 x 50 nm in length, width and height, respectively, compared with a slightly depressed ellipsoid approximately 90 x 80 x 50 nm, in *P. haustor*. It lies lateral to the wedge-shaped, anteriorly tapered tip of the nucleus, and contains moderately electron dense material which, in *S. plicata*, accumulates on the side facing its inner membrane with strands extending to the outer membrane. The outer membrane of the acrosome is in close contact with the overlying plasmalemma in the region where fuzzy extracellular material is absent. The inner and outer nuclear membranes make close contact with each other to form a pedestal for the acrosome. There is a gap, crossed by connections and considered to be a subacrosomal space, between the acrosome and the nuclear membrane. In both species the acrosome vesicle appears to form by fusion of two smaller vesicles in the spermatid. The small size of the acrosome vesicle of the sperm suggests that acrosomal lysins for penetration of the egg chorion are intercalated into the plasmalemma enclosing the sperm head as is actually observed in *Perophora annectens* (q.v.) (Fukumoto, 1983).

Subsequently, Fukumoto (1985, 1986) has demonstrated a putative acrosomal vesicle, said to resemble that of *Ciona intestinalis*, containing moderately electron dense material at the apex of mature spermatozoa in *Styela clava*, *Cnemidocarpa finmarkiensis* and *Botryllus schlosseri* (family Styelidae); in *Boltenia villosa* and *Herdmania momus* (family Pyuridae) and in *Molgula manhattensis* (family Molgulidae). The outer membrane of the vesicle in all of these species is in close contact with the overlying plasmalemma and the here contiguous inner and outer nuclear membranes form a pedestal for the vesicle. A fuzzy material decorating the external surface of the plasmalemma is apparently not restricted to the apex (though this is claimed for *Perophora annectens*, Fukumoto, 1984). The development of the apical vesicle in the sperm of *Molgula manhattensis* is illustrated and described by Fukumoto (1985) (Fig. 3.11).

An estimated acrosomal volume in the six species and in *S. plicata* and *Pyura haustor* ranges from 0.0004 μm^3 in *Pyura haustor* to 0.001 μm^3 in the two *Styela* species, in *Cnemidocarpa finmarkiensis* and in *Herdmania momus*, compared with volumes of 0.03-0.16 μm^3 in the sperm of several echinoderms. The acrosome volume relative to the volume of the sperm head ranges from a minute 0.05-0.20% in the ascidians compared with 2.1-16% in the echinoderms (Fukumoto, 1986).

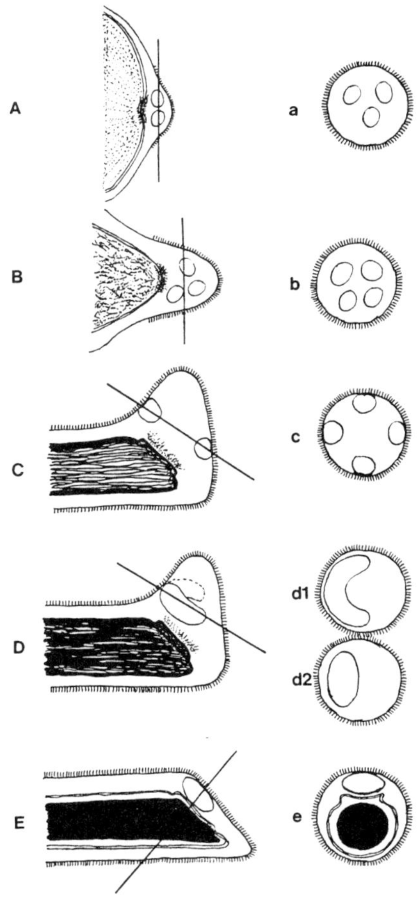

Fig. 3.11. *Molgula manhattensis*. Schematic illustration showing differentiation of the axoneme. In early spermatids (A, B), the plasma membrane at the apex expands to form a blister which is covered with fuzzy extracellular material. Three or four moderately electron dense vesicles (50-60 nm in diameter) are present in the blister (a, b). Midway through spermiogenesis (C, D), these vesicles attach to the inner surface of the plasmalemma enclosing the blister (c). The vesicles then fuse with each other along the inner surface of the plasmalemma and form a horseshoe-shaped acrosomal vesicle (d1) which transforms into a sphere (d2). In mature spermatozoa (E), the acrosome is a slightly depressed sphere positioned at the apex (e). From Fukumoto M. (1985). *Journal of Ultrastructure Research* **92**, 158-166. Fig. 23.

Although a similar, Golgi-derived acrosomal

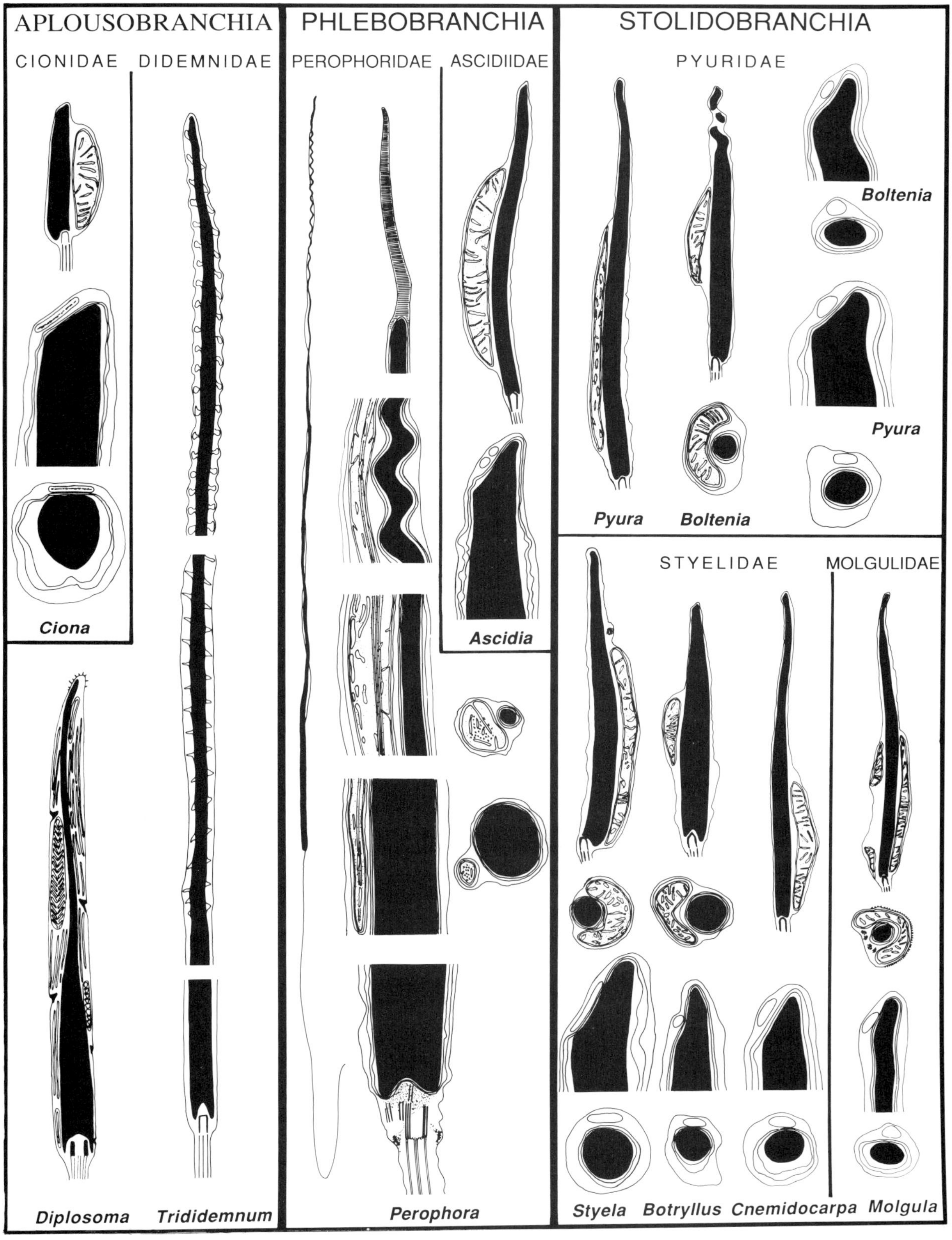
APLOUSOBRANCHIA
CIONIDAE
DIDEMNIDAE
Ciona
Diplosoma
Tridideminum
PHLEBOBRANCHIA
PEROPHORIDAE
ASCIDIIDAE
Ascidia
Perophora
STOLIDOBRANCHIA
PYURIDAE
Boltenia
Pyura
Pyura
Boltenia
STYELIDAE
MOLGULIDAE
Styela
Botryllus
Cnemidocarpa
Molgula

vesicle is vestigial at maturity in *Halocynthiaroretzi*, the nuclear envelope at the tip has thickened; the chromatin is not closely attached to the envelope, unlike other regions of the nucleus, so that an intranuclear space is formed. The narrow space between the thickened nuclear envelope and the plasma membrane is filled with amorphous material (Kubo *et al.*, 1978). There is biochemical evidence for production of typical acrosome products in fertilization of *Halocynthia roretzi* although their location is not known. Two proteases, acrosin and spermosin, have been demonstrated in sperm extracts of this species by Sawada *et al.* (1982, 1984).

In light of the above, supposed absence (Villa, 1981) of an acrosome in *Molgula impura* must be doubted. Finally, a cap-shaped structure at the anterior end of the nucleus occurs, at least in the late spermatid, of *Microcosmus* sp. (Villa and Tripepi, 1982).

Nucleus and mitochondrion. The reported length of the stolidobranch nucleus varies from 5 to 14 µm. In *Styela plicata* the sperm is of the typical ascidian type. The head is 8 µm long, relatively short for an ascidian as appropriate to external fertilization, and the nucleus is a slightly tortuous tapering rod. The mitochondrion extends for more than half the length of the nucleus and embraces it as a C-shape (Fukumoto, 1983).

The sperm of the pyurid *Boltenia villosa* is of the basic ascidian type but the head has a corkscrew-like tip which includes the anterior end of of the nucleus (Cloney and Abbott, 1980; Fukumoto, 1986). The nuclear envelope resembles that of *Ascidia callosa*. The anterior and posterior parts are in contact with the chromatin; the posterior two thirds are loosely arranged and convoluted. At the anterior end the inner and outer layers are more parallel, the lumen is filled with dense amorphous material and there are nuclear pores. The mitochondrion is wrapped about halfway around the nucleus. The cristae are elongate and triangular in cross section and are often packed tightly in hexagonal arrays. As in the phlebobranch *A. callosa*, the glycocalyx is prominent on the head but very thin or absent on the tail (Cloney and Abbott, 1980).

The head of the sperm of another pyurid, *Halocynthia roretzi*, is only about 5 µm long but it is relatively elongate as it is only 0.3 µm wide; the tip is weakly tapered and bends ventrally. The mitochondrion occupies the posterior two thirds of the length of the head (Kubo *et al.*, 1978).

In *Pyura haustor*, the head is 14 µm long (Fukumoto, 1983). In *Pyura vittata* longitudinal tube-like structures, 300 Å wide, are present in the matrix of spermatid mitochondria exclusively along the inner membrane adjacent to the nucleus. During elongation of the mitochondrion, which extends finally throughout the posterior two thirds of the nucleus, the number increases, reaching 14-16 in the late spermatid and mature sperm (Fukumoto, 1979). Similar intramitochondrial tubules are seen in spermatids of *Pyura haustor* (Fukumoto, 1983) and in rat pinealocytes, rat interscapular brown adipose tissue, and in renal tubules of the snake *Elaphae quadrivirgata* (references in Fukumoto, 1979). It is deduced that, in *Pyura* at least, they contribute to

Fig. 3.12. (Opposite). Summary of sperm ultrastructure in the three suborders of the Ascidiacea. Within the Aplousobranchia, *Ciona* shows basic features of the ascidiosperm (including a reduced acrosomal vesicle). Internally fertilizing aplousobranchs show interesting modifications of the ascidiosperm. The helical structure in *Trididemnum* is a nuclear flange accompanied by gyres of the spiral mitochondrion (Holland, pers. comm.). Among Phlebobranchia, internally fertilizing perophorids differ significantly from *Ascidia*, notably by the presence of a non-membrane bound striated apical structure (acrosomal vesicle in *Ascidia*). Within the Stolidibranchia, few modifications of the ascidiosperm have been described (helical nuclear tip in *Boltenia*). as. striated apical structure. dc. distal centriole. er. endoplasmic rods. f. flagellum. fs. fibrous structures within mitochondrion. g. helical groove. m. mitochondrion. pc. possible proximal centriole. From micrographs by Clony, R.A. and Abbott, L.C. (1981). *Cell and Tissue Research* **206**, 261-270. Figs. 1, 3 (*Ascidia callosa*); Fukumoto, M. (1981). *Journal of Ultrastructure Research* **77**, 37-53. Figs. 1b, 4a-c, 5b, c, 6a,b, 8. (*Perophora formosana*); Fukumoto, M. (1983). *Development, Growth and Differentiation* **25**, 503-515. Figs. 10, 11b, 12a (*Pyura haustor, Styela plicata*); Fukumoto, M. (1984a). *Journal of Ultrastructure Research* **87**, 252-262. Figs. 1-3 (*Ciona intestinalis*); Fukumoto, M. (1986). *International Journal of Invertebrate Reproduction and Development* **10**, 335-346.Figs. 6, 7, 18, 19 (*Boltenia villosa, Botryllus schlosseri, Cnemidocarpa finmarkiensis, Molgula manhattensis*); Tuzet, O. *et al.* (1974). *Bulletin de Biologie de France et Belgique* **108**, 151-167. Figs. 19, 20, 22b (*Tridideminum cereum*); *Diplosoma listerianum*, redrawn from Burighel *et al.* (1985). *Cell and Tissue Research* **241**, 513-521. Fig. 30D.

mitochondrial elongation. They are not simple microtubules but have a solid core and a wall with a fairly regular transverse banding suggestive of a helix or stack of rings (Fukumoto, 1979).

In *Molgula impura*, the nucleus, 6 μm long and 0.3 μm wide, is rodlike though anteriorly slightly pointed. The mitochondrion, with convoluted cristae, appears U-shaped in a micrograph of a transverse section with the nucleus between the tips of the arms, thus bending of the mitochondrion around its own longitudinal axis appears somewhat independent of the nucleus, as, indeed, seems the case in other ascidian species with C-shaped or other crescentic mitochondria. A glycocalyx occurs on the plasma membrane surrounding the mitochondrion (Villa, 1981).

Fukumoto (1986) has confirmed or extended these observations with demonstration of a single long mitochondrion laterally applied to the nucleus, and embracing it in a C-shaped configuration in cross section, in the styelids *Styela clava*, *Cnemidocarpa finmarkiensis* and *Botryllus schlosseri*, the pyurids *Boltenia villosa* and *Herdmania momus* and the molgulid *Molgula manhattensis*.

Centriole and flagellum. A single centriole, 0.3 μm long and 0.15 μm wide, is present in a small concave basal fossa in *Halocynthia roretzi*; it has nine triplets anteriorly but doublets posteriorly (Kubo *et al.*, 1978). Presence of only the distal centriole is confirmed for *Botryllus schlosseri* (Burighel *et al.*, 1982). The 9+2 flagellum of *Molgula impura* is about 28 μm long (Villa, 1981).

CLASS THALIACEA

Diagnosis. Planktonic. Branchial and atrial apertures at opposite ends of the body. Water current used for locomotion in addition to feeding and respiration. Life cycle complex; sexually reproducing stages derived as buds (blastozooids) from vegetative individuals. Three orders: Pyrosomatida; Doliolida and Salpida.

Sperm structure. Sperm of the Pyrosomatida, Doliolida and Salpida have been examined ultrastructurally. All are internally fertilizing and live-bearing.

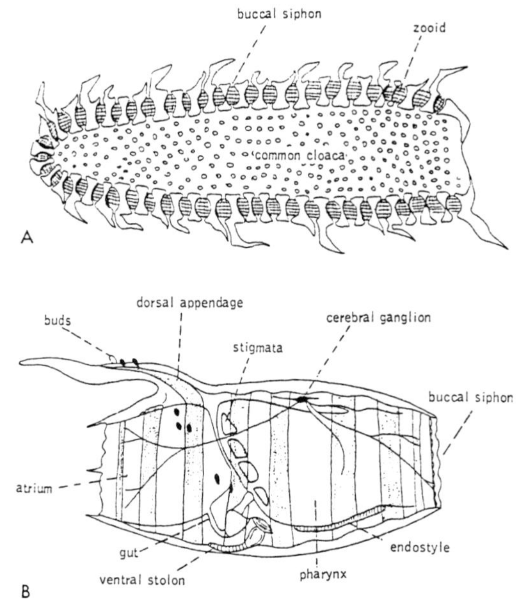

Fig. 3.13. Thaliacea. A. *Pyrosoma*. The colony, in longitudinal section. B. Oozooid of the solitary thaliacean *Doliolum*. Buds develop on the ventral stolon and migrate to the dorsal process where they attach and grow. From Barnes, R. S. (1980). *Invertebrate Zoology*. 5th. Edition. Saunders, Philadelphia. Fig. 20-20.

Order Pyrosomatida

Diagnosis. Compact colonies with numerous individuals embedded in the wall of a gelatinous tube, which is usally long, more or less cylindrical, and closed at one end. The zooids, which are virtually independent, are held together by the common test, and have their branchial apertures at the outer surface of the tube. The atrial apertures open into the common cloacal cavity. The individual water currents fuse into a powerful stream emerging from the open, posterior end of the tube and effecting jet propulsion. Strongly phosphorescent. Vegetatively produced blastozooids progressively added to the colony. No alternation of generations. No tailed larvae (Berrill, 1950).

***Pyrosoma atlanticum* sperm**. Spermatozoal ultrastructure has recently been investigated in *Pyrosoma* by Holland (1990) (Fig. 3.14).

Acrosome. The spermatozoon lacks an acrosome but at the tip of the spermatid there are several 50 nm proacrosomal vesicles. These disappear leaving no trace in early elongating spermatids. A 35 μm long acrosome identified by light microscopy (Franzén, 1958, 1983) was clearly nucleus and a spiral cytoplasmic extension observed ultrastrucutrally.

Nucleus and spiral extension. The head is 35 μm long. The nucleus is differentiated into a bulbous posterior portion 5 μm long × 1 μm wide, a thin anterior portion 25 μm 1ong × 0.2 μm wide, and a very thin anterior extension 5 μm long × 0.1 μm wide. At the start of elongation, the anterior extension begins to form just lateral to the proacrosomal vesicles as a spiral projectlon comprising part of the nucleus covered by a thin sheath of cytoplasm. This sheath of cytoplasm undergoes a complex differentiation. Ultimately, the nucleus in the anterior extension is overlain by two membrane bound sheaths of cytoplasm connected by a spiral flange of cytoplasm. Between these two sheaths is a spiral space, open to the exterior through a subteminal pore near the sperm tip. A manchette of microtubules transiently encircles the thin anterior portion of the nucleus during the last phase of elongation.

Mitochondria. In early spermatids the mitochondria fuse into a single mitochondrion which remains lateral to the nucleus. The cristae become modified late in spermatogenesis. Throughout elongation of the spermatid there are patches of dense material between the nucleus and mitochondrion. In the spermatozoon the mitochondrion, which has reticulate cristae, spirals a few times about the nucleus and extends from the junction between the bulbous portion and the thinner anterior portion of the nucleus to the junction between the thinner anterior portion and the nuclear extension.

Centriole. The proximal centriole disappears in the early spermatid. Its remains apparently persist in the spermatozoon as dense material adjacent the distal centriole. The latter gives rise to the axoneme.

Sperm phylogeny. Spermatogenesis in *Pyrosoma atlanticum* compared to that in other tunicates, most closely resembles that in colonial ascidians, and is considered by Holland to support the majority view that pyrosomes arose from aplousoboanch ascidians that lost their attachment to the substratum. Pyrosome sperm are more highly derived than doliolid sperm, which are plesiomorphic in having an acrosome that is probably capable of exocytosis. When salp and pyrosome sperm are compared both are seen to be

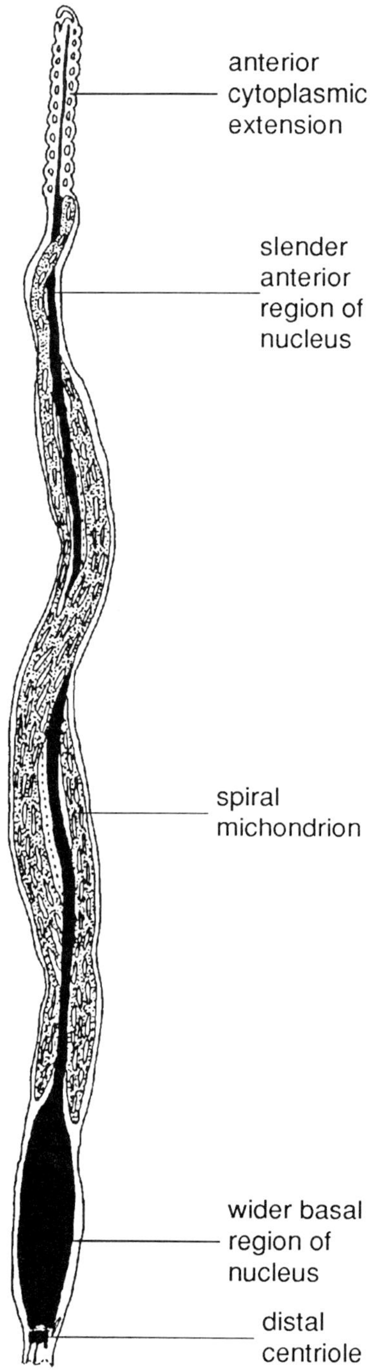

Fig. 3.14. *Pyrosoma atlanticum*. Spermatozoon. After Holland, L. (1990). *Marine Biology* (In press). Fig. 3.

highly derived but neither shows any apomorphies with the other that it does not share with at least one other tunicate order. Thus sperm morphology does not support the majority view that pyrosomes gave rise to doliolids and neither confirms nor denies the idea that pyrosomes are intermediate between aplousobranch ascidians and salps (Holland, 1990). The paucity of apomorphies presented by tunicate and specifically thaliacean sperm, and therefore of clear phylogenetic affinities, possibly does not warrant Holland's conclusion, provisionally accepted in Fig. 3.20, that the class Thaliacea is polyphyletic, with doliolids arising very early from the ascidian lineage and with salps and pyrosomes arising somewhat later.

Order Doliolida

Diagnosis. Body barrel-shaped, completely surrounded with 8 or 9 hooplike muscles. Long parallel stigmata at posterior end of branchial sac. Alternation of solitary sexual gonozooid (producing tailed larvae) with polymorphic, colonial, vegetative generation (Kott, pers. comm.).

Sperm literature. The spermiogenesis and a few details of the mature spermatozoon of *Dolioletta gegenbauri* and spermiogenesis to the mid-spermatid of *Doliolum nationalis* (zooids, Fig. 3.13B, 3.15), have been examined ultrastructually by Holland (1989). The mid-spermatids of the two species are virtually identical. Only the spermatozoon of *Doliolum gegenbauri* is described here. Holland's description is augmented from accompanying micrographs.

***Dolioletta gegenbauri* sperm**. The spermatozoon (possibly a very late spermatid) is about 70 µm long, including the head which is 9 to 10 µm long and about 1 µm wide.

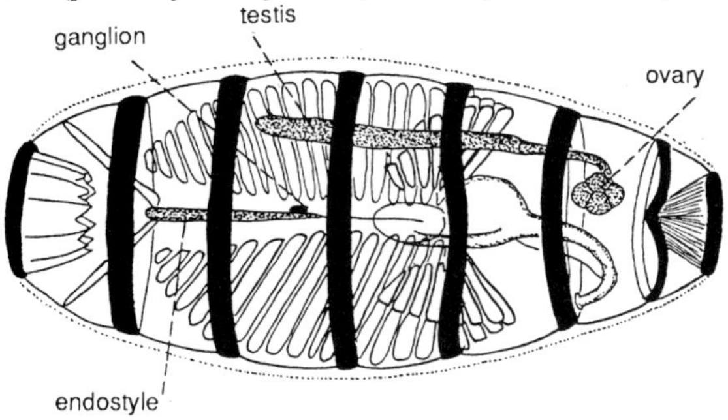

Fig. 3.15. *Doliolum nationalis*. Gonozooid. After Berrill, N.J. (1950). *The Tunicata*. Ray Society, London. Fig. 102A. After Borgert.

Acrosome. The acrosome, present in the spermatid, has completed exocytosis. The fibrous material derived from the acrosome remains loosely associated with the inner surface of the acrosomal membrane, which has become the outer surface of the plasma membrane at the tip of the sperm and back about 300 nm along its sides. In the spermatid the acrosome is present as a small vesicle, about 250 nm in diameter by 50 nm thick with its long axis parallel to the plasma membrane from which is separated by a narrow space about 15 nm wide. It has dense contents with a denser plate of material in the centre parallel to its long axis. It lies in a lens shaped area of cytoplasm between the thickened anterior portion of the nuclear envelope and the apical plasma membrane.

Nucleus and mitochondrion. The nucleus is rod-shaped, gradually narrowing anteriad with a blunt, rounded tip. A thin sheath of "mitochondria" surrounds the anterior two thirds of the nucleus. It is stated that they remain numerous but Holland questions the full maturity of the gametes investigated. From micrographs it appears possible that some, if not all of them, fuse to form a single structure, C-shaped in cross section, in the mature sperm. At the posterior third of the sperm head, the plasma membrane is closely applied to the nucleus. At the posterior end of the head, the nuclear envelope has an indentation about 0.5 µm deep (implantation fossa) into which fits the centriole.

Centriole and flagellum. A proximal centriole is absent but, in the spermatid at least, there are two parallel profiles which appear to be short (150 nm) microtubules extending at about 45° from the proximal end of the distal centriole (basal body) which may be the remnant of the proximal centriole. There are no specializations of the plasma membrane at the base of the flagellum or along the axoneme; this is of the 9+2 type.

Fertilization biology. All doliolids are hermaphroditic and produce ripe eggs and sperm simultaneously. Fertilization is probably internal, at least in *Doliolum*. Eggs are apparently released from the ovary into the atrial cavity of the mother where they are fertilized by sperm entering with the incoming current of water and soon thereafter are shed into the external seawater (the conditions for recognition of ent-aquasperm in the terminology of Rouse and Jamieson, 1987). Eggs of doliolids, like those of solitary ascidians, have test cells in the perivitelline space and follicle cells external to the vitelline layer [chorion] but the follicle cells are more widely separated than in ascidians (references in Holland, 1989). Holland (1989) therefore suggests that access of doliolid sperm to the vitelline layer should be somewhat easier than for ascidian sperm. Although the acrosome was seen to undergo exocytosis in the late spermatid while still in the testis, the possibility was recognized that this might be a fixation artefact and that the acrosome reaction (exocytosis, though with no acrosome filament) might in nature occur while the sperm were exterior to the vitelline layer. Whether the reaction occurred at this point or in the testis, it was hypothesized that the dense fibrous acrosomal contents released by exocytosis might briefly bind the sperm to the vitelline layer while other components of the acrosome might create a hole for passage of the sperm.

Order Salpida

Diagnosis. Body transparent cylindrical or prism-shaped body, surrounded with hooplike muscle bands interrupted ventrally. Stigmata short, in an oblique line across posterior end of the branchial sac. Alternation of solitary vegetative oozoid with colonial (aggregated) sexual blastozooids. No tailed larva (Thompson, 1948; Berrill, 1950).

Sperm literature. A light microscope description have been given for the sperm of *Thalia* (=*Salpa*) *democratica* by Franzén (1958) and for that of *Heterosalpa virgula* by Pictet (1891). Spermiogenesis in the salps *Thalia democratica* and *Cyclosalpa affinis* has been examined by Holland (1988). The following account refers to *Thalia democratica* but Holland (1988) states that the fine structure of the male germinal cells in the testes *T. democratica* and *Cyclosalpa affinis* is identical and that the general description for *H. virgula* by Pictet (1891) is confirmed. The spermatozoa are, again, ascidiosperm.

***Thalia democratica* sperm.** Mature sperm of *Thalia* (=*Salpa*) *democratica* are 54 µm (Franzén, 1958) to 58 µm (Holland, 1988) long. The rod-shaped head, 0.8 µm wide at the flagellar end and about 140 nm at the tip, has a narrow, elongate nucleus with a length of 18 µm (Holland, 1988). Presence of a 3 µm long acrosome (Franzén, 1958) has not been confirmed; indeed, Holland (1988) states that at no stage in spermiogenesis of *T. democratica* is there an active Golgi apparatus or any clearly derived Golgi-derived vesicle. An unusual "cytoplasmic" spiral around the nucleus (Franzén, 1958), has been shown by Holland to be the mitochondrion.

Nucleus and mitochondrion. A midpiece is absent. The mature mitochondrion is a single tube spiralled 45 times around the entire length of the nucleus. The nucleus tapers, from about 0.7 µm anterior to the centriole to about 50 nm at the tip. The chromatin is fully condensed in the anterior two thirds of the sperm and at the periphery of the posterior third. A cone of less condensed chromatin extends anteriorly from the centriole into the axis of the flagellum. "Fuzzy material" may be apparent exterior to the plasma membrane of the head, particularly near the tip. No microtubules are involved in the shaping of the head. This accords with the view that the forces required to shape the sperm nucleus and the spiral mitochondion are generated within these organelles and that the shape of the head is caused by a genetically determined pattern of chromatin condensation. It is further suggested that dense material within the mitochondrion may be responsible for its elongation which occurs after, and presumably independently, of nuclear elongation.

Plasmalemmal groove. A deep indentation of the plasma membrane encircles the base of the flagellum. This groove is possibly homologous with the dense groove of didemnid sperm.

Centriole and flagellum. The distal centriole inserts into a pocket at the posterior end of the nucleus. There is no proximal centriole. The tail is 40 µm long. The axoneme is of the 9+2 type but has dense material peripheral to each of the doublets; it has a "fuzzy coat" 20 nm thick.

Remarks. Because *H. virgula* and *C. affinis* are in the family Cyclosalpinae whereas *T. democratica* is in the Salpinae, it seems likely that spermiogenesis follows the same pattern among all salps (Holland, 1988).

CLASS APPENDICULARIA (LARVACEA)

Diagnosis. Planktonic. Adults retaining many characteristics of the ascidian tadpole and possibly neotenic. Mouth located at the anterior end; intestine opening directly to the outside ventrally. Only one pair of pharyngeal clefts, opening directly to the exterior; no atrium. Tail held at right angles to the body. Epidermis typically secreting a delicate mucoid "house"; tunic absent. Reproduction sexual only, giving a tadpole larva which undergoes metamorphosis without settling and retains the tail (Fig. 3.16).

Sperm literature. *Oikopleura dioica*, Flood and Afzelius (1978); Holland *et al*. (1988).

***Oikopleura* sperm**. The sperm of *Oikopleura dioica* (Fig. 3.3, 3.17) is about 30 µm long, with a spherical to cylindrical, about 1 µm wide, head, a 3 µm long and 1 µm wide midpiece, and 25 µm long free tail with a tapered 2 to 3 µm long endpiece (Flood and Afzelius, 1978; Holland *et al*., 1988).

The well developed acrosome and discrete midpiece are notable differences from the Ascidiacea and Thaliacea and are considered plesiomorphic conditions.

Acrosome. The long axis of the acrosome is slightly oblique to the cell axis. The acrosome contains an acrosomal vesicle and has a diminutive subacrosomal space. The vesicle is oval or oblong, 0.3 µm long and up to 0.2 µm wide (Flood and Afzelius, 1978, Holland *et al*., 1988). Its volume, 0.008 μm^3, is only about 10% of that of the holothurian acrosome. It is inserted into a 0.3 µm deep anterior nuclear fossa; thus the anterior and posterior poles of the nuclear envelope are closely apposed and form a thin septum between the centriole and acrosome (erroneously termed the axoneme). The posterior pole of the acrosome is narrowly indented and a small evagination of the thick nuclear envelope inserts into the indentation; in midsagittal sections no subacrosomal space is visible. The indented acrosomal and nuclear membranes appear to blend into a dense cord extending the length of the acrosome (Holland *et al*., 1988). This core corresponds with the central fibrous strand surrounded by a less dense zone described by Flood and Afzelius. It seems possible to the writer that this axial zone and strand (perforatorium?) is in fact subacrosomal in a deep basal invagination of the vesicle.

Acrosome reaction. On encountering the egg, the sperm undergoes an acrosome reaction involving exocytosis of the acrosome vesicle and production of an acrosomal tubule (Fig. 3.19).

The acrosomal contents bind the sperm to the vitelline layer (follicle cells are absent) and the posterior part of the acrosomal membrane and the anterior portion of the nuclear envelope evaginate together to form an acrosomal tubule which fuses with the egg plasma membrane to form a fertilization cone. By 45 s after insemination, the sperm nucleus, centriole,

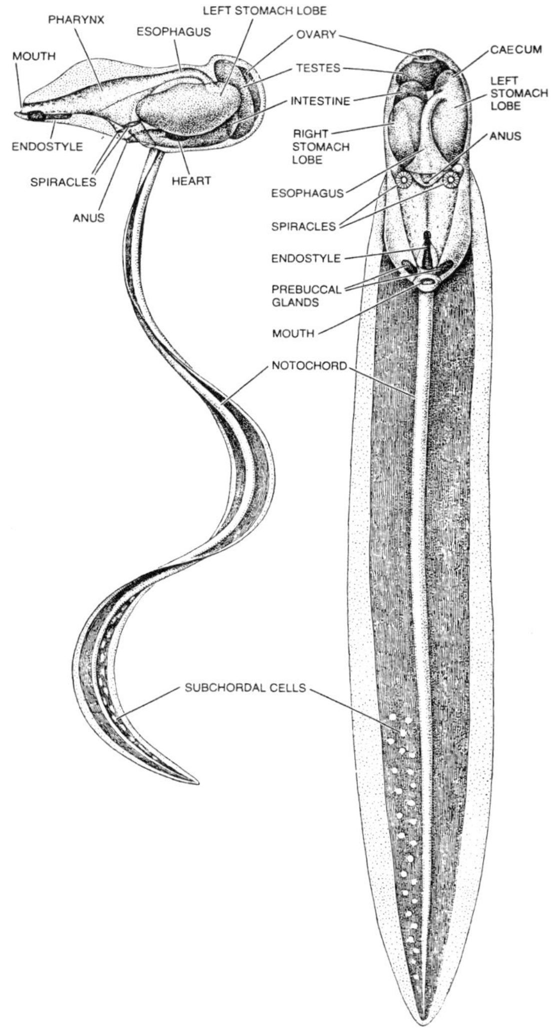

Fig. 3.16. *Oikopleura albicans*. Lateral and dorsal view of the tadpole-like adult. From Alldredge A. (1976). *Scientific American* **23**, 95-102. Fig. 1.

mitochondrion and at least the anterior portion of the axoneme are within the fertilization cone. By 60 s sperm entry is complete (Holland *et al.*, 1988). These authors hold that *O. dioica* resembles echinoderms and enteropneusts (we might add cephalochordates and fish) in having eggs with a cortical reaction and an acrosome reaction but differs from ascidians in these respects. However, there is some evidence for an, albeit reduced, acrosome reaction in ascidians (see above) though never with the classicial acrosomal tubule seen in *Oikopleura*.

Nucleus. The nucleus is 1 µm long and 0.6 µm wide (Holland *et al.*, 1988). The chromatin volume of the nucleus is only about 0.1 µm^3 (against 8 µm^3 for nuclei of internal epithelial cells), indicating only about 9 times the DNA content of the *E. coli* haploid genome (Flood and Afzelius, 1978), compared with the still low figure of 35 times in *Ciona intestinalis* (and *Drosophila*) and hundreds of times in most gametes of multicellular eukaryotes (references in Atkins and Ohno, 1967; Laird, 1971). An anterior inpocketing of the nucleus containing a compact but well developed acrosome is almost contiguous with a posterior inpocketing which contains the basal body (a normal 9+0 centriole with anterior triplets) and the proximal part of the 9+2 flagellum. The chromatin is thus completely penetrated axially, forming an annulus, but, as already indicated, the nuclear membranes are continuous across the gap between acrosome and basal body (Flood and Afzelius, 1978). Most of the intranuclear material is condensed chromatin, except for a small amount of diffuse fibro-granular material adjacent to the anterior fossa.

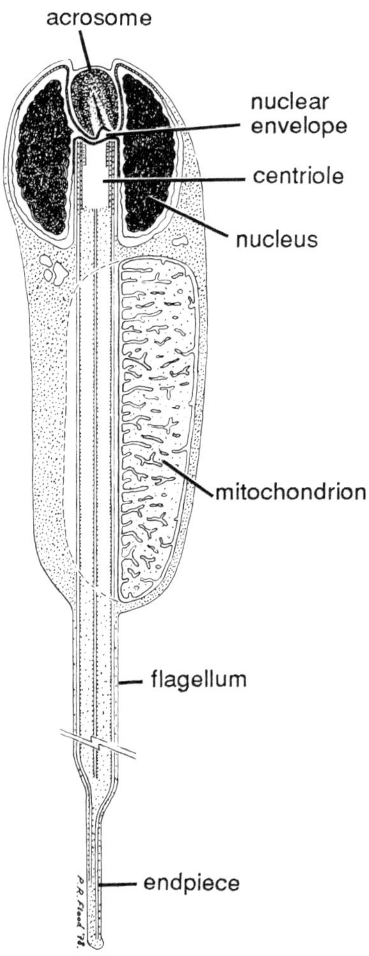

Fig. 3.17. *Oikopleura dioica*. Longitudinal section of spermatozoon. After Flood, P.R. and Afzelius, B.A. (1978). *Cell and Tissue Research* **191**, 23-37. Fig. 13.

Mitochondrion. Unlike that of the Ascidiacea, the mitochondrion is located behind the nucleus, giving a distinct midpiece, but it is similar in being single with a C-shaped profile, though here embracing the axoneme (Flood and Afzelius, 1978). A major difference from other tunicates is that the mitochondrion enters the egg (Holland *et al.*, 1988).

Centriole and flagellum. A further similarity to the Ascidiacea (with the possible exception of *Ciona intestinalis*, Cotelli *et al.*, 1980) is the absence of a pericentriolar anchoring apparatus (Flood and Afzelius, 1978). As usual in the latter order a proximal centriole is absent. Holland *et al.* (1988) add that the posterior nuclear fossa into which the axoneme inserts is 0.7 µm deep, most of the length of the nucleus.

The small size of the nuclei is thought probably to allow the storage of large numbers of sperm for expulsion in the short time that the animal is outside its jelly coat.

Sperm phylogeny in tunicates. A phylogeny of the Urochordata based on general anatomy is given in Fig. 3.18 and may be contrasted with relationships which might be deduced from sperm morphology (Fig. 3.20).

Sperm morphology suggests that tunicates are monophyletic. Thus the ascidiosperm, with reduced

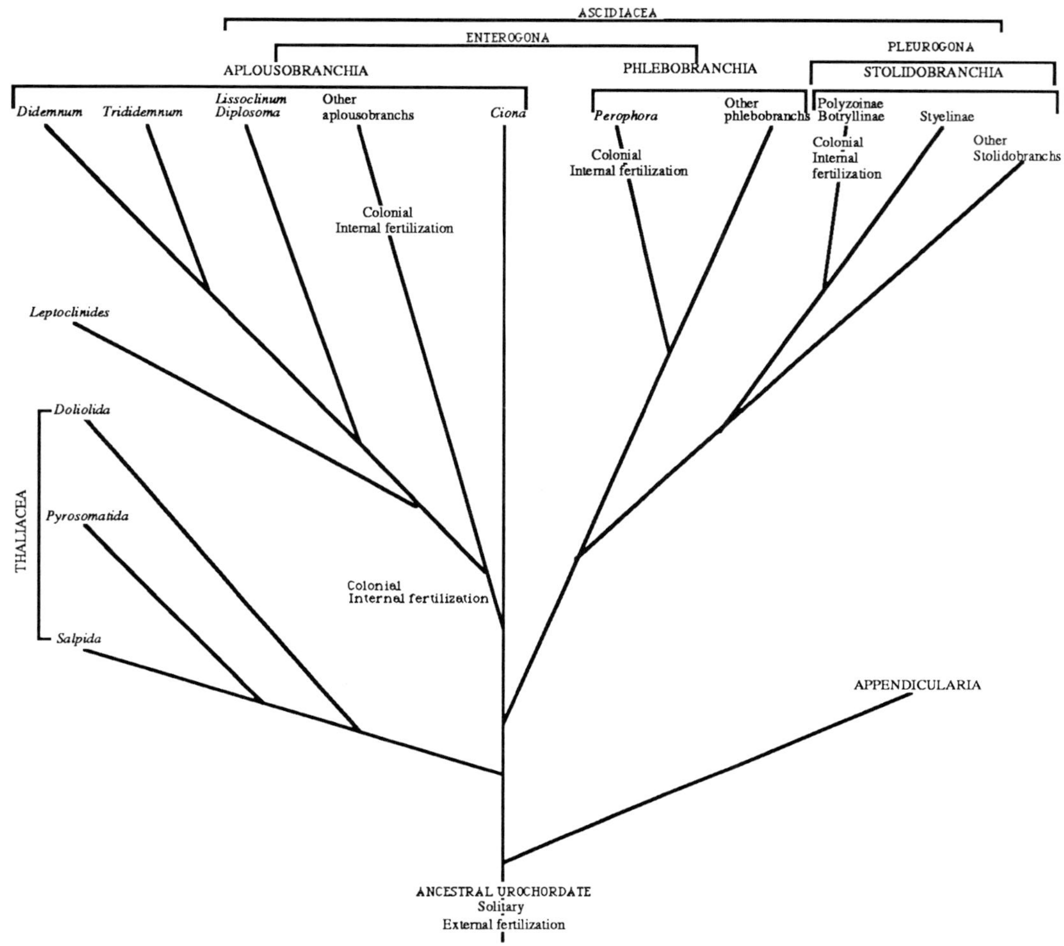

Fig. 3.18. Phylogeny of the major groups of the tunicates, including genera discussed here, based on general anatomy. Courtesy of Dr. Patricia Kott (pers. comm.)

acrosome, and a mitochondrial mass lateral to the nucleus (questionably with many separate mitochondria in doliolids) is seen in Ascidiacea and Thaliacea (doliolids, salps and pyrosomatids) while the more plesiomorphic spermatozoon of the appendicularians also has a lateral mitochondrion. Plesiomorphic features of the appendicularian sperm include the para-axonemal location of the mitochondrion and presence of a well developed and reactive acrosome. A major link between the three classes is the presence of a tadpole at some stage of the life history. Loss of the proximal centriole (questionable in *Ciona intestinalis*) to which Holland (1989) has drawn attention is a weak tunicate synapomorphy as it occurs in many phyla as a sporadic variant, for instance, in cnidarians. Neither this nor the reduction of the acrosome space is unique or autapomorphic but they are nevertheless unifying trends in the Urochordata.

In Fig. 3.20 the chief features of the sperm of urochordates which have been investigated ultrastructurally are arranged hierarchically in a tree. Although this tree is based partly on taxonomic preconceptions of relationships, it may be regarded as a tentative spermatologically derived phylogeny of the Urochordata. Though having points of similarity, it is considerably, at variance with the phylogeny (itself one of many conflicting phylogenies which have been proposed) based on general anatomy of adults and larvae shown in Fig. 3.18.

Holland (1989) rightly considers spermiogene-

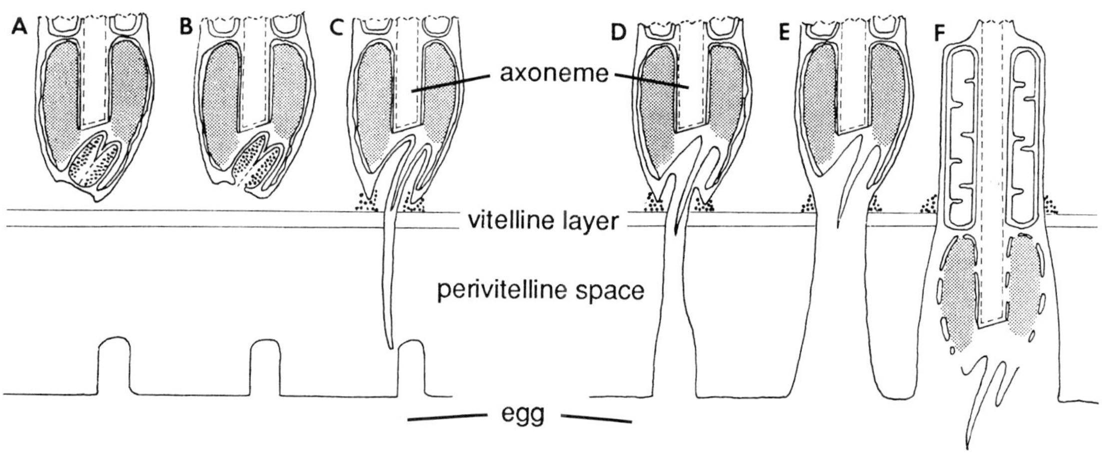

Fig. 3.19. *Oikopleura dioica* . Schematic drawing of sperm-egg interaction. Sperm chromatin is indicated by fine stippling and the contents of the acrosome vesicle by coarse stippling. A. Unreacted sperm approaching the egg vitelline layer. The space between the nuclear envelope and acrosomal membrane is exaggerated. B. The acrosome reaction is initiated. At the anterior end of the sperm the acrosomal membrane and sperm plasma membranes fuse. C. The acrosome completes exocytosis. The acrosomal contents bind the sperm to the vitelline layer of the egg. The posterior portion of the acrosomal membrane and the anterior portion of the nuclear envelope evaginate together to form the acrosomal tubule and this penetrates the vitelline layer. D. The acrosomal tubule fuses with the egg plasma membrane to form a fertilization cone, at about 30-40 s after insemination. E. The fertilization cone widens as it fills with egg cytoplasm. F. The sperm moves through the fertlization cone into the egg. The anterior portion of the nuclear envelope moves away from the chromatin, at about 45 s after insemination. After Holland, L.Z., Gorsky, G. and Fenaux, R. (1988). *Zoomorphology* ***108***, 229-243. Fig. 5.

sis, and the spermatozoon, in doliolid thaliaceans when compared with other tunicates to be closest to that in solitary members of the Ascidiacea, except that in the latter the sperm mitochondria fuse and the acrosome appears to be incapable of exocytosis. In contrast, salps, also within the Thaliacea, and colonial didemnid ascidians have acrosomeless sperm [though the dense groove in didemnids may be a highly modified acrosome-derivative] with a spiral mitochondrion. A further similarity of salps and ascidians is their asynchronous spermatogenesis while that of appendicularians is synchronous (Holland, 1988). By outgroup comparison with echinoderms and acraniates, it is deduced (Holland, 1989) that appendicularian sperm are plesiomorphic within the Tunicata. Thus, it was argued, gamete morphology indicates that (1) solitary ascidians and doliolids had a common ancestor, (2) the popular idea that doliolids gave rise to appendicularians is incorrect, and (3) that Thaliacea are polyphyletic, doliolids having arisen very early from the ascidian lineage and salps having arisen later. These postulates appear acceptable but polyphyly of the Thaliacea remains controversial.

Holland (1990) has since demonstrated similarity of the thaliacean *Pyrosoma* to salps in possession of a spiral mitochondrion with modified cristae. Holland (1988) has drawn attention to similarities of the sperm of salps and those of dideminid ascidians, particularly *Trididemnum cereum*, which suggest that salps and didemnids may be closely related and that, because their developing spermatids both have a single lateral mitochondrion, they are both derived from a sperm like that of modern ascidians with external fertilization such as *Ascidia callosa* with a single lateral mitochondrion with unmodified cristae. This "ancestral sperm" was envisaged as having a rudimentary acrosome which was lost in the evolution of didemnids and salps. In the light of recent demonstration (Holland, pers. comm.) that the sperm of *Trididemnum cereum* are less similar to those of salps than was predicted by Holland (1988), notably in that *Trididemnum* does not have the spiral tubular acristate form of mitochondrion seen in salps, and that the aplousobranch *Clavelina* has a spiral mitochondrion of the didemnid type, a special relationship between salps and didemnids is no longer proposed but the affiinity of salps with aplousobranchs is upheld.

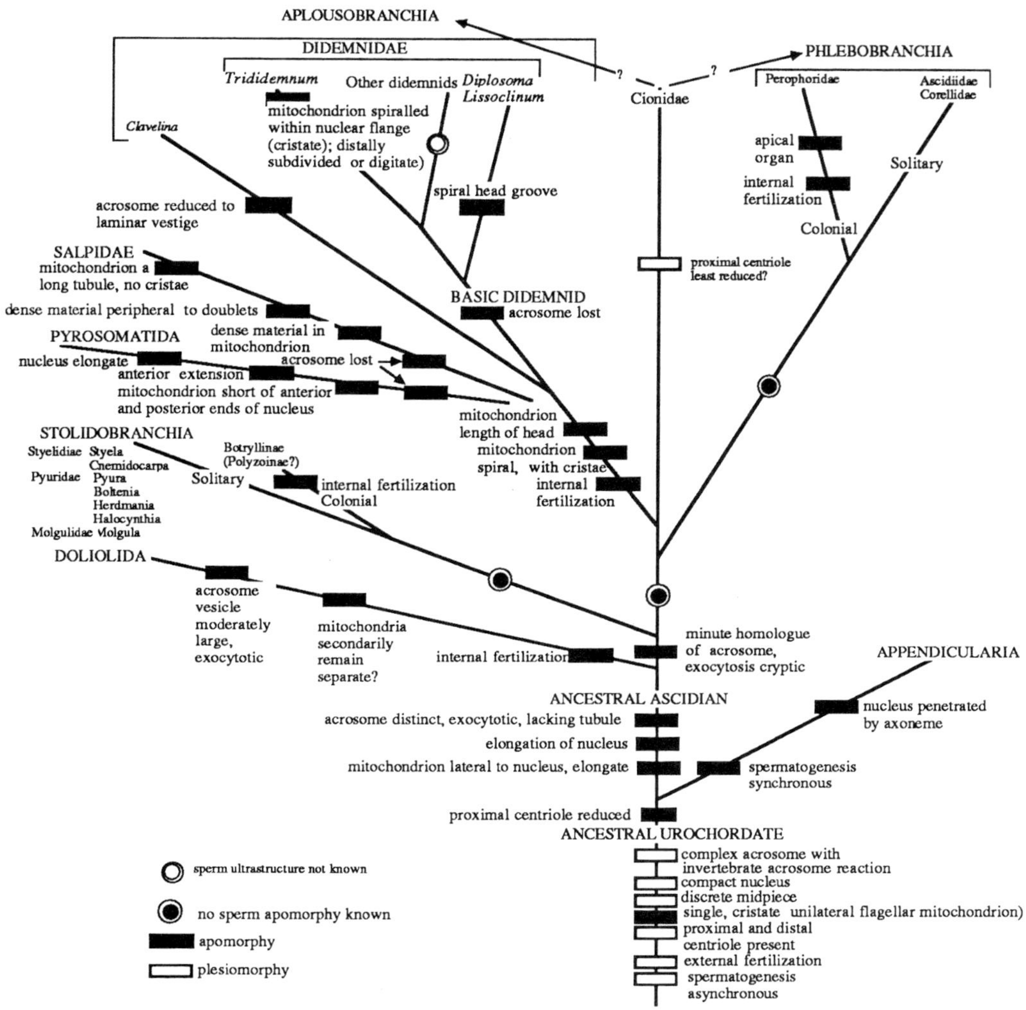

Fig. 3.20. Tentative phylogeny of the Urochordata based on sperm ultrastructure. (Original, incorporating the ideas of Holland). Explanation in text.

The further suggestion (Holland, 1989) that it is likely that appendicularians gave rise to both the 'thaliaceans' and the ascidians has profound implications. It would suggest that the tunicate tadpole, as manifested in the appendicularians is not a neotenous form of a doliolid tadpole larva which had been secondarily interpolated into an ascidian life cycle (the theory of Garstang, 1928; Berrill, 1950) but that it may represent the original adult form of the tunicates. The writer concurs with these views and envisages (Fig. 3.20) that the ancestral urochordate sperm was like that of appendicularians, though the nucleus probably was not penetrated by the axoneme, but derives the assemblage consisting of ascidians and the debatable thaliacean-didemnid assemblage from a common ancestor shared with appendicularians, a view with which Holland (1990) latterly concurs.

Holland (1989) accepts the controversial view that solitary ascidians have evolved from colonial ascidian ancestors which had internal fertilization. She therefore proposes that external fertilization has been derived from internal fertilization in ascidians. This is considered to explain why solitary, externally fertilizing ascidians have an elongate sperm head and no midpiece, features attributed to internal fertilization. A similar explanation, secondary development of external from internal fertilization, has been examined by Jamieson and Rouse (1989) to explain the complex form of the sperm of "lower" fish but is rejected, though with reservation, in the present work. Derivation of external from internal fertilization in tunicates is contraidicated by evidence of Kott (1969) that the solitary, externally fertilizing (aplousobranch?) *Ciona intestinalis* is a very plesiomorphic ascidian and with further evidence (Kott, 1985) that the colonial mode has originated independently several times in tunicates from the solitary condition. In *Ciona intestinalis* the nucleus is the shortest known among the ascidians and reduction of the proximal centriole appears to be less than in other tunicates, both clearly plesiomorphic conditions. To regard the solitary ascidians as advanced is also unparsimonious with regard to the view of Holland (1989), endorsed here, that the appendicularian sperm, with its compact nucleus and well developed acrosome which exhibits a complete invertebrate acrosome reaction, is plesiomorphic in these respects. The lack of parsimony lies in then considering the apparently plesiomorphic conditions of the *Ciona* sperm as secondary.

To summarise Fig. 3.20, the sperm of the ancestral urochordate is envisaged as having the following plesiomorphies: a complex acrosome capable of undergoing an acrosome reaction with extrusion of an acrosomal filament; a compact, subspheroidal or perhaps slightly cylindrical nucleus; a discrete midpiece, containing the mitochondrial material, situated behind the nucleus and encircling the base of the flagellum; a proximal and distal centriole; and external fertilization. Apomorphic conditions relative to other invertebrates were fusion of the spermatid mitochondria to form a single mitochondrion and its asymmetrical location in the mature sperm lateral to and partly embracing the flagellum. As this condition of the mitochondrion is seen in cephalochordates, it is conceivable that it was inherited from a common ancestor and that, though apomorphic for invertebrates in general, it is plesiomorphic for urochordates. It is difficult to determine whether the asynchronous spermatogenesis of salps and ascidians or the synchronous spermatogenesis of appendicularians should be regarded as plesiomorphic and which of these conditions should be attributed to the ancestral urochordate. Asynchrony appears to be widely distributed in invertrebrates and is taken to be plesiomorphic. I have therefore regarded the synchrony in *Oikopleura* as an apomorphy, perhaps correlating with the relatively oligopyrene condition in ensuring maximal sperm production for the brief reproductive period when the animal is out of its house. The hypothetical ancestor of ascidians and appendicularians is envisaged as differing from the above only in reduction of the proximal centriole.

A second apomorphy of appendicularians, in addition to sychonous spermiogenesis, is the penetration of the nucleus by the axoneme (excepting a thin diaphragm of nuclear envelope). Embedment of the acrosome in the nucleus in *Oikopleura* is possibly apomorphic but could be a plesiomorphy and a simultaneous synapomorphy with echinoderms.

In agreement with Holland (1988, 1989), the sister group of the Appendicularia is seen as the Thaliacea - Ascidiacea assemblage, and the Thaliacea are tentatively regarded as paraphyletic. Hence doliolids have the least modified of the non-appendicularian sperm in retaining the complete invertebrate acrosome reaction. Salps and pyrosomatids are dissociated from the doliolids not merely because they have lost the acrosome, a weak synapomorphy with ascidians which could conceivably have occurred convergently, but because they also share the spiral mitochondrion with modified cristae, and the type of internal fertilization, seen in non-cionid aplousobranchs. Doliolids have internal fertilization in the cloaca whereas the internal fertilization of these aplousobranchs, and of salps and pyrosomes occurs either in the oviduct (some aplousobranchs) or in the ovary (other aplousobranchs, salps and pyrosomes) (Holland, pers. comm.).

Apomorphies of the sperm of the ancestor of the 'Thaliacea' - Ascidiacea assemblage (non-appendicularian tunicates) which distinguished it from the common ancestor shared with the appendicularians are deduced to have been: some reduction of the acrosome which remained distinct and exocytotic but had lost the tubule; some elongation of the nucleus which, although possibly reaching 3 µm in length, was considerably thinner than long; and, attributable to it with some confidence, movement of the unilateral mitochondrion from the tail to a paranuclear position. It was thus a basic ascidiosperm. This sperm type is preserved in *Ciona intestinalis* which has alternatively been placed in the Aplousobranchia or in the Phlebobranchia, and is seen, with slight further

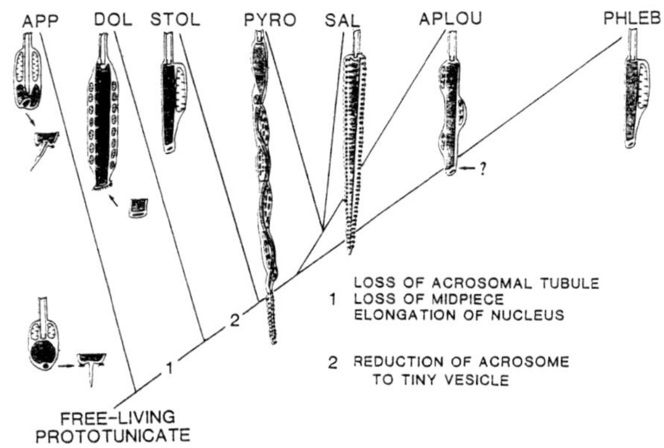

Fig. 3.21. Evolution of tunicate sperm as envisaged by Holland (1990). The phylogenetic significance of tunicate sperm morphology. In Proceedings of the VIth International Congress of Spermatology. Siena, Italy. (ed. B. Baccetti). (In press).

elongation of the nucleus, in stolidobranchs. Doliolids are somewhat arbitrarily given a basal position immediately below the stolidobranchs in conformity with the views of Holland (1990) (see Fig. 3.21). If the single report of persistence of the proximal centriole, that for *Ciona intestinalis* by Cotelli *et al.* (1980), which is indubitable for the spermatid, were not confirmed for the mature sperm, reduction of this centriole would be upheld as a monophyletic event further characterizing the basic asciodiosperm. The non-appendicularian tunicate assemblage is here named the Ascidiospermia.

We have seen that the doliolid thaliaceans appear plesiomorphic members of the Ascidiospermia in having an ascidiosperm in which the acrosome vesicle is larger than in other non-appendicularians and is capable of gross exocytosis. If the supposed persistence of many separate mitochondria in the mitochondrial mass of doliolids is a true plesiomorphy, this feature would have to be ascribed to the ancestral ascidiospermian. It is possible, though, merely a "neotenous" retention of a normal feature of ascidian spermatids. The non-doliolid thaliaceans (salps and pyrosomes) are seen as offshoots (possibly monophyletically) of an aplousobranch clade, sharing with the latter the spiral mitochondrion with modified cristae, though whether this, like the loss of the acrosome, is a true synapomorphy remains open to debate. *Pyrosoma atlanticum* is apomorphic in great elongation of the nucleus, and development of a peculiar anterior cytoplasmic extension (Holland, 1990).

In the Phlebobranchia, the Ascidiidae and Corellidae have a basic ascidiosperm but the Perophoridae, as represented by *Perophora formosana* and *P. annectens*, have added a cross striated possibly perforatorial "apical structure" behind the minute acrosomal vesicles. Whether the "helical string" in *P. annectens* is a parallelism (homoplasy or even paramorphy) with the dense groove of advanced didemnid sperm cannot at present be determined. *Ciona intestinalis*, in having a simple cristate lateral mitochondrion, lacking both the spiral structure and modification of other aplousobranchs, and possibly in persistence of both centrioles, cannot readily be accommodated in the Aplousobranchia, as it is here, and a phlebobranch status, usually recognized, is spermatologically more acceptable. The relatively unmodified condition of its sperm and its ambivalent taxoomic position are indicated in Fig. 3.20.

In the aplousobranchs, *Diplosoma* and *Lissoclinum* are distinguished by a spiral plasmalemmal groove extends throughout the head which may have an acrosomal role (Fig. 3. 6). *Trididemnum* has become specialized in developement of a regular spiral nuclear flange with, between its coils, a spiral mitochondrion which in this case is divided anteriorly into many (conjoined?) vesicles.

External fertilization, here attributed to the ancestral urochordate, persisted in the Appendicularia, and Cionidae, but facultative or obligate internal fertilization developed in non-cionid didemnids and many other Ascidiospermia.

Sperm structure in tunicates is summarized in a phylogeny of the group based mainly on non-spermatozoal evidence in Fig. 3.21, drawn from Holland (1990).

Chapter 4

SUBPHYLUM CEPHALOCHORDATA

Diagnosis. Cephalochordates are small marine, fishlike animals (up to 8 cm long) (Fig. 4.1)which lie buried in the substrate with buccal cirri exposed, feeding by straining minute organisms from the water.

posteriorly at the atriopore and is covered by lateral folds of the body wall, the metapleural folds. The feeding current is maintained by cilia of a "wheel organ" in the oral hood; and by cilia of the gill bars and of the endostyle. Near the centre of the wheel organ

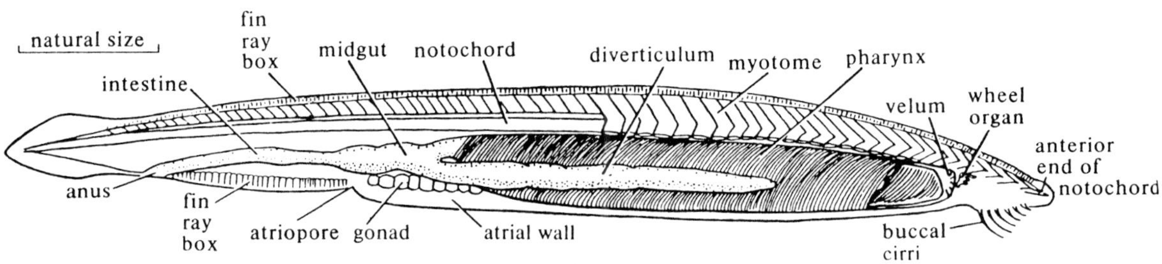

Fig. 4.1. *Branchiostoma*, the amphioxus or lancelet. The body wall and atrial wall have been removed on the right side to show the pharynx, midgut with "liver diverticulum", and intestine. The oral hood has been cut away on the right, leaving the buccal cirri, wheel organ and velum. From Young, J.Z. (1981). *The Life of Vertebrates*. Third Edition. Clarendon Press, Oxford.Fig. 2.1.

The notochord extends to the anterior end of the body, in front of the brain. A cranium is absent. There is a cerebral vesicle but no true brain. The fibres of the peripheral nerves differ from those of vertebrates in lacking a myelin sheath. Unlike the vertebrates, peripheral muscles send processes to the nerve cord (as in echinoderms but also in nematodes). There are no vertebrae; no cartilage or bone; and no red blood corpuscles. The long dorsal fin and shorter ventral fin are each supported by a fin ray box. The heart is simple, consisting of a contractile vessel. The epidermis has only a single layer of cells. The excretory system is protonephridial, with solenocytes. The alimentary canal has a liver diverticulum and a pharyngeal endostyle which contains iodine fixing cells and is considered to be the homologue of the vertebrate thyroid. Anteriorly a series of buccal cirri forms a sieve around the opening of an oral hood. The numerous gill slits open from the pharynx into a chamber, the atrium (Fig. 4.2), which opens to the exterior

opens a groove, Hatschek's pit, which represents the opening of the left, first coelomic sac to the exterior, an opening which also occurs in the oligomerous phyla, including echinoderms. Sensory velar tentacles are present within the hood. The anus is well anterior to the hind end of the body, leaving a definite postanal tail as is typical of chordate metamerism. Sexes are separate (Wickstead, 1975; Young, 1981; Nelson, 1984).

Cephalochordates and (at least as embryos) vertebrates share attributes indicative of a close relationship: a notochord; a dorsal tubular nerve cord; paired lateral gill slits; an hepatic portal system; and an endostyle or it homologue the thyroid.

There is a single order, the Amphioxiformes (lancelets), exclusively marine, in the Atlantic, Indian and Pacific oceans.

There are two families: Branchiostomidae, with one genus, *Branchiostoma* (Fig. 4.1) with about 15 species; and Epigonichthyidae, containing *Epigon-*

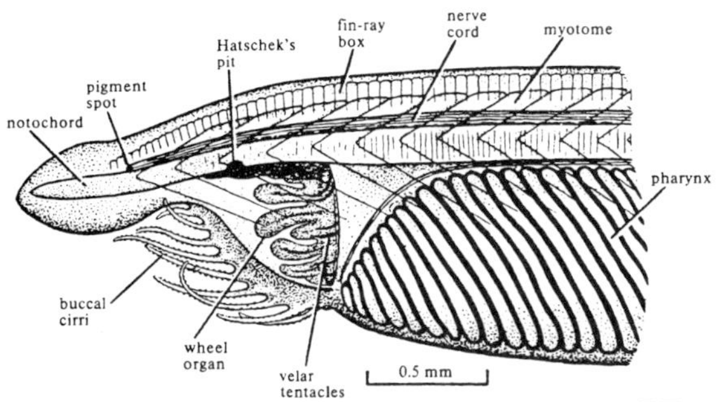

Fig. 4.2. *Branchiostoma*. Oral region. From Young, J.Z. (1981). *The Life of Vertebrates*. Third Edition. Clarendon Press, Oxford.Fig. 2.8.

ichthys (=*Asymmetron*, *Heteropleuron*) with about 5 species, differing from *Branchiostoma* in having the gonads on the right side only (Nelson, 1984).

Sperm literature. Ultrastructural studies on cephalochordate sperm have been limited to *Branchiostoma lanceolatum* (Baccetti *et al.*, 1972; Wickstead, 1975); *B. belcheri* (Lin *et al.*, 1987); *B. floridae* (Holland and Holland, 1989) and *B. moretonensis* (Jamieson, 1984). Wickstead gives a valuable account of reproduction and related morphology in the Cephalochordata.

***Branchiostoma* sperm. General.** The spermatozoon of *Branchiostoma* (Fig. 4.3; 4.4A) approximates to the primitive type *sensu* Franzén (1956) or ect-aquasperm (Rouse and Jamieson, 1986; Jamieson, 1986) in having a cap-like acrosome surmounting a compact nucleus, mitochondrion incorporated in the head behind the nucleus and not elongated, two mutually perpendicular centrioles with triplet microtubules and a long flagellum with the 9 + 2 configuration of microtubules, but it differs notably from the basic ect-aquasperm in possessing only a single mitochondrion and in the form of this. The head (including the acrosome, nucleus and mitochondrion) is approximately 3 µm long.

Acrosome. The acrosome (Fig. 4.3A-D) in all species consists solely of a bell-shaped acrosome vesicle (0.3 - 0.4 µm long and maximally, near the posterior rim, 0.7-0.8 µm wide in *B. moretonensis*; 0.5 µm long and basally 0.7 µm wide in *B. lanceolatum*) deeply invaginated posteriorly by the acrosomal space so that its anterior and posterior bounding membranes are only narrowly separated at the apex, though well separated at the posterior rim.

The subacrosomal invagination (Fig. 4.3A-D) contains unstructured, somewhat diffuse material

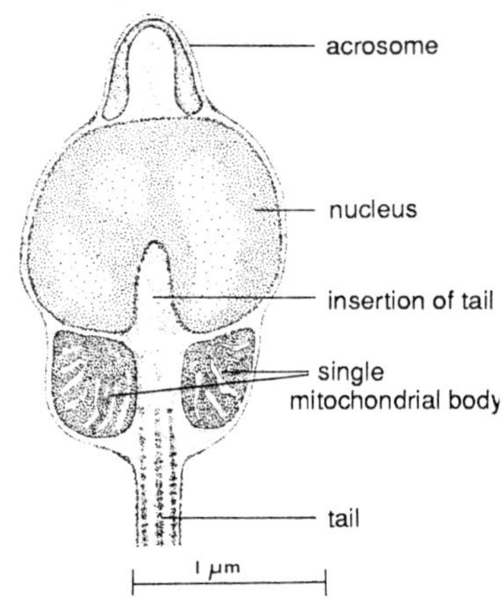

Fig. 4.3. *Branchiostoma lanceolatum*. Semidiagrammatic longitudinal section of the spermatozoon. After Wickstead, J.H. (1975). In *Reproduction of marine invertebrates. II. Ectoprocts and lesser coelomates* (eds A.C. Giese and J.S. Pearse), pp. 283-319. Academic Press, New York. Fig. 6.

Fig. 4.4. (Opposite). *Branchiostoma moretonensis*. Spermatozoal ultrastructure. A.B. Longitudinal section (LS) of the heads of two spermatozoa, showing the bell-shaped acrosome vesicle (see also inset, Fig. A), subacrosomal material, nucleus with tubular posterior fossa and single mitochondrion enclosing the two mutually perpendicular centrioles. C. Spermatozoon with strongly eccentric axoneme. D. Transverse section (TS) of acrosome. E. TS of the 9+2 flagellar axoneme. F. L.S. centriolar region showing a TS of the proximal and an LS of the distal centriole (basal body) (detail from H). The proximal centriole has a core and a spur (striated rootlet) which extends into the nuclear fossa.G. LS further sperm showing rootlet. Note that in this plane the mitochondrion is seen only on one side of the centrioles, the other side being occupied by the vesicle of the perinuclear cisterna, and that the flagellum is at an angle to the nuclear axis. H. LS sperm, showing the centriioles and narrow arms of the C-shaped mitochondrion. Inset: TS of a distal centriole, showing anchoring apparatus. I. LS centriolar region approximately at right angles to F. J. TS distal centriole, showing the C-shaped mitochondrion, in the opening of which is a portion of the vesicle of the perinuclear cisterna. av. acrosome vesicle. dc. distal centriole. f. flagellum. m. mitochondrion. n. nucleus. pnf. posterior nuclear fossa. pc. proximal centriole. pm. plasma membrane. sm. subacrosomal material. sp. spur-like process (striated rootlet) of proximal centriole. v. vesicle of perinuclear cisterna. After Jamieson, B.G.M. (1984b). *Zoologica Scripta* **13**, 223-229. Figs. 1-10.

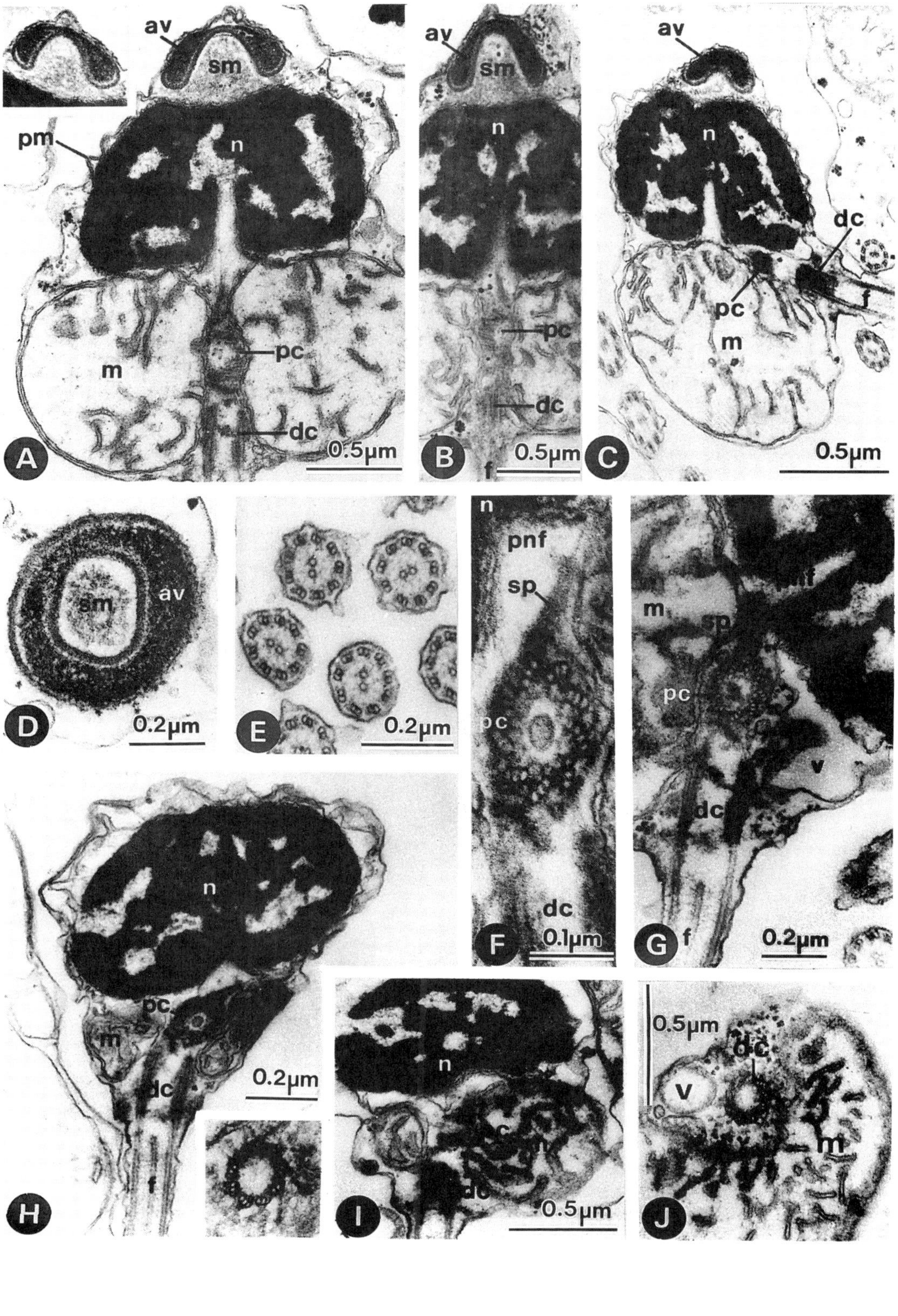
av
sm
pm
n
pc
m
dc
0.5μm
A
B
f
C
D
0.2μm
E
F
0.1μm
G
0.2μm
pnf
sp
v
H
I
J
0.5μm

which is less electron dense than the contents of the vesicle. This material extends for about 0.1 µm posteriorly, separating the vesicle from the nucleus. In *B. moretonensis* the plasma membrane covering the acrosome projects as a small apical cone or button; this is a common artefact in various phyla, but possibly reflects some underlying differentiation (Jamieson, 1984). The acrosome vesicle originates, as is normal, from the Golgi apparatus but there is no evidence for such an origin of the subacrosomal material (Holland and Holland, 1989).

Nucleus. The nucleus (Fig. 4.4A-C), subovoid and wider (1.8 µm) than long (1.0 µm), is shallowly concave anteriorly and posteriorly is penetrated for about half its length by a narrow, anteriorly tapering, axial nuclear fossa (Fig. 4.4A-C, F, G). This fossa or canal is circular in cross section and is about 0.12 µm wide at its midlength. A narrow perinuclear space (cisterna) is present between the inner and outer nuclear membranes and (Figs. 4.4G and J) expands as a unilateral vesicle in the mitochondrial region. It is possible that this vesicle is at least partly artefactual, however. The contents of the nucleus consist of electron dense homogeneous material, with some evidence of a flocculent nature, but a considerable proportion of its volume consists also of tortuous lacunae, lacking chromatin, some of which extend to its surface (Figs. 4.4A-C, G-I).

Midpiece. The mitochondrion is closely juxtaposed to, but does not indent, the nucleus (Figs. 4.4A-C, H-I). It has numerous, turtuous cristae, but apart from its shape appears unmodified. In *B. moretonensis* and *B. floridae* the mitochondrion is C-shaped in the transverse plane, with terminally tapering arms extending from a wide central region. The two centrioles are embraced by the arms, the proximal centriole being at midlength, the distal centriole at the posterior end of the mitochondrion (Figs. 4.4A, B, H, I). The C-shape is seen in cross section in Fig. 4.4J, in which the axial distal centriole and a peripheral vesicle of the perinuclear cisterna are shown. it appears that the opening in the C is nearly, but not exactly, opposite one end of the transverse, proximal centriole.

The mitochondrion in *B. lanceolatum*, though wider on one side, forms a complete ring surrounding the centrioles (Baccetti *et al.*, 1972), as confirmed by Wickstead (1975) in micrographs of cross sections.

Centrioles. Transverse sections of the proximal (Fig. 4.4F, G) and distal centriole (Fig. 4.4H inset, J) reveal the normal structure of 9 skewed triplet but the proximal centriole, with its long axis transverse to the long axis of the sperm shows unusual features. Most distinctive in *B. moretonensis*, but not apparently in other species, is a discrete central core (Fig. 4.4F-H) which, though not uncommon in basal bodies of sperm in various animal groups, is perhaps unprecedented in proximal centrioles. Less peculiar, but nevertheless remarkable, is a spur-like process, oblique to the long axis of the sperm, extending from the anterior border of the proximal centriole into the base of the posterior nuclear fossa; the spur shows weak transverse or slightly oblique striations; the posterior border of the centriole is also fringed by satellite-like structures (Fig. 4.4F-H) which abut on the distal centriole and there is some suggestion of continuity of the spur with these and with the distal centriole. A projection towards the distal centriole is also seen in *B. floridae*. The pointed anterior tip of this spur appears to attach to the nuclear membrane on one side of the basal region of the posterior nuclear fossa. The distal centriole lying in the long axis of the axoneme, which originates from it, and at right angles to the proximal centriole has satellite rays (presumably 9) radiating from the triplets. Each ray is inclined at about 45° to the radius passing through its triplet and ends peripherally in a swelling, though no terminal branches have been detected (Fig. 4.4H inset, J). The distal centriole is surrounded by many putative glycogen granules and these are sparsely present between other organelles and under the plasma membrane (Jamieson, 1984). No equivalent of the core to the proximal centriole has been described for, nor is it visible in published micrographs of, the centriole of *B. lanceolatum*; and it is absent from *B. floridae* (Holland and Holland, 1989) and probably from *B. belcheri*. There is some suggestion of the spur extending from the proximal centriole into the nuclear fossa in *B. lanceolatum* (Fig. 2 of Baccetti *et al.* 1972), and it is clearly visible in micrographs of *B. floridae* and it is probably to be regarded as a characteristic of cephalochordate sperm.

Flagellum. The axoneme, arising from the distal centriole, has 9 peripheral doublets, each with two dynein arms and two central singlets. All microtubules are hollow. In some transverse profiles a bridge between doublets 5 and 6 is apparent (Fig. 4.4E). There are no accessory microtubules. In longitudinal sections passing transversely through the proximal centriole (Fig. 4.4G) the flagellum appear strongly tilted relative to the axis of the posterior nuclear fossa. This is more evident in advanced spermatids when they are loosely connected in tetrads than in mature sperm, but is sometimes still pronounced in the latter (Fig. 4.4C) (Jamieson, 1984). The flagellum is 40 µm long in *B. lanceolatum* in which it is tilted at 135° relative to the axis of the endonuclear canal; its last 4 µm is less than 0.1 µm wide owing to loss first of the nine doublets and then of the two central singlets (Baccetti *et al*., 1972). In the terminal few micrometres of the axoneme there is a reduced number of microtubules (Holland and Holland, 1989).

Cephalochordate sperm: summary. Close resemblance of the spermatozoon of *Branchiostoma moretonensis*, Jamieson (1984) to that of *B. lanceolatum*, Baccetti *et al.* (1972); Wickstead (1975), *B. belcheri*, Lin *et al.* (1987) and *B. floridae*, Holland and Holland (1989), allows us to recognize a cephalochordate type in so far as this can be established on four species. The characteristics of this are: a slight modification of the primitive form; a bell-shaped acrosome; diffuse subacrosomal material not structured as an acrosome rod; subovoidal nucleus with shallow anterior concavity, deep posterior fossa ("endonuclear canal" of Baccetti *et al.* 1972) and condensed but lacunate chromatin; single asymmetrical, postnuclear ring-shaped or C-shaped mitochondrion encircling the centrioles; mutually perpendicular proximal and distal centrioles of the triplet type, with the distal forming the basal body of the flagellum; flagellum tilted relative to the longitudinal axis (and endonuclear canal) of the nucleus; a 9 + 2 axoneme with hollow tubules and both dynein arms present on the doublets; and scattered glycogen granules, numerous around the distal centriole.

Sperm and lophophorate-deuterostome evolution

A protostome stock ancestral to the lophophorates probably had an ect-aquasperm little modified, if at all, from the plesiosperm morphology, with the usual complement of cap-like acrosome, subspheroidal nucleus, several discrete rounded mitochondria, two centrioles, an anchoring satellite apparatus arising from the distal centriole, and a 9+2 flagellum. The Bryozoa have ent-aquasperm little modified from the plesiosperm (lacking nuclear membranes and an acrosome) in *Bugula* or highly modified in others. Phoronids have highly modified sperm, probably as a modification for expulsion to the exterior in a spermatophore.

Further lophophorates, brachiopods, retain a primitive fertilization biology, with ect-aquasperm, or (*Terebratulina*) acquire ent-aquasperm. Their aquasperm foreshadow special features of those of enteropneusts, echinoderms and cephalochordates. These similarities are presumably symparamorphies indicative of close relationships but the possibility exists that brachiopods have played a more central role in the origin of deuterostomes than has been formerly supposed. Certainly sperm structure supports the view that the brachiopods originated close to the branching point between the protostome and deuterostome lines (Afzelius and Ferraguti, 1978). In the inarticulate brachiopods, which as the plesiomorph sister-group would be structurally nearer this branching, the sperm of *Lingula* shows remarkable deuterostome features: it approaches the cephalochordate *Branchiostoma* in shallow depression of the anterior face of the nucleus and well developed subacrosomal material, while resembling the enteropneusts (*Saccoglossus*, *Saxipendium*) in the inflated, roughly heart-shaped form of the acrosome, embedment in of this in periacrosomal material (both apomorphies or perhaps symparamorphies) and the plesiomorph retention (in *Saccoglossus*, at least) of separate mitochondria. Another inarticulate, *Crania*, has a caplike acrosome with abundant subacrosomal material, much as in *Branchiostoma*, but differs from the latter in the independent mitochondria. Finally, *Terebratulina*, in an order of the articulate brachiopods which developed latest in the fossil record of the phylum (Williams and Rowell, 1965), has an ent-aquasperm which appears plesiomorph in most features, but, as a symparamorphy with the echinoderms and cephalochordates, has a single, annular mitochondrion.

In the enteropneusts, we saw above a sperm which has features foreshadowed in *Lingula*. The heart-shaped acrosome with periacrosomal material well developed laterally but less so subacrosomally, resting on a broad depression of the nucleus is strikingly precursory to echinoderm and appendicularian (larvacean) urochordate sperm with which the separate mitochondria are a plesiomorph contrast. In pterobranchs, which are generally considered representative of the lophophorate ancestry of echinoderms and of protochordates, the only investigated species, *Rhabdopleura normani*, has highly modified sperm (Lester, 1988) which cannot be construed as basic for the lophophorate-deuterostome lineages (see Chapter 1).

In a protoechinoderm derived from a putative hemichordate-like ancestry shared with the protochordates, we may envisage a spherical acrosome embedded in periacrosomal material and underlain by (but not enclosed in) an anterior nuclear (subacrosomal) fossa, which was developed as an adaptation primarily for housing the actomere; a slight posterior nuclear (centriolar) fossa may have been present and, as a symparamorphy with some cephalochordates and with *Terebratulina*, an annular mitochondrion. Fusion of mitochondria, though as a C-shape, is also seen in cephalochordates and in the urochordates.

Within the echinoderms (see Chapter 1), the emergent acrosome of echinoid sperm, separated from the nuclear fossa, is here considered a plesiomorph retention from the protoechinoderm but a striking echinoid apomorphy is the elongation of the nucleus, a modification usually but not in this case attributable to internal fertilization. In the crinozoids, holothuroids, asteroids and ophiuroids deep embedment of the acrosome and periacrosomal material in the anterior nuclear fossa, with or without presence of a separate subacrosomal depression occurred giving a distinctive aquasperm conveniently, if unfelicitously, termed the echinosperm by Jamieson (1985; see Echinodermata, below). The highly modified sperm of *Xyloplax* shows remarkable symparamorphic resemblance to the *Phoronis* sperm.

The hemichordate ancestry (cephalodiscus-like but with simpler sperm) of the echinoderms is envisaged as giving rise to a sister-group of the echinoderms, those deuterostomes with gill slits, including extant enteropneusts and rhabdopleuran pterobranchs, the urochordates and the cephalochordates. Appendicularian sperm show a remarkable and close symparamorphy with echinoderms in embedment of the acrosome (otherwise heart-shaped as in enteropneusts) in the anterior nuclear fossa and similarly have fused the mitochondria, though as a C-shape, and we have envisaged a urochordate ancestor with a sperm resembling that of Appendicularia except in the larger nucleus (apomorphically oligopyrene in Appendicularia), shorter centriolar fossa and, as in enteropneusts, cephalochordates and the ascidiacean *Ciona* (spermatids and apparently spermatozoa), two centrioles. In the Ascidiacea the mitochondrion has become perinuclear and the acrosome vesicle is reduced.

Branchiostoma sperm give no clear indications of the commonly accepted origins of the cephalochordates from a stock shared with the urochordates after emergence of the echinoderms. It shows resemblances to the sperm of both the brachiopods and the echinoderms.

In fact, as we have seen, the cephalochordate sperm falls well within the known variation for brachiopods: having a caplike acrosome with subacrosomal material (as in *Crania*), and an annular mitochondrion (as in *Terebratulina*). This is in striking contrast with the inflated acrosome, embedded in periacrosomal material, seen in echinoderms, hemichordates and appendicularian urochordates. From a purely spermatological standpoint it would therefore be tempting to derive cephalochordates from a protocephalodiscus below the branching point of the echinoderms. This would necessitate the unorthodox propositions that pharyngotremy in notochordates (cephalochordates and, if monophyletic with them, craniates) is an independent symparamorphy compared with enteropneusts and urochordates or, alternatively, that echinoderms have lost pharyngeal perforations during the profound modifications which they have clearly undergone from the presumed cephalodiscus ancestry (see Jefferies, 1981 a,b). If we accept the conventional view of a close affinity of cephalo- and other chordates with the enteropneust-urochordate assemblage, i.e. monophyly of chordates *sensu lato*, it will be necessary to assume that the inflated, embedded acrosome has occurred in parallel (by symparamorphy) in enteropneusts and appendicularians relative to echinoderms or that this condition is basic to echinoderms and chordates and that cephalochordate sperm have secondarily approached the plesiosperm condition of a caplike acrosome. Greater secondary simplification is well demonstrated by the teleosts and the ectoprocts, at least one species of *Bugula* lacking an acrosome, but a functional reason for deflation of the *Branchiostoma* acrosome, if secondary, is elusive.

Branchiostoma sperm resemble those of some echinoderms in a number of respects beyond mere similarity of primitive sperm. These resemblances include well developed periacrosomal material, though this is subacrosomal in cephalochordates, as noted by Afzelius (1977); the commonly asymmetrical ring-shaped or sometimes (*Cucumaria lubrica*, Atwood and Chia 1974; *Branchiostoma moretonensis*, *B. floridae*) C-shaped single mitochondrion; and the rootlet arising from the proximal centriole and extending into the nuclear fossa (shared with holothuroids) and the often tilted emergence of the flagellum.

It is noteworthy in view of the above apomorphies shared between cephalochordate and echinoderm sperm that Jefferies (1979, 1981b) considers that cephalochordates are less closely related to vertebrates than are urochordates. In vertebrates and urochordates the locomotory muscles are innervated by nerves from the dorsal nerve cord or the brain (a shared derived character according to Jeffries) while in cephalochrodates and echinoderms, as we have seen, skeletal muscles send processes to the dorsal nerve cord or its equivalent. Even if the latter is a symplesiomorphic condition it distances cephalochordates from the urochordate-vertebrate assemblage in this respect.

If urochordates have an especially close relationship with vertebrates we might speculate that the modified sperm typical of basal gnathostomes (only the Actinopterygii having sperm with the "primitive" sperm morphology) have been inherited from a common urochordate-vertebrate ancestor or that the complex form represents a common genetic and phylogenetic propensity independently realized in each group. Against the first proposition is the view developed in Chapter 3, on the evidence of *Oikopleura*, that the ground plan for the urochordate spermatozoon is not complex and that complexity of the sperm in "lower" fish could not be a retention of a urochordate condition. The spermatozoon of the ancestral urochordate is there envisaged as having the following plesiomorphies: a fully developed acrosome; a compact, subspheroidal or perhaps slightly cylindrical nucleus; a discrete midpiece, containing the mitochondrial material, situated behind the nucleus and encircling the base of the flagellum; a proximal and distal centriole; and external fertilization. Fusion of the spermatid mitochondria to form a single mitochondrion and its asymmetrical location in the mature sperm lateral to and partly embracing the flagellum is seen as an apomorphy of the basal urochordate but, as this condition of the mitochondrion is seen in cephalochordates (and echinoderms), it is conceivable that it was inherited from a common ancestor.

We may, alternatively, regard similarities of *Branchiostoma* sperm to those of brachiopods and echinoderms as symparamorphies, parallelisms by virtue of relationship, which nevertheless do not necessitate a closer relationship of cephalochordates to these groups than to vertebrates. If also we accept the by no means unequivocal view that the sperm of the ancestral vertebrates were ect-aquasperm of a simple type, we may attempt to reconstruct its morphology and compare this with that of the *Branchiostoma* sperm. Holland and Holland (1989) correctly note similarities of the *Branchiostoma* acrosome (bell-shaped with granular subacrosomal material) to that of the hagfish (Jespersen, 1975), and we may attribute these features to the ancestral vertebrate ect-aquasperm. A rounded nucleus, as in *Branchiostoma*, and unlike the elongate nucleus of agnathans, chondrichthyans and Chondrostei may be envisaged. The midpiece can be ascribed several, separate mitochondria, as parsimoniously demanded by the multiple mitochondria of these groups, fusion as a single annulus in, for instance, salmoniforms being a symparamorphy of these with *Branchiostoma* and of this with echinoderms, urochordates and some brachiopods, and not a plesiomorphic state for vertebrate sperm. Two centrioles, as in most fish sperm, and a satellite apparatus which is present at least as a propensity in fish (and strangely realized most fully in advanced neopterygians, the atheriniforms) and a simple axoneme are other attributes of the basic vertebrate sperm. While a plesiosperm origin of the sperm of fish and protochordates is likely, we will nevertheless consider below the alternative possibility that ancestral fish had modified ect-aquasperm or, less likely, internal fertilization.

Chapter 5

INTRODUCTION TO FISH SPERMATOZOA AND THE MICROPYLE

by B.G.M. Jamieson and L. K.-P. Leung

Fish - a paraphyletic group

Pisces, a term used here for all types of fish, are aquatic vertebrates which are (1) jawless and lack paired limbs or (2) have jaws with paired limbs which are not polydactylous. The term Pisces is often, as in Nelson (1984), used only for vertebrates of type 2.

The term "fish" is therefore applied here to the Agnatha together with the jawed (gnathostome) fish while recognizing, as is seen in Fig. 5.1, that the total assemblage is paraphyletic, as also are the agnathans and gnathostome fish themselves. The Agnatha appears to be a paraphyletic grouping of two separate groups, the Myxinidae (hagfish) and Petromyzontidae (lampreys), artifically united by the symplesiomorphic absence of jaws.

Gnathostome fish represent several lineages arising independently and in apparent succession from the basal gnathostome line which led to the tetrapods. Successive gnathostome lineages are: the Chondrichthyes (elasmobranchs and holocephalans); the Actinopterygii, which is an apparently monophyletic

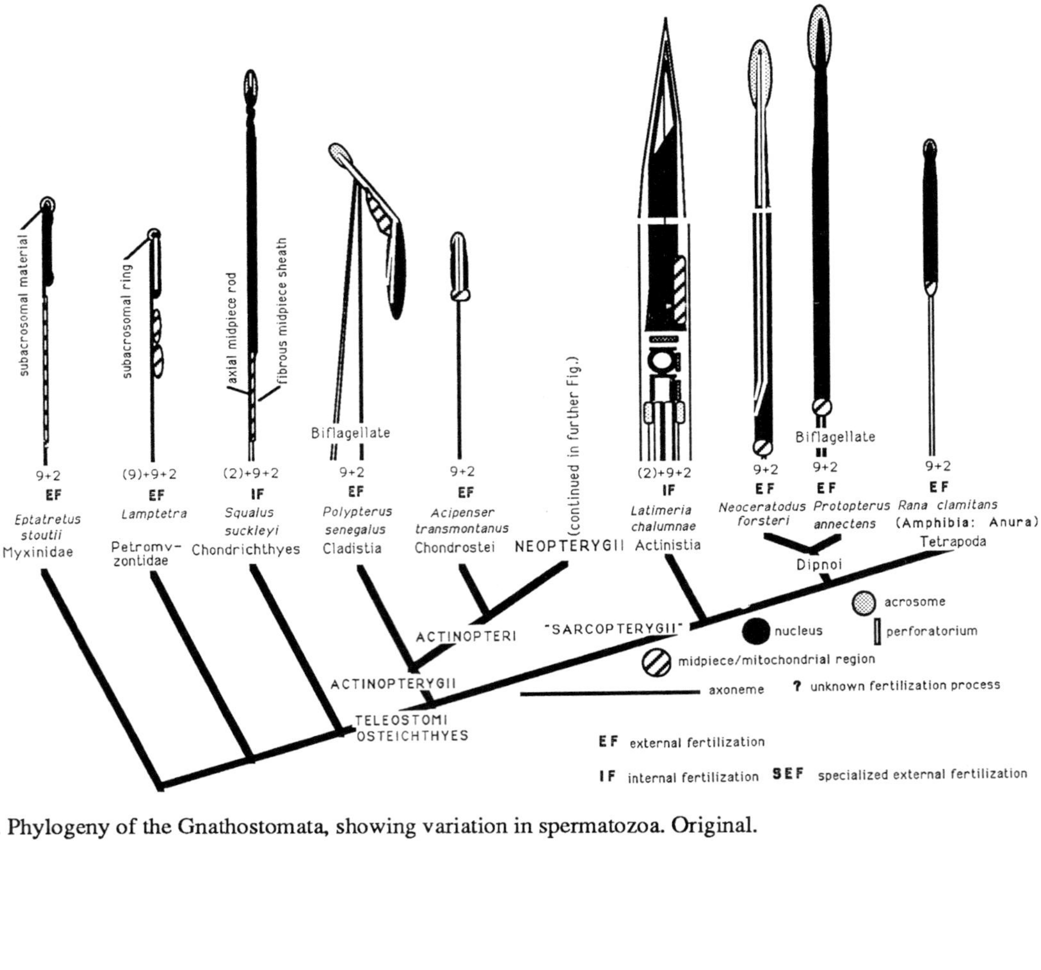

Fig. 5.1. Phylogeny of the Gnathostomata, showing variation in spermatozoa. Original.

group containing the Cladistia (*Polypterus*), Chondrostei (sturgeons) and Neopterygii (the vast assemblage of modern fish); the crossopterygian Actinistia (coelacanths); the Dipnoi (lung fish) and the tetrapods. Actinistia, Dipnoi and Tetrapods comprise the Sarcopterygii.

The gnathostome assemblage can be regarded as monophyletic only if the tetrapods are included in addition to the jawed fish. Tetrapod sperm are represented in Fig. 5.1 by that of the anuran *Rana* (Poirier and Spink, 1971).

Primitive and advanced sperm in fish

Classification of sperm as primitive or advanced (Retzius, 1904, 1905; Franzén 1956, 1970) becomes equivocal in classifying fish sperm as all "primitive" sperm belong to the more advanced fish groups (teleosts), and most of the "advanced" sperm are present in the "lower" groups (agnathans, chondrichthyans, cladistians, actinistians and dipnoans). These observations raise questions with regard to the evolution of fish sperm. In fish are "primitive" sperm secondary? Are their "advanced" sperm, although derived in relation to the primitive sperm of invertebrates and most lower deuterostomes, nevertheless basic to lower fish groups and have they re-evolved in the few higher fish groups which display them with the redevelopment of internal fertilization? If "advanced" sperm are in fact basic to fish does this mean that internal fertilization is also basic to the group?

To anticipate, we will propose that modifications of the sperm of externally fertilizing fish are related to aspects of external fertilization, including, probably, the nature of the egg envelopes, and are not evidence of former internal fertiization. Fish eggs are relatively large, usually exceeding 1 mm in diameter, and have complex and thick envelopes. Possibly it was only the development of a micropyle in teleost eggs which permitted reduction in size of the sperm and reversion to a "primitive" sperm facies. Multiple micropyles of sturgeon eggs, coinciding with elongate, relatively "advanced" sperm may represent a transitional stage before secondary simplification of the sperm in teleosts. The occurrence of a micropyle in myxinid eggs is elusive of explanation in view of the considerable modification of the spermatozoa. In chapter 1 we have gone so far as to attempt to morphologically define the ancestral fish sperm.

We must, nevertheless, examine arguments which can be presented in favour of the view that fertilization in fish was primitively internal. To explain the occurrence of "advanced" sperm in externally fertilizing fish the hypothesis that their ancestors had internal fertilization which was subsequently lost, while the modified sperm structure was retained, has been invoked by Afzelius (1970) for the lamprey and by Baccetti (1978, 1985) for teleosts. However, Baccetti (1986) depicts internal fertilization in fish as secondary.

Irrespective of whether internal fertilization is plesiomorphic or is a derived condition in fish, the basic spermatozoon of the Chordata was presumably an aquasperm (*sensu* Jamieson, 1986a,b) as this is seen in the presumed sister-group of the agnathan-gnathostome assemblage, the Cephalochordata (Fig. 4.3, 4.4), and appears to be basic to the two other protochordate subphyla and to invertebrate deuterostomes, the Echinodermata (see, however, alternative propositions in Chapter 4). The single, ring or C-shaped mitochondrial mass of *Branchiostoma* is, however, to be regarded as an apomorphy (albeit paramorphic with echinoderms and other taxa mentioned in Chapter 1) and the basic chordate stock may be attributed several separate mitochondria as in most invertebrate and some lophophorate and hemichordate aquasperm. The issue of plesiomorphy versus apomorphy for internal fertilization in ancestral fish will be further discussed after a listing of the categories of fish sperm.

The chief types of sperm in fish

Species for which spermatozoal ultrastructure has been described are listed in Table 5.1. The spermatozoa fall into the following categories:

1. **Aquasperm** (freed into water; in fish always externally fertilizing)
    - a. Acrosomal aquasperm (with an acrosome)
    - b. Anacrosomal aquasperm (lacking an acrosome)
        - i. Uniflagellate (simple, i.e. ectaquasperm-like, or complex, including elopomorph sperm)
        - ii. Biflagellate (simple, i.e. ectaquasperm-like, or complex)

iii. Aflagellate

2. **Introsperm** (Not freed into ambient water; internally fertilizing sperm)

a. Acrosomal introsperm (with an acrosome)

b. Anacrosomal introsperm (lacking an acrosome)

i. Simple (plesiosperm-like)

ii. Complex

These chief sperm types of fish will now be briefly discussed. The discussion will be followed by a taxonomic and phylogenetic treatment of the spermatozoa of fish.

(1) Aquasperm. Externally fertilizing sperm

a. Acrosomal aquasperm. In all "lower" fish, other than chondrichthyans and coelacanths, from Agnatha to Dipnoi, external fertilization occurs with possession of a spermatozoon with a more or less complex, introsperm-like morphology, notably in elongation of the nucleus. Whether this external fertilization is a secondary development from former internal fertilization or is a retention of a primitive condition again remains undecided. Such sperm are here termed *acrosomal aquasperm* as opposed to the usual neopterygian aquasperm which has no recognizable acrosome and may be designated the *anacrosomal aquasperm.* Thus lampreys and, it is believed, hagfish have external fertilization yet their sperm show features normally associated with internal fertilization: elongate nucleus; long perforatorium in *Lampetra* (also seen in many invertebrate ect- and ent-aquasperm, however); and elongate mitochondrial sheath around the axoneme, with, in *Lampetra*, accessory axonemal fibres. Afzelius (1970) has suggested that these features have been retained from cyclostome ancestors which were internally fertilizing but this hypothesis requires further support and need not necessarily include ancestral fish. Elongation and great modification are not exclusive to introsperm but are features of a large proportion of known entaquasperm of polychaetes (see Jamieson and Rouse, 1989) and are seen in ect- and ent-aquasperm of urochordates (see Chapter 2). We may thus regard the modified sperm of cyclostomes and other non-neopterygian fish as resembling not only introsperm but also entaquasperm.

Micropyle-perforatorium correlation. Co-existence of an egg micropyle with acrosomal sperm in the hagfish *Eptatretus burgeri* (Fernholm, 1975) is rare in fish (see also the sturgeon), development of the micropyle being accompanied in teleosts by loss of the acrosome. In *Eptatretus* and, despite the presence of endonuclear canals, in the sturgeon, there is no perforatorium whereas in the lamprey (lacking a micropyle) the perforatorium appears to develop from the endonuclear canal. It therefore appears that the presence of a micropyle correlates with absence of a perforatorium and not specifically with absence of an acrosome as this may coexist with it. An acrosome with one or more intranuclear canals also occurs in the externally fertilizing sperm of *Polypterus*, and *Neoceratodus* but there the acrosome reaction has not been observed.

Accessory fibres. The development of nine accessory fibres around the 9+2 basic axonemal structure, restricted to lamprey sperm and, independently developed, in the Osteoglossomorph, *Pantodon*, is seen in many animal groups which, like *Pantodon*, have internal fertilization (Nicander, 1970). There are two elements lateral to the axoneme in the Chondrichthyes, Actinistia, Dipnoi and Amphibia (see Mattei, 1988). The 9 accessory fibres may be an adaptation for movement in viscous fluids. The same may also be true of the radial periaxonemal rods of the sperm of the exocoetoid *Hemirhamphodon* (see Chapter 17). The significance of accessory fibres in lampreys, with their external fertilization, is obscure and, we have seen, has been thought by Afzelius (1970) to persist from former internal fertilization.

Acrosome in Neopterygii. In the Neopterygii, typified by anacrosomal aquasperm, a putative acrosome has been reported in the spermatid of the ent-aquasperm of *Salmo gairdneri* (Billard, 1983b) as an anterior vacuole similar to that visible at the beginning of acrosome formation in mammals (Mattei and Mattei, 1978); and one with an apparent perforatorium occurs transiently in the spermatid of the lophiiform *Neoceratias spinifer* (Jespersen, 1984) which appears to have an intimate and modified form of external fertilization. One or more anterior vesicles at the normal location of an acrosome are reported in the present work for the ect-aquasperm of gymnotoid *Sternarchus*, the percichtyid *Lates*, and the atheriniform genus *Melanotaenia*. Differentiation and densification of the nuclear envelope which might play a part in membrane fusion during fertilization (Billard, 1983b) have been observed in the anterior part of spermatids and spermatozoa in trout (Billard, 1983a) and are interpreted as remnants of the location of a former acrosome in the spermatid of the gobiesociform *Lepadogaster lepadogaster* (Mattei and Mattei, 1978a) in which it possibly has some relationship to penetration of neigbouring cells in the testis which has been observed in this species. The acrosome shows its fullest development for teleosts in the introsperm of the salmoniform *Lepidogalaxias salamandroides* (see Salmoniformes below) in

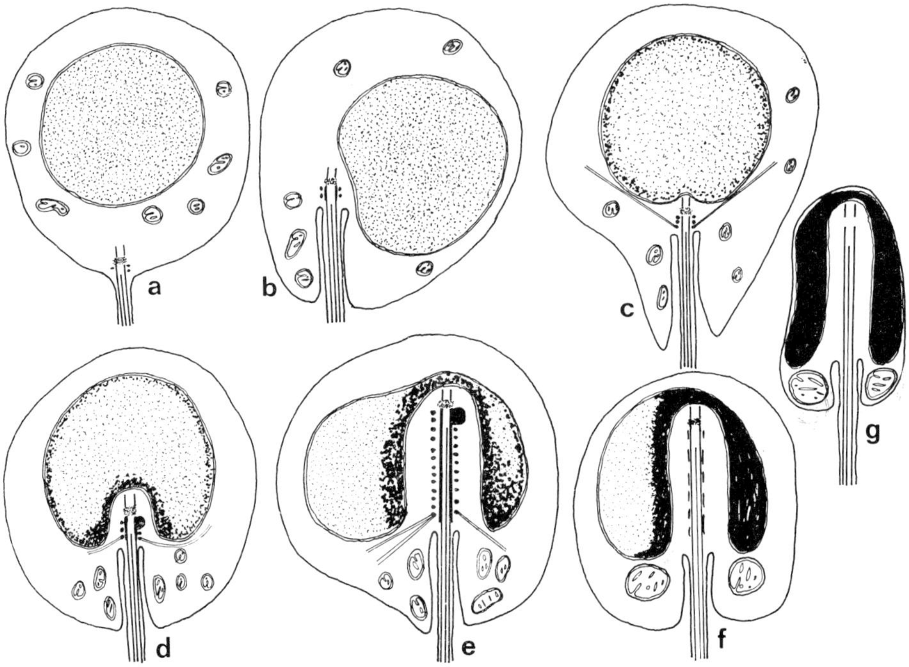

Fig.5.2. Type I teleost spermiogenesis*sensu* Mattei, exemplified by *Upeneus prayensis*. From Mattei, X. (1970). In *Comparative Spermatology* (ed. B. Baccetti), pp. 59-69. Academic Press, New York. Fig. 5.

which an egg micropyle is absent.

b. Anacrosomal aquasperm

i. Uniflagellate anacrosomal aquasperm. We will recognize here a basic teleostean aquasperm which will be referred to as the *teleostean aquasperm* or as a *simple anacrosomal aquasperm*. This is the "primitive spermatozoon" defined for fish by Mattei (1988) and described morphologically and developmentally below. It is overwhelmingly the most common type of spermatozoon in Neopterygii and therefore in fish as a whole and is already established (but with a single mitochondrion by fusion) in the holostean *Lepisosteus*. Some acrosomeless aquasperm have a more complex structure and may be termed *complex uniflagellate anacrosomal aquasperm*; these represent one form of the the "evolved spermatozoon" *sensu* Mattei (1988). Mattei also refers the sperm of *Lepisosteus* to the evolved category because of the mitochondrial fusion. Mattei (1970) recognizes two types of such uniflagellate anacrosomal sperm with regard to course of their late spermiogenesis. Recognized initially for perciforms they are useful, though not all-inclusive, concepts for other groups also and are described below.

Type I sperm. Teleostean (and Neopterygian) aquasperm typically have a small rounded or ovoid nucleus, approximately 2-3 µm in diameter, but no

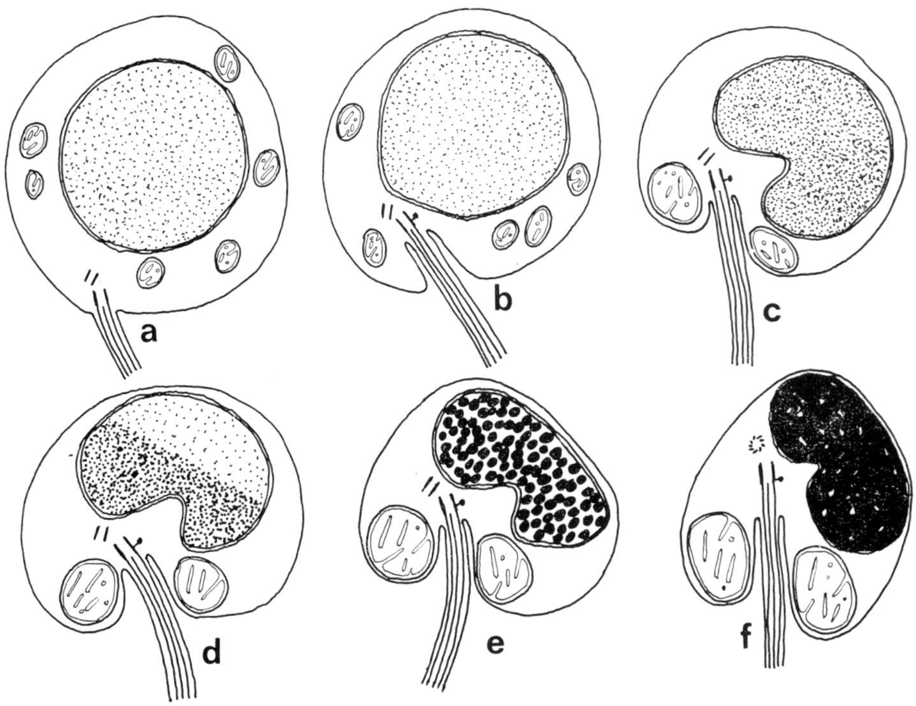

Fig. 5.3. Type II teleost spermiogenesis *sensu* Mattei, exemplified by *Parapristipoma octolineatum*. From Mattei, X. (1970). In *Comparative Spermatology* (ed. B. Baccetti), pp. 59-69. Academic Press, New York. Fig. 6.

acrosome; two centrioles each consisting of 9 triplet microtubules are present distal to the nucleus; the proximal centriole is often at right angles to the distal centriole which forms the basal body of the flagellum; one or both of the centrioles may or may not be located in a basal fossa of the nucleus if, as is frequent, a fossa is present; small cristate mitochondria are present around the centrioles and the base of the flagellum and are commonly situated in a cytoplasmic collar which extends around the base of the flagellum and is separated from this by a periaxonemal space loosely termed the cytoplasmic canal; the region containing the mitochondria forms a midpiece, usually indistinctly demarcated, less than 1 µm long; the flagellum contains an axoneme of classical form with 9 peripheral doublet microtubules (each consisting of an incomplete B microtubule apposed to a complete A microtubule) and two central singlet microtubules; the plasma membrane of the flagellum extends as 1, 2 or 3 longitudinal flattened processes (sometimes absent), here termed fins.

Mattei (1970, 1988) has shown that what is regarded as the most basic manifestation of this "primitive" spermatozoon results from a simple spermiogenesis: in the young spermatozoon the diplosome (centriolar apparatus) is close to the plasma membrane, to which the future distal centriole is linked. The distal centriole then gives rise to the flagellum while the diplosome moves into the cytoplasm to the presumptive base of the nucleus. Because the distal centriole is linked to the cell membrane, the membrane is drawn inwards to form the cytoplasmic canal. A rotation then causes the flagellar axis to become perpendicular to the base of the nucleus (i.e. with its length in the longitudinal axis of the nucleus). The previously scattered mitochondria aggregate around the base of the flagellum from which they remain separated by the cytoplasmic canal. This spermatozoal morphology, presumed to be primitive, shows variations: the plasma membrane of the flagellum may extend as fins; the centrioles may sink more or less deeply into a nuclear fossa.

A spermatozoon developing in this way is termed by Mattei (1970) a Type I sperm and is exemplified by

Upeneus prayensis, spermiogenesis of which is illustrated in Fig. 5.2 and is further discussed in Chapter 15 (Mullidae).

Type II sperm. The above type of spermiogenesis contrasts with that of a type II spermatozoon in which rotation of the flagellar axis in relation to the nucleus does not take place. In the type II spermatozoon the flagellum remains parallel to the base of the nucleus. Although a depression (fossa) is usually found at this point, the centrioles remain outside it (Mattei, 1970, 1988). The cytoplasmic canal may be absent if the midpiece is very small, for instance in *Maccullochella* (Chapter 15). In the axoneme the A tubules of doublets 1, 2, 5 and 6 exhibit an intratubular differentiation, the lumen appearing dense. This type of spermatozoon and flagellar modification is found only in the more evolved teleosteans, the acanthopterygians, and there is recognized only in the order Perciformes where 25 out of 39 investigated families are stated to have this model (Mattei *et al.*, 1979, Mattei, 1988) though of utility, we will see, in other groups. Type II spermiogenesis is exemplified by *Parapristipoma octolineatum* (Mattei, 1970) (Fig. 5.3) and is further discussed in Chapter 15 (Haemulidae).

Anacrosomal aquasperm morphology is adapted to external fertilization (Franzén, 1970) but effects internal fertilization in anacrosomal introsperm of goodeids (category 2b, above).

Elopomorph sperm. Elopomorphs show a distinctive form of the anacrosomal aquasperm (Chapter 11), complex in several respects, which merits separate treatment, as also recognized by Mattei (1988). Two constant features (autapomorphies) of the Elopomorpha (including the Elopiformes, Notacanthiformes and Anguilliformes) are a 9+0 flagellum and division of the proximal centriole into two elongate bundles of 4 and 5 triplets here collectively termed the pseudoflagellum whether or not this projects freely from the nucleus. The 4 triplet bundle is transformed in 4 doublets by opening of the B subfibre. The nucleus is usually crescentic and there is typically a protuberant centriolar rootlet. There is no acrosome. A supposedly advanced elopomorph group, the Muraenidae, has, however, the appearance of a typical teleostean aquasperm, with no rootlet, but indicates its affinities in retaining the divided centriole and the 9+0 axoneme (Billard and Ginsburg, 1973; Çolak and Yamamoto 1974a,b; Gibbons *et al.*, 1983; Ginzburg and Billard, 1972; Mattei, 1970; Mattei and Mattei, 1972, 1973, 1974; Todd 1976).

The typical elopomorph spermatozoal features may be specific adaptations to elopomorph-specific peculiarities of external fertilization in aquasperm, a view apparently embraced by Mattei (1988) and here considered probable.

If, alternatively, these features were derived from ancestral introsperm, the question would arise as to whether these were primary or secondary introsperm and the answer to this would depend on the phylogenetic position of the group. Elopomorphs are regarded (as the Elopoidei and excluding anguilloids) as the lowest of the teleosts by Breder and Rosen (1966), as maintained by Patterson (1973, p. 235), who terms *Elops* "the most archaic teleost". The elopid sperm, and here also the anguilloid sperm, could therefore reasonably be regarded as derived from a primary introsperm, if internal fertilization were regarded as basic to fish (if not derived from a simple aquasperm). If the elopmorphs are given a higher position in the teleosts, tentatively above the Osteoglossomorpha (Nelson, 1984, and in the present account) a precursory introsperm would be envisaged as having originated from a neopterygian aquasperm. Neither derivation from an introsperm can be selected over the other with any confidence (for further discussion see Chapter 11). Origin from a simple teleostean aquasperm has been proposed above.

ii. Biflagellate aquasperm. A variant on the anacrosomal teleostean aquasperm is the biflagellate sperm. These spermatozoa utilize both centrioles as basal bodies for the development of flagella. Biflagellate sperm have evolved independently at least six times in fish (see Figs. 5.1, 18.1, 18.4 and 18.5), (1) in *Polypterus* (Mattei 1970) (Cladistia), (2) in *Ictalurus* (Poirier and Nicholson 1982) and a malapterurid (Mattei, 1988) (Siluriformes), (3) in *Lampanyctus* sp. (Mattei and Mattei 1976a) (Myctophiformes), (4) in *Opsanus tau* (Casas *et al.* 1981) and *Porichthys notatus* (Stanley 1965b) (Batrachoidiformes), (5) in *Lepadogaster lepadogaster* (Mattei and Mattei, 1978a,b) (Gobiesociformes), and (6) in an apogonid, (Perciformes) (Mattei, 1988), all of which are externally fertilizing. They also occur in various invertebrate groups. Biflagellate sperm are the predominant sperm type in platyhelminthes in which, in contrast, sperm are exclusively internally fertilizing. Again, we may recognize simple (*Ictalurus*, *Opsanus*) and complex categories within this type (both categories occurring in the Batrachoidiformes). The simple type differs insignificantly from the basic teleostean aquasperm apart from being biflagellate.

iii. Aflagellate aquasperm. Aflagellate sperm have evolved in osteoglossomorphs: the Mormyridae (Mattei *et al.* 1972) and Gymnarchidae (Mattei *et al.* 1967a) (Fig. 18.2). Derivation from forms with anacrosomal aquasperm is supported by occurrence of the latter in the sister group of the Mormyroidei, the notopteroids (*Papyocranus afer*, Matttei, 1970), and by intermediacy in the form of the aflagellate sperm of mormyrids from this to the more modified, gymnarchid type (Mattei, 1988). These aflagellate sperm are known to be externally fertilizing, and therefore a type of aquasperm, in *Gymnarchus* in which they possess microtubules and are amoeboid. Fertilization, though presumed to be external, has not been observed in mormyrids. There is also an unconfirmed report of aflagellate sperm for the myctophyforms *Lampanyctodes hectoris* and *Diaphus danae* (Young *et al.*, 1987).

(2) Introsperm (Internally fertilizing sperm)

Internally fertilizing sperm are termed introsperm in accordance with the terminology of Rouse and Jamieson (1987). They, too, fall within the category of the "evolved spermatozoon" *sensu* Mattei (1988).

Internal fertilization, with introsperm, seen in the Chondrichthyes and coelacanths is considered apomorphic for the Metazoa but we have considered, but found equivocal, the possibility that internal fertilization is a basic plesiomorphy for fish as a whole (see discussion below). Although external fertilization is generally, and probably correctly, regarded as the original mode in the Metazoa, Jamieson and Rouse (1989) caution against facile dismissal of internal fertilization, constant for platyhelminthes, as the primitive metazoan mode.

a. Acrosomal introsperm. The introsperm of chondrichthyans and *Latimeria*, taxa widely held to be phylogenetically primitive, could conceivably be retentions of a primitive internally fertilizing sperm type. If so they could be considered *primary introsperm.* They have been examined ultrastructurally in the several species of elasmobranchs, in two species of the Holocephali, and in *Latimeria chalumnae* (Table 5.1). The acrosome, the nucleus, the midpiece and/or the tail are elongated, but no satisfactory explanation of the adaptive significance of such elongation has been given. In polychaetes elongation is frequently correlated with transdermal impregnation (Jamieson and Rouse, 1989) but this phenomenon is unknown in fish. It might be considered adaptive to better motility in narrow and tortuous ducts of the internal reproductive system but occurrence of similar elongation (see below) in externally fertilizing fish cannot have this function. Where externally fertilizing species exhibit elongation of the sperm (in acrosomal aquasperm) the eggs of these fishes are usually covered with viscous material and/or have no micropyles (e.g. lung fish), suggesting that elongation is adapative to penetration of egg envelopes. Nevertheless great elongation occurs in myxinids, and in a few internally fertilizing neopterygians, in which an egg micropyle is present.

In the light of sporadic occurrence of an acrosome in ect-aquasperm and introsperm noted above, it would appear that a plesiomorphic genetic propensity to produce an acrosome has been retained in neopterygians though rarely expressed Presence of the acrosome in *Lepidogalaxias* may reasonably be regarded as a secondary development correlated with the fact (Pusey and Stewart, 1989) that the egg is exceptional for a teleost in lacking a micropyle.

It is unlikely that occurrence of the acrosome and absence of the micropyle are plesiomorphies and that the *Lepidogalaxias* sperm is a primary introsperm (derived from introsperm which may have been basic to Pisces) as it is not a plesiomorphic fish and is deduced to have descended through forms with a micropyle and anacrosomal sperm.

b. Anacrosomal introsperm. All teleostean introsperm which lack an acrosome appear to be secondarily derived from externally fertilizing aquasperm, as is also suggested (from "primitive sperm") by Mattei (1988). They may have undergone little or no further modification from the simple morphology of the teleostean aquasperm or may have become relatively complex.

Simple anacrosomal introsperm. A few cases in which a sperm type with aquasperm structure is associated with internal fertilization (simple secondary introsperm) are known. These are the viviparous cyprinodonts *Ameca splendens, Ataenobius toweri, Characoden lateralis* and *Xenotoca eiseni* (Goodeidae) and the perciform *Zoarces viviparus*, the viviparous blenny (Zoarcidae). It seem incontestable, as indicated in the previous paragraph, that internal fertilization in these fishes has developed recently, in evolutionary terms, from external fertilization involving aquasperm. (For an excellent review of morphological and physiological adaptations for viviparity, see Webb and Brett, 1972).

Complex anacrosomal introsperm Complex, although acrosomeless, introsperm with elongate nucleus and midpiece occur in viviparous Poeciliidae, Jenynsiidae and Anablepidae among the cyprinodontoids (Breder and Rosen, 1966; Parenti, 1981). They are, again, here considered to have evolved secondarily from simple teleostean aquasperm (see also Mattei, 1988). The spermatozoa show subtle differences between these groups (see Chapter 17). Although undescribed spermatologically, the Horaichthyidae also have internal fertilization. This sperm type has also developed independently in the osteoglossomorph *Pantodon buchholzi*. In a few percomorph groups, also, the spermatozoa are elongated, though again derivable from the aquasperm, and again differ between groups (e.g. the scorpaeniforms *Oligocottus maculosus* and *Sebastiscus marmoratus*, and the perciform *Cymatogaster aggregata*). The sperm of *Cottus gobio* is intermediate in structure between the simple and complex anacrosomal introsperm types, and, interestingly, whether its mode of fertilization is internal or external is disputed.

The plesiomorphic mode of fertilization in fish

It is doubtful that the fact that the spermatozoa of the more plesiomorphic extant fish groups are of the "advanced" type implies that fertilization was plesiomorphically internal in fish.

In so far as Fig. 5.1 may be accepted as the true phylogeny of fish and Amphibia, it would appear that if external fertilization were basic (plesiomorphic), internal fertilization in the groups represented (omitting viviparous teleosts) would have originated independently twice, once in the Chondrichthyes and once in the Actinistia (*Latimeria*) unless, as held by some, see Chapter 9, the two groups have a common origin. On the other hand, if internal fertilization were assumed to be plesiomorphic in fish in general and in Chondrichthyes and Actinistia in particular, external fertilization would have originated independendently 4 or 5 times, (1) in the myxinids (although the mode of fertilization is not certainly known in these), (2) in the petromyzontids, (3) in the presumed common ancestor of the cladistian through neopterygian assemblage, and (4) in the common ancestor of the Dipnoi and Amphibia or in each of these independently. (Other possibilities, such as plesiomorphic retention of internal fertilization in chondrichthyes but apomorphic reversal to this in Actinistia exist). From the two chief alternatives it can be seen that it is more parsimonious to assume that external fertilization is plesiomorphic for fish as a whole and that the immediate chordate ancestor of the fish had, like the cephalochordate *Branchiostoma*, an ect-aquasperm ("primitive sperm" *sensu* Franzén).

However, some consideration should be given to evidence which might be thought to indicate that a basic internal fertilization has been independently relinquished in favour of external fertilization in a number of lines. Thus, although lampreys are externally fertilizing, they possess a penis (cloacal tube), a fact which suggests loss of former internal fertilization. (We cannot, however, exclude the possibility that this form of fertilization is merely a refinement of external fertilization which never proceeded in lampreys to true internal fertilization.) In *Petromyzon marinus* the cloacal tube of the male is fitted closely to the female pore and the eggs are fertilized as they leave the body in what appears to be neither truly external nor internal fertilization (Breder and Rosen, 1966). More cogently perhaps, the ultrastructure of the "advanced" sperm of the lower fish, the myxinid through dipnoan assemblage in Fig. 5.1, is unlike that of the ect-aquasperm of invertebrates and the lower chordates (e.g. *Branchiostoma*, Jamieson, 1984). It might be a retention of former adaptations to internal fertilization rather than a modification for external fertilization in view of the evident suitability of the simpler ect-aquasperm structure for the latter. Evidence that the telostean aquasperm has indeed undergone some simplification is seen in the transient and/ or sporadic occurrence of an acrosome in the species noted above, suggesting the occurrence of acrosome-bearing (though not necessarily internally fertilizing) sperm in the immediate ancestry of the Neopterygii.

If internal fertilization were basic to fish while the ancestral chordates had ect-aquasperm it would be necessary to explain why external fertilization, clearly so effective in many invertebrates and lower chordates (as in teleosts), would have been replaced with internal fertilization in early fish. No convincing answer can be given to this question but it is at least possible to envisage one evolutionary "bottle-neck" which could have led to internal fertilization in ancestral fish. Internal fertilization is often correlated with production of small numbers of large eggs as in small, sediment-living or interstitial animals. It is seen as an adaptation maximizing the likelihood of fertilization and production of sufficient progeny. It is conceivable that ancestors of fish had developed a tendency to enter the sediments to the extent of becoming obligate burrowers. This tendency is seen in *Branchiostoma* (without suggesting that cephalochordates, although generally accepted as the sister-group of the vertebrates, were directly ancestral to fish). The eel-like form of the Agnatha might, in this view, have been derived from that of a burrowing ancestor. In support of this thesis, the ammocoete larva of lampreys, possibly partly recapitulatory of ancestral adult stages of petromyzontiforms, burrows in the muddy substrate of streams and rivers, filter feeding, like cephalochordates, by means of an endostyle. This highly hypothetical explanation (Jamieson, unpublished), whether or not it represents the true selective pressure originally favouring internal fertilization in fish, does illustrate how a "bottle-neck" of internal fertilization might have been interposed in the main line of evolution of fish from an externally fertilizing chordate ancestor. Berrill (1955), too, saw the ammocoete as "the original kind" of larva at least for freshwater vertebrates but he did not suggest that it is recapitulatory of the adult also. Rather, he saw Amphioxus, with its 'abortive head" as 'degenerate' and also [presumably because of persistence in the adult of the endostyle] neotenic.

The alternative view, that early fish were externally fertilizing, with ect-aquasperm, and that internal fertilization in the Chondrichthyes and coelacanths are early specializations which did not involve the main line of fish evolution remains cogent, however. It is possible that the complex sperm of lampreys, lungfish and other lower gnathostomes represent specializations for peculiar aspects of their external fertilization, much as the ent-aquasperm of polychaetes are modified relative to the "primitive" facies (Rouse and Jamieson, 1987), and do not indicate previous internal fertilization. In this context the size of the egg and configuration of egg envelopes in fish requires investigation in relation to sperm structure.

Some support for regarding external fertilization as primitive, at least for the Osteichthyes, is here considered to lie in the presence of a pair of lateral fins on the flagellum in disparate bony fish groups: Dipnoi (*Neoceratodus*, *Protopterus*); Chondrostei (*Acipenser*); Ginglymodi (*Lepisosteus*); Euteleostei including and above the Protacanthopterygii. Fins must parsimoniously be considered plesiomorphic for this assemblage. Afzelius (1978) reviews the independent occurrence of flagellar fins in two echinoderm, *Ophiocoma*, species and four nereid polychaetes, all externally fertilizing. In fish they do not appear to be consistent with former internal fertilization. Internally fertilizing groups, including the Chondrichthyes, lepidogalaxids, and live-bearing hemiramphids, lack fins or (poeciliids) show signs of their reduction and it is unlikely that fins have been independently reevolved from internally fertilizing precursors, lacking fins, in the diverse lineages with finned sperm. Where fins coexist with internal fertilization this is clearly derived from relatively recent external fertilization, as in poeciliids. In this interpretation of the significance of fins, advanced (complex) sperm retaining acrosomes may well be adaptations in Dipnoi, Cladistia and Chondrostei for peculiarities of fertilization biology, probably features of the egg envelopes in the absence initially (Dipnoi, Cladistia) of micropyles rather than evidence of former internal fertilization.

Table 5.1. Sources of information on the ultrastructure of fish spermatozoa.

(See opposite page). Some references to light microscopy[lm] are included. Scanning electron microscopy to the exclusion of transmission electron microscopy is also indicated[sem]. Taxonomic ranks listed are chiefly those of Nelson (1984). Although most groupings are considered cladistically sound it is not to be expected that all nested series and rank names would survive a cladistic parsimony analysis, at least in terms of hierarchical equivalence.

Superclass Agnatha
Class Myxini

Order Myxiniformes - Myxinidae: *Eptatretus deani, E.* sp., *E. stoutii*, Jespersen, 1975; *Myxine glutinosa*, Alvestad-Graebner and Adam, 1977; Nicander, 1968.

Class Cephalaspidomorphi

Order Petromyzontiformes - Petromyzontidae: *Lampetra japonica*, Jaana and Yamamoto, 1981; *Lampetra fluviatilis*, Nicander, 1968, 1970; Nicander and Sjoden, 1968, 1971; *L. planeri*, Follenius, 1965; Stanley, 1967; *Mordacia mordax* ; *M. praecox*, Hughes and Potter, 1969.

Superclass Gnathostomata; Grade Pisces; Class Chondrichthyes

Subclass Holocephali

Order Chimaeriformes - Chimaeridae: *Chimaera phantasma*, Hara and Tanaka, 1986; *Hydrolagus colliei*, Stanley, 1965a, 1983; *Neoharriotta pinnata*, Mattei, 1988.

Subclass Elasmobranchii

Superorder Selachimorpha

Order Hexanchiformes - Chlamydoselachidae: *Chlamydoselachus anguinesus* Hara and Tanaka, 1986.

Order Lamniformes (Galeoidea) - Carcharinidae: *Prionace glauca*, Hara and Tanaka, 1986; Scyliorhinidae: *Scyliorhinus caniculus*, Gusse and Chevaillier, 1978; *Scyliorhinus* sp., Stanley, 1971, 1983.

Order Squaliformes - Squalidae: *Centrophorus atromarginatus* (spermatogenesis), Tanaka *et al.* 1978; *Centroscymnus owstonii*, Hara and Tanaka, 1986; *Squalus suckleyi*, Stanley, 1964, 1965a, 1971.

Superorder Batidoidimorpha (Hypotremata)

Order Rajiformes - Dasyatidae: *Dasyatis kuhlii, Dasyatis garouensis*; Hara and Tanaka, 1986; Rajidae: *Raja* sp., Stanley, 1971; Rhinobatidae: *Rhinobatos cemiculus*, Boisson *et al.*, 1968b; Mattei, 1970.

Class Osteichthyes

Subclass Crossopterygii

Superorder Coelacanthimorpha (Actinistia)

Order Coelacanthiformes - Latimeriidae: *Latimeria chalumnae*, Tuzet and Millot, 1959[lm]; Mattei *et al.*, 1988.

Subclass Dipnoi (Dipneusti)

Order Lepidosireniformes - Protopteridae: *Protopterus annectens*, Boisson, 1963; Boisson *et al.*, 1967; Mattei, 1970; *Protopterus aethiopicus*, Purkerson *et al.*, 1974.

Order Ceratodontiformes - Neoceratiidae: *Neoceratodus forsteri*, Jespersen, 1971.

Subclass Actinopterygii

Infraclass Cladistia (Brachiopterygii): a separate subclass in Nelson (1984).

Order Polypteriformes - Polypteridae: *Polypterus senegalus*, Mattei, 1970, 1988.

Infraclass Chondrostei

Order Acipenseriformes - Acipenseridae: *Acipenser guldenstadti*, Ginzburg, 1968; *Acipenser stellatus*, Ginzburg, 1968; *Acipenser transmontanus*, Cherr and Clark, 1984; *Huso huso*, Ginzburg, 1968.

Infraclass Neopterygii

Division Ginglymodi

Order Lepisosteiformes - Lepisosteidae: *Lepisosteus osseus*, Afzelius, 1978; Mattei *et al.*, 1981; Mattei, 1988.

Division Halecostomi

Subdivision Halecomorphi

Order Amiiformes - Amiidae: *Amia calva*, Retzius, 1905[lm].

Subdivison Teleostei

Superorder Osteoglossomorpha

Order Osteoglossiformes - Suborder Osteoglossoidei; Pantodontidae: *Pantodon buchholzi*, van Deurs, 1973, 1974; van Deurs and Lastein, 1973. **Suborder Notopteroidei**; Notopteridae: *Papyocranus afer*, Mattei, 1970. **Suborder Mormyroidei**; Mormyridae: *Gnathonemus niger*; *G. senegalensis*; *Hyperopisus bebe*; *Mormyrus rume*; *Petrocephalus bovei*, Mattei *et al.*, 1972. Gymnarchidae: *Gymnarchus niloticus*, Mattei *et al.*, 1967a.

Infradivision Elopomorpha

Order Elopiformes - Suborder Elopoidei; Elopidae: *Elops lacerta*, Mattei and Mattei, 1974; Suborder Albuloidei; Albulidae: *Albula vulpes*, Mattei and Mattei, 1972, 1973, 1974; *Pterothrissus belloci*, Mattei and Mattei, 1974.

Order Anguilliformes - Anguillidae: *Anguilla anguilla*, Baccetti *et al.*, 1979a, 1981; Billard and Ginsburg, 1973; Gibbons *et al.*, 1983; Ginzburg and Billard, 1972; *A. australis schmidtii*, Todd, 1976; *A. dieffenbachii*, Todd, 1976; *A. japonica*, Çolak and Yamamoto, 1974a,b. Congridae: *Congermuraena bertini*, *Cynoponticus ferox*, *Paraconger notialis*. Undetermined Congridae, Mattei and Mattei, 1974. Ophichthyidae: *Echelus myrus*; *Ophichthus ophis*, *Pisodonophis semicinctus*; Muraenidae: *Lycodontis afer*, *Muraena helena*, *Muraena robusta*, *Mystriophis rostellatus*; Heterenchelydae: *Pythonichthys microphthalmus*, Mattei, 1970, Mattei and Mattei, 1974.

Order Notacanthiformes - unnamed species, Mattei, 1988.

Infradivision Clupeomorpha

Order Clupeiformes - Clupeidae: *Clupea harengus*, Retzius, 1905[lm]; *Ethmalosa fimbriata*, Mattei, 1970; Mattei *et al.*, 1981; Engraulidae: *Anchoa* (*=Engraulis?*) *guineensis*, Mattei, 1970; Mattei *et al*, 1981; *Sardinella aurita*, Mattei, 1970; Mattei *et al.*, 1981.

Infradivision Euteleostei

Superorder Esocomorpha

Order Esociformes - Esocidae: *Esox lucius*, Billard, 1970a; Stein, 1981.

Superorder Ostariophysi

Order Cypriniformes - Cyprinidae: *Abramis brama*, Stein, 1981; *Alburnus alburnus alborella*, *Barbus barbus plebejus*, Baccetti *et al.*, 1984; *Blicca bjorkna*, Stein, 1981; *Brachydanio rerio*, Wolenski and Hart, 1987; *Carassius auratus*, Baccetti *et al.*, 1984; Fribourgh *et al.*, 1970; Munoz-Guerra *et al.*, 1982; *Chondrostoma toxostoma*, Baccetti *et al.*, 1984; *Cyprinus carpio*, Billard, 1970a; Fujimura *et al.*, 1957; Kudo, 1980; Stein, 1981; *Leuciscus cephalus*, Baccetti *et al.*, 1984; Stein, 1981; *Leuciscus souffia*, Baccetti *et al.*, 1984; *Leuciscus leuciscus*; *Phoxinus phoxinus*, Stein, 1981; *Rhodeus ocellatus*, Ohta and Iwamatsu, 1983; *Rutilus rubilio*, Baccetti *et al.*, 1984; *Rutilus rutilus*, Stein, 1981; *Scardinius erythrophthalmus*, Stein, 1981. Cobitidae: *Acanthophthalmus semicinctus*, Jamieson (unpublished) .

Order Characiformes - Characidae: *Paracheirodon* (=*Hyphessobraycon*) *innesi*, Jamieson (unpublished).

Order Siluriformes - Suborder Siluroidei; Ictaluridae: *Ictalurus punctatus*, Jaspers *et al.*, 1976; Poirier and Nicholson, 1982; Yasuzumi, 1971; Clariidae: *Clarias senegalensis*, Mattei, 1970; Pimelodidae: *Rhamdia sapo*, Maggese et al., 1984[sem]. Suborder Gymnotoidei; Apteronotidae: *Sternarchus albifrons*, Jamieson (unpublished).

Superorder Protacanthopterygii

Order Salmoniformes - Suborder Argentinoidei; Alepocephalidae: *Xenodermichthys* sp., Mattei *et al.*, 1981. Searsidae: *Searsia* sp., Mattei *et al.*, 1981. **Suborder Salmonoidei**; Superfamily Salmonoidea; Salmonidae: *Coregonus wartmanni*; *Hucho hucho*, Stein, 1981; *Oncorhynchus gorbuscha*, Drozdov *et al.*, 1981; *Oncorhynchus keta*, Kobayashi and Yamamoto, 1987; *Oncorhynchus kisutch*, Lowman, 1953; *Oncorhynchus tshawytscha*, Zirkin, 1975; *Salmo gairdneri*, Billard, 1970a; 1983a; Fribourgh and Soloff, 1976; Mattei *et al.*, 1981; Stein, 1981; *Salmo salar*, Nicander, 1968; *Salmo trutta*, Furieri, 1962; *Salmo trutta fario*, Billard, 1983a; Stein, 1981; *Salvelinus alpinus*, Stein, 1981; *Salvelinus fontinalis*, Fribourgh, 1978; Stein, 1981; *Thymallus thymallus*, Stein, 1981. Superfamily Osmeroidea; Plecoglossidae: *Plecoglossus altivelis*, Kudo, 1983. Superfamily Galaxioidea; *Galaxias olidus*, Marshall, 1989. **Suborder Leipidogalaxioidei**; Lepidogalaxiidae: *Lepidogalaxias salamandroides*, Leung, 1988a.

Superorder Scopelomorpha

Order Aulopiformes - Synodontidae: *Trachinocephalus myops*, Mattei, 1970.

Order Myctophiformes - Myctophidae: *Lampanyctus* sp., Mattei and Mattei, 1976a; *Diaphus danae*; *Lampanyctodes hectoris*, Young *et al.*, 1987[sem].

Superorder Paracanthopterygii

Order Gadiformes - Gadidae: *Lota vulgaris*, Retzius, 1905[lm].

Order Ophidiiformes - Ophidiidae: *Ophidion* sp., Mattei *et al.*, 1989.

Order Batrachoidiformes - Batrachoididae: *Opsanus tau*, Casas *et al.*, 1981; Hoffman, 1963[lm]. *Porichthys notatus*, Stanley, 1965b.

Order Gobiesociformes - Gobiesocidae: *Lepadogaster lepadogaster*, Mattei and Mattei, 1978a,b.

Order Lophiiformes - Antennaridae: *Antennarius senegalensis*, Mattei, 1970; *Neoceratias spinifer*, Jespersen, 1984.

Superorder Acanthopterygii

Series Percomorpha

Order Zeiformes - Zeidae: *Zeus faber*, Mattei, 1970.

Order Syngnathiformes - Fistulariidae: *Fistularia tabacaria*, Mattei, 1970.

Order Dactylopteriformes - Dactylopteridae: *Dactylopterus* (=*Cephalacanthus*) *volitans*, Boisson *et al.*, 1968a; Mattei, 1970.

Order Scorpaeniformes - Cottidae: *Cottus gobio*, Stein, 1981; *Oligocottus maculosus*, Stanley, 1966, 1969. Scorpaenidae: *Scorpaena angolensis*, Mattei, 1970; *Sebastiscus marmoratus*, Mizue, 1968.

Order Perciformes

Suborder Percoidei; Centropomidae: *Lates calcarifer*, Leung, unpublished; Percichthyidae: *Maccullochella macquariensis*, *Maccullochella peeli*, *Macquaria ambigua*, *Macquaria australasica*, *Macquaria novemaculeata*, Leung (unpublished); Serranidae: *Plectropomus leopardus* (=*oligocanthus*), Jamieson (unpublished); Kuhlidae: *Nannoperca oxleyana*, Marshall, 1989; Apogonidae: *Paronocheilus* sp., Mattei and Mattei, 1984; Percidae: *Perca fluviatilis*, Retzius, 1905[lm]; Stein, 1981; Carangidae: *Vomer setapinnis*, Mattei *et al.*, 1979; Haemulidae: *Parapristipoma octolineatum*, Mattei, 1970; Sparidae: *Boops boops*, Mattei, 1970; Centracanthidae: *Spicara chryselis*, Carrillo and Zanuy, 1977; Mullidae: *Upeneus prayensis*, Boisson *et al.*, 1969; Mattei, 1970.

Suborder Mugiloidei; Mugilidae; *Liza aurata* Brusle, 1981; *Liza dumerili* van der Horst, 1976; van der Horst and Cross, 1978; Polynemidae: *Galeoides decadactylus*, Mattei, 1970.

Suborder Labroidei; Zoarcidae: *Zoarces viviparus*, Retzius, 1905[lm]; Pomacentridae: *Pomacentrus leucostictus*, Mattei and Mattei, 1976; Cichlidae: *Hemichromis fasciatus*; *Tilapia nilotica*, Mattei,

1970; *Oreochromis niloticus*, Guha *et al.*, 1988; Embiotocidae: *Cymatogaster aggregata*, Gardiner, 1978a, b

Suborder Trachinoidea; Mugiloididae: *Parapercis* sp., Jamieson (unpublished).

Suborder Blennioidei; Clinidae: *Clinus nuchipinnis*, Mattei, 1970; Blenniidae: *Blennius cristatus*, *Blennius vandervekeni*, *Ophioblennius atlanticus*, Mattei, 1970; *Blennius pholis*, Silveira *et al.*, 1990

Suborder Gobioidei; Gobiidae: *Gobius niger*, Retzius, 1905[lm]; *Periophthalmus papilio*, Mattei, 1970; Eleotridae: *Hypseleotris galii*, Jamieson (unpublished).

Suborder Scombroidei; Trichiuridae: *Trichiurus lepturus*, Mattei and Mattei, 1976b.

Order Pleuronectiformes - Pleuronectidae: *Platichthys flesus* Retzius, 1905[lm]; Jones and Butler, 1988. Soleidae: *Pegusa triophthalmus*, Mattei, 1970.

Order Tetraodontiformes - Tetraodontidae: *Gastrophysus hamiltoni*, Hansford and Jamieson (unpublished). Balistidae: *Balistes forcipatus*, *Chilomycterurus antennatus*, Mattei, 1970; *Pseudobalistes fuscus*, Jamieson (unpublished).

Series Atherinomorpha

Order Cyprinodontiformes - Cyprinodontidae: *Fundulus heteroclitus*, Yasuzumi, 1971; Brummett and Dumont, 1979; Selman and Wallace, 1986; *Cyprinodon variegatus*, Yasuzumi, 1971; *Fundulosoma thierryi*, *Epiplatys ansorgei*, *E. bifasciatus*, *E. chaperi E. fasciolatus*, (flagellar structure only) Thiaw *et al.*, 1986. Goodeidae: *Ameca splendens*, *Ataenobius toweri*, *Characodon lateralis*, Grier *et al.*, 1978. Jenynsiidae: *Jenynsia lineata*, Dadone and Narbaitz, 1967. Poeciliidae (*additional data provided in this volume): **Gambusia affinis*, Grier, 1975; **Poecilia latipinna*, Mizue, 1969; Grier, 1973a,b; **Poecilia reticulata*, Asai, 1971; Billard, 1970b; Mattei, 1970; Mattei and Boissin, 1966; Porte and Follenius, 1960; **Xiphophorus helleri*, Jonas-Davies *et al.*, 1983; *Xiphophorus* (=*Poecilia*) *maculatus*, Russo and Pisano, 1973; *Xenotoca eiseni*, Grier *et al.*, 1978. Aplocheilidae: *Epiplatys senegalensis*, Mattei, 1970, *Aphyosemion gardneri*, Jamieson (unpublished); *Aphyosemion guignardi*, *A. herzogi*, *A. nigrifluvi*, *A. riggenbachi*, *A. splendopleure*; *Aplocheilichthys lamberti*, *A. normani*, *Aplocheilus lineatus*, *Cynolebias wittei*; and *Notobranchius steinforti*, (flagellar structure only) Thiaw *et al.*, 1986.

Order Atheriniformes - Atherinidae: *Craterocephalus stercusmuscarum*, *C. helenae*, *C. marjoriae*, *Querichthys stramineus*. Melanotaenidae: *Cairnsichthys rhombosmoides*, *Iriatherina werneri*, *Melanotaenia duboulayi*, *M. maccullochi*, *Pseudomugil mellis*, *P. signifer*, *P. tenellus*, all Marshall, 1989; Marshall and Jamieson (unpublished).

Order Beloniformes - Suborder Adrianichthyoidei; Oryziidae: *Oryzias latipes*, Grier, 1976; Sakai, 1976; Iwamatsu and Ohta, 1981[sem]. **Suborder Exocoetoidei**; Exocoetidae: *Fodiator acutus*, Mattei, 1970; Hemiramphidae: *Arramphus sclerolepis*, Jamieson (unpublished); *Hemirhamphodon pogonognathus*, Jamieson, 1989.

Apomorphic features of fish sperm.

In investigating the congruence, or dissonance, of fish sperm ultrastructure with phylograms deduced from other, chiefly somatic characters apomorphic features of the sperm will be indicated on the phylograms. These features will be numbered as indicated in the following list (Table 5.2).

Table 5.2
Apomorphies of fish sperm relative to Branchiostoma
(only apomorphies are numbered; plesiomorphies unnumbered)

Acrosome
present
caplike
1. conical to elongate
2. with spiral ridge
3. absent
4. with endonuclear canal
5. with postacrosomal ring
6. with perforatoria
7. acrosomoid present

Nucleus
8. elongate, >3 µm (give length) - reaching 70 µm in *Neoceratodus* ;
9. spiral
10. chromatin only in anterior part of nucleus
11. with anterior peg
12. basal fossa subterminal
13. very deep basal fossa
14. pointed with "ventral" fossa
15. locally thin chromatin layer
16. crescentic

Midpiece

short

17. spiral
18. elongate, >3 μm
19. axial rod of rhizoplast origin
20. cytoplasmic canal present
22. with fenestrated sheath
23. with submitochondrial net

Mitochondria

several small

24. numerous
25. 1 in nonbasal nuclear concavity
26. 9 helical derivatives
27. 2 elongate
28. 6-10 elongate
29. end to end in 3 to 5 longitudinal columns
30. end to end bilaterally
31. anterior to basal body
32. around nucleus
33. fused as single ring or C-shape
34. single unilateral (occasionally 2 in *Macculochella peeli*)

Centrioles

proximal perpendicular to distal

35. angle >90°
37. proximal and distal serial coaxial (orthogonal)
38. proximal pseudoflagellum divided into 4 + 5 doublets
39. striated centriolar rootlet
40. proximal centriole reduced
41. proximal lost
42. basal body prenuclear
43. triplet centrioles absent
44. external to the fossa or eccentric to nucleus (Type II)

Axoneme

45. 1 fin
46. 2 fins
47. fins lost
48. 9+0
49. biflagellarity
50. aflagellate
51. 9 accessory fibres
52. accessory longitudinal columns, at doublets 3 & 8
53. septate doublets 1 2 6 7
54. septate doublets 1 2 5 6 7
55. septate doublets 1 2 3 5 6 7
56. septate doublets 1 3 5 6 7
57. outer dynein arms absent
58. large amount of glycogen

Other features

59. retronuclear body
60. lamellated body
61. 9 midpiece fibres
62. many cytoplasmic tubules
63. osmiophilic body
64. testis telogonic
65. spermatozeugmata with parallel sperm
66. spermatozeugmata with centrifugal sperm heads
67. one accessory axonemal column, at 3
68. microtubular sheath around midpiece

Egg

69. micropyles: many
70. micropyle: single
71. nucleus: secondarily rounded

Fish sperm and phylogeny

An inferred phylogeny of all living fish groups is illustrated in Fig. 5.1. Phylogenies of the individual groups will be given in subsequent chapters and are summarized in Chapter 18. The cladograms are based chiefly on somatic structure of adults and larvae according to the sources cited, original and in the literature. Construction of cladograms solely from the ultrastructure of spermatozoa which has proved so effective in other groups (e.g. oligochaetes, Jamieson *et al.*, 1987) has less extensive application in fish, at least in the present state of our knowledge, but cases will be presented where sperm synapomorpies used alone give highly heuristic phylogenies. A striking example is the incontrovertible demonstration of relationship between the Anguilliformes and Elopiformes from their highly apomorphic sperm structure which is supported by co-occurrence of the leptocephalus larva (Mattei and Mattei, 1974).

The significance of the micropyle in fertilization

Egg micropyles have evolved in Actinopteri (*sensu* Patterson, 1981, the Actinopterygii above the

Cladistia) although also seen in myxinids. The Neopterygii (Actinopteri above the Chondrostei) are distinguished from other fish in possession of an ect-aquasperm, of remarkable simplicity, lacking even the acrosome, whereas the sperm of all other fish are of a modified type whether ect-aquasperm or introsperm albeit plesiomorphic in retaining an acrosome. Possession of the neopterygian aquasperm, probably as a result of secondary simplification from a more complex sperm type, is coincident with presence of a single egg micropyle prefigured in the numerous micropyles of sturgeons (Chondrostei) with their perforatorium-possessing* and somewhat elongate sperm. Specifically, coevolution appears to have occurred in proto-neopterygians for development of a single micropyle and a block to polyspermy on the one hand and loss of the acrosome and particularly its perforatorium on the other. Precursory to this was the development of multiple micropyles with no blocks by the egg to polyspermy as in sturgeons.

The egg envelopes are also penetrated by numerous pore canals which function in transporting nutrients during oogenesis. It is not known whether these pores remain open and fully traverse the egg envelopes after spawning (Stehr and Hawkes, 1979).

As most of the present work is devoted to the ultrastructure of the neopterygian sperm it is appropriate to pay some attention to the structure and role of the micropyle. It is not proposed, however, to give a detailed review of investigations on this subject. Whatever the genetic-ontogenetic events which initially caused the development of micropyles, one adaptive advantage which may have led to their retention by selection would seem to be that they provide a specialized passage through the egg membranes which allows the spermatozoon to come into contact with the oolemma at fertilization without penetrating outer egg membranes. The acrosome, which serves to penetrate the egg membranes has become unimportant for penetration although transit of the oolemma still has to occur (Ginzburg, 1968, p 115). The claim that egg-binding has been lost in many actinopterygians nevertheless requires substantiation.

The micropyle is typically situated above the animal pole of the egg. The structure of the external micropylar opening permits recognition of three types of micropyles (Riehl and Götting, 1974; Mikodina, 1987). The first type has a deep pit or funnel-like depression in the chorion surface leading into a short canal which opens on the inner surface of the chorion. The second has a shallow saucer-like pit leading to a long canal. The third has the canal opening directly onto the chorion surface. A fourth type, with the canal opening externally on the summit of a conical elevation of the chorion described for *Oryzias latipes* by Hart *et al.* (1984) was not recognized by Iwamatsu and Ohta (1981) for the same species. The area surrounding the external micropylar opening is usually differentiated from the remainder of the chorion. It is frequently devoid of pore canals which typify other regions of the chorion or canals if present have enlarged plugs. Attaching filaments and other surface structures are often absent around the micropyle or, conversely, are elaborated (see review by Laale, 1980 and other references herein).

The open canal of the micropyle, penetrating the chorion, connects the exterior directly to the plasma membrane of the oocyte (oolemma). The internal opening is always smaller than the exterior opening. The canal walls are ribbed or ridged (Kuchnow and Scott, 1977; Stehr and Hawkes, 1979; Kobayashi and Yamamoto, 1981; Hart and Donovan, 1983; Mikodina, 1987), probably reflecting the lamellar structure of the internus of the chorion. The oolema and egg surface are depressed immediately under the micropyle, conforming in shape with the depression in the chorion formed by the micropylar funnel (Brummett and Dumont, 1979; Brummett *et al.*, 1985; Hart and Donovan, 1983; Iwamatsu and Ohta, 1981; Kobayashi and Yamamoto, 1981, 1987; Kudo, 1982, 1983; Kudo and Sato, 1985; Ohta and Iwamatsu, 1983; Ohta, 1985a,b; Onitake and Iwamatsu, 1986; Stehr and Hawkes, 1979; Wolenski and Hart, 1985; Yamamoto and Kobayashi, 1988; Yanagamachi, 1957). Most reports indicate that species-specific elaboration of the oolemma is located directly under the internal opening of the micropyle. In *Rhodeus ocellatus*, *Cyprinus carpio*, *Tribolodon hakoensis* and *Brachydanio rerio*, this takes the form of a group or tuft of microvilli which project into the micropylar canal (Hart and Donovan, 1983; Kudo, 1980, 1982; Kudo and Sato, 1985; Ohta, 1985a,b; Ohta and Iwamatsu, 1983; Wolenski and Hart, 1987, 1988). Alternatively, in *Plecoglossus altivelis*, *Onchorhynchus keta*, *Oryzias latipes*, *Platichthys stellatus* and

*The contents of the three endonuclear canals in *Acipenser* and here regarded as perforatoria although this is disputed.

Fundulus heteroclitus, a rounded or slightly flattened protuberance on the oocyte surface extends into the proximal part of the micropylar canal. This protuberance is often surrounded by, though not bearing, microvilli (Kudo, 1982, 1983; Kobayashi and Yamamoto, 1981, 1987; Stehr and Hawkes, 1979) or microvilli are absent, as in *Fundulus heteroclitus* (Brummett *et al.*, 1985). It has been demonstrated that either type of protrusion into the micropyle is the site of sperm contact with the ovum. Such sperm entry sites have been identified in most eggs which have been investigated.

It might be expected that the sperm entry site is differentiated during oogenesis but this is contraindicated by the observation of Ohta (1985b) that on experimental removal of the chorion the specialized structure of the sperm entry site in *Rhodeus ocellatus*

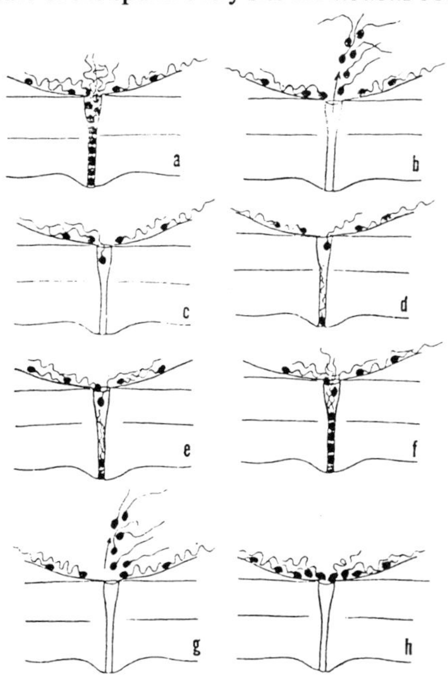

Fig.5.4. *Clupea pallasii.* Semidiagrammatic drawings showing the manner of sperm entry into the egg (at 10 ° C). Numbers in the figure represent the time after insemination in minutes. After Yanagimachi, R. (1975). *Zoological Magazine* (Tokyo) **66**, 218-233. Fig. 6.

eggs became virtually indistinguishable from the surrounding surface of the egg. A single egg formed many fertilization cones on contact with spermatozoa. Other workers had previously shown that dechorionated eggs undergo polyspermy (references

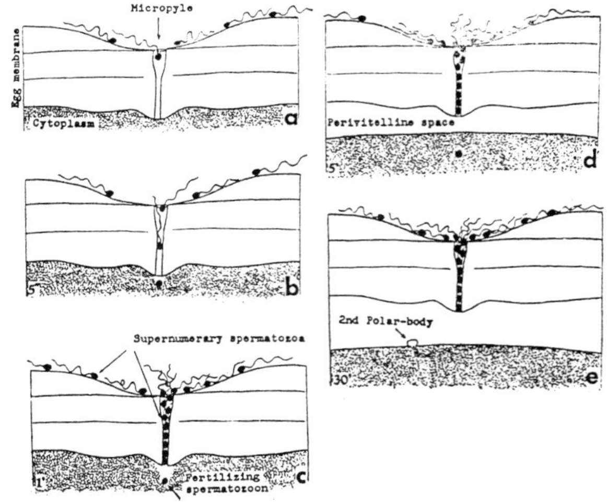

Fig. 5.5. *Clupea pallasii.* Extrusion of supernumerary sperm from the micropyle. A. 30 min after fertilization. B. 35 min, the first extrusion. C-D. 36-38 min, further advance of sperm into the micropyle. E-F. 40 min, micropyle filled with several spermatozoa. G. 45 min, the second extrusion. H. 55 min, further attempts of spermatozoa to enter the micropyle have ceased (all at 10° C). From Yanagimachi, R. (1975). *Zoological Magazine* (Tokyo) **66**, 218-233. Fig. 7.

in Yamamoto, 1961). This suggests that the integrity of the sperm entry site, in this species, is maintained at least partly by the chorion. Acceptance of sperm thus appears to be a property of the entire oolemma which is restricted by the chorion to the micropyle. That the successful sperm, that which brings about fertilization of a monospermic egg, is merely the first to reach the bottom of the micropyle must not be assumed, however. There is a possibility that expulsion of supernumerary sperm reflects an underyling ability of the egg to "choose" which sperm fuses with it. The possibility of egg "selection" of sperm by an electrochemical mechanism compatible with Paterson's (1982) specific mate recognition system is discussed by Peterson (1989).

Contact between the male and female gametes usually occurs within a few seconds of spawning. The apical region of the spermatozoon contacts the oolemma at the specialized site of entry at the base of the micropylar canal. Immediately after gamete fusion the oolemma rises and forms a protrusion, the so-called fertilization cone, enclosing the spermatozoon and extending into the canal. Initially this protrusion takes the form a collar extending around the apical portion of the spermatozoon and into the micropylar canal, as shown for *Plecoglossus altivelis* (Kudo, 1983), *Fundulus heteroclitus* (Brummett and Dumont, 1979; Brummett *et al.*, 1985), *Oryzias latipes* (Onitake and Iwamatsu, 1986) and *Oncorhynchus keta* (Kobayashi and Yamamoto, 1987). This response is apparently absent in *Brachidanio rerio* (Wolenski and Hart, 1985, 1987). A second fertilization cone forms, usually after the spermatozoon has been incorporated into the ooplasm and formation of the pronucleus has been completed. The second cone rises under the first, which it exceeds in size, thus forming a tiered fertilization complex. This cone complex may be transient, subsiding back to a normal membrane (Brummett and Dumont, 1979; Brummett *et al.*, 1985; Iwamatsu and Ohta, 1981; Kobayashi and Yamamoto, 1987; Kudo, 1983; Kudo and Sato, 1985; Lessman and Huver, 1981; Wolenski and Hart, 1985, 1987) or has been reported to be permanent and to remain attached to the micropylar region of the chorion (Ohta, 1985a,b; Ohta and Iwamatsu, 1983).

In addition to providing a route through the egg membranes, the micropyle provides a potential mechanism for preventing polyspermic penetration of the egg. It would be wrong, however, to regard the micropyle simply as a filter ensuring monospermy by virtue of its narrowness. Exceptionally, as in *Cyprinus carpio*, the terminal micropylar opening is sufficiently large to potentially allow the passage of more than one spermatozoon yet monospermy is maintained (Kudo and Sato, 1985). Nevertheless, in most investigated species the inner opening of the micropyle is too narrow to allow passage of more than one spermatozoon at a time and thus simultaneous polyspermy is prevented (e.g. *Clupea pallasii*, Yanagamachi, 1957, Fig. 5.4, 5.5; *Salmo salar*, *S. trutta*, Ginsburg, 1963; *Fundulus heteroclitus*, Kuchnow and Scott, 1977; *Oryzias latipes*, Iwamatsu and Ohta, 1981; *Onchorhynchus keta*, Kobayashi and Yamamoto, 1981; *Brachydanio rerio*, Hart and Donovan, 1983; *Plecoglossus altivelis*, Kudo, 1983; and *Rhodeus ocellatus*, Ohta and Iwamatsu, 1983). Entry of the first spermatozoon through the oolemma is concomitant with protrusion of the fertilization cone into the micropyle. This event is known to be accompanied by expulsion of supernumerary sperm (Kudo, 1980; Iwamatsu and Ohta, 1981; Kobayashi and Yamamoto, 1981; Kudo and Sato, 1985; Ohta and Iwamatsu, 1983). In fertilization of the egg of *Clupea pallasii*, Yanagamachi (1957) reports two seprate expulsions of supernumerary sperm (Fig. 5.5).

It is probable that the supernumerary sperm do not fuse to the fertilization cone because the oolemma no longer contains binding sites for spermatozoa, possibly because of incorporation of sperm plasma membrane into the oolemma of the cone. Although formaton of the cone appears necessary to prevent polyspermy it is not necessary for development of the fertilized egg. Thus eggs of *Brachydanio rerio* can be successfully fertilized, and continue development, if formation of the fertilization cone is experimentally inhibited (Wolenski and Hart, 1988). In *Oryzias latipes* the earliest fertilization cone was recognized at the region of sperm egg-membrane fusion 10-15 seconds after insemination. At the same time a ball-like stalked protrusion containing a number of sperm was observed on the outer opening of the micropylar canal and was considered to be concerned with trapping the supernumerary sperm and excluding them from the micropylar canal (Onitake and Iwamatsu, 1986).

Final structural blockage against sperm correlates in some species with egg hydration. The increased internal pressure on the chorion in conjunction with hardening of the chorion constricts or completely occludes the inner opening of the micropyle (Laale, 1980; Kobayashi and Yamamoto, 1981; Wolenski and Hart, 1987; Yamamoto and Kobayashi, 1988). Chemicophysical blockage to polyspermy has also been proposed. Changes have been observed in electrical potential over the egg surface which appear to be linked to calcium flux immediately following fertilization. These are followed by a phase corresponding to discharge of cortical alveoli of the egg (Ginsburg, 1961). A free calcium wave, traversing the activating egg, has been demonstrated using aequorin injection for the Medaka, *Oryzias latipes* (Gilkey *et al.*, 1978). However, as suspected by Ginsburg, passage of the fertilization potential over the egg is not necessary to block polyspermy as monospermy is maintained when passage of the potential is experimentally blocked with voltage clamps

(Nuccitelli, 1980). In contrast, the wave of exocytosis of cortical alveoli into the perivitelline space (following the wave of calcium flux and the fertilization potential) is widely believed to sustain monospermy. The resulting perivitelline fluid has a strong agglutinating action on spermatozoa (Brummett and Dumont, 1981; Ginsburg, 1961; Iwamatsu and Ohta, 1981; Kobayashi and Yamamoto, 1981; Kudo, 1980; Kudo and Sato, 1985; Ohta and Iwamatsu, 1983; Wolenski and Hart, 1987). Furthermore, it is the pressure created in the perivitelline space which squirts the fluid and any supernumerary spermatozoa out of the micropyle (Sakai, 1961; Iwamatsu and Ohta, 1981; Yamamoto, 1952; Kobayashi and Yamamoto, 1981, 1987). The perivitelline fluid may in turn give rise to the amorphous micropylar plug which has been observed to form after this expulsion (Brummett and Dumont, 1981; Iwamatsu and Ohta, 1981; Sakai, 1961; Wolenski and Hart, 1987; Yamamoto, 1952).

As a result of the formation of the first fertilization cone, the spermatozoon is engulfed and penetrates into the egg cortex through plasma membrane fusion. The head and midpiece of the spermatozoon penetrate more rapidly than the tail and the head is exposed to the cortical cytoplasm (Brummett and Dumont, 1979; Iwamatsu and Ohta, 1981; Kobayashi and Yamamoto, 1987; Ohta and Iwamatsu, 1983). The sperm nuclear envelope breaks down through vesiculation caused by fusion of its inner and outer membranes (Ohta and Iwamatsu, 1983; Wolenski and Hart, 1985). This breakdown is essential for pronucleus formation in *Carassius auratus langsdorfii* (Yamashita *et al.*, 1988). After the sperm has penetrated, the naked nuclear material decondenses. The nascent pronucleus is surrounded by smooth vesicles, of supposed ovigerous origin, which unite to form a new pronuclear envelope (Lessman and Huver, 1981; Ohta and Iwamatsu, 1983).

There is some disagreement as to whether the fertilization cone forms independently of (though normally coincident with) insemination. In parthenogentically developing eggs, Kobayashi and Yamamoto (1987) found its formation to be suppressed whereas Wolenski and Hart (1987), in an SEM and TEM study of *Brachydanio rerio*, state that development of the cone, formation of the second polar body and exocytosis of cortical granules regularly occurred, indicating that these surface rearrangements did not require sperm binding and/or fusion. Because only the micropylar canal is entered by and presumably attractive to spermatozoa and because only one sperm is enabled to penetrate the oolemma despite multiple entry of sperm into the canal, it thus appears that the micropylar complex (canal and "fertilization cone") acts as much as a control mechanism, governing sperm entry, as a means of easy access of the spermatozoon to the oolemma.

Chapter 6

SUPERCLASS AGNATHA ('CYCLOSTOMATA')

Diagnosis. The Agnatha are the jawless vertebrates. A biting apparatus not derived from gill arches is present in some fossil forms (Nelson, 1984). It is probable that it is a paraphyletic group as indicated in Figs. 5.1 and 6.2 which show their presumed relationships with gnathostomes.

General. Lampreys (Fig. 6.1B) and, it is believed, hagfish (Fig. 6.1A) have external fertilization yet their sperm show features often associated with internal fertilization: long endonuclear perforatorium in *Lampetra* (but also in ect-aquasperm in polychaetes); elongate nucleus; in hagfishes especially, an elongate mitochondrial sheath around the axoneme; and, in lampreys, nine accessory fibres around the axoneme. Afzelius (in Nicander, 1970) has suggested that these features have been retained from "cyclostome" ancestors which were internally fertilizing. In possible support of this view, lampreys have a penial tube in the male (Fig. 6.7) and copulation (Breder and Rosen, 1966; Hardisty and Potter, 1971) although fertilization is not internal (see also Chapter 5). It thus appears plausible that the ancestors of agnathans, or even of gnathostomes as a whole, had internal fertilization. Lampreys are unique among vertebrates in having no male ducts, discharging the sperm, like the ova, into the coelom for egress through pores to the exterior. One might speculate that the penis-like structure in the male is a substitute for a ductus ejaculatorius or other muscular component of the normal vertebrate male duct and is not evidence of former internal fertilization. From comparison with other Metazoa with accessory axonemal fibres which are generally internally fertilizing, the presence of accessory fibres in the lamprey sperm axoneme remains inconsistent with external fertilization, however.

The sperm of lampreys and hagfish show no similarities which demand recognition of relationship between the two groups. This supports the view (above) that the Cyclostomata is an artifical group unified by the symplesiomorphy jawlessness. On the other hand remarkable homogeneity of sperm structure is seen within *Lampetra* (but see *Mordacia*, below) and, of a different nature, within the hagfish *Eptatretus*. The current view favours the hypothesis that the morphological and physiological similarities shared between lampreys and gnathostomes, but not hagfishes are due to common ancestry and not convergent evolution. These similarities are: highly differentiated kidney tubules, absence of a persistent pronephros, more than one semicircular canal, large exocrine pancreas, photosensory pineal organ, vertebral elements, histological structure of the adenohypophysis, and composition of the body fluid (Nelson, 1984).

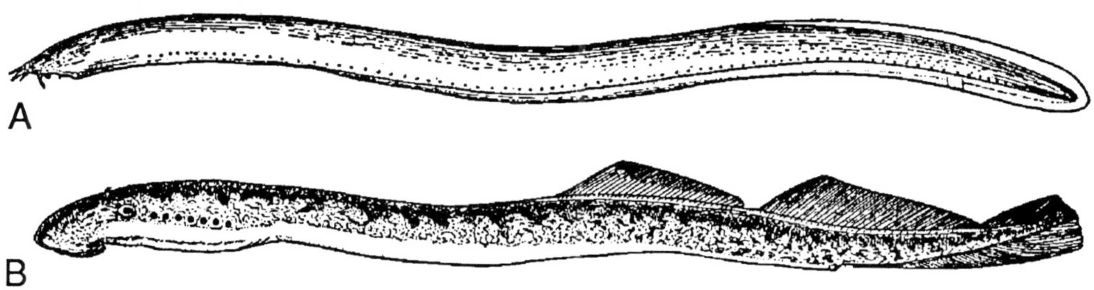

Fig. 6.1. A. The hagfish, *Myxine*. B. The lamprey, *Petromyzon*. From Romer, A.S. and Parsons, T. (1977). *The Vertebrate Body*. W.B. Saunders Company, Philadelphia. Fig. 16B, C. After Dean.

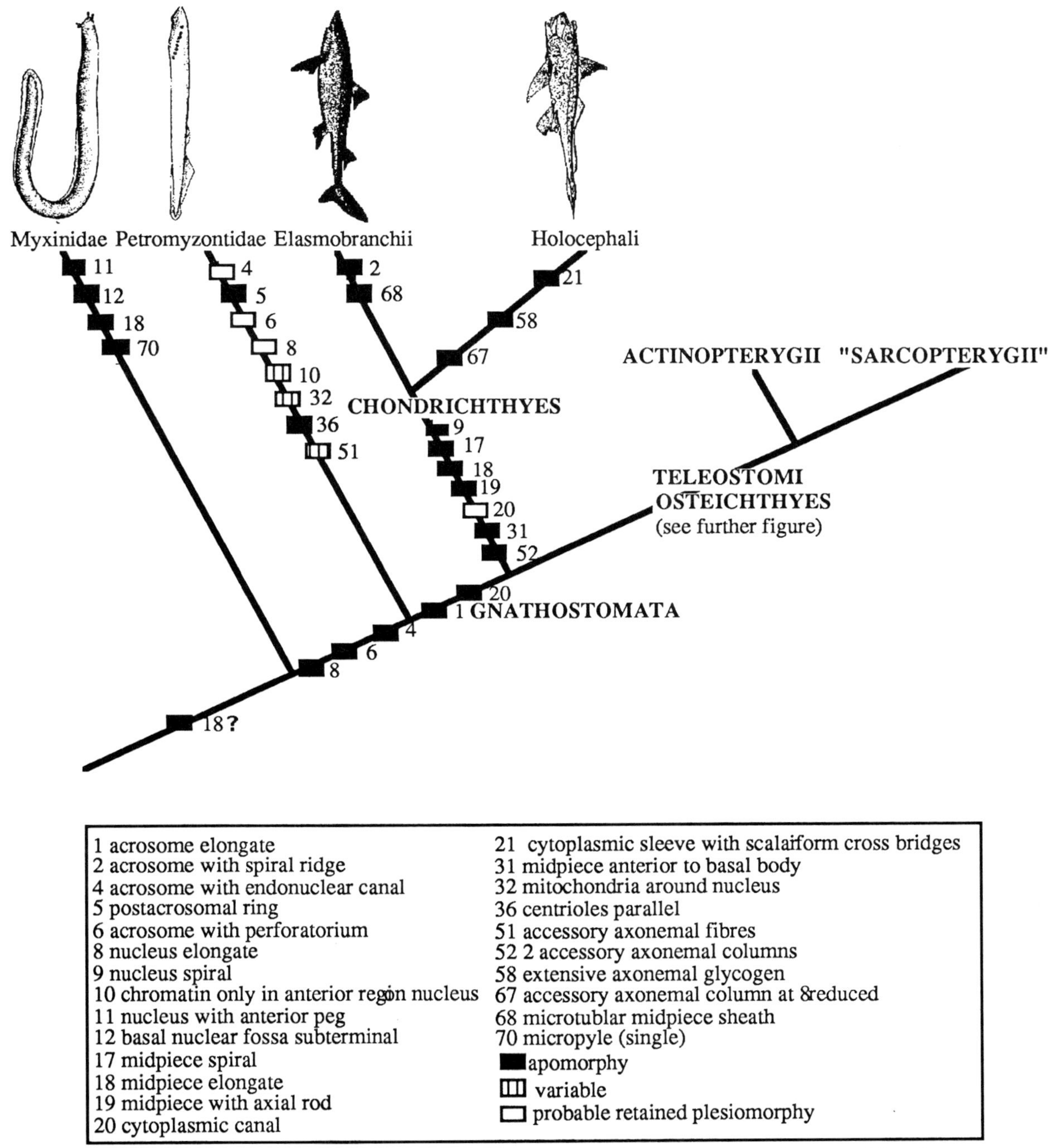

Fig. 6.2. Phylogeny of the Craniata (Vertebrata) tentatively adopted here. Major spermatozoal apomorphies of agnathans and chondrichthyans are indicated. Some features which may be plesiomorphic retentions are indicated for their descriptive value in this and subsequent phylograms.

We will deal in some detail with the sperm of agnathans in view of the significance of the lampreys and hagfishes to an understanding of basal gnathostome sperm structure.

CLASS MYXINI

Diagnosis. As for the order Myxiniformes.

Order Myxiniformes

Diagnosis. Unlike lampreys and their fossil relatives, myxinoforms have only one semicircular canal on each side. Like the lampreys they lack bone, though bone is present in cephalaspidiform relatives of the latter. The single family, Myxinidae, consists of the hagfish, marine in the temperate zones and the Gulfs of Mexico and Panama (Nelson, 1984). See Heintz (1963) for phylogeny.

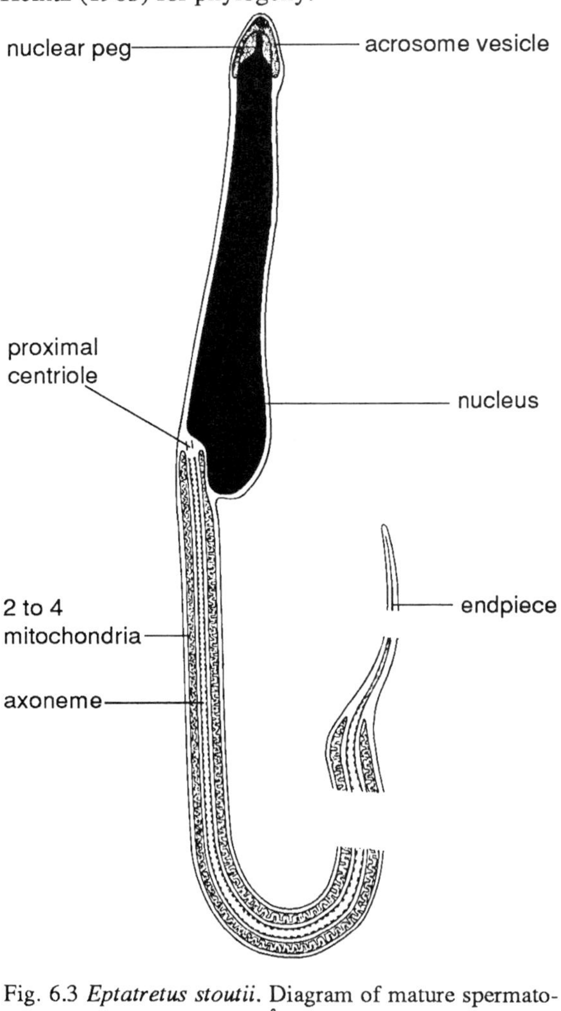

Fig. 6.3 *Eptatretus stoutii.* Diagram of mature spermatozoon. From Jespersen, Å. (1975). *Acta Zoologica (Stockholm)* **56**, 189-198. Fig. 19.

Family Myxinidae

Sperm literature. Spermatogenesis to the spermatid stage has been described ultrastructurally for the Atlantic, or European, Hagfish *Myxine glutinosa* (Alvestad-Graebner and Adam, 1977) and some data on the mature sperm of this species have been given by Nicander (1968, 1970). Spermiogenesis to the mature spermatozoon has been described for eastern Pacific hagfish, *Eptatretus stoutii, E. deani* and *E.* sp. (Jespersen, 1975).

***Eptatretus stoutii.* General form.** The mature sperm of *Eptatretus stoutii*, the Pacific Hagfish (Fig. 6.3), is an elongate structure, about 40-60 μm long. It consists of a head (acrosome and nucleus), a midpiece and an endpiece.

Acrosome. The acrosome (Fig. 6.4A), apical on the nucleus, has the form of a conical cap about 1 μm long enclosing subacrosomal material into which projects a central cylinder which is a narrow peglike extension of the nucleus. This anterior process is also seen in *E. deani* and *E.* sp. (Fig. 6.4B, C (Jespersen, 1975).

Nucleus. The nucleus is lanceolate, 12 μm long and maximally, near it base, about 1.5 μm wide, tapering anteriorly and posteriorly and with highly condensed chromatin. On one side, for about 1 μm from the posterior end, the nucleus is indented to receive the centriole and the base of the flagellum.

Midpiece and flagellum. The midpiece, which is about 1 μm wide and at least 20 μm long, consists of a 9+2 flagellum, lacking accessory fibres outside the doublets, surrounded by 2, 3 or 4 long and slightly twisted irregular mitochondria. Behind this is the free flagellum (narrowed to form the endpiece) of undetermined length (Jespersen, 1975).

Other Eptatretus species. Sperm structure in the other two species of *Eptatretus* is said to be similar but the distribution of electron dense material in the acrosome vesicle of *E.* sp. differs in that instead of being uniform it is chiefly apical, with pale flocculent material filling the remainder (Fig. 6.4C) (Jespersen, 1975).

Myxine. The account of spermatogenesis for *Myxine glutinosa* by Alvestad-Graebner and Adam (1977) contains no data on the mature sperm but an acrosome illustrated for the late spermatid and by Nicander (1970) for the mature sperm (Fig. 6.5)

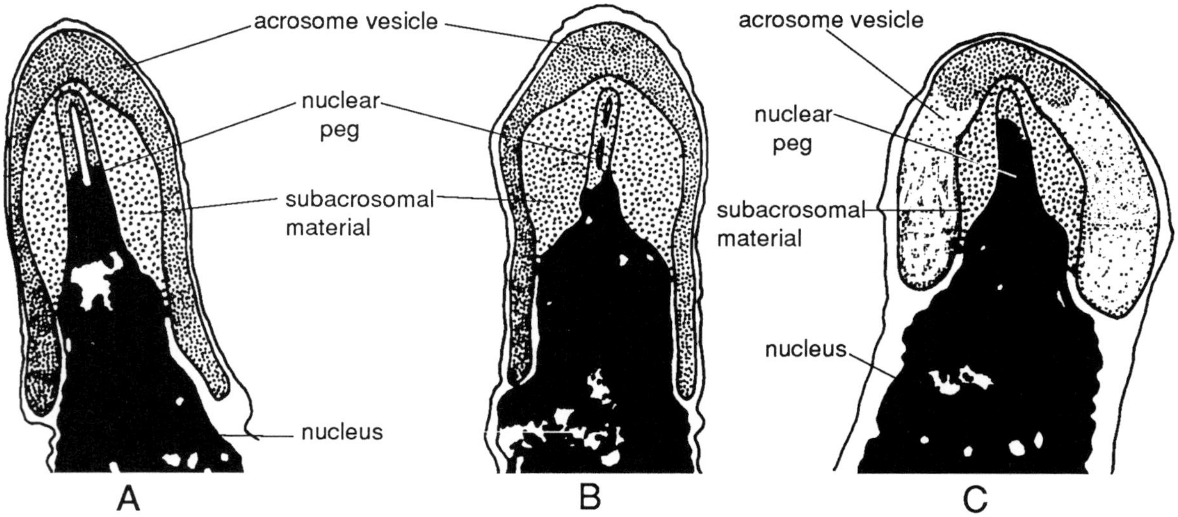

Fig.6.4. *Eptatretus.* Longitudinal sections of the acrosome and nuclear peg. A. *E. stoutii.* B. *E. deani.* C. *E.* sp. After micrographs by Jespersen, Å. (1975). *Acta Zoologica* (Stockholm) **56,** 189-198. Figs. 14-16.

resembles that of the *Eptatretus* sperm, including the presence of a central nuclear peg embedded in subacrosomal material.

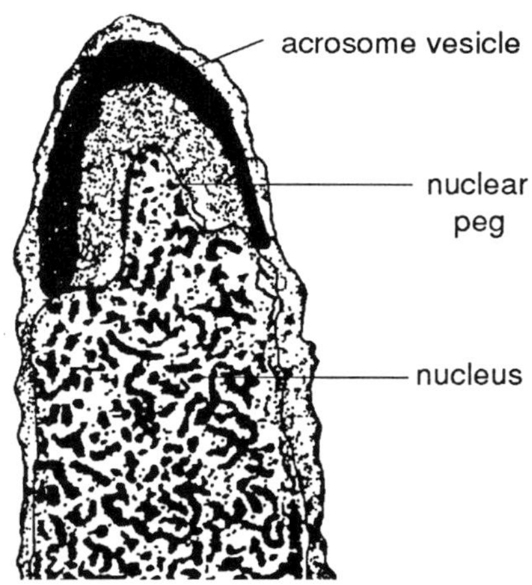

Fig. 6.5. *Myxine glutinosa*. Acrosome. After a micrograph by Nicander, L. (1970). In *Comparative Spermatology* (ed. B. Baccetti), pp. 47-55. Academic Press, New York. Fig. 2.

Alvestad-Graebner and Adam note an association of the chromatoid body with the inception of the acrosome. Jespersen (1975) refers to unpublished work indicating similarity of *Myxine* and *Eptatretus* sperm. Nicander (1968) refers to a true mitochondrial sheath in the sperm of *M. glutinosa* and gives a micrograph of a transverse section showing five round mitochondrial profiles symmetrically distributed around the axoneme (Fig. 6.6); groups of microtubules are present peripheral to the mitochondria in what is regarded as a mature spermatozoon.External fertilization is deduced from the absence of copulatory organs (Walvig, 1963).

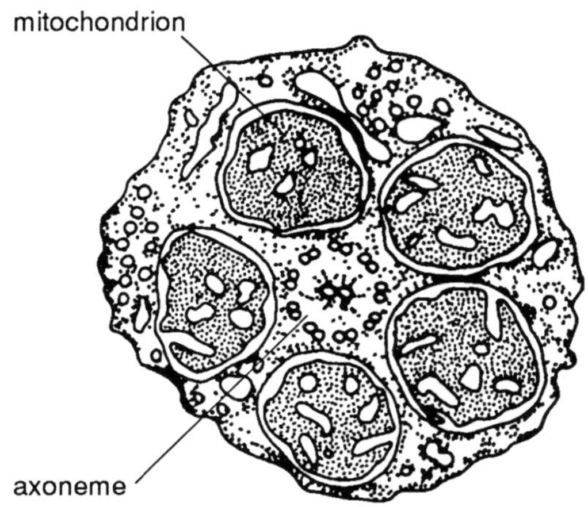

Fig. 6.6. *Myxine glutinosa* . Cross section of sperm showing midpiece with mitochondrial sheath and microtubules. After a micrograph by Nicander, L. (1968). *Proceedings of the 6th International Congress on Artificial Animal Reproduction Paris* 1: 89-107. Fig. 13.

Eptatretus **micropyle.** The egg of the hagfish *Eptatretus burgeri*, unlike that of the lamprey, has a micropylar funnel. This is surrounded by eight radiating capitate filaments. At the bottom of the funnel is a honeycomb pattern of about 3,500 'cells' (chambers) of which only one is open at the bottom. The successful sperm presumably is that which enters this cell which, at 3.5 µm, is wide enough for only a single sperm cell (Fernholm, 1975). We contend that there is a high positive correlation betwen absence of a perforatorium and presence of egg micropyles in fish, of which hagfishes are the first example in the present study.

CLASS CEPHALASPIDOMORPHI

Diagnosis. Differing from Myxini and bony fish in having two semicircular canals on each side and a single median nostril (nasohypophysial) opening between the eyes with a pineal eye behind. The orders Anaspidiformes and Cephalaspidiformes are known only as fossils (Nelson, 1984).

Order Petromyzontiformes

Diagnosis. Resembling Myxini in lacking bone. Differing from gnathostome fish, among other respects, in having two, not three, semicircular canals and in lacking paired fins. Anadromous and freshwater in cool regions (Nelson, 1984).

Family Petromyzontidae

Sperm literature. The spermatozoon of the non-parasitic lamprey *Lampetra planeri* (Fig. 6.7) has been described ultrastructurally by Follenius (1965) (Fig. 6.8) and Stanley (1967) (Fig. 6.9) and that of the parasitic *L. fluviatilis*, which possibly represents its ancestral stock, by Nicander (1968, 1970) and Nicander and Sjödén (1968, 1971). That of *Lampetra japonica* has been described by Jaana and Yamamoto (1981).

Some data on spermatogenesis of the parasitic lamprey *Mordacia mordax* and the sympatric, and possibly descendant, non-parasitic *M. praecox* are given in an interesting paper comparing gametogenesis in these 'paired' species by Hughes and Potter (1969).

***Lampetra*.** The sperm of *Lampetra* has a length of 130 µm (*L. japonica*) or about 140 µm (*L. planeri*). The length of the rod-shaped head is about 8 µm (*L. japonica*) or 14-16 µm (*L. planeri*), with a maximum diameter, posteriorly, of 1.0 µm.

Acrosome. The acrosome vesicle is subovoidal with its greatest width transverse. Apposed to the posterior face of the acrosome vesicle there is a small ring of very dense material (apical corpuscle of Follenius, 1965) which, in *L. planeri* at least, is attached by short finger-like extensions to the nuclear membrane. From the centre of this postacrosomal ring (subacrosomal ring, Nicander, 1968, 1970; Jaana and Yamamoto, 1981) a long 'central fibre' (subacrosomal fibre, Nicander, 1970), extends posteriorly throughout the length of an endonuclear canal to where this is terminated posteriorly by the indented

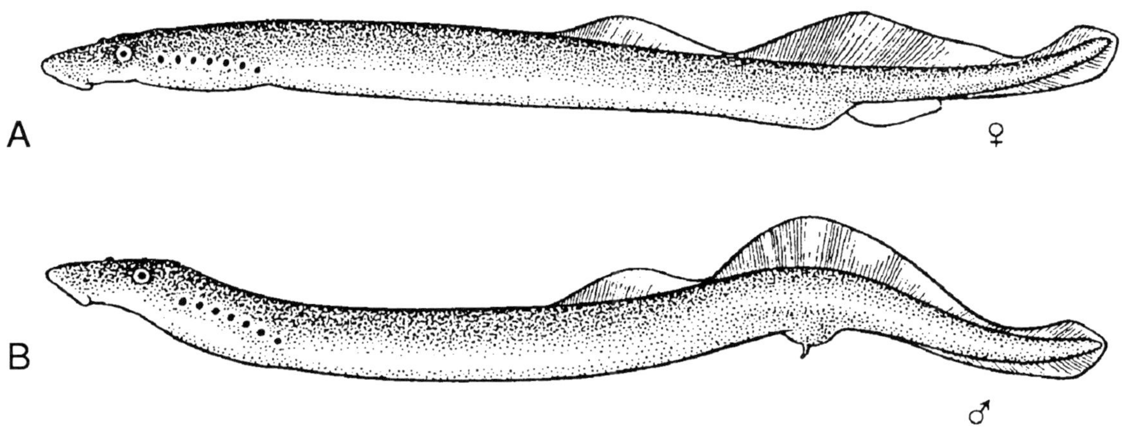

Fig. 6.7. *Lampetra planeri*. The Brook Lamprey. A. Ripe female with anal fold. B. Ripe male, showing copulatory papilla with penis-like structure. After Young, J.Z. (1981). *The Life of Vertebrates*. Third Edition. Clarendon Press, Oxford. Fig. 4.1.

nuclear membranes, well into the tail (*L. planeri*, *L. japonica*).

The central fibre is clearly a perforatorium and

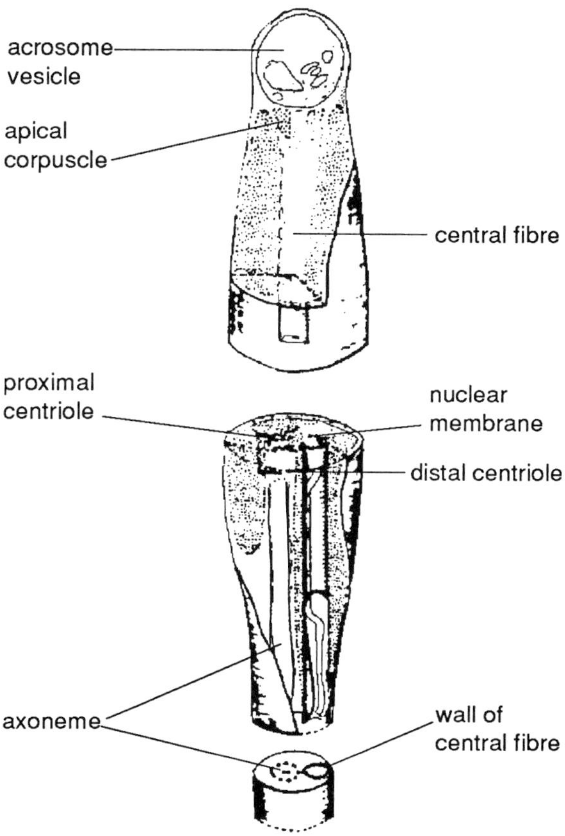

Fig. 6.8 *Lampetra planeri.* Diagram of the spermatozoon. The middle part of the head has been omitted. From Follenius, E. (1965). *Journal of Ultrastructure Research* **13**, 459-468. Fig. 1.

has been shown to be capable of extrusion as a 50 µm long "head filament" in ultrastructural investigation of *L. fluviatilis* (Afzelius and Murray, 1957) and in an optical study of this species and *L. planeri* (Kille, 1960).

It is termed an "acrosomal filament" by Jaana and Yamamoto (1981). In sperm apparently reacted by fixation, the nuclear membranes of the canal, and the acrosome vesicle, disappear (Stanley, 1967), the posterior acrosomal membrane being pushed into contact with the anterior membrane which in turn fuses with the plasma membrane (Jaana and Yamamoto, 1981) whereas the plasma membrane is drawn out into a slender sheath containing the central fibre and projecting anteriorly through the postacrosomal ring (Stanley, 1967).

The central fibre appears to be a fully formed acrosomal filament undergoing no observable change on extrusion (Follenius, 1965). It is possibly extruded by a spring-like action as it has an undulating course before reaction (Stanley, 1967). Similarly, in the river lamprey, *Lampetra fluviatilis*, spermatozoal ultrastructure of which agrees (Nicander, 1968,

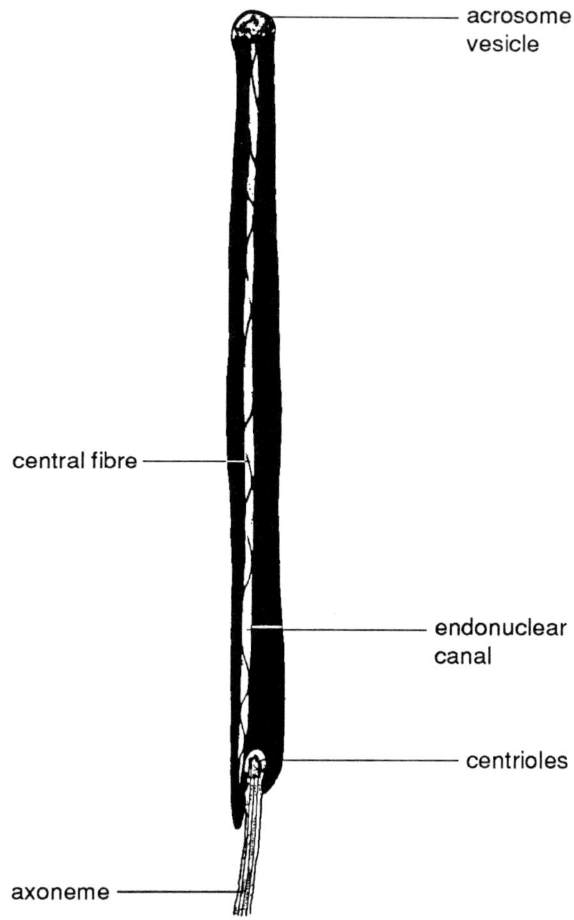

Fig. 6.9. *Lampetra planeri.* Longitudinal section through the head of the spermatozoon. The small vesicular acrosome covers the anterior tip of the nucleus, and the basal body and flagellum are inserted into a pit at the posterior end of the nucleus (at bottom). A portion of the nuclear canal and central fibre lie within the section anteriorly. After a micrograph by Stanley, H.P. (1967). *Journal of Ultrastructure Research* **19**, 84-99. Fig. 1.

1970) with that of *L. planeri*, sperm in the egg coatings show a true acrosome reaction: the acrosome vesicle bursts and the central fibre is extended, surrounded by a membrane continuous with the plasma membrane. Thus an acrosomal tubule is formed and penetrates the egg envelopes to reach the egg surface; the membrane covering the fibre is 60-70 Å thick compared with 90-100 Å for the plasma membrane in testicular spermatozoa (Nicander, 1968; Nicander and Sjödén, 1968). In lampreys, unlike hagfishes, there is no micropyle in the egg envelopes (Jaana and Yamomoto, 1981). As noted by Follenius (1965), the apical corpuscle and long perforatorium have remarkable parallels in the sperm (ect-aquasperm) of *Limulus*.

Nucleus. The chromatin-containing part of the rod-like nucleus is 7 µm (*L. japonica*) or 14 µm long (*L. planeri*) but the nuclear membranes, bounding the nucleus externally and lining the canal, extend at least 6 µm-34 µm, respectively, behind this as does the central fibre (Stanley, 1967; Jaana and Yamamoto, 1981). Within the nucleus the canal is loosely helical with, in *L. planeri*, about 15 to 17 turns (Follenius, 1965). Behind the chromatin the central fibre is thus surrounded by four concentric nuclear membranes (Follenius, 1965; Stanley, 1967; Jaana and Yamomoto, 1981).

Mitochondria. In *L. planeri* the elongate mitochondria, cristate with dense matrices, are arranged longitudinally along the proximal portion of the tail although few lie adjacent to the central fibre-endonuclear canal complex. More distally they are numerous, seven or eight sometimes being observed in a single cross section (Stanley, 1967). In *L. japonica* the mitochondria are either at the beginning of the flagellum or further posterior (Jaana and Yamomoto, 1981).

Cytoplasmic vesicles. In *L. planeri* posteriorly to the central fibre complex one or more large, often elongate, cytoplasmic vesicles are present. Longitudinally beaded filaments lie between the mitochondria or vesicles and the plasma membrane. In some posterior regions of the tail flattened vesicles partly or completely surround the axoneme (Stanley, 1967).

Centrioles and flagellum. Independently of the endonuclear canal, the chromatin cylinder in the sperm of *L. planeri*, *L. fluviatilis* and *L. japonica* has a deep eccentric posterior invagination, the implanta-

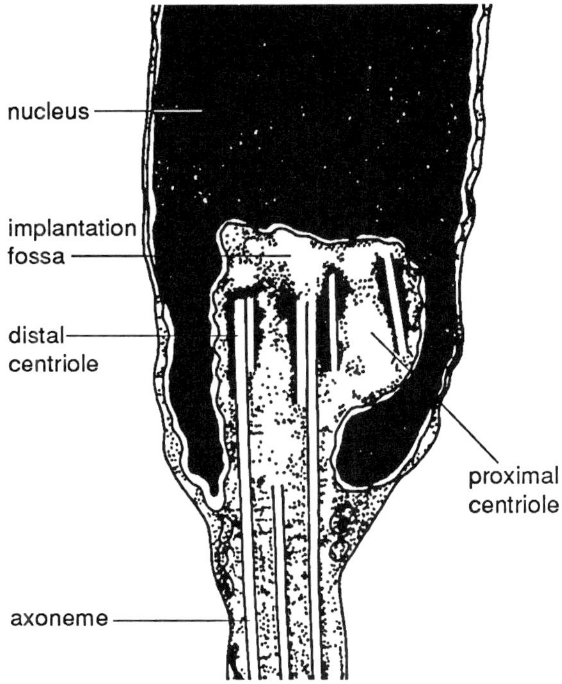

Fig. 6.10. *Lampetra planeri*. Longitudinal section near the posterior end of the nucleus, showing the implantation fossa containing the basal body (distal centriole) and "accessory" (proximal) centriole. After a micrograph by Stanley, H.P. (1967). *Journal of Ultrastructure Research* **19**, 84-99. Fig. 7.

tion fossa (Fig. 6.11), containing two parallel centrioles of triplet construction. One of these, the basal body, gives origin to the typical 9+2 flagellum which has 9 accessory fibres (Follenius, 1965; Nicander, 1970; Jaana and Yamamoto, 1981). As shown for *L. planeri* by Stanley (1967), each axonemal doublet has hollow subtubules and two [dynein] arms. The accessory fibres, small dense rods (outer dense fibres) adherent to the outer surface of each doublet appear to be narrow continuations of dense material which surrounds each triplet of the basal body and is in turn continuous with an "epicentriolar body" which caps the basal body. Fine filaments extend from this body to the nuclear membrane which lines the implantation fossa. These, with the epicentriolar body, are probably an anchoring apparatus, as suggested by their absence from the proximal (accessory) centriole. The dense fibres, in their small size and close proximity to the doublets, resemble those of reptilian rather than mature avian or mammalian sperm although similar to those of spermatids of these latter two groups (Stanley, 1967). The occurrence of accessory fibres, known elsewhere only in internally fertilizing sperm, affords some support for the view that lampreys were formerly internally fertilizing. This would not necessitate regarding internal fertilization as basic to fish as whole.

At the posterior tip of the flagellum, in *L. planeri*, the axonemal elements terminate in a consistent proximal to distal order: the central two filaments, subtubule B, subtubule A of the doublets, the outer dense fibres and finally granular material of medium density central to and binding together and preventing disarray of the doublets (Stanley, 1967). Whether triads of microtubules paralleling the axoneme in *L. fluviatilis* (Nicander, 1968) indicate immaturity or are present in the definitive sperm is uncertain.

Mordacia. The mature sperm of *Mordacia praecox* have not been observed but the structure of the advanced spermatid described by Hughes and Potter (1969) is presumably definitive.

General ultrastructure. The nucleus of the advanced *Mordacia* spermatid superficially bears at least three deep longitudinal folds or ridges which are somewhat spirally twisted. In the grooves between these are located mitochondria and extensive vacuoles, beneath the plasma membrane, a notable difference from the periaxonemal location of the mitochondria in *Lampetra* and other vertebrates. The small acrosomal cap is separated from the nucleus by a complex of vacuoles. As in *Lampetra*, there is a deep implantation fossa containing two parallel triplet centrioles (a notable synapomorphy of *Mordacia* and *Lampetra*) and an endonuclear ('intranuclear') canal containing a central fibre which penetrates the tail. No mention is made of extension of the nuclear membranes around the fibre posterior to the chromatin. Reported absence of outer dense fibres accompanying the doublets, if confirmed, would be an interesting but presumably symplesiomorphic agreement with the hagfishes *Myxine* and *Eptatretus* (see above).

Sperm phylogeny. Lampreys represent the first occurrence of an endonuclear canal and perforatorium in vertebrates. As these structures are present in most major groups up to and including Lissamphibia and Chelonia, it seems reasonable to conclude that they were acquired in the common ancestor of lampreys and gnathostomes and therefore that they are plesiomorphic for gnathostomes, as indicated in Fig. 18.7.

Chapter 7

SUPERCLASS GNATHOSTOMATA

Diagnosis. The jawed vertebrates. Jaws, derived from gill arches, are present (Nelson, 1984). For an analysis of the interrelationships of the main groups of the Gnathostomata see Chapter 8 and Fig. 8.2. Fig. 7.1 shows the internal relationships of gnathostomes recognized by Forey (1980) and supported here.

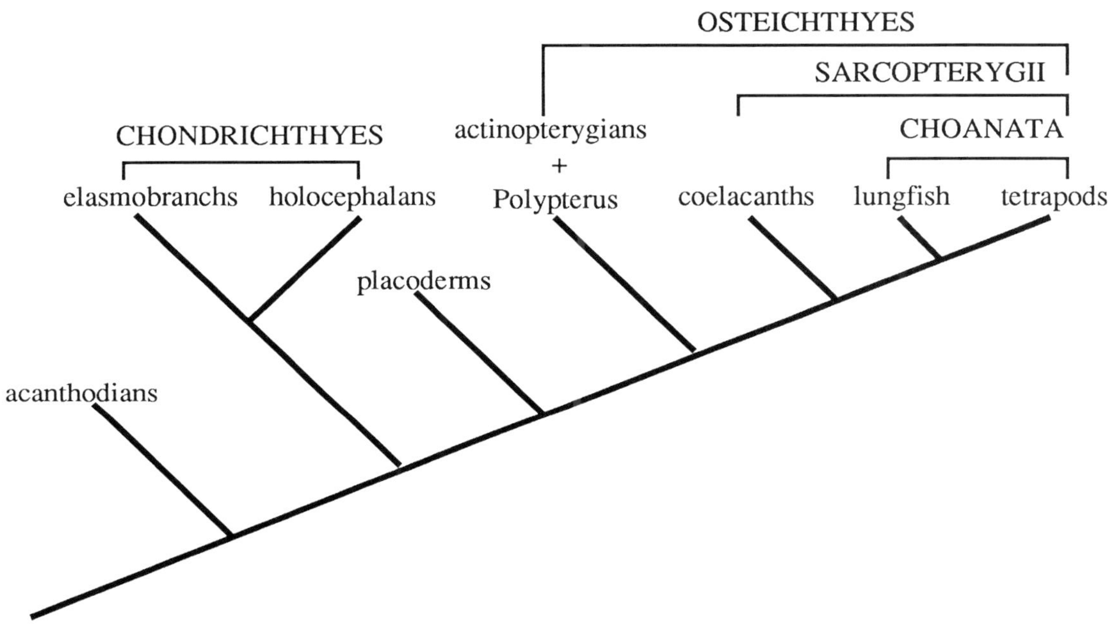

Fig. 7.1. Phylogeny of Recent and fossil gnathostomes, excluding osteolepiforms and porolepiforms. After Forey, P.L. (1980). *Proceedings of the Royal Society of London* B **208**, 369-384. Figs. 1 and 2. (See character set there listed).

Class Chondrichthyes

Diagnosis. The cartilagenous fish. The cartilagenous skeleton is often calcified but is seldom if ever ossified. Teeth not usually fused to the jaws; serially replaced. Horny, soft fin rays are unsegmented and epidermal in origin. Nasal openings on each side usually single; more or less ventral. Biting edge of the upper jaw formed from the palatoquadrate. Swim bladder and lung absent. Intestinal spiral valve present. Modern forms with internal fertilization, the male with pelvic claspers. Embryo encapsulated in a leatherlike case (gestation the longest in vertebrates, up to two years). High blood urea and trimethylamine oxide (Nelson, 1984). Monophyly of chondrichthyans is corroborated by the presence of coracobranchial muscles of hypobranchial origin (Wiley, 1979). There are two subclasses, the Holocephali and the Elasmobranchii (Nelson, 1984). Examples of the three chief types of organization in the Chondrichthyes: Order Chimaeriformes (rat fish and chimaeras), and superorders Selachimorpha (sharks) and Batidoidimorpha (rays) are shown in

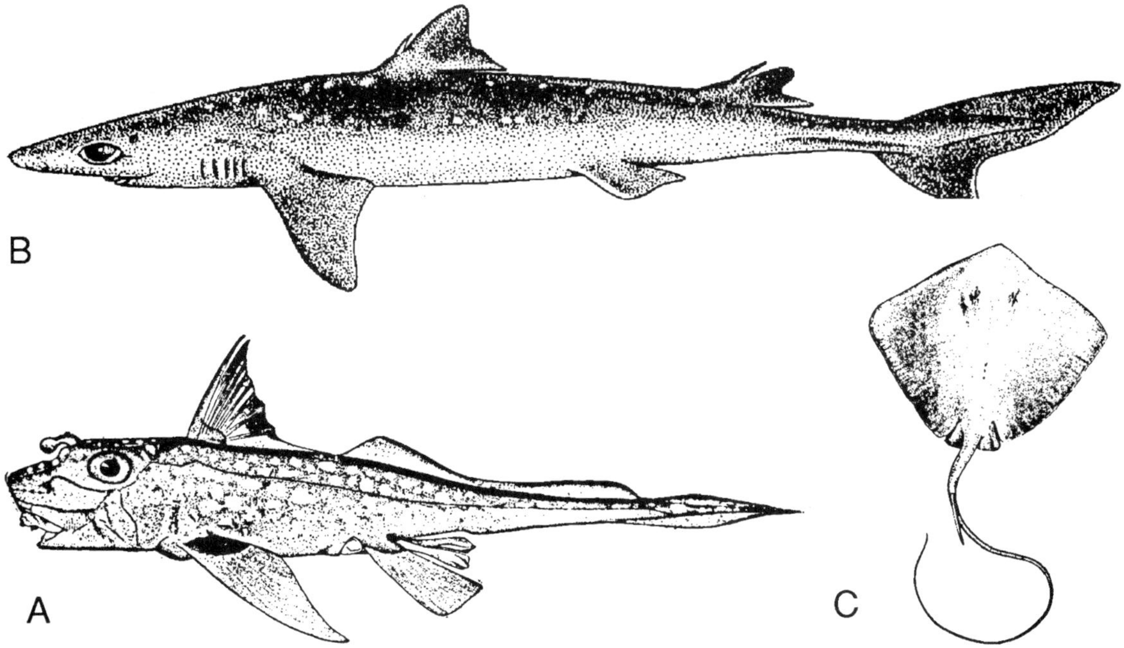

Fig. 7.2. Three chondrichthyans. A. *Hydrolagus* (=*Chimaera*) *colliei*. After Jordan, D.S. (1907). *Fishes*. Henry Holt, New York. Fig. 158. B. *Squalus acanthias*. After Jordan, D.S. (1907). *Fishes*. Henry Holt, New York. Fig. 144. C. *Dasyatis*. After Romer, A.S. and Parsons, T.S. (1977). *The Vertebrate Body*. W.B. Saunders Company, Philadelphia. Fig. 24D. After Garman.

Fig. 7.2.

SUBCLASS HOLOCEPHALI

Diagnosis and relationships. Gill cover over the four gill openings, giving one opening on each side; palatoquadrate (upper jaw) usually fused to cranium (holostyly or holocephaly); branchial basket mostly below the neurocranium; no spiracle; tooth replacement slow (hence an alternative name, Bradyodonti, for the subclass); teeth as a few grinding plates in extant and many fossil forms; separate anal and urinogenital openings, no cloaca; skin in extant forms naked; no stomach; no ribs; males with clasping organ on head, in addition to pelvic claspers. With a single extant order, the Chimaeriformes (Nelson, 1984).

Relationship of Holocephali with the Elasmobranchii has been questioned but Nelson (1984) considers that the two groups form a monophyletic entity, the Chondrichthyes. He considers that these had a single origin from the placoderms. Monophyly of the Chondrichthyes is supported from the evidence of brain traits by Northcutt (1989) in a cladistic analysis of an intuitive nature, not using parsimony programs, in which an alternative nomenclature is advanced (Fig. 7.3).

We will see that sperm ultrastructure confirms the unique relationship of the two subclasses and indicates that autapomorphies such as the two axonemal elements and interpolation of the mitochondria between the nucleus and basal body were present in the common ancestor (placoderm?). The mutuality of sperm structure and internal fertilization in the holocephalan-elasmobranch assemblage further suggests that this ancestor was internally fertilizing as has been proposed (Nelson, 1984) for the extinct holocephalan-like placoderm order Ptyctodontiformes in which the male had claspers (see Fig. 7.5).

Accounts of chondrichthyan sperm ultrastructure are often fragmentary. They are not, therefore, reviewed for individual species. A combined account for Holocephali and Elasmobranchii is given below but is chiefly drawn from accounts of the well-described sperm of two species, the holocephalan

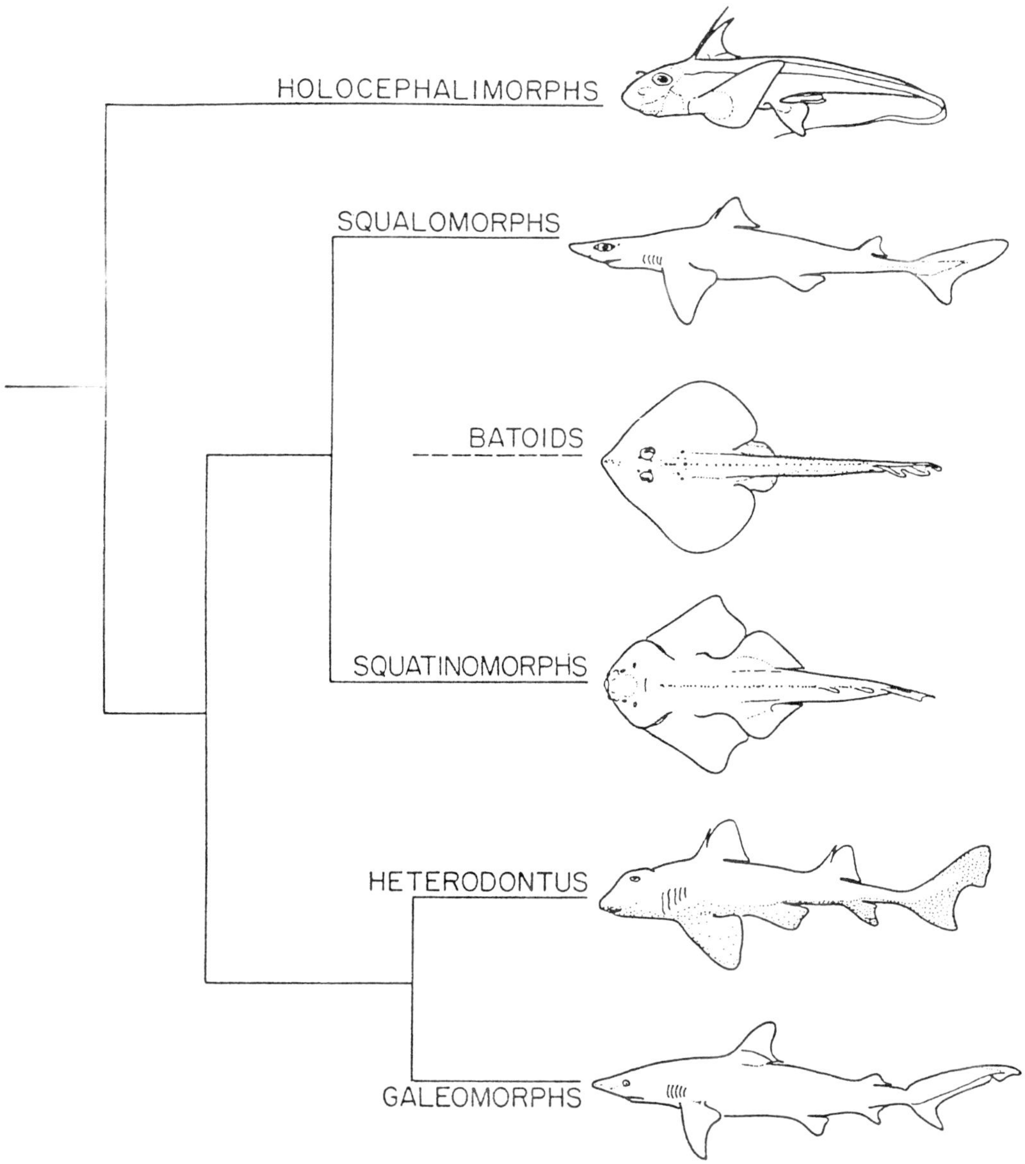

Fig. 7.3. Cladogram of the suspected interrelationships of the living groups of chondrichthyan fish (alternative nomenclature) based on brain anatomy. From Northcutt (1989). *Journal of Experimental Zoology Supplement* **283**, 83-100.

Hydrolagus colliei (Stanley 1983) and the elasmobranch *Squalus suckleyi* (Stanley, 1971).

The following holocephalan species have been examined for sperm ultrastructure:

Order Chimaeriformes: Chimaeridae - *Chimaera phantasma*, Hara and Tanaka, 1986; *Hydrolagus colliei*, Stanley, 1965a, 1983; *Neoharriotta pinnata*, Mattei, 1988.

SUBCLASS ELASMOBRANCHII

Diagnosis and relationships. Five to seven separate gill openings on each side; dorsal fin(s) and spines, if present, rigid; males lack claspers on head; dermal placoid scales often present; palatoquadrate not fused to cranium (suspension amphistylic or hyostylic); branchial basket mostly behind neurocra-

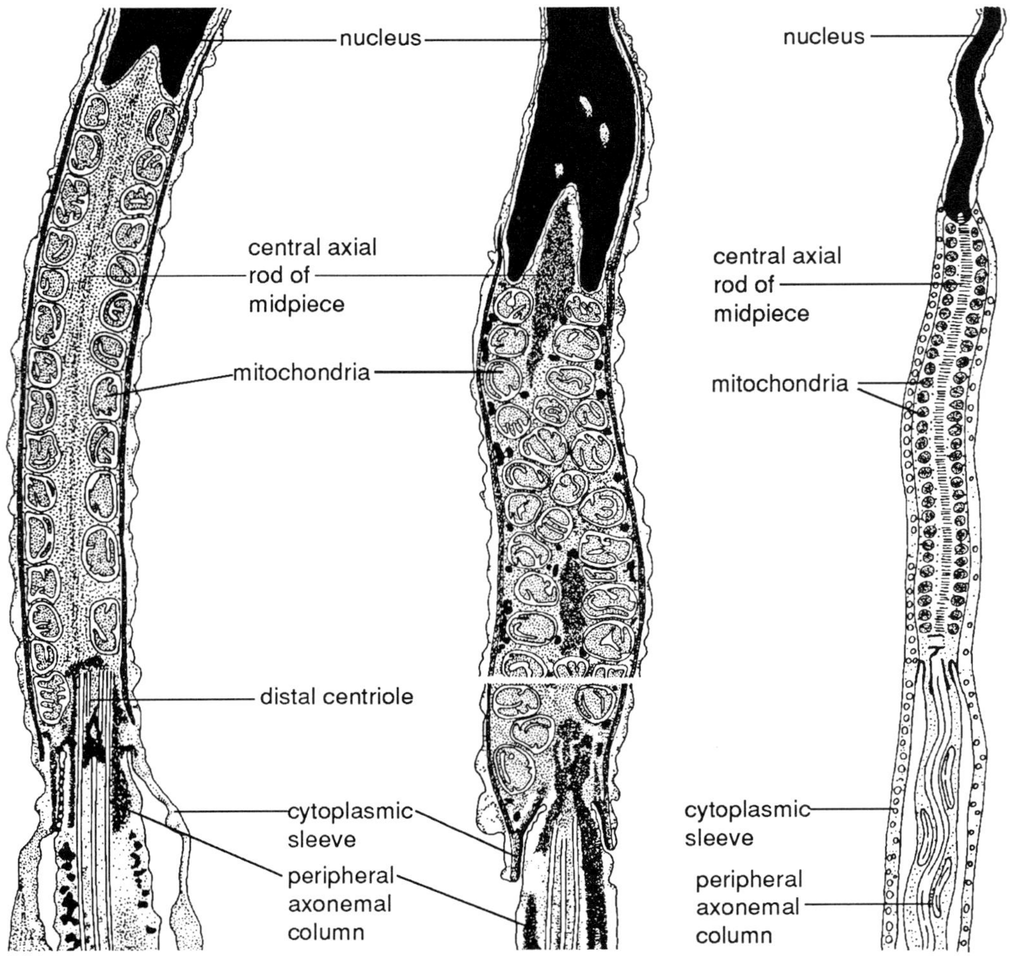

Fig. 7.4. Sperm of three chondrichthyans. A. *Hydrolagus colliei*. Longitudinal section through the midpiece with the posterior end of the nucleus at the top and the basal body of the flagellum at the bottom. After a micrograph by Stanley, H.P. (1983). *Journal of Ultrastructure Research* **83**, 184-194. Fig. 11. B. *Squalus suckleyi*. Longitudinal section through the nucleus-midpiece junction of a spermatozoon from the ampulla ductus deferentis. The striation pattern on the axial midpiece rod is no longer visible. Mitochondria interspersed with glycogen granules are tightly compressed about the midpiece axis, restricted by the closely apposed fibrous midpiece sheath. After a micrograph by Stanley, H.P. (1971). *Journal of Ultrastructure Research* **36**, 103-118. Fig. 18. C. *Rhinobatus cemiculus*. Longitudinal section of spermatozoon from posterior region of nucleus to anterior region of midpiece. Redrawn from Mattei (1970) (After Boisson *et al.*, 1968b. In *Comparative Spermatology* (Ed B. Baccetti) , pp 59-69. Academic Press, New York. Fig. 1h.

nium; tooth replacement relatively rapid; teeth numerous; some ribs usually present; spiracle (remains of hyoidean gill slit) usually present.

Typically, predacious fish; strongly reliant on the sense of smell. With five extant orders, the Hexanchiformes, Lamniformes, Squaliformes, comprising the sharks (Superorder Selachimorpha or Pleurotremata) and the Rajiformes, containing the rays and sawfish (Superorder Batidoidimorpha or Hypotremata) (Nelson, 1984). Northcutt (1989) (Fig. 7.3) regards the sharks as a paraphyletic group one branch of which contains the batoids.

Sperm ultrastructure has been examined in the following species:

Superorder Selachimorpha

Order Hexanchiformes: Chlamydoselachidae. *Chlamydoselachus anguineus*, Hara and Tanaka, 1986.

Order Lamniformes (Galeoidea): Carcharinidae. *Prionace glauca*, Hara and Tanaka, 1986; Scyliorhinidae. *Scyliorhinus caniculus*, Gusse and Chevaillier, 1978; *Scyliorhinus* sp., Stanley, 1971.

Order Squaliformes: Squalidae. *Centrophorus atromarginatus* (spermatogenesis), Tanaka *et al.*, 1978; *Centroscymnus owstoni*, Hara and Tanaka, 1986; *Squalus suckleyi*, Stanley 1964, 1965a, 1971.

Superorder Batidoidimorpha (Hypotremata).

Order Rajiformes: Dasyatidae. *Dasyatis kuhlii*; *Dasyatis garouensis*, Hara and Tanaka, 1986; Rajidae. *Raja* sp., Stanley, 1983; Rhinobatidae. *Rhinobatos cemiculus*, Boisson *et al.* 1968b; Mattei, 1970.

Comparative ultrastructure of chondrichthyan sperm

General. The ultrastructure of the spermatozoa of a holocephalan, a shark and a ray is illustrated in Fig. 7.4. Mature sperm of *Hydrolagus colliei*, in the ampulla ductus deferentis, are clustered into spermatozeugmata. All of the sperm in a spermatozeugma, each approximately 143 µm long, are orientated in the same direction and they are adherent chiefly by the, albeit transient, cytoplasmic sleeves at the anterior end of the flagella (Stanley, 1983).

Acrosome. The length of the acrosome is 3.5 µm in *Hydrolagus colliei* and 4.7 µm in *Squalus suckleyi.*

In *Squalus* (Stanley, 1971), as in *Hydrolagus* (Stanley, 1983), the acrosome has a deep posterior indentation which fits over the pointed tip of the nucleus. This contains a subacrosomal rod in *Squalus*. In *Hydrolagus* it is eccentric and contains a heterogeneous assemblage of medium electron density in which the denser material consists of several longitudinally orientated strands tentatively identified as a perforatorium. In addition, a 'ring of dense material' occupies the region between the acrosome and the nuclear tip.

The acrosome in *Squalus* is helical and in transverse section shows a unilateral shelflike expansion also with a spiral course. The posterior end of the acrosome is slanted, as in *Hydrolagus*. The external surface of the plasma membrane over the acrosomal region is covered by a series of low ridges orientated at about 12° to the long axis of the spermatozoon (Stanley, 1971).

In *Hydrolagus* the anterior half of the acrosome is bent at about 30° from the sperm axis and this anterior region again bears parallel arrays of extracellular fibrous material (Stanley, 1983).

Nucleus. In *Raja*, *Scyliorhinus* and *Squalus*, as in *Hydrolagus*, the nucleus consists of a central core and a less dense "parachromatin" sheath.

In *Squalus suckleyi* the nucleus attains the great length, compared with most spermatozoa in the Animal Kingdom, of 37 µm. It has the form of an attenuated cone, narrowly pointed at the anterior end. Posteriorly it is rounded and has a depression which accommodates the tip of the axial midpiece rod. It is helical and has highly condensed, electron dense chromatin. At maturity the nuclear envelope is closely applied to the nucleus but parachromatin material, gray in appearance, forms an apparently continuous sheath over the outer surface of the chromatin. Nuclear pores, believed to be redeveloped in place of those present in the earlier spermatid, are present in the posteriormost part of the nuclear envelope (Stanley, 1971).

In *Hydrolagus colliei* the nucleus is shorter than that of the *Squalus* sperm although still long, at 18 µm. It again forms a loose helix, here of three to four gyres. The moderately dense parachromatin is clearly differentiated from the dense chromatin. Anteriorly, in transverse section of the nucleus, the parachromatin forms a crescent, only partially surrounding the chromatin but posteriorly it forms a thin ring around the whole circumference (Stanley, 1983).

Midpiece. The midpiece seems always to consist of many approximately isodiametric mitochondria; its length is 3.5 µm in *Hydrolagus colliei* (Fig. 7.2A) and 12 µm in *Squalus suckleyi* (Fig. 7.2 B). There are approximately 70 mitochondria in the midpiece of *Hydrolagus* in which they are pressed together to form small polyhedral units with mostly concentric cristae. In both species the midpiece has a central axial rod which fits into an indentation at the posterior end of the nucleus whereas its posterior end inserts on the basal body (distal centriole) of the flagellum.

In *Squalus* the central rod which, like the entire midpiece is helical in register with the nucleus, is formed by fusion of two elements, a bundle of cytoplasmic microfilaments and a transversely striated filamentous band originating from the centriolar complex. The band, at least, is considered homologous with a ciliary rootlet (Stanley, 1964, 1971); its dual nature, as a transient amorphous rhizoplast and a persistent cross striated rhizoplast, is also seen in *Rhinobatos cemiculus* (Boissin *et al.*, 1968b; Mattei, 1970). Presence in the axial core of the midpiece in *Hydrolagus* of an inner core and outer shell may also relate to the dual origin of the structure (Stanley, 1983).

According to (Stanley, 1971), for *Squalus*, the mitochondria are formed by fragmentation of elongate mitochondria but in *Rhinobatos cemiculus* they are discrete throughout spermatogenesis (Boissin *et al.*, 1968b; Mattei, 1970).

In *Squalus* the mitochondria are interspersed with clusters of 200 Å putative glycogen granules, this region giving a PAS positive reaction which is mostly absent after diastase treatment. The mitochondrial sheath is surrounded by a fibrous sheath which overlaps the posterior end of the nucleus by 2 or 3 µm. Posteriorly the fibrous sheath attaches to the granular layer of the transient cytoplasmic sleeve (Stanley, 1971). A fibrous midpiece sheath, consisting of spirally orientated filaments also occurs in *Hydrolagus colliei* (see Stanley, 1983) and is reported for *Rhinobatos cemiculus* by Boissin *et al.* (1968b) and for *Centrophorus atromarginatus* by Tanaka *et al.*, (1978). In *Rhinobatos* it consists of microtubules extending anteriorly from a satellite of

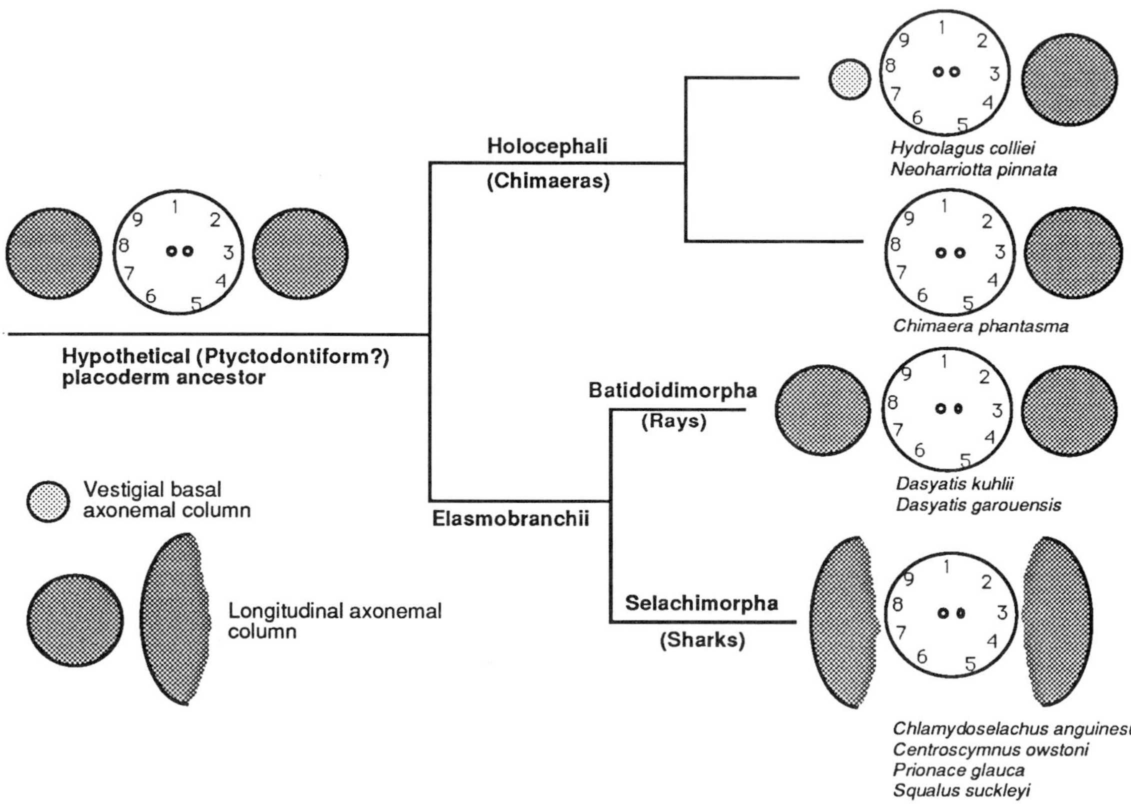

Fig. 7.5. Phylogram of known arrangements of longitudinal axonemal columns in Chondrichthyes. It is inferred that a circular cross section is plesiomorphic for the columns and that the column at doublet 8 in Holocephali is undergoing reduction. Original, based on data of Stanley (1971), Hara and Tanaka (1986), and Mattei (1988).

the distal centriole.

Cytoplasmic sleeve. The cytoplasm of the midpiece is reflected as a sleeve around the proximal portion of the tail. The sleeve is about 6.5 μm long in *Squalus suckleyi and* 10 to 16 μm long in *Hydrolagus colliei.*

The contents of the sleeve in *Hydrolagus* consist of smooth membranes, coated invaginated vesicles, and masses of dense material up to 0.2 μm in diameter. On the side of the flagellum nearest doublets 4 through 8, broad, shelflike connections extend, in a series, from the outer surface of the plasma membrane at the flagellar base to the medial surface of the plasma membrane of the remnant sleeve. In some longitudinal sections the series of 9 or 10 transverse bands appears ladder-like but resemblance to a septate junction noted by Stanley (1983) must be regarded as superficial.

Mature sperm in the ampulla ductus deferentis have mostly lost this cytoplasmic remnant; the sleeve is pinched off at the site of junction of the midpiece sheath fibres and the granular layer (Stanley, 1971) and eventually slips off the terminal portion of the sperm tail (Stanley, 1983).

Centrioles. The distal centriole is in the longitudinal axis of the sperm in *Squalus* or is at 15° to this in *Hydrolagus*. It has 9 triplets. In *Squalus*, as in *Hydrolagus*, from the basal body arise satellite rays which appear to attach to the plasma membrane at the point of its reflection from the flagellum to the inner surface of the cytoplasmic sleeve.

In *Squalus*, just posterior to the reflection, the basal body is attached to the plasma membrane by 9 Y-links. Anteriorly the 9+2 axoneme is surrounded by a dense ring which subdivides to form the two longitudinal accessory columns. A proximal centriole lies anteriorly and almost perpendicularly to the distal centriole. From a micrograph it is seen that the two centrioles intrude slightly into posterior region of the mitochondrial sheath, the proximal being entirely within it (Stanley, 1971). In *Rhinobatos* a proximal centriole is present in the spermatid (Mattei, 1970) but it is not clear whether this persists in the mature spermatozoon

In *Hydrolagus* no proximal centriole has been observed. Dense material adheres to the periphery of the basal body and almost fills its interior. The proximal ends of the two central singlets are embedded in a plate of dense material [basal plate] at the basal body-flagellar junction. A ring of fibrous material attached to the anterior end of the basal body extends posteriad as a truncated cone to line a reflected portion of the plasma membrane which is continuous with the inner remnant sleeve membrane. This is considered by Stanley (1983) to be probably homologous with the annulus of mammalian sperm.

Nucleus, midpiece and tail elements are helical in accordance with the observed rotation of the sperm about its long axis which allows anteriorward or posteriorward locomotion. Lateral undulation is minimal (Stanley, 1964, 1971).

Flagellum. The tail, originating from the posterior end of the midpiece, contains the usual 9+2 axoneme but this is accompanied by two accessory longitudinal columns external to doublets 3 and 8, both of which are fully developed in *Squalus suckleyi.* The two columns (discussed below) thus lie in a line passing approximately through the two central singlets. The axoneme as a unit appears straight but the axoneme and its two columns are helical, the columns describing a double helix (Stanley, 1971) whereas in *Hydrolagus* the entire flagellum is helical (Stanley, 1983).

Axonemal columns. According to Hara and Tanaka (1986) the spermatozoa of the sharks *Chlamydoselachus anguineus*, *Centroscymnus owstoni*, and *Prionace glauca*, like *Squalus*, have two longitudinal columns in the tail, oval in cross section with a flattened interior surface; in rays (*Dasyatis kuhlii*, *D. garouensis*), the two columns are rounded in cross section while in the holocephalan *Chimaera phantasma* only a single, rounded rod is present. In *Hydrolagus colliei* (Stanley, 1983), and *Neoharriotta pinnata* (Mattei, 1988) the column close to doublet 3 is similar in diameter and length to those of elasmobranchs; the other, close to doublet 8, is smaller in diameter and very short while in *Chimaera phantasma* (Hara and Tanaka, 1986) it is apparently lost. In *Hydrolagus colliei*, at least, glycogen is apppsed to the axoneme throughout its length. The phylogenetic significance of these arrangements, within the Chondrichthyes, is indicated in Fig. 7.5. Similarly located elements in the Actinistia and Dipnoi are discussed under those taxa.

The accessory axonemal columns are described in some detail for *Hydrolagus colliei* (Fig. 7.6). The smaller column, at doublet 8, is approximately circular in cross section, 0.09 μm in diameter and about 1.5 μm long. The larger column, at doublet 3, is about 22 × 27 μm in diameter anteriorly and slightly ovoid in cross section. Both columns are similar in structure. A cylinder of dense material surrounds a less dense core consisting of a ribbon attached to the inner surface of the cylinder by its medial edge and containing material of moderate electron density on either side. The outer surface of the cylinder has a thin layer adherent to but distinct from it. The larger column appears attached, at least intermittently, to doublet 3 by a short bridge. This column and the axoneme form

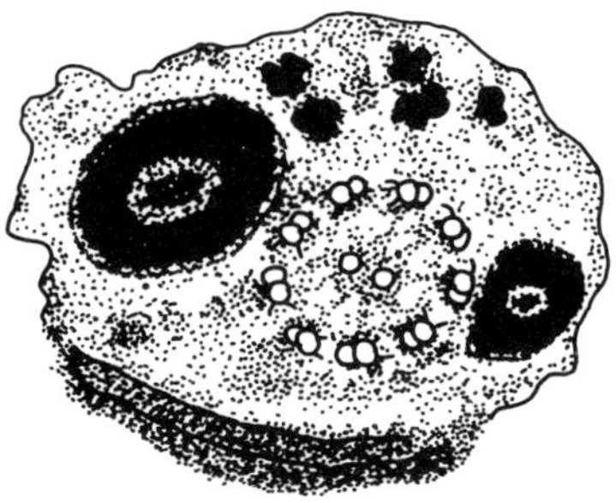

Fig. 7.6. *Hydrolagus colliei*. Transverse section of the axoneme at the level of the cytoplasmic sleeve remnant, showing the two accessory longitudinal columns. The column at doublet 8 is reduced and posteriorly disappears. On the side of the flagellum nearest doublets 4-7, a bridge connects the axoneme to the sleeve in which the annulus lies. After a micrograph by Stanley, H.P. (1983). *Journal of Ultrastructure Research* **83**, 184-194. Fig. 17.

a double helix with a short pitch: a 180° turn is made in every 4 μm of the flagellar length. It extends to within 2.0 μm of the flagellar tip, gradually tapering to a terminal diameter about equal to that of the smaller column. Glycogen granules are present along the entire length of the tail, excepting the terminal 1 μm, lying in the cytoplasm at right angles to the line through the central singlets.

In *Squalus suckleyi* the two columns are similar to each other (Fig. 7.7) and extend nearly to the posterior tip of the flagellum from their origin from the ring which surrounds the axoneme anterior to the commencement of the doublets. Each column is composed of a cylinder of dense material with lateral ridges of less dense material which give its cross section an elliptical if axially flattened profile.Each cylinder has a lightly staining centre in which a longitudinal membrane extends from the side nearest the axoneme into the interior; attached to each side of this membrane are rows of granules, about 100 Å in diameter (Stanley, 1971), an arrangement similar to that described above for *Hydrolagus*.

Spermatozoal synapomorphies of Holocephali and Elasmobranchii.

The Holocephali and Elasmobranchii are unified by a suite of spermatozoal characters: small [in fact moderately elongate] conical apical acrosome; long

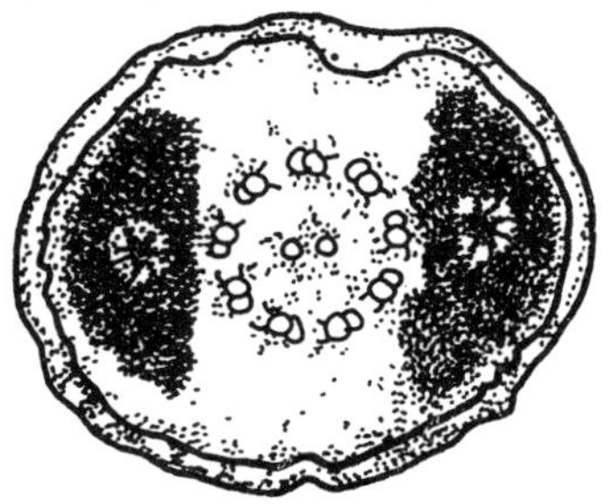

Fig. 7.7. *Squalus suckleyi*. Transverse section of the flagellum through the two accessory axonemal columns. After a micrograph by Stanley, H.P. (1971). *Journal of Ultrastructure Research* **36**, 103-118. Fig. 7f.

helical nucleus; long midpiece, composed of many subspherical mitochondria, with fibrous axial core or rod of rhizoblast origin; basal body situated behind the midpiece mitochondria [a very rare condition, seen also, with no phylogenetic construction, in Onychophora and Euclitellata]; location of two longitudinal columns in the axoneme opposite doublets 3 and 8; and sloughing of much of the cytoplasm by the formation of a remnant sleeve which eventually detaches from the spermatozoon over a ring-shaped area at the posterior end of the midpiece (Stanley, 1965a, 1983). All of these appear to be apomorphies relative to the ect-aquasperm of invertebrates and lower chordates.

Distinctions between holocephalan and elasmobranch sperm.

General. Holocephali (Fig. 7.4 A) show several distinct modifications of the common structural plan outlined above when compared with elasmobranchs

(Stanley, 1983). It is, however, uncertain whether these differences are constant for the Holocephali. The acrosome in *Hydrolagus* is oval in cross section and bent in its long axis, instead of being straight with a spiral ridge as in *Squalus*. The mitochondrial rod differs in being differentiated into a core and outer shell of more distinctly staining densities, and the proximal centriole, known for *Squalus*, has not been observed in *Hydrolagus*. Only in Holocephali (*Hydrolagus*) (Fig. 7.4 A) is the cytoplasmic sleeve joined to the flagellum by ladder-like cross connections. The cross bridges remain with the sperm after sloughing of the sleeve. The longitudinal column at doublet 8 is very short in *Hydolagus* and *Neoharriotta* and possibly absent in *Chimaera* (Fig. 7.5), rendering the axoneme asymmetrical, and not only do the microtubules of the tail follow an helical course (as in sharks) but the axonemal cylinder as a whole is spiral. *Hydolagus* has glycogen distributed along more than 100 μm of the tail while elasmobranchs show a small number of glycogen granules, among the mitochondria of the midpiece (Stanley, 1983).

Chapter 8

CLASS OSTEICHTHYES

Composition of Osteichthyes. All the remaining fish are members of the class Osteichthyes, which, as rightly recognized by Lauder and Liem (1983), in strict cladistic terms contain also the tetrapods (the amphibian - reptilian - avian - mammalian assemblage). If the tetrapods are arbitrarily excluded and the Osteichthyes are confined to the true bony fish, as is customary (Nelson, 1984), a paraphyletic group is obtained.

Nelson (1984) divides the class into four subclasses: the Dipneusti (lungfish); Crossopterygii (including, among others, the order Coelacanthiformes or Actinistia); the Brachiopterygii or Cladistia (including *Polypterus*); and the Actinopterygii (Ray-finned fish). We will, however, adopt, and will cladistically verify, the classification of Rosen *et al.* (1981) which is as follows:

Class Osteichthyes
- Subclass Actinopterygii
  - Infraclass Cladistia
  - Infraclass Actinopteri
    - Series Chondrostei
    - Series Neopterygii
- Subclass Sarcopterygii
  - Infraclass Actinistia
  - Infraclass Choanata
  - Series Dipnoi
  - Series Tetrapoda

Diagnosis. Inclusion of tetrapods gives a monophyletic Osteichthyes. This is defined on a large number of basal synapomorphies though these are further modified in the Sarcopterygii and, within these, the choanates. These synapomorphies, indicated in Fig. 8.2, include: presence of transverse ventral gill arch muscles; ventral interarcual muscles; hypohyal bones in the hyoid arch; medial insertion of the adductor mandibulae complex in the mandibular fossa; pleural ribs; lepidotrichia (fin rays); and a unique ossification pattern in the dermal shoulder girdle (suprascapular, supracleithrum, cleithrum and clavicle) (Rosen *et al.*, 1981; and other references in Lauder and Liem, 1983).

Additional characters, though some are consequential or even plesiomorphic, are given for the Osteichthyes *sensu strictu* by Nelson (1984): the skeleton, in part at least, has true bone (hence the name of the group); the skull has sutures; the teeth are usually fused to the jaws; the soft fin rays are dermal; nasal openings on each side are usually double and more or less dorsal; and the biting edge of the upper jaw is usually formed by dermal bones, the maxilla and premaxilla; a swim bladder appears for the first time; an intestinal spiral valve is present only in a few lower forms; internal fertilization is rare [and secondary?] and a pelvic copulatory device is seen only in phallostethoids; embryos are not encapsulated in a case; and except in *Latimeria* and dipnoans, the blood concentration of urea and trimethylamine oxide is low.

Position of Actinopterygii and Actinistia. In terms of living groups, the Osteichthyes *s. lat.* are widely accepted as the apomorph sister-group of the Chondrichthyes (Lauder and Liem, 1983) (see also Fig. 8.1). There is wide agreement that the Actinoperygii (ray-finned fish) form the plesiomorph sister-group of the remaining Osteichthyes (see, for instance, Forey, 1980; Lauder and Liem, 1983) but the internal constitution and relationships of the latter has been the subject of much debate.

Some workers do not recognize a basal position for the Actinopterygii. Thus Wiley (1979), from a consideration of gill arch muscles, considers actinopterygians to be the the terminal and most highly evolved members of a group which he terms the Euosteichthyes (Fig. 8.1). This consists of Dipnoi and tetrapods (grouped in a restricted Sarcopterygii) and ray-finned fish but excludes the Actinistia. Wiley restricts the Choanata to tetrapods but we will use the term in the more usual sense to embrace the Dipnoi also.

There are other dissenters from the view that actinistians are part of a monopyletic Osteichthyes. Some (Lagios, 1975; Lemire and Lagios, 1975; Løvtrup, 1977) go so far as to suggest that *Latimeria* and other actinistians must not only to be excluded from the Osteichthyes and Sarcopterygii but are the sister-group of the Chondrichthyes. We will examine this claim below.

Crossopterygian-tetrapod relationship. Nelson (1984) does not support a chondrichthyan relationship for the Actinistia when he contends that the difference between two osteichthyan subclasses, Crossopterygii and Dipnoi (the first two subclasses in his system) is not as great as generally thought, being bridged by a Devonian fossil *Youngolepis*. Although recognizing a monophyletic Osteichthyes, Nelson favours the Crossopterygii (including the Acinistia) over the Dipnoi as the ancestral group [more properly plesiomorph sister-group] of the tetrapods. In support of this contention, it has been

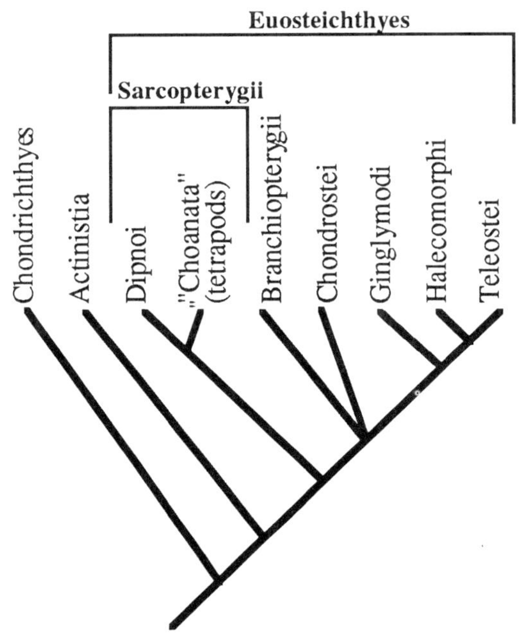

Fig. 8.1. Interrelationship of Recent gnathostomes as envisaged by Wiley. After Wiley, E.O. (1979). *Zoological Journal of the Linnean Society* **67**, 149-179. Fig. 8.

suggested that the basilar papilla in the inner ear of the actinistian *Latimeria* is a synapomorphy with that of tetrapods (Fritzsch, 1987). Using 28S ribosomal DNA sequences in PAUP analyses, Hillis and Dixon (1989) found that the most robust findings placed *Latimeria* with the tetrapods. The only other fish used in the analyses was the actinopterygian *Notropis* but the demonstration that the actinopterygian was the sister-taxon of the *Latimeria*-tetrapod assemblage at least suggests that relationship of *Latimeria* would not be found to lie with the Chondrichthyes if the latter were included in the analyses. Panchen and Smithson (1988), consider that the Actinistia may well be the sister group of the tetrapods, though not dismissing the alternative that that the Dipnoi warrant this status.

Dipnoan-tetrapod relationship. In contrast, Gardiner (1980), Lauder and Liem (1983) and Forey (1980, 1988) (see Chapter 9) see the Dipnoi and Tetrapoda as sister-groups, with the Actinistia (of living forms) as their plesiomorph sister-group, within the Sarcopterygii (Actinistia - Dipnoi - Tetrapoda), as demonstated here by parsimony analysis in Fig. 8.2. A very large number of synapomorphies can be demonstrated between Dipnoi and Tetrapoda, including the co-occurrence of a choana (see legend to Fig. 8.2 and Chapter 6). *Latimeria*, on the other hand, does not even possess a choana.

Cladistic analysis of gnathostomes. The phylogeny of the living groups of the Gnathostomata (Chondrichthyes + Osteichthyes) has been reconstructed in Fig. 8.2 (Jamieson, unpublished) using the PAUP program of Swofford (1989), and apomorphic features recognized by Lauder and Liem (1983), Løvtrup (1977), Lagios (1979), Lemire and Lagios (1979) and Forey (1980), with the addition of some spermatozoal and other reproductive characters demonstrated by Jespersen (1971), Stanley (1983), Mattei (1988b) and in the present work. The analysis is preliminary in that apomorphies ascribed by Lauder and Liem (1983) to the ground plan for each group (e.g. the Osteichthyes) have been attributed to all members of that putative group irrespective of loss (reversal) or alteration in some members of the group. A more critical analysis is in preparation. The present analysis is, nevertheless, a more rigorous evaluation of current hypotheses of relationship than has been possible without application of parsimony criteria.

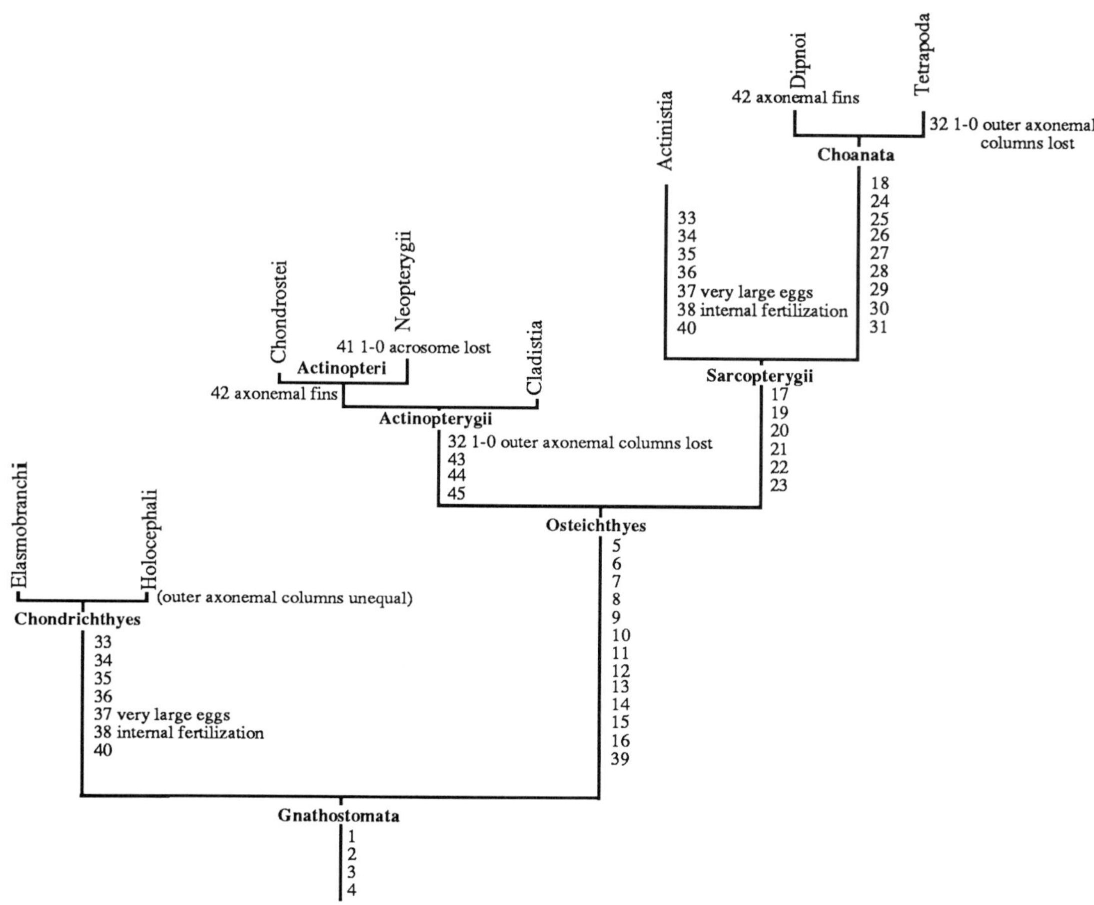

Fig. 8.2. Phylogram derived by the PAUP program of Swofford, representing the relationships of the major groups of living vertebrates, using somatic and some reproductive characters. Branch and bound sorting. A single tree of 54 steps. Consistency index 0.816. Original, from Jamieson, unpublished. All character states listed were scored as apomorphic in this analysis with the exception of 32 and 41 which gave greater parsimony if considered plesiomorphic. Characters: 1, the presence of three semicircular canals (placoderms also share this feature); 2, the presence of a ventral otic fissure between the embryonic trabecular and parachordal segments of the chondrocranium; 3, the presence of a lateral occipital braincase fissure; 4, five other characters including branchial arches consisting of basibranchial, hypobranchial, ceratobranchial, epibranchial and pharyngobranchial elements, and internal supporting girdles for the pectoral and pelvic appendages; 5, ossified dermal opercular plate(s) covering the gills laterally; 6, presence of an interhyal bone in the hyoid arch; 7, branchiostegal rays present; 8, mandibular depression primarily mediated by posteroventral rotation of the hyoid apparatus (inferred from the similarity of the hyoid-palatoquadrate relationships to that in osteichthyans and a mandibulohyoid ligament is inferred to have been present); 9, sclerotic ring present; 10, transversiventrales gill arch muscles present; 11, interarcuales ventrales muscle present; 12, hypohyal bones present in the hyoid arch; 13, medial insertion of the adductor mandibulae complex in the mandibular fossa; 14, pleural ribs present; 15, lepidotrichia present; 16, unique ossification pattern in the dermal shoulder girdle (suprascapular, supracleithrum, cleithrum, and clavicle); 17, true enamel present on the tooth surface; 18, double articulation of the hyomandibula with the neurocranium; 19, unique supporting skeleton in paired fins; 20, presence of an endoskeletal urohyal; 21, last gill arch articulates with base of preceding arch; 22, muscular lobes form the base of pelvic and pectoral appendages; 23, anocleithrum subdermal; 24, presence of a choana; 25, structure of the pelvic girdle; 26, the presence of multiple pharyngoclaviculari muscles; 27, numerous other features of soft anatomy such as partially divided conus arteriosus, an atrial septum, and ciliation of larval forms; 28, ciliation of larva; 29, dermal bone pattern covering the braincase; 30, loss of

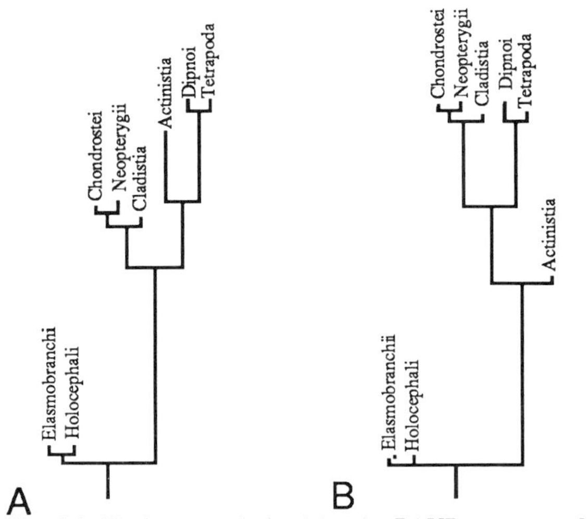

Fig. 8.3. Phylograms derived by the PAUP program of Swofford, representing the major groups of living vertebrates, showing the effect of scoring supposed apomorphies of the Chondricthyes and Actinistia (*Latimeria*) as plesiomorphies. Two topologies are obtained, one identical with that in Fig. 8.2, the other placing the Actinistia below the Actinopterygii. Characters as in Fig. 8.2 but with the following plesiomorphies: 32, 2 peripheral axonemal elements; 33, osmoregulation by uremia (plesiomorphic for all but Chondrostei); 34, rectal gland; 35, endocrine cells around pancreatic ducts; 36, large renal glomeruli; 37, very large eggs; 38, internal fertilization; 39, oil droplets in cones; 40, first nerves telencephalic. Original, from Jamieson, unpublished.

Polarity of characters, plesiomorphy versus apomorphy, has been determined with relation to a hypothetical ancestor (not shown), used as the outgroup, displaying the presumed plesiomorphic condition for each character. The effect of altering the polarity of characters for which this is less confidently deduced has been extensively investigated (Jamieson, unpublished). These characters will be discussed below after portrayal in Fig. 8.2 of the phylogeny and the character-states from which it is derived.

The phenomenon of osmoregulation by retention of urea in the blood (uremia) merits early mention as its polarity, and distribution, is particularly debatable. It is scored in Fig. 8.2 as an apomorphy of only Chondrichthyes and *Latimeria*, appearing on parsimony analysis to be homoplasically derived. It is, however, also reported for aestivating *Lepidosiren and Protopterus*, for the teleost *Periophthalmus*, for some Amphibia, and for *Testudo hermanni* (Forey, 1980). If, consequently, uremia is included in the anlaysis as an apomorphy of all the gnathostome groups to which these belong an identical tree is obtained. If, as Forey (1980) proposes, uremia is regarded as a plesiomorphic condition for gnathostomes, (excepting Chondrostei, in which it is not reported) the tree is unchanged though a slightly higher consistency index is obtained.

Fig 8.2 shows that even when the suite of Chondrichtyan-actinistian character states which Løvtrup (1977) and Lagios (1979) contend are synapomorphies are so treated, the Acinistia do not associate with the Chondricthyes but remain in the Sarcopterygii. The same tree topology is obtained whether these states, and other characters of particularly debatable polarity, here listed, are made apomorphic or plesiomorphic:

32, 2 peripheral axonemal elements; 33, osmoregulation by uremia (plesiomorphic for all but Chondrostei); 34, rectal gland; 35, endocrine cells around pancreatic ducts; 36, large renal glomeruli; 40, first nerves telencephalic. The tree is slightly less parsimonious (by one evolutionary step) if character 32 is considered apomorphic.

If 38, internal fertilization, is made plesiomorphic *in addition* to these characters (there is no effect on topology if it alone is made plesiomorphic), two, alternative topologies are obtained. One (Fig. 8.3A) has the same topology as Fig. 8.2 but the other (Fig. 8.3B) places the Actinistia below the Actinopterygii. It is somewhat ironic that only by reversing Løvtrup's decision that the several features (33-36, 40) are synapomorphies and regarding them as plesiomorphies do the Actinistia come to occupy a position near

interhyal; 31, structure of the pelvic and pectoral appendage (Characters 1-31 from Lauder and Liem, 1983); 32, 2 peripheral axonemal elements at doublets 3 and 8 (Jespersen, 1971; Stanley, 1983; Mattei, 1988); 33, osmoregulation by urea retention (Løvtrup, 1977); 34, rectal gland (Millot and Anthony, 1958); 35, endocrine tissue of pancreas around exocrine ducts (Lagios, 1979); 36, large renal glomeruli (Lemire and Lagios, 1979); 37, very large eggs (Løvtrup, 1977; Lagios, 1979; Anthony, 1980); 38, internal fertilization (Smith *et al.*, 1975); 39, oil droplets in retinal cones (Forey, 1980); 40, first pair of nerves in telencephalic vesicle (Løvtrup, 1977); 41, acrosome present (plesiomorphic state); 42, axonemal fins (Jamieson, unpublished); 43, autosphenotic and large opisthotic bone in braincase; 44, acrodin caps on teeth; 45, pelvic plate present.

Chondrichthyes, though never as their sister group, and that such plesiomorphy, advocated by Forey (1980), has the effect of splitting the Actinistia from the Sarcopterygii in which Forey included them. This reminds us that intuitive estimates may give results incompatible with strict Hennigian parsimony analysis. Nevertheless, it appears that the phylogeny proposed by Forey (1980) (Fig. 7.1), which is congruent with that derived here in Fig. 8.2, though based on a different if overlapping character set, may be correct, at least for living forms.

Since the above was written, Ahlberg (1989) has further confirmed the sarcopterygian nature of the Actinistia. He has produced maximum parsimony phylograms based on the structure of the paired fin skeletons which differ from those of Forey and the present work in showing the Actinistia as the sister-group of the Dipnoi; the tetrapods are seen as the sister-group of the Actinistia + Dipnoi. If fossil forms are taken into account, the porolepiforms intervene

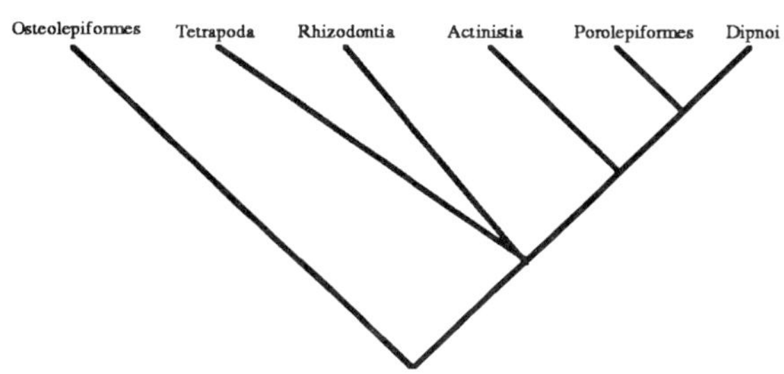

Fig. 8.4. Maximum parsimony cladogram of the Sarcopterygii based on characters of the paired fin skeletons. After Ahlberg, P.E. (1989). *Zoological Journal of the Linnean Society* **96**, 119-166. Fig. 16B.

between the Actinistia and Dipnoi as the immediate sister-group of the Dipnoi while the Osteolepiforms are the sister-group of the entire tetrapod through dipnoan assemblage (Fig. 8.4). Clearly, a holomorphological approach, including as many characters as possible is required to evaluate the various preliminary analyses presented and reviewed here, but Ahlberg's suggestion that tetrapods are a basal rather than terminal group of the Sarcopterygii is of great interest. It finds some support in the relatively simple organization of the sperm of some Amphibia, for instance those of *Xenopus laevis* described by Bernardini *et al.* (1986), when compared with Actinistia and would make the co-occurrence of two outer axonemal columns in Actinistia and Dipnoi a synapomorphy rather than a homoplasy*.

Subclass Actinopterygii (Cladistia, Chondrostei + Neopterygii)

Diagnosis and relationships. The Actinopterygii (ray-finned fish) are so named because of the presence of dermal, segmented, raylike supports within the fins. Among living forms, they may be the sister-group of the Sarcopterygii (Actinistia - Dipnoi - Tetrapoda assemblage) (see Figs. 7.1 and 8.2). Inclusion of the Cladistia (e.g. *Polypterus*) in the Actinopterygii, as advocated by Wiley (1979) or close relationship with these is, however, the subject of debate.

In one view cladistians are considered to be allied to the Dipnoi (Mok, 1981; G. Nelson, 1969). The second, more prevalent view is that cladistians are the plesiomorph sister-group of the Actinopterygii or that they are to be included in the latter as its most primitive members (see Wiley, 1979, and discussion in Nelson, 1984). In a third proposal, Bjerring (1985) argues for exclusion of *Polypterus* from the Actinopterygii and placement in a teleostome group of its own.

Some seven synapomorphies in support of a monophyletic Actinopterygii, including the Cladistia as its most basal living group (see references in Lauder and Liem, 1983) (Fig. 8.5), are listed here: a unique, ganoid, scale, with an outer lamellar layer (ganoine), a central dentinous layer with vascular canals, and a deep layer of spongy bone; an anteriodorsal peglike process on the scale; presence of acrodin, a dentinous tissue, which forms a cap on the teeth, including those of *Polypterus*; a unique pectoral fin structure, with an expanded protopterygial element of the pectoral fin base and an extensive articular surface with the endoskeletal shoulder girdle; pelvic girdle with greatly expanded metapterygium which supports the fin radials. The ontogenetic fusion of internal pelvic cartilages into a larger adult structure is confined to ray-finned fish; general presence of a single dorsal fin (although secondarily divided or otherwise modified in some) whereas acanthodians, actinistians (*Latimeria*) and lungfish all primitively have two dorsal fins; median fins supported internally by paired segmental dermal rays (lepidotrichia) which are attached to an internal skeleton which does not extend into the fin at the

*However, I concur here with Lee (pers. comm.) who considers the paraxonemal columns of Lissamphibia to be the homologues of these columns in Dipnoi and Actinistia, as shown in Fig. 18.7.

ACTINOPTERIGII

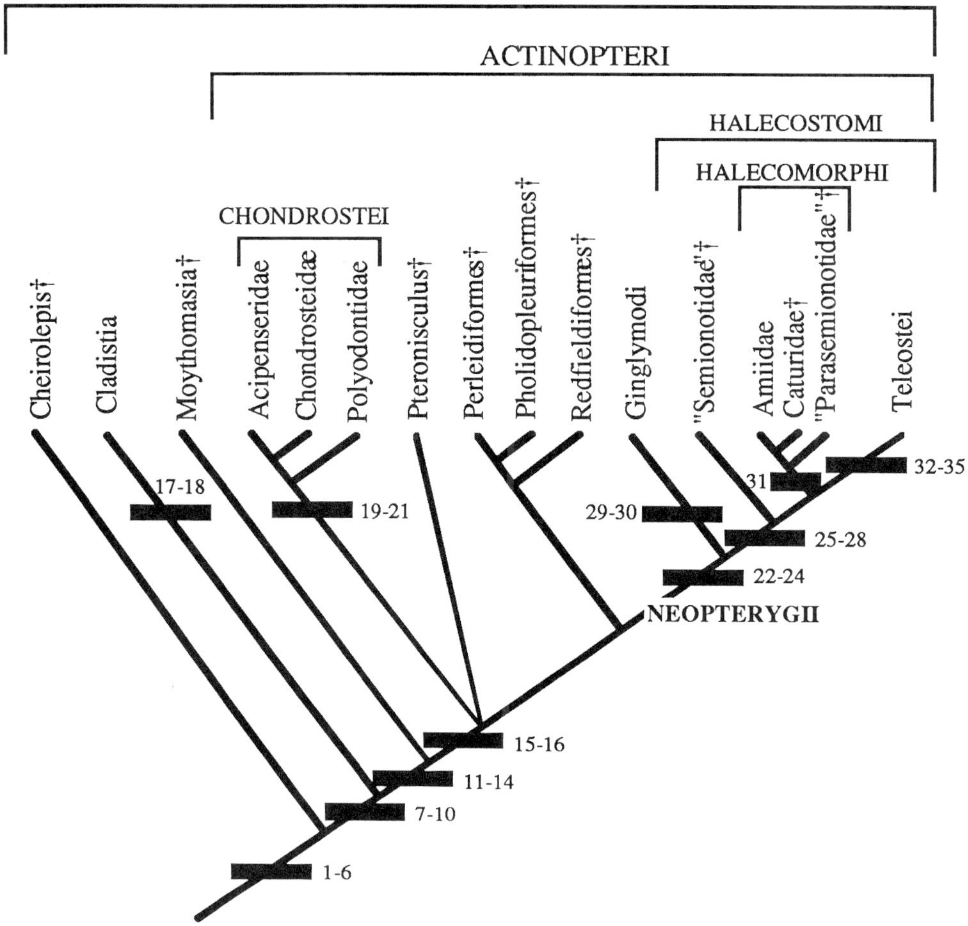

Fig. 8.5. Phylogeny of the main actinopterygian subgroups showing one hypothesis of the relationships between the main actinopterygian subgroups. Taxa with no living representatives are indicated with a dagger. The characters are: 1, presence of a single dorsal fin; 2, a pectoral propterygium ; 3, ganoin, 4, anterodorsal peglike process on the scales; 5, jugal pitlines; 6, mandibular sensory canal enclosed in the dentary bone; 7, autosphenotic and large opisthotic bone in the braincase; 8, acrodin caps on teeth; 9, pelvic plate present ; 10, numerous features of soft anatomy including brain development, jaw muscles, and gill arch muscles; 11, a perforated propterygium; 12, basal fulcra on dorsal caudal margin; 13, supra-angular bone present in lower jaw; 14, hemopoietic organ above the medulla oblongata; 15, fringing fulcra on fins; 16, spiracular canal; 17, dorsal finspines; 18, ontogenetic fusion of infraorbitals with the maxilla; 19, absence of myodomes; 20, fusion of premaxillae, maxillae, and dermopalatines; 21, anterior palatoquadrate symphysis; 22, fin rays equal in number to their supports in the dorsal and anal fins; 23, upper pharyngeal dentition consolidated; 24, clavicle lost or reduced to small plate lateral to cleithrum; 25, mobile maxillary bone in the cheek; 26, interopercular bone present; 27, median neural spines; 28, quadratojugal lost or fused with quadrate; 29, opisthocoelous centra; 30, a series of toothed infraorbital bones; 31, both the symplectic bone and the quadrate contribute to the jaw articulation; 32, the presence of uroneurals (elongated ural neural arches); 33, unpaired basibranchial toothplates; 34, internal carotid foramen enclosed in the parasphenoid. Patterson (1977) also mentions two other features as tentative teleostean features: seven epurals and a pectoral propterygium fused with the first pectoral fin ray. Living teleosts share many features in the jaw musculature, including loss of the anterior (suborbital) jaw adductor component. After Lauder, G.V and Liem, K.F. (1983). *Bulletin of the Museum of Comparative Zoology* **150**, 95-197. Fig. 6. See that work for sources of characters, many of which were abstracted from Patterson (1982).

finbase; and an unsegmented actinotrich composed of elastoidin between the distal ends of each pair of lepidotrichs. Actinotrichs are also seen in coelacanths but if actinistian relationships deduced here are correct cannot be true homologues of chondrichthyan ceratotrichs.

We provide cladistic evidence above (Fig. 8.2) that cladistians (*Polypterus*) are the plesiomorph sister-group of the Neopterygii + Chondrostei which jointly comprise the Actinopteri of Rosen *et al.* (1981) and which with the Cladistia comprise their Actinopterygii. The Cladistia and Chondrostei will now be discussed but treatment of the great radiation of the Neopterygii will be deferred to follow description of crossopterygian (actinistian) and dipnoan sperm.

Osteichthyes, the Branchiopterygii. More recently, however, evidence has been presented (Patterson, 1982; Daget, 1986; Gardiner and Schaeffer, 1989) for considering them to be highly specialized survivors of primitive Actinopterygii and for their inclusion in the latter (Fig. 8.5). Preliminary molecular phylogenetic work based on 28 S RNA sequence analysis appears to have confirmed this view. Some of the trees nevertheless associate the tetrapod *Xenopus* with *Polypterus* and the two show striking identical "signature" portions of the molecule. It has been concluded that such cases probably reflect an ancient origin of *Polypterus* at a time period close to that of amphibians (Le *et al.*, 1989). Major autapomorphies are the dorsal finspines and ontogenetic

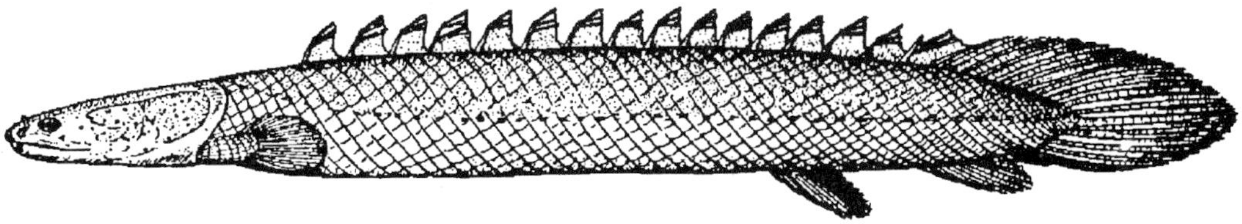

Fig. 8.6. *Polypterus.* After Romer, A.S. and Parsons, T.S. (1977). *The Vertebrate Body.* W.B. Saunders Company, Philadelphia. Fig. 32B. After Dean.

fusion of the infraorbitals with the maxilla. In view of the widespread occurrence of ventral lunglike diverticula in primitive osteichthyan groups, the presence of lungs and probably the occurrence of external gills in young may be plesiomorphic features of actinopterygians which have been retained in *Polypterus* (Lauder and Liem, 1983).

Infraclass Cladistia (Brachyopterygii)

Order Polypteriformes

Diagnosis and relationships. These are the bichirs, in the single family Polypteridae with two genera, *Polypterus* (Fig. 8.6) and *Erpetoichthys*, restricted to tropical African freshwaters.

Among their characteristics are: rhombic ganoid scales; presence of a spiracle; dorsal fin with 5-18 finlets, each with a single spine to which is attached one or more soft rays; pectoral fin rays supported by numerous ossified radials linked to the scapula; spiral valve present; lungs partially used in respiration (Nelson, 1984). There are two extant genera, both in the family Polypteridae, *Polypterus* and the eel-like *Calamoichthys* (=*Erpetoichthys*). Brainerd *et al.* (1989) have shown that polypterids have aspiration breathing utilizing deformation and recoil of the bony-scaled integument.

Placed earlier in the Sarcopterygii, polypterids were later recognized as a separate subclass of the

Sperm literature. The ultrastructure of the sperm of *Polypterus* has been briefly described by Mattei (1970) in an account chiefly referring to the spermatid.

Polypterus senegalus. The sperm of the African lungfish *Polypterus senegalus* (Fig. 8.7) are biflagellate a curious resemblance, but the only notable spermatological one, to those of the dipnoan *Protopterus.* Biflagellarity is also seen in a few teleosts: ictalurids, malapterurids, myctophids, batrachoids and gobiesocids, and apogonids, as in some Amphibia.

General sperm ultrastructure. The *Polypterus* sperm nucleus appears from a diagram (Fig. 8.7) tear-shaped, with a small acrosome on the narrow anterior end.

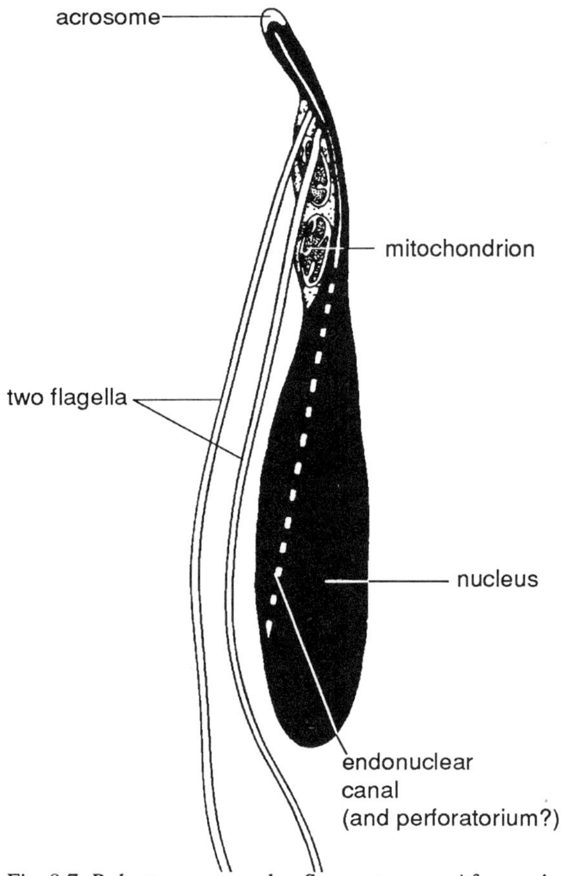

Fig. 8.7. *Polypterus senegalus*. Spermatozoon. After a micrograph by Mattei, X. (1970). In *Comparative Spermatology* (ed. B.Baccetti), pp. 59-69. Academic Press, New York. Fig. 3.

An endonuclear canal [and perforatorium?] penetrates the nucleus; the canal is axial in the anterior region of the nucleus; posteriorly it crosses the nucleus obliquely and ends against the nuclear membrane near the posterior end of the nucleus. The chromatin is compact. A microtubular manchette, present in the spermatid, disappears by maturity. The two centrioles are parallel, located at the junction of acrosome and nucleus, and each bears a flagellum. Mitochondria are associated with the centrioles and are lodged in a concavity of the nucleus. Migration of the mitochondria towards the centrioles which generally occurs late in spermiogenesis occurs earlier in *Polypterus*, before the centrioles have elaborated their flagella (Mattei, 1970; 1988).

Significance of biflagellarity. Although biflagellarity has clearly evolved more than once in fish sperm, sharing of this condition between cladistians and dipnoans does constitute weak support for the above-mentioned hypothesis of a special (sister-group) relationship between cladistians and the Dipnoi. The prenuclear implantation of the flagella in *Polypterus* is a notable difference from the plesiomorphic, and usual, postnuclear implantation in Dipnoi but such a variation (albeit of a single flagellum) can occur in unified subgroups, as between ascothoracican and cirripedian maxillopod Crustacea. On the other hand, the occurrence of a single flagellum in the sperm of *Neoceratodus* may indicate that biflagellarity in *Protopterus* is an independent development.

We have adhered to the classifications (e.g. Patterson, 1982) which place the polypteriforms in, and at the base of, the Actinopterygii but consider that the spermatological information justifies reinvestigation of their affinities.

Absence of the egg micropyle. It was seen in Chapter 5 that egg micropyles, absent in *Polypterus*, have evolved in their putative apomorph sister-group the Actinopteri (*sensu* Patterson, 1981, the Actinopterygii above the Cladistia) although also seen in myxinids.

Infraclass Chondrostei

Diagnosis and relationships. Synapomorphies include: absence of myodomes; fusion of premaxillae, maxillae and dermopalatines; presence of anterior palatoquadrate symphisis; and absence of an interoperculum. Usual presence of a spiracle is symplesiomorphic (references in Lauder and Liem, 1983; Nelson, 1984). Living members are the northern hemisphere anadromous and freshwater sturgeons (Fig. 8.8) and the paddlefish of freshwaters of N. America and China (Nelson, 1984). Despite the fact that several synapomorphies are reductive, the Chondrostei appear to comprise a monophyletic group. Among living forms they are the sister-group of the Neopterygii (Lauder and Liem, 1983) (Fig. 8.5 see also Fig. 8.2).

In accordance with this relationship, chondrostean gametes represent a transitional stage towards the neopterygian condition of a single micropyle on

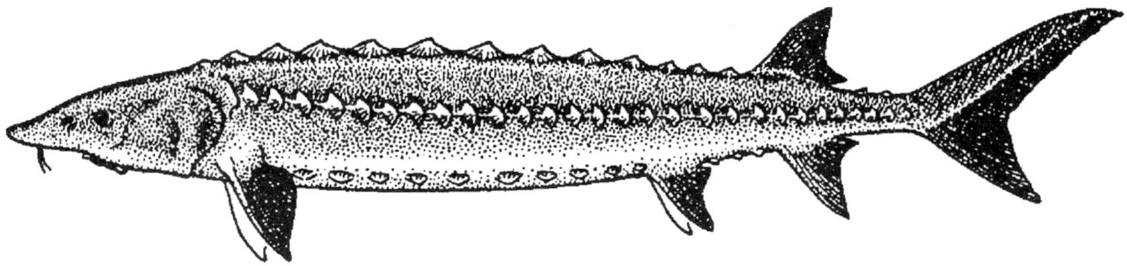

Fig. 8.8. *Acipenser sturio*. After Norman, J.R. (1937). *Illustrated Guide to the Fish Gallery. British Museum (Natural History)*. Trustees of the British Museum, London. Fig. 16.

the egg and no acrosome on the sperm. The sturgeons have not only egg micropyles but also sperm acrosomes. The egg has a thick impenetrable envelope perforated by numerous micropyles, the number of which varies within and among individual fish, and between species. This allows polyspermic fertilization which, if it occurs, is pathological (Ginzburg, 1968, p. 61; p. 183).

Order Acipenseriformes

Diagnosis. As for the infraclass Chondrostei.

Sperm literature. Sperm ultrastructure has been investigated in four species of the Acipenseridae: *Acipenser guldenstadti*, *A. stellatus*; and *Huso huso*, (Ginzburg 1968) and *Acipenser transmontanus* (Cherr and Clark 1984b).

Acipenser transmontanus. The sperm of *Acipenser transmontanus* (Fig. 8.9), externally fertilizing, has an elongate head (acrosome and nucleus) 7 µm long and maximally 1.5 µm wide.

Acrosome. The acrosome (Fig. 8.9), 1 µm long, consists of a scalloped bell-shaped cap, the acrosome vesicle, 0.2 µm thick, underlain by a thick zone of subacrosomal material consisting of 6 nm wide actin filaments. Laterally, between the inner vesicle membrane and the nuclear mebrane, this material appears more granular. The subacrosomal material passes posteriorly through three membrane-lined 40 nm wide channels [endonuclear canals] which terminate at the posterior nuclear fossa and form a triple helix.

Acrosome reaction. Although the natural inducer of the sturgeon sperm acrosome reaction and its ionic controls have been described by Cherr and Clark (1984b), the role of the acrosome during actual fertilization in these animals remains unknown. The egg envelope consists of four layers, the outermost layer being an adhesive jelly coat (Cherr and Clark, 1984a).

This jelly does not contain acrosome reaction inducing activity. However, an acrosome reaction is experimentally induced by ionophore A23187 with obligate presence of Ca^{2+} and Mg^{2+}; high Ca^{2+}; high pH, optimally 9.0; or egg water, of which the active component is a 66 K D glycoprotein (Cherr and Clark, 1984b). Detlaf and Ginzburg (1963) also reported that Ca^{2+} is required for the acrosome reaction in sperm of sturgeon and other species; their finding that that Mg^{2+} could be substituted for Ca^{2+} is not confirmed for *A. transmontanus*. The acrosome undergoes exocytosis and a process aproximately 10 µm long and 0.1 µm wide is produced from the subacrosomal material. Some of the filamentous material, presumably actin, extends through the subacrosomal region and into the acrosomal process. No observable change occurs in the nuclear channels as a result of the acrosome reaction (though the acrosomal process is said to originate from the nuclear channels in addition to the subacrosomal region). However, the lateral margins of the subacrosomal region exhibit a decrease in the thickness of the granular region which lies lateral to the subacrosome and below the acrosome vesicle. It is suggested by Cherr and Clark that H^+ influx (internal alkalinization) occurs during this acrosome reaction.

Presence of actin. The major protein isolated from the sperm head comigrates on gels with rabbit muscle actin and has a molecular weight of 43 K Daltons. In conjunction with the occurrence of 6 nm microfilaments this indicates that actin present in sperm heads is the protein responsible for formation of the acrosomal process. Following the acrosome

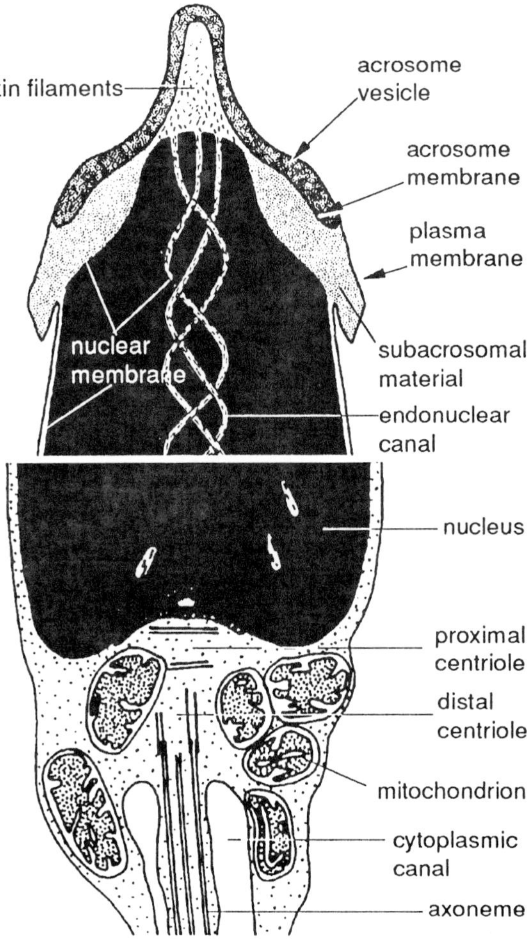

Fig. 8.9. *Acipenser transmontanus*. Diagram of the spermatozoon. The long middle region of the nucleus is omitted. Adapted from Cherr, G.M. and Clark, W.H. (1984b). *The Journal of Experimental Zoology* **232**, 129-139. Figs. 3, 7 and 10.

reaction, the material in the nuclear channels remains closely associated with the midpiece and the centriolar extensions. Cherr and Clark (1984b) consider that this may ensure that a centriole is incorporated into the egg after sperm-egg fusion and it is therefore uncertain that they regard the material in the nuclear canals as perforatorial.

Nucleus. From a micrograph, the nucleus is an elongate, anteriorly tapering cylinder, 5.2 μm long, with dense homogenous chromatin. A broad, basal, nuclear fossa with an anterior indentation, partly includes the proximal centriole.

Midpiece. The midpiece, 1 μm long [scarcely elongated compared with an ect-aquasperm] contains numerous mitochondria but forms an 8 μm long collar, into which mitochondria extend, which surrounds the base of the flagellum.

Centrioles. Proximal and distal centrioles are mutually perpendicular. Both have apparently microtubular extensions into a basal nuclear fossa as also shown in garpike sperm by Afzelius (1978).

Flagellum and moltility. The flagellum, 30-40 μm long, is of the 9+2 type. Two lateral extensions of the axonemal plasma membrane [fins] are present in the plane of the two central microtubules. The sperm remain motile in freshwater for 5 minutes, compared with less than 30 seconds for sperm of other freshwater fish. This is attributed to a very impermeable plasma membrane as inferred from difficulty in solubilizing the sperm for electrophoresis and poor penetration of fixatives (Cherr and Clark, 1984b).

Remarks. The fertilization biology of *Acipenser* poses some enigmas. Co-occurrence of an acrosome and, on reaction, an acrosome process with presence of micropyles is inconsistent with fertilization in other fish where presence of a micropyle usually excludes occurrence of an acrosome. An exception exists in the myxinids, but there, although an acrosome coexists with a micropyle, no acrosomal process is produced. An acrosomal process occurs in lampreys, sharks, and lungfish, none of which possesses an egg micropyle. The neopterygian *Lepidogalaxias*, unique in possessing an acrosome (Leung, 1986), lacks a micropyle.

A further peculiarity is that, although *Acipenser* sperm are spawned in freshwater, the ionic controls of the acrosome reaction appear to Cherr and Clark (1984b) to be similar to those reported in the sperm of marine invertebrates.

Chapter 9

Crossopterygians and Dipnoi

SUBCLASS CROSSOPTERYGII

SUPERORDER COELACANTHIMORPHA (ACTINISTIA)

Relationships. The relationships of the Actinistia, and specifically *Latimeria*, have been analyzed in Chapter 8 in the context of gnathostome evolution. It was shown that two, albeit preliminary, parsimony analyses on different though partly overlapping data indicate that coelacanths are the plesiomorph sister-group of the Choanata (Dipnoi + tetrapods), the triplet comprising the Sarcopterygii. The alternative view, that coelacanths are the sister-group of the Actinopterygii + Choanata, was supported only on the assumption that internal fertilization and two lateral axonemal columns, seen in chondrichthyans and *Latimeria,* are plesiomorphic for gnathostomes and support was equivocal as an alternative tree supported the first view.

If we accept the sarcopterygian status of the Actinistia, the relationship of coelacanths with the rhipidistians (the extinct crossopterygian superorder Osteolepimorpha) emerges as a point of interest. It is noteworthy that Lauder and Liem (1983), who also place the Actinistia at the base of the Sarcopterygii, give no synapormorphies that are not shared with the osteolepiform *Eusthenopteron* which they see as the the plesiomorph sister-taxon of the Sarcopterygii. Furthermore, they see the osteolepimorph order Porolepiformes as originating in common with the Actinistia, within the Sarcopterygii. *Eusthenopteron*-sarcopterygian synapomorphies include: true enamel on the tooth surface found also in rhipidistians but differing from that of actinopterygians (Smith, 1978); double articulation of the hyomandibular with the neurocranium (Gardiner, 1973); a unique type of supporting skeleton in the paired fins; presence of an endoskeletal urohyal (Patterson, 1977a); articulation of the last gill arch

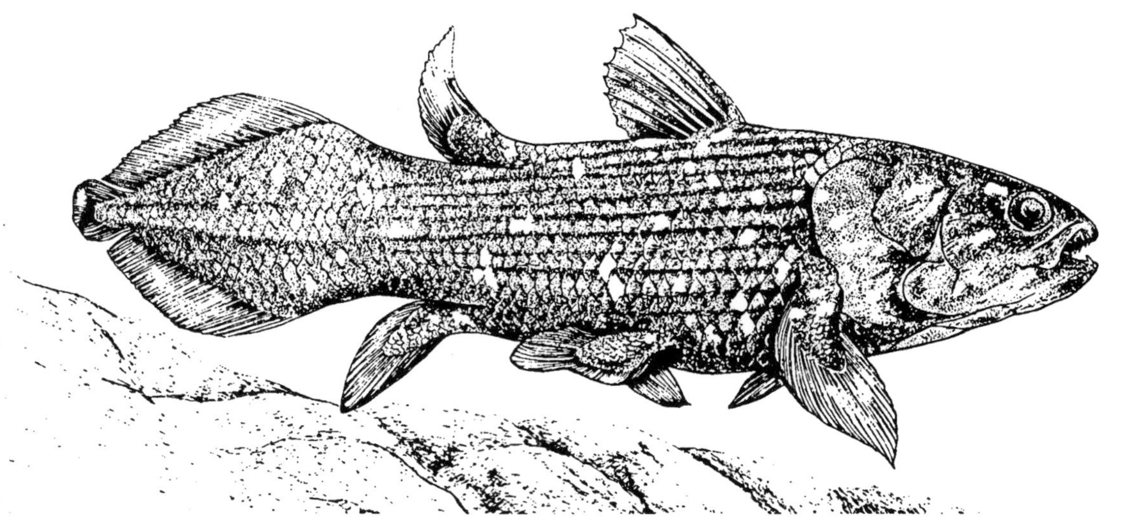

Fig. 9.1. The living coelacanth, *Latimeria chalumnae*. Adult. From Greenwood, P.H. (1989). *Biologist* **36**, 15-19. Fig. 1.

with the base of the preceding arch; muscular lobes forming the base of the pelvic and pectoral appendages (Rosen *et al.*, 1981); and a subdermal anocleithrum (Rosen *et al.*, 1981; see Lauder and Liem, 1983). A monophyletic Tetrapoda rooted in the Crossopterygii is also envisaged by Szarski (1977) and Schultze (1987). In terms of simple gestalt, the similarity of *Latimeria* and the porolepiforms (Fig. 8.3) is striking.

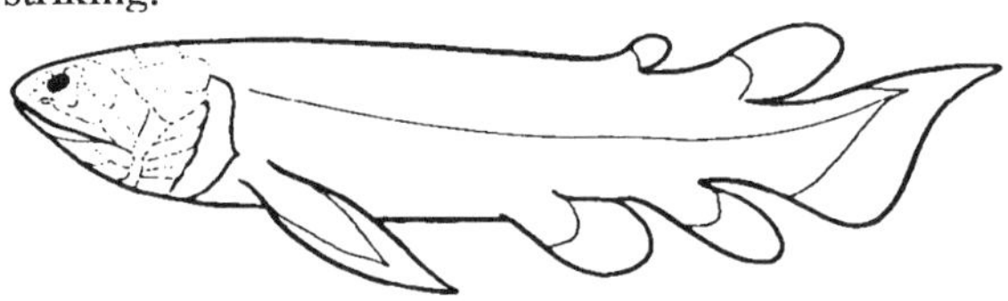

Fig. 9.2. A porolepiform rhipidistian (Osteolepimorpha). From Nelson, J.S. (1984). *Fishes of the World*. 2nd edition, John Wiley and Sons, New York.p. 75.

Order Coelacanthiformes

Diagnosis. Caudal fin diphycercal, consisting of three lobes; external nostrils, no choana; branchiostegals absent; lepidotrichia never branched; lepidotrichia in tail equal to number of radials or somewhat more numerous; anterior dorsal fin in front of centre of body. Length reaching 1.8 m, in *Latimeria*. Several Devonian to Cretaceous families. One family (Latimeriidae) with a living representative (Nelson, 1984).

Sperm literature. The spermatozoon of *Latimeria chalumnae*, was known until recently only from light microscopy (Tuzet and Millot, 1959) (Fig. 9.3) but TEM examination of formalin-fixed material has yielded a commendably complete description of its ultrastructure (Mattei, 1988; Mattei *et al.*, 1988) (Fig. 9.4).

Latimeria chalumnae. The spermatozoon of *Latimeria* is filiform (Tuzet and Millot, 1959). The ultrastructure of sperm in the testis is the same as that in the renal canal (Mattei *et al.*, 1988) (Fig. 9.4).

Acrosome. The gently curved acrosome is 2.8 µm (Tuzet and Millot, 1959), confirmed as about 3 µm long (Mattei *et al.*, 1988). Presence, almost constantly, of a "siderophilic" granule, questionably thought to correspond with a centrosome, at its anterior extremity (Tuzet and Millot, 1959) has not been confirmed. The anterior end of the acrosome is prolonged into a slender point, 0.2 µm in diameter, and

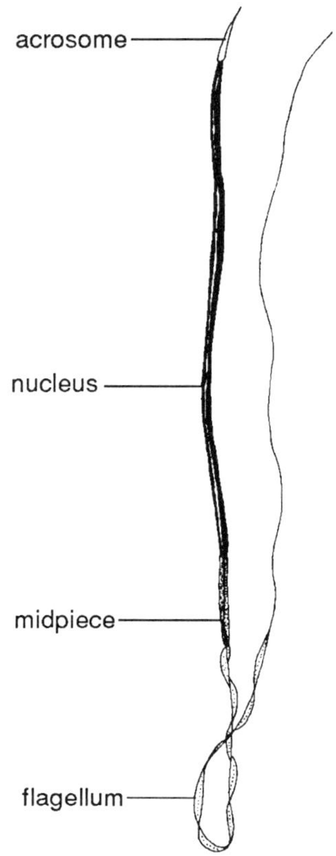

Fig. 9.3 *Latimeria chalumnae*. Light microscope drawing of spermatozoon. After Tuzet, O. and Millot, J. (1959). *Annales des Sciences Naturelles Zoologie* 12 Ser **1**, 61-69. Fig. 3C.

its base is situated laterally on the tip of the nucleus. The acrosome contains three longitudinal rodlets which extend as far as its anterior extremity (Mattei *et al.*, 1988).

Nucleus. The head [acrosome + nucleus] is 26 µm long and is penetrated by an "intranuclear filament" (Tuzet and Millot, 1959). A length of approximately 25 µm is reported for the nucleus by Mattei *et al.* (1988). The contents of the nucleus are dense and compact, though in the aged spermatid concentric laminae are visible. The anterior extremity of the nucleus is drawn out into a bevel and penetrated by a central cavity, like a hypodermic needle. This intranuclear canal penetrates the entire length of the nucleus but does not perforate its base.

The diameter of the nucleus is 0.4 µm at the base of the acrosome, 0.6 µm in its midregion, and 0.3 µm for a posterior region of about 5 µm in length. The

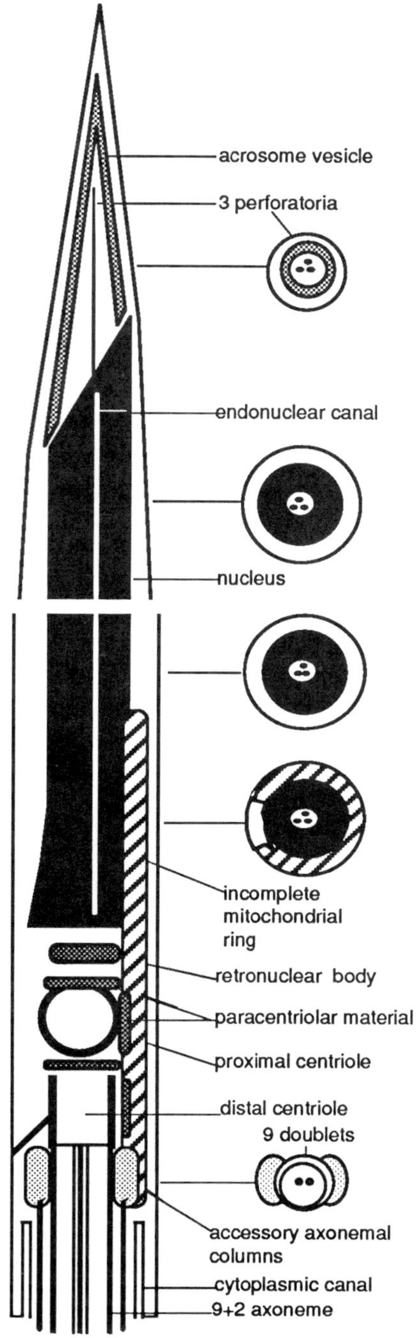

Fig. 9.4. *Latimeria chalumnae*. Diagrammatic reconstruction of a longitudinal section of the spermatozoon. Original, based on micrographs by Mattei *et al.* (1988). *Journal of Ultrastructure and Molecular Structure Research* **101**, 243-251.

posterior extremity of the nucleus is enlarged, as is the endonuclear canal, in this region.

Perforatoria. The transverse section of the intranuclear canal is trilobed as it contains the three rodlets (here regarded as perforatoria) which are prolonged into the acrosome. The rodlets are each 50 nm in diameter and are very regularly arranged. Two giant nuclei were observed, among the very many spermatozoa observed, in one of which the intranuclear canal had an ovoid cross section and contained only two rods and the other of which had three canals and four rods.

Perinuclear sleeve. The posterior region of the nucleus is surrounded by a cylindrical sleeve (manchon périnucléaire) which is usually not complete. This sleeve appears to be homogeneous and to be limited by a double membrane. It is absent from some sperm, probably because of its fragility. It is implied that the sleeve is mitochondrial* as spermatozoa observed by light microscopy after staining for mitochondria show a nuclear length of only 20 µm and an intermediate piece 5 µm long (Mattei *et al.*, 1988). This corresponds with the report by Tuzet and Millot (1959) of an elongated midpiece which, with the two posterior "centrosomes", constitutes an intermediate segment 5 µm long.

Centrioles. The 'neck', about 0.8 µm long, contains two centrioles and paracentriolar structures which are disposed between the proximal centriole and the nucleus and around the two centrioles. The proximal centriole, consisting of 9 triplets, is at right angles to the distal centriole which latter forms the basal body of the flagellum. The distal centriole is provided with 9 satellite rays (Mattei, 1988; Mattei *et al.*, 1988). A "paracentriolar" mass anterior to the proximal centriole closely resembles the retronuclear body of *Protopterus* sperm.

Membranous collar. The region of the neck is prolonged as a membranous collar ('manchon membranaire'), 0.6 to 1 µm long. It is connected to the distal centriole by the satellite rays. Nine dense granules are present at this level (Mattei *et al.*, 1988). The collar appears in micrographs to be separated from the axoneme by a space equivalent to the so-called cytoplasmic canal typical of the sperm of gnathostome fish.

Flagellum. The flagellum begins within the membranous collar. It is cylindrical for a length of

* In a micrograph, what appears to be an indubitable mitochondrion extends from below the centriolar complex, reaching almost to the beginning fo the cytoplasmic canal, and extends anteriorly up the side of the basal region of the nucleus.

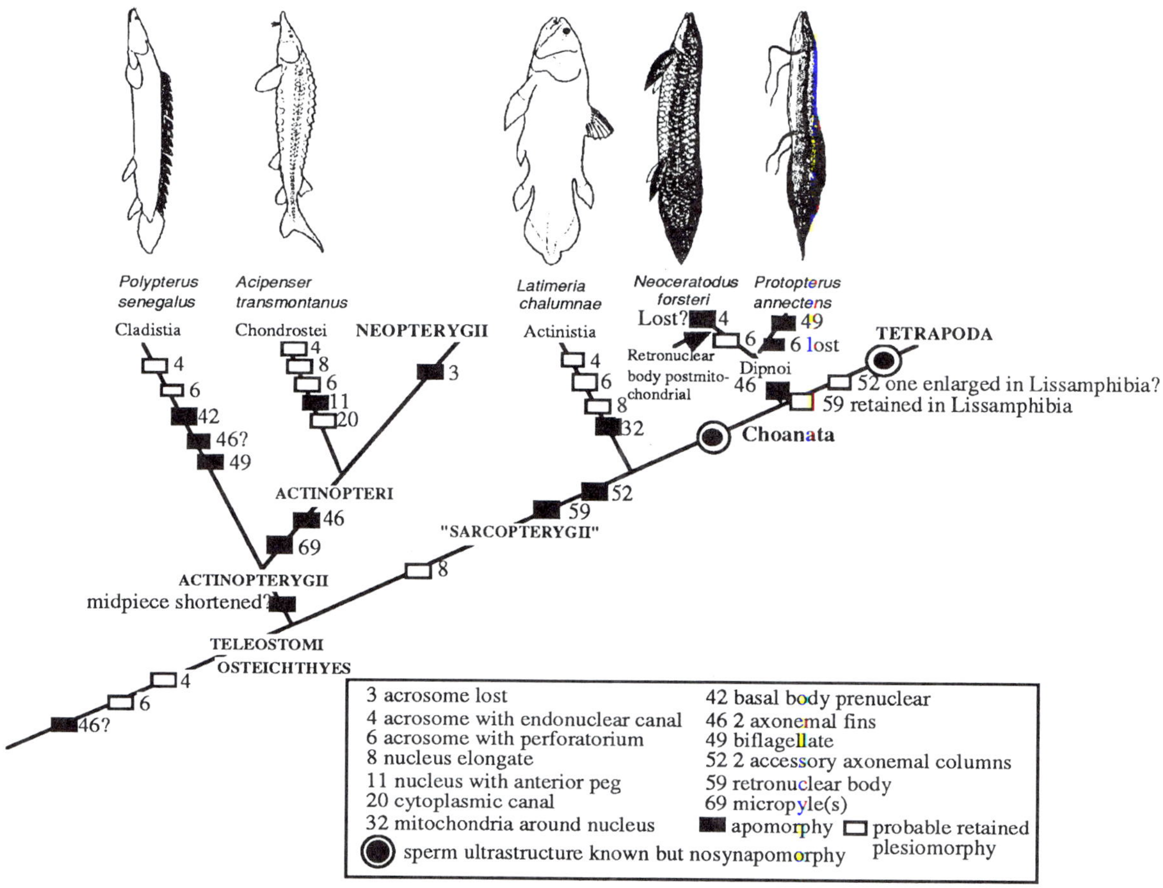

Fig. 9.5. Phylogeny of the Osteichthyes, chiefly after Lauder and Liem (1983), and references therein, with spermatozoal apomorphies added. Fish after Nelson and Norman.

0.1 µm after which two lateral elements are associated with the axoneme (Mattei, 1988; Mattei *et al.*, 1988). On the free flagellum these elements are not as wide as at the level of the collar. We note, from micrographs, that within the collar each element is applied to almost half of the periphery of the axoneme whereas on the free flagellum it is not much greater in diameter than an axonemal doublet. The two elements are associated with doublets 3 and 8; most sections of axonemes lack the elements (Mattei *et al.*, 1988) which presumably, therefore, extend through only a small portion of the flagellum. The elements clearly correspond with the undulating membrane recognized by Tuzet and Millot (1959).

The axoneme has the 9+2 pattern. The doublets have two arms and rays extend to the two central singlets (Mattei, 1988; Mattei *et al.*, 1988).

Remarks. Mattei *et al.* (1988) state that location of the mitochondria around the posterior region of the nucleus has been observed in a nemertean. We add, perhaps more pertinently, that it is also seen in the Australian lamprey *Mordacia praecox* (Hughes and Potter, 1969) and is also an independent apomorphy of non-appendicularian urochordates.

As noted by Mattei *et al.* (1988), and we add to their account, one or more intranuclear, or endonuclear, canals are described for the spermatozoon of several vertebrates: the lampreys, *Lampetra fluviatilis* (Afzelius and Murray, 1957), *L. planeri* (Follenius, 1965; Nicander, 1968, 1970), and *L. japonica* (Jaana and Yamamoto, 1981); the cladistian *Polypterus senegalus* (Mattei, 1970); the dipnoan *Neoceratodus forsteri* (Jespersen, 1971), but not *Protopterus* (Boissin *et al.*, 1967); the chondrostean *Acipenser stellatus* (Ginzburg, 1977) and the amphibian *Pleurodeles waltlii* (Picheral, 1967). In these canals, the number of rods is one (*Lampetra*, in which it arises from a subacrosomal ring, *Polypterus* and *Pleurodeles*), two (*Neoceratodus*) or three as in *Latimeria* (*Acipenser*). The rods are restricted to the nucleus in *Lampetra*, *Polypterus* and *Pleurodeles* whereas they reach the anterior extremity of the acrosome in *Latimeria*, *Neoceratodus* and *Pleurodeles*.

The acrosome reaction is accompanied by extrusion of the rod(s) in *Lampetra* (Afzelius and Murray, 1957; Jaana and Yamamoto, 1981) and *Acipenser* (Ginzburg, 1977; Cherr and Clark, 1984b) whereas in *Pleuodeles* it remains intranuclear (Picheral, 1977). The fate of the rods during fertiization is not known in *Latimeria*, *Polypterus* and *Neoceratodus*.

We may also add that a purely subacrosomal (neither acrosomal nor nuclear) rod occurs in sharks, for instance *Squalus suckleyi* (Stanley, 1971), and that a less distinct subacrosomal perforatorium is present in the holocephalan *Hydrolagus colliei* (Stanley, 1983).

Satellite rays, arising from the distal centriole, are common in the aquasperm of invertebrates but are rare in fish, being seen, as Mattei *et al.* (1988) note, in dipnoans (Boissin *et al*, 1967; Jespersen, 1971) and, we may add, in some perciforms and atheriniforms.

Presence of two lateral elements on the axoneme appears on first consideration to be an important similarity and possible synapomorphy with chondrichthyan sperm, especially as these elements lie opposite doublets 3 and 8, as in sharks. However, we have seen in Chapter 5 that even if co-occurrence of these elements in *Latimeria* and sharks is taken as a true homology and synapomorphy, synapomorphies between *Latimeria* and choanates so far outnumber those between it and chondrichthyans that sister-group relationship of *Latimeria* and choanates, within the Sarcopterygii, advocated by Forey (1980) is unassailed. In addition, the agreement in position, opposite doublets 3 and 8, seems less impressive if it is recognized, as here proposed, that these doublets lie approximately in the plane of the two central singlets in all 9+2 axonemes and that the location of the two elements is therefore merely in a given latitudinal plane (one cannot rightly say plane of symmetry) of the axonemal cross section. This is the same plane in which fins develop in actinopterygian sperm and, although this is probably an unrelated development, it perhaps indicates that this is the only plane in which lateral elements may be positioned for efficient functioning of the flagellum unless multiple, circumferential elements (9 fibres as in several fish and in amniotes) are present. Nor is this correspondence between *Latimeria* and the Chondrichthyes unique; location of two lateral elements opposite doublets 3 and 8 is also reported for *Neoceratodus*, *Triturus vulgaris* and several other urodeles, for the anuran *Bufo boreus* and for the rat. In urodeles the two columns are disproportionate, the column opposite doublet 3 developing earlier and attaining a larger size (Stanley, 1970). In *Neoceratodus* (Jespersen, 1971) the columns are restricted to the portion of the axoneme within the cytoplasmic collar and differ from those of chondrichthyans in having dense rather than pale cores. There is some indication, despite formalin fixation, that the cores are dense in *Latimeria* also.

In *Hydrolagus*, *Neoharriotta* and *Bufo*, apparently as specializations, the column opposite doublet 8 is eliminated entirely except for a basal remnant, while in *Chimaera* it may be entirely lost (Hara and Tanaka, 1986; Mattei, 1988; see also Chapter 7). The small diameter of the lateral elements in *Latimeria* further casts doubt on their homology with the large elements of chondrichthyans.

In short, two lateral axonemal elements, in the plane of the central singlets, are widespread in vertebrates and, although it is possible that they represent a facultative expression of the same gene or genes there is no compelling reason to consider them a synapomorphy of *Latimeria* and the Chondrichthyes. If they represent a gene complex which may be switched on or off they constitute a symparamorphy

Fig. 9.6. *Latimeria chalumnae*. Two eggs, among the largest known in the non-amniote vertebrates. From Anthony, J. (1980). *Proceedings of the Royal Society of London* **B 208**, 349-367. Plate 2.

in the terminology of Jamieson (1984). The possibility remains that they are merely analogous structures. Examination of better fixed material is necessary to determine the degree of ultrastructural correspondence of the elements in the Chondrichthyes, Actinistia and Dipnoi.

In view of the suggestion of Ahlberg (1989) that the Actinistia are the sister-group of the Dipnoi, it is possible that the two outer axonemal columns and the retronuclear bodies are synapomorphies of *Latimeria* and Dipnoi. Alternatively, they may be synapomorphies of the Actinistia - Dipnoi -Tetrapoda assemblage which have been modified in the tetrapods (Fig. 9.5, 18.7).

Reproductive biology. *Latimeria* is a special type of live-bearer of a type called a matrotrophous oophage or is possibly an adelphophage. This energetically efficient type of viviparity is also known in most mackerel sharks (Lamniformes); the large and dense yolk facilitates the early development of a definitive phenotype within the oviduct. This allows the large young to begin oral feeding on other less advanced siblings and ova in the same oviduct. Thus one or a few large, fully developed, urea-retaining predatory young are born (see review by Balon *et al.*, 1988). The eggs are among the largest known in the non-amniote vertebrates*, exceeding 9 cm in diameter (Anthony, 1980) (Fig. 9.6). Ovoviviparity was established by Smith *et al.* (1975) (Fig. 9.7).

SUBCLASS DIPNOI (DIPNEUSTI)

Diagnosis and relationships. The three extant genera of the Dipnoi are illustrated in Fig. 9.8. The old view (e.g. Bischoff, 1840), that dipnoans are more closely related to tetrapods, specifically the Amphibia, than are any other fish has been revived by several workers, notably Patterson (1980); Gardiner (1980); Rosen *et al.* (1981); Lauder and Liem (1983) (see also Fig. 9.4); Northcutt (1986) and Forey (1980, 1986, 1988). It hinged on the conclusion that the internal excurrent nostril of Recent lungfish is a true choana such as exists in tetrapods.

As Patterson (1980) states, the synapomorphies between lungfishes and tetrapods listed by Bischoff in 1840 appear uncontested and to them may be added (*inter alia*) about 34 derived characters of the 57 listed by Kesteven (1951), the gill-arch muscles

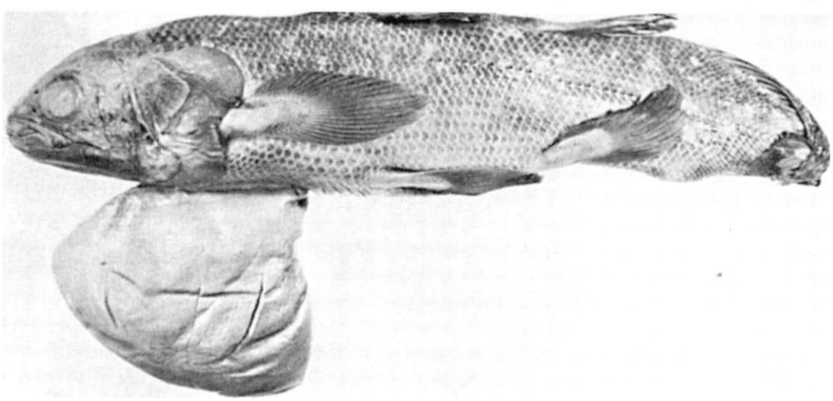

Fig. 9.7. *Latimeria chalumnae*. Yolk-sac young. Note fins and tail compressed by oviducal wall and morphological similarity to the adult. From Smith, C.L., Rand, C.S., Schaeffer, B. and Atz, J.W. (1975). *Science* **190**, 1105-1106. Fig. 2. Copyright 1975 by the American Association for the Advancement of Science.

*Eggs of some sharks reach 10 cm or more in diameter.

discussed by Wiley (1979) and ciliation of the larvae described by Whiting and Bone (1980). Fig. 8.2 summarizes these major synapomorphies between Dipnoi and Tetrapoda. They also include the presence of the choana; the structure of the pelvic girdle (Rosen *et al.*, 1981); and numerous further features of the soft anatomy such as the partially divided conus arteriosus, an atrial septum; the dermal bone pattern covering the braincase; loss of the interhyal; and the structure of the pelvic and pectoral appendage Rosen *et al.* (1981). Embryologically Dipnoi resemble amphibians and differ from other fish in that cleavage is total and gastrulation produces a yolk plug. The presence in the larvae of *Lepidosiren* and *Protopterus* of a sucker and external gills is a further resemblance to the Amphibia (Young, 1981). We have seen that Nelson (1984), nevertheless, favours the Crossopterygii as the ancestors of the Amphibia, and by implication the Tetrapoda.

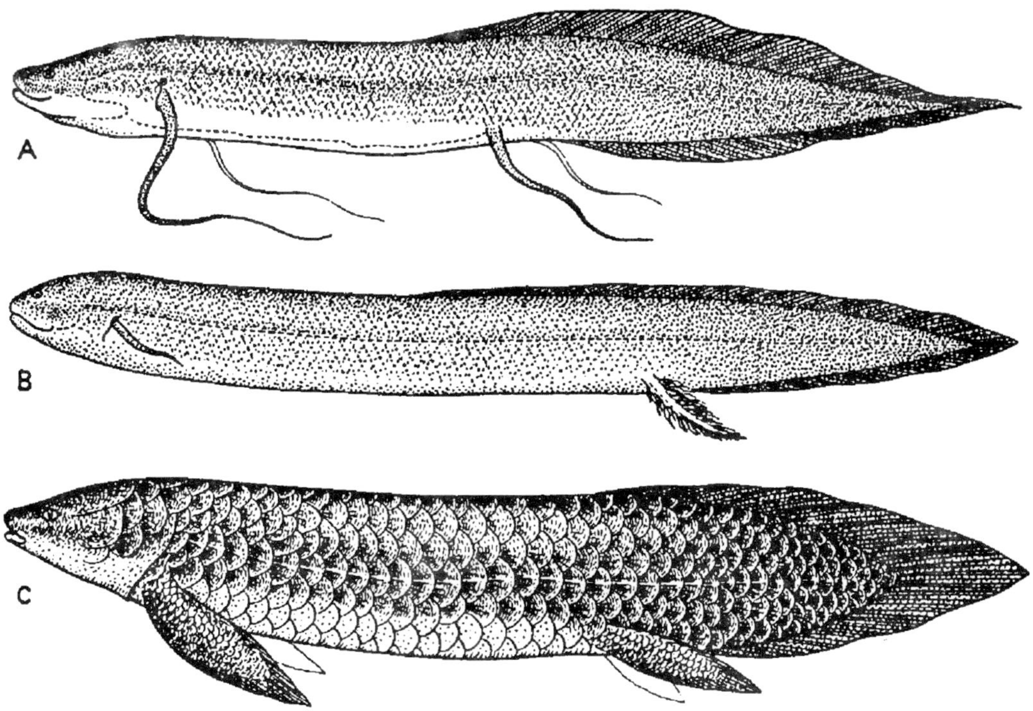

Fig. 9.8. Adult Dipnoi. A. *Protopterus aethiopicus* (Africa). B. *Lepidosiren paradoxa* (S. America). C. *Neoceratodus forsteri* (Australia). From Norman, J.R. (1937). *Illustrated Guide to the Fish Gallery. British Museum (Natural History).* Trustees of the British Museum, London. Fig. 50.

The sperm of three species of Dipnoi are now well studied ultrastructurally. Sperm ultrastructure reinforces separation of *Neoceratodus* from *Protopterus* in separate families of the Dipnoi (see Nelson, 1984) while offering an autapomorphy (retronuclear body or its derivative) linking the two families. We have seen that this body and the two outer axonemal columns are possible synapomorphies of Dipnoi and Actinistia.

Order Lepidosireniformes

Diagnosis. Pectoral and pelvic fins filamentous, without rays. Larvae with external gills. S. American and African lungfish (Nelson, 1984).

Protopterus. Both examined species of Protopteridae, *Protopterus*, *P. annectens* (Boissin *et al.*, 1967; Mattei, 1970) and *P. aethiopicus* (Purkerson *et al.*, 1974), have biflagellate sperm

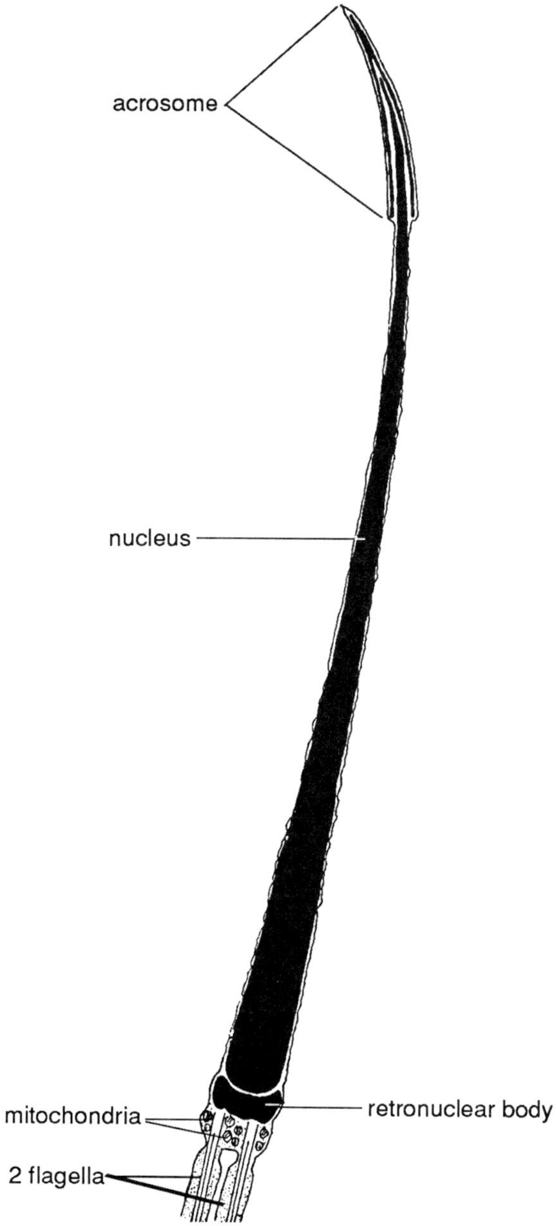

Fig. 9.9. *Protopterus annectens.* Mature spermatozoon. After Boisson*et al.* (1967). *Institut Fondamental d'Afrique Noire. Bulletin* Série A. (Sciences Naturelles) **29**, 1097-1121. Fig. 1f.

whereas that of *Neoceratodus* is uniflagellate. The spermatozoa of the two *Protopterus* species are uniform.

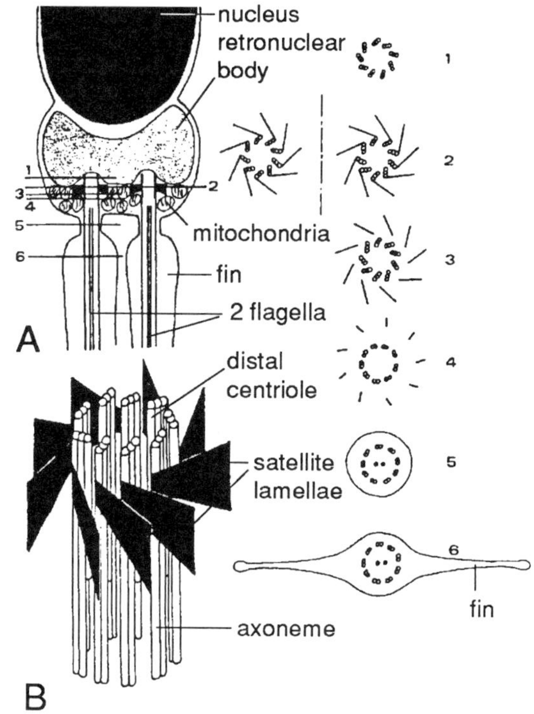

Fig. 9.10. *Protopterus annectens.* Mature spermatozoon. A. Intermediate piece. B. Diagrammatic reconstruction of a centriole. Right column: transverse sections through levels indicated by numerals in A. After Boisson *et al.* (1967). *Institut Fondamental d'Afrique Noire. Bulletin* Série A. (Sciences Naturelles) **29**: 1097-1121. Fig. 2.

Acrosome and nucleus. The very long tapering nucleus is capped, and deeply penetrates, a pointed acrosome (Fig. 9.9) which (Boisson *et al.*, 1967; Mattei, 1970) arises not from Golgi secretions, as in Chondrichthyes, but apparently from a single Golgi saccule. According to Purkerson *et al.* (1974) the nucleus remains filamentous but the other workers indicate final condensation.

Midpiece or intermediate piece. This consists of the retronuclear body, the mitochondria and the two centrioles. The midpiece is approximately a twentieth of the length of the nucleus from which it is delimited by a circumferential groove.

Retronuclear body. Between the base of the nucleus and the two parallel centrioles there is a large, structure, the retronuclear body. In *P. aethiopicus* this is said to display excessively dense regions some of which are in striated array just proximal to what is unaccountably termed the "distal" centriole. Purkerson *et al.* (1974) consider the retronuclear body to be strikingly comparable with the striated columns of mammalian sperm. In *P. annectens*, although there are concentric rings during development, the mature body later consists of spherules disseminated in a less osmiophilic mass and finally becomes a compact formation (Boissin *et al.*, 1967).

Mitochondria. The remainder of the arbitrarily defined midpiece or intermediate segment is short, containing numerous small cristate mitochondria associated with the two centrioles.

Centrioles. The two triplet centrioles are parallel and longitudinal, separated by 1 µm. Each bears a flagellum. Their proximal extremities are lodged in the centre of an osmiophilic cylinder. Extending from each cylinder are satellites (Fig. 9.10A, B) consisting of 9 thin lamellae, 0.2 µm long, situated in the perpendicular plane relative to the axis of the centriole and inserted between the triplets (Boissin *et al.*, 1967).

Flagella. In both species each 9+2 flagellum (inexplicably stated by Mattei, 1988, to be of 9+0 construction) is circular in cross section on leaving the midpiece but further distally has two lateral expansions [fins] giving a total flagellar width of 50 µm (Boissin, 1963). As is frequent in fish, the fins lie in the plane of the two central singlets. In *P. annectens*, at least, the fins are pointed in the proximal region of the flagellum, where they are about 1 µm wide, but in the distal region, where they are about 2 µm wide, are terminated by a vesiculate expansion; the vesicles correspond with longitudinal tubules. Fine trabeculae unite the two surfaces of each fin (Boissin, 1963). Scanning electron microscopy reveals that the flagella are about twice the length of the head (Purkerson *et al.*, 1974).

In vivo the flagella are seen to beat slowly relative to flagella of most animal sperm and this is ascribed by Boissin (1963) to the presence of the fins, an interesting contrast with the questionable view that from the standpoint of fluid-dynamic efficiency it is virtually immaterial whether the cross sectional shape of the sperm tail is circular or of another shape (Flower, 1967).

Peculiarities of spermiogenesis. Boissin *et al.* (1967) list a number of pecularities which seem specific to spermiogenesis and the mature sperm in *Protopterus*: 1. the germinal cells are immense although the nuclei, mitochondria and Golgi apparatus are of normal size; 2. the cytoplasm of the spermatid is frothy, being invaded by innumerable vacuoles; 3. the Golgi apparatus, well developed in spermatocytes I and II, resolves in the spermatid into small vesicles of secretion; only one of these survives as a

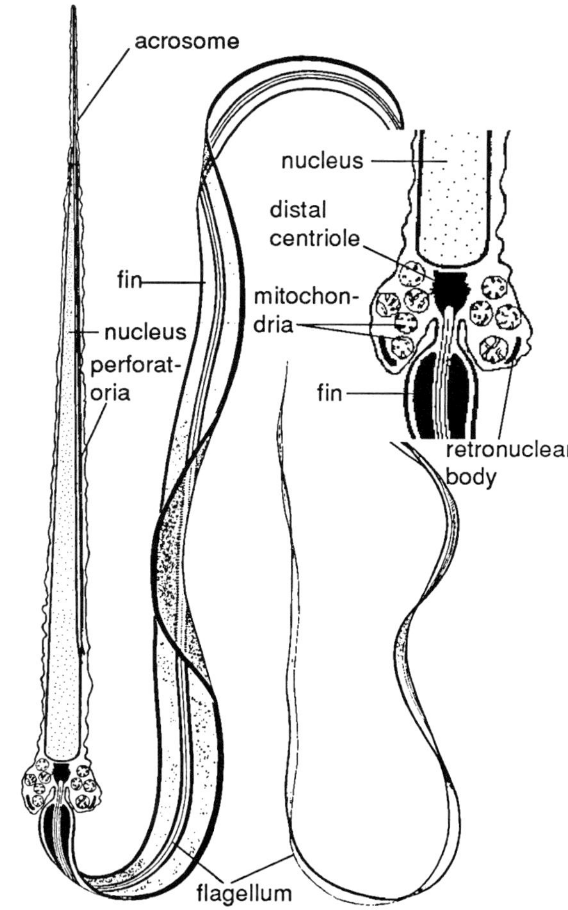

Fig. 9.11. *Neoceratodus forsteri.* Diagram of the mature spermatozoon. Inset shows detail of midpiece region. After Jespersen, Å. (1971). *Journal of Ultrastructure Research* **37**, 178-185. Fig. 18.

saccule in contact with the nuclear membrane and becomes the somewhat reduced acrosome; 4. each of the two centrioles develops a flagellum; 5. a retronuclear body is present; this forms from osmiophilic granules originating from the cell membrane; 6. the two flagella confer bilateral symmetry on the spermatozoon.

Order Ceratodontiformes

Diagnosis. Pectoral and pelvic fins flipperlike. Larvae without external gills. The Australian lungfish (Nelson, 1984) (Fig. 9.8C).

Neoceratodus. The spermatozoon of the Australian Lungfish, *Neoceratodus forsteri*, (Fig. 9.11) has been described ultrastructurally by Jespersen (1971).

Head. The head is about 70 µm long, to our knowledge by far the longest known in fish. Its proportions appear similar to those of *Protopterus* for which lengths are not available and it similarly tapers very gradually to the pointed tip, being 2 µm wide at its greatest width, near its posterior end.

Acrosome. A slender elongate acrosome is present on, but unlike that of *Protopterus*, does not appear to ensheath, the tip of the nucleus. Two rod-shaped structures [perforatoria], absent in *Protopterus*, extend from the tip of the acrosome through about four fifths of the nucleus and end abruptly in the thin layer of cytoplasm coating the sides of the nucleus; through most of this length they run in endonuclear canals; they are close together in the acrosome and anterior region of the nucleus but gradually separate posteriorly. The acrosome again forms, as in *Protopterus*, from a single Golgi vesicle in close contact with the nuclear membrane.

Nucleus. The 70 µm long nucleus is lanceolate with condensed chromatin.

Midpiece. The short, almost globular midpiece is about 2 µm long. It contains numerous irregularly arranged spherical mitochondria and, axially, two triplet centrioles differing from those of *Protopterus* in being mutually perpendicular. A "button" extends from the proximal centriole into the basal nuclear fossa. Behind the longitudinal distal centriole (basal body) the midpiece surrounds a fossa, deeper than that in *Protopterus*, containing the base of the 9+2 axoneme. This fossa is, clearly, the equivalent of the cytoplasmic canal of teleost sperm.

Retronuclear body. At the posterior limit of the midpiece there is a dense, granular, ring-shaped body. In the spermatid this surrounds the centrioles and Jespersen appears correct in equating it with the retronuclear body of the *Protopterus* sperm.

Flagellum. The tail is a remarkable 200 µm long. Where the flagellum is still within the cytoplasmic canal it is bordered by two oblong dense bodies. As in *Protopterus*, after an initial strictly cylindrical region where the the flagellum emerges from the midpiece, the cell membrane is expanded as two equal sized "wings" or fins.

Remarks. As noted by Jespersen (1971) the sperm of *Neoceratodus* and *Protopterus* show notable similarities: the heads are almost equally large; the acrosome and nucleus are of similar proportions; and the acrosome develops from a single Golgi saccule. Distinguishing features are the uniflagellate condition of *Neoceratodus*; the absence of perforatoria in *Protopterus*; and the postmitochondrial rather than premitochondrial location of the retronuclear body in *Neoceratodus*. Homology of this body in the two genera seems assured by its initial premitochondrial location in the spermatid of *Neoceratodus*.

Presence in Dipnoi of axonemal fins, seen also in Neopterygii, is computed as an independent apomorphy in Fig. 8.2 and is so regarded in the work of Forey (1980) and Ahlberg (1989). Such homoplasic origin is acceptable as fins are known spasmodically in polychaete and echinoderm sperm. Nevertheless, the alternative that fins are retainedin Dipnoi from the ancestor of the Osteichthyes and were lost in the Actinistia in not improbable and is canvassed in Figs. 9.5 and 18.7.

CHAPTER 10

INFRACLASS NEOPTERYGII

HOLOSTEANS THROUGH OSTEOGLOSSOMORPHA

Diagnosis and relationships. The Neopterygii contain two major groups, the Ginglymodi containing only the family Lepisosteidae (gars) and the Halecostomi (*Amia* and the teleosts) (Fig. 10.1; see also actinopterygian phylogeny, Fig. 8.5).

Neopterygians have dorsal and anal fin rays equal in number to their supports (plesiomorphically the number of fin rays was greater than the number of supports); consolidation of the upper pharyngeal dentition and clavicle lost or reduced to a small plate

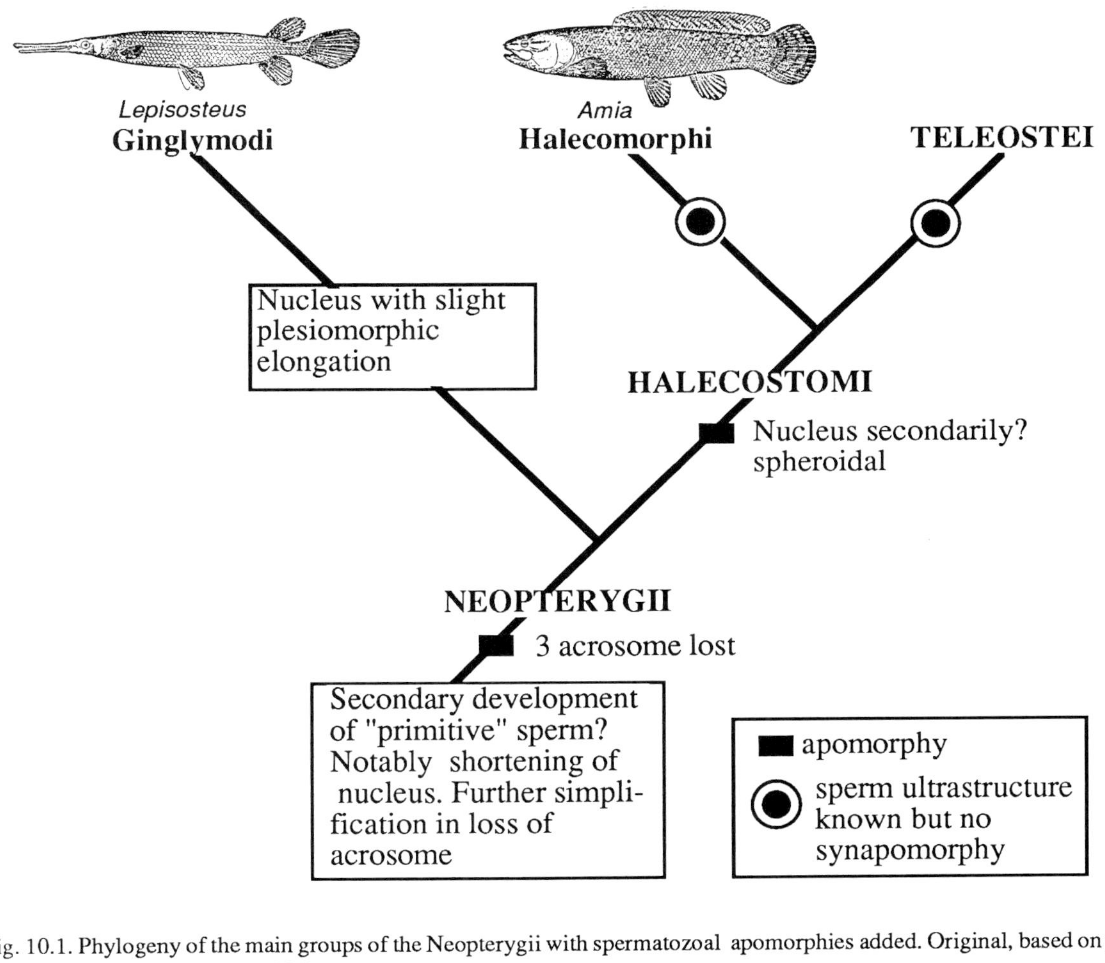

Fig. 10.1. Phylogeny of the main groups of the Neopterygii with spermatozoal apomorphies added. Original, based on phylogeny of Lauder and Liem (1983).

lateral to the cleithrum; premaxilla with internal process lining the anterior part of the nasal pit; symplectic developed as an outgrowth of the hyomandibular cartilage. It is generally agreed that the Neopterygii constitute a monophyletic group (Patterson, 1973; Wiley, 1976; Bartram, 1977; Patterson and Rosen, 1977; Lauder and Liem, 1983; Nelson, 1984).

Gamete features. Irrespective of whether internal fertilization is primitive or derived in fish as a whole, neopterygians are spermatologically diagnosed from the Agnatha through Dipnoi in having lost the acrosome, giving an anacrosomal aquasperm.

Only one micropyle is present in the neopterygian egg, presumably as an adaptation preventing polyspermy (Ginzburg, 1968, p 62; p 200). Divergence from the simple anacrosomal sperm type, though only in *Lepidogalaxias* with redevelopment of the acrosome, is found in the few neopterygians which have redeveloped internal fertilization (poeciliids, hemiramphids etc.) or have external fertilization of a presumably specialized type.

An environmental change from sea water to freshwater has been suggested as a possible cause in the loss of the acrosome in teleost spermatozoa (Baccetti, 1979, 1985). This hypothesis is not supported by the fact that fishes of Petromyzontidae, Cladistia, Chondrostei and Dipnoi spawn in freshwater and yet possess sperm acrosomes (Leung, unpublished).

Halecomorphi and Ginglymodi, the "holosteans"

Status of Holostei. The category Holostei is now considered paraphyletic and therefore obsolete (Fig. 10.1). The Halecomorphi, represented by *Amia*, the Bowfin, are regarded as the plesiomorph sister-group of the Teleostei. The Halecomorphi and Teleostei comprise the Halecostomi. The Ginglymodi, again with only one extant genus, *Lepisosteus*, the Garpike (Fig. 10.2), are seen as the sister-group of the halecomorph-teleost assemblage and therefore as the most primitive living neopterygians (Lauder and Liem, 1983; Nelson, 1984; Patterson, 1973) (Fig. 10.1). Accordingly the sperm of *Lepisosteus*, with that of *Amia*, is the first example of an externally fertilizing, anacrosomal aquasperm in extant fish.

Division Ginglymodi

Diagnosis. Interoperculum absent; two or more supratemporal lobes on each side; maxilla small and immobile; supramaxilla absent; a series of toothed infraorbital bones present; myodome absent; centra opisthocoelous (see Wiley, 1976, for many other characters). A single order and family (Lauder and Liem, 1983; Nelson, 1984).

Order Lepisosteiformes

Diagnosis. As for the Ginglymodi. One extant family (Lepisosteidae) and genus (*Lepisosteus*) with seven species, the garpikes. Freshwater, occasionally brackish, very rarely in marine water; N. and Central America and Cuba (Nelson, 1984).

Lepisosteus osseus. The spermatozoon of the Garpike *Lepisosteus osseus* (Fig. 10.3, 10.4) is rightly described by Afzelius (1978) as being of the "primitive" type like that of most teleosts, that is an anacrosomal aquasperm.

Nucleus. The bullet-shape of the nucleus, 2.5 μm long and 1.1 μm wide, is, however, a slight departure from the more rounded form of the nucleus in most teleostean aquasperm. It has a posterior indentation which houses the proximal centriole.

Midpiece. The ring- or C-shaped mitochondrion (Afzelius, 1981), described as a single annular mitochondrion by Mattei (1988), is not usual for neopterygians; it is known in only five teleost families (see Clupeomorpha, below); in *Branchiostoma* and, *inter alia*, in echinoderms, cnidarians and some polychaetes (see Jamieson, 1984) and has doubtful cladistic significance. In some *L. osseus* sperm the midpiece also contains a lipid droplet. A thin sleeve projects posteriorly for 0.6 μm from the midpiece around the anterior end of the flagellum (Afzelius, 1981).

Centrioles. The two triplet centrioles have the mutually perpendicular arrangement considered plesiomorphic for the animal kingdom (Afzelius, 1979). Along one side of the proximal centriole there is a cross-striated fibrous body (Fig. 10.4) which appears to be attached to the centre of the basal nuclear fossa by an extension consisting of several thin strands. Afzelius equates this with a body [the "button"] seen in the sperm of *Neoceratodus* but it has alternatively been regarded as a short flagellar root (Mattei *et al.* 1981; Mattei, 1988). The two views may be compat-

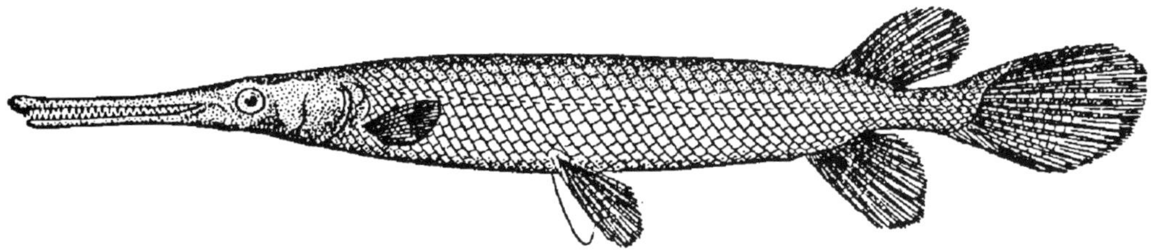

Fig. 10.2. *Lepisosteus osseus*. The Garpike. After Norman, J.R. (1937). *Illustrated Guide to the Fish Gallery.* British Museum (Natural History. Trustees of the British Museum, London. Fig. 19.

ible. Each centriole has a doublet-containing distal extension, each doublet having a hook-like projection. In addition, the distal centriole has a satellite apparatus. The doublet portion of the proximal centriole has been interpreted as an abortive flagellum (Mattei *et al.* 1981; Mattei, 1988) but there seems no reason to accept the implication that biflagellarity is plesiomorphic for actinopterygians.

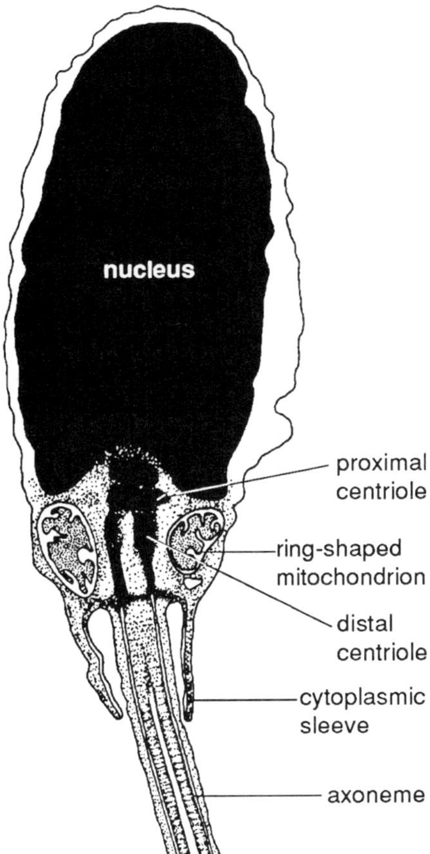

Fig. 10.3. *Lepisosteus osseus*. Longitudinal section through head and midpiece of spermatozoon. An acrosome is absent. After a micrograph by Afzelius, B.A. (1978). *Journal of Ultrastructure Research* **64**, 309-314. Fig. 1.

Flagellum. The 9+2 flagellum, 45 µm long, has a pair of lateral fins (Fig. 10.5). Each doublet appears to have two dynein arms (Afzelius, 1981). The doublets are not all identical. The A tubule of doublets 1, 2, 6 and 7, contains a dense substance whereas the other A

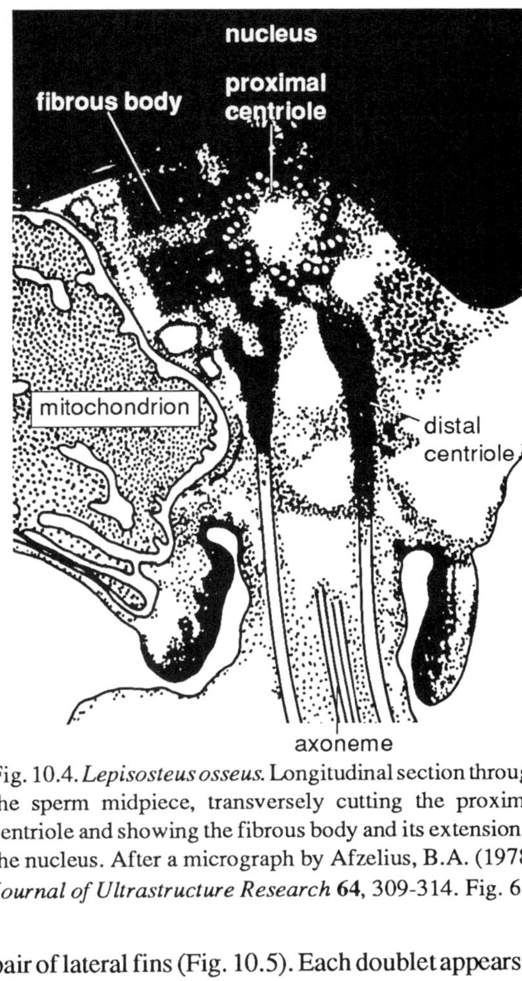

Fig. 10.4. *Lepisosteus osseus.* Longitudinal section through the sperm midpiece, transversely cutting the proximal centriole and showing the fibrous body and its extension to the nucleus. After a micrograph by Afzelius, B.A. (1978). *Journal of Ultrastructure Research* **64**, 309-314. Fig. 6.

tubules and the B tubules appear empty. This densification is due to the presence of an intratubular differentiation (ITD) which has the form of a partition,

Fig. 10.5. *Lepisosteus osseus*. Cross section through the axoneme, showing the pair of lateral fins. After a micrograph by Afzelius, B.A. (1978). *Journal of Ultrastructure Research* **64**, 309-314. Fig. 9.

inside the tubule, in line with the outer arms of the doublet (Mattei, 1988).

DIVISION HALECOSTOMI

Diagnosis. *Amia* (Halecomorphi) and the teleosts. Interopercular bone present; one supratemporal bone on each side; mobile maxillary bone in the cheek; supramaxilla and large myodome present; quadratojugal lost or fused with the quadrate; median neural spines (Patterson and Rosen, 1977; Lauder and Liem, 1983; Nelson, 1984) (Fig. 8.5, 10.1).

Subdivision Halecomorphi

Diagnosis. Symplectic forms an auxillary articulation with the lower jaw (Lauder and Liem, 1983; Nelson, 1984).

Order Amiiformes

Diagnosis. As for the Halecomorphi. Include fossil forms and, in the suborder Amioidei, family Amiidae, the single species *Amia calva*, the Bowfin, in freshwater in eastern N. America (Nelson, 1984).

Optical structure. The anacrosomal aquasperm of *Amia calva* (Amiidae) has been described optically by Retzius (1905). It differs from that of *Lepisosteus* in the spheroidal nucleus but little further comparative information is available.

Subdivision Teleostei

The Teleostei, with an estimated 20,000 species, is by far the most diverse group of the Actinopterygii. Teleosts are distinguished by elongation of the ural neural arches into uroneurals which function to stiffen the upper tail lobe and to support a series of dorsal fin rays. The hypurals have become expanded. Other features indicating the monophyly of teleosts are: unpaired basibranchial toothplates; a mobile premaxilla; internal carotid foramen enclosed in the parasphenoid; seven epurals and a pectoral propterygium fused with the first pectoral fin ray. Living teleosts share many features in the jaw musculature, including loss of the anterior (suborbital) jaw adductor component (Patterson, 1977a,b; Lauder and Liem, 1983).

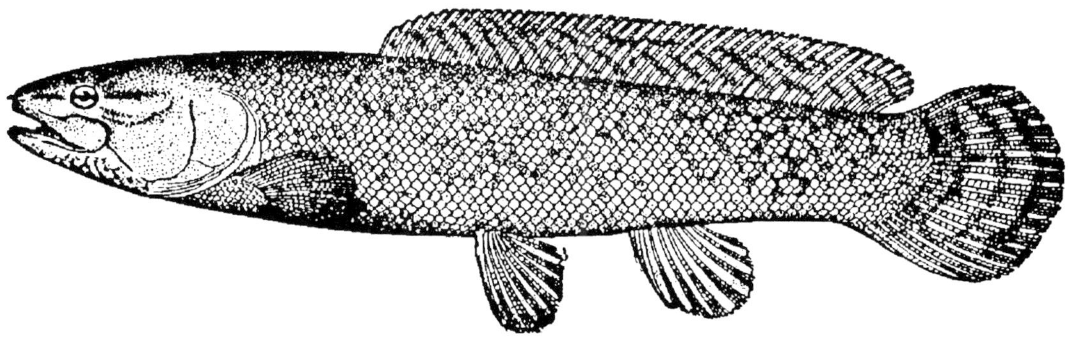

Fig. 10.6. *Amia calva*. After Jordan, D.S. *Fishes*. Henry Holt, New York. Fig. 198.

Superorder Osteoglossomorpha

Order Osteoglossiformes

Diagnosis. Parasphenoid and tongue bones usually with well developed teeth; "shearing bite" between the basihyal teeth and lateral pterygoquadrate teeth; small premaxilla firmly fixed to the skull;

supramaxillae lost; connection of the sternohyoideus to hypobranchial two; intestine passes posteriorly to left of oesophagus and stomach (in most other gnathostomes it passes to right); with a distinctive caudal skeleton in which one or more epurals are fused with neural arches of the caudal verebrae to form "neurepurals"; parapophyses fused with vertebral centra; other, more questionable characters proposed for the Osteoglossomorpha include reticulate scales, 16 or fewer branched caudal rays, and a ventral process at the base of the second gill arch (Gosline, 1960; Greenwood *et al.*, 1966; G. Nelson, 1972; Taverne, 1979; Lauder and Liem, 1983; Nelson, 1984). The Osteoglossomorpha are termed an Infradivision by Nelson (1984).

Relationships. Internal relationships of the Osteoglossomorpha summarized by Lauder and Liem (1983) are shown in Fig. 10.9. With regard to wider relationships, Greenwood (1973) considers that the Osteoglossomorpha and Clupeomorpha share a common ancestry and together form the plesiomorph sister-group of all other teleosts, i.e. of the Elopomorpha and their apomorph sister-group the Euteleostei

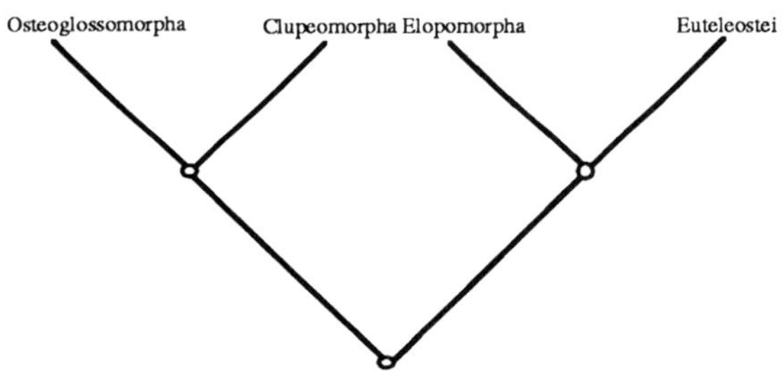

Fig. 10.7. Phylogenetic relationships of the Osteoglossomorpha and the Clupeomorpha according to Greenwood, P.H. (1973). In *Interrelationships of Fishes* (eds. P.H. Greenwood, R.S. Miles & C. Patterson). *Journal of the Linnean Society of London* Zoology 53 (suppl 1), pp. 307-332. Academic Press, New York. Fig. 7.

(Fig. 10.7). The Clupeomorpha are alternatively considered to be the sister-group of the Euteleostei (Fig. 11.1).

Thus Smith (1984), chiefly on the basis of larval morphology, considers the Clupeomorpha to be the plesiomorph sister-group of the Euteleostei; the Elopomorpha to be the plesiomorph sister group of these two; and the Osteoglossomorpha to be the plesiomorph sister-group of all three (Fig. 10.8).

This view accords with that of Lauder and Liem

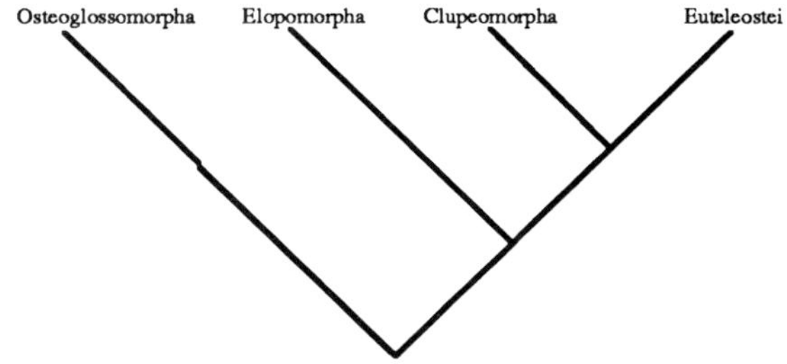

Fig. 10.8. Hypothesis of relationships between major groups of Teleostei. From Smith, D.G. (1984). In *Ontogeny and Systematics of Fishes*. Special Publication Number 1, p. 94-102. American Society of Ichthyologists and Herpetologists. Fig. 53.

(1983) (Fig. 11.1). These workers recognize four major groups of living teleosteans, the Osteoglossomorpha, Elopomorpha, Clupeomorpha and Euteleostei. The oesteoglossomorphs are regarded as the most primitive group of living teleosts (Gosline, 1980; Lauder and Liem, 1983) (but contrast Patterson, 1973, who regards *Elops* as the most primitive living teleost). The other three groups, sometimes termed the elopocephalans, are united by the presence of only two uroneurals.

Evidence that the osteoglossomorphs are monophyletic seems good though not unequivocal. The group is known from the Upper Jurassic and branching from the elopocephalan lineage must have been considerably older than this. Living or fossil osteoglossomorphs are known from every continent except Europe. The osteoglossomorph fish take their name from the "tongue bite" in which the basihyal, covered by a massive toothplate, "bites" against the roof of the mouth which bears large teeth. They are also supposedly united by the presence of paired bony rods or processes at the base of the second gill arch. However, *Heterotis* and *Gymnarchus* lack parasphenoid (roof of mouth) teeth. These teeth may be plesiomorphic for teleosts and therefore not an autapomorphy of osteoglossomorphs. Specific connection of the sternohyoideus to hypobranchial two does appear autapomorphic. A further unifying and distinctive feature of osteoglossomorphs is the fact that the gut passes to the left of the oesophagus and stomach and one ro two pyloric caeca are consistently present. In contrast, in all primitive actinopterygians and in the higher teleosts the anterior part of the intestine passes to the right. Other synapomorphies of the Osteoglossomorpha and internal groups are given

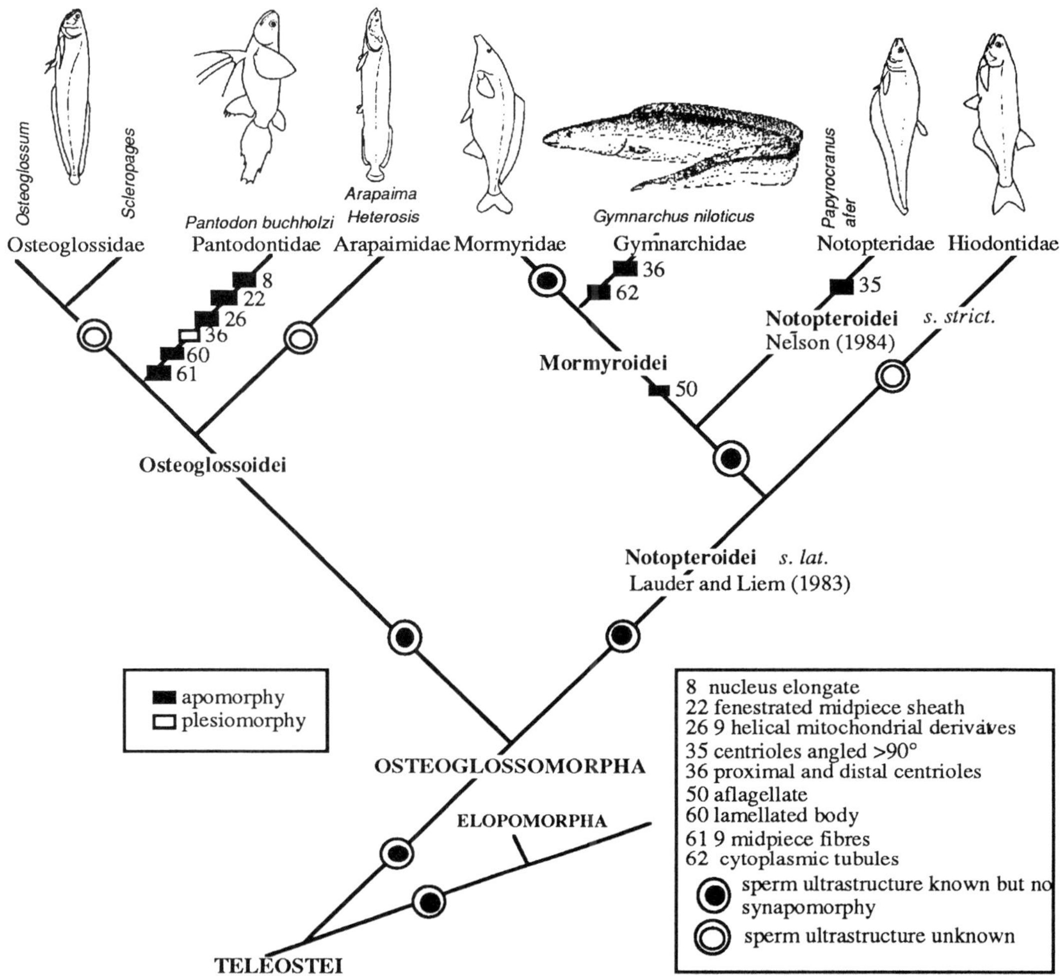

Fig. 10.9. Phylogenetic relationships of the Osteoglossomorpha. Chiefly after Lauder and Liem (1983), with spermatozoal apomorphies superimposed.

by Lauder and Liem (1983). Spermatozoal apomorphies are superimposed on a slight modification of their phylogeny in Fig. 10.9. The phylogeny is modified to separate the Gymnarchidae from the Mormyridae but branching relationships are unchanged.

Internal relationships. Taxonomic interrelationships within Osteoglossomorpha are controversial (see Greenwood, 1973; Lauder and Liem, 1983; Nelson, 1984).

Investigation by Greenwood (1973) of features of the branchial skeleton, hyoid musculature, the inner ear and the nature of the swimbladder-ear connection support monophyly of the Osteoglossomorpha. Specializations of these character complexes, especially those of the inner ear, indicate that the Notopteridae and Mormyridae (including there the gymnarchids) are the closest living relatives of each other (see also Lauder and Liem, 1983, and Fig. 10.9). For example, only these two families have an inner ear in which the utriculus and semicircular canals are physically unconnected with the sacculus and lagena. The Hiodonotidae show few specialized characters and most of these are unique. The few characters known to be shared with other groups suggest a closer relationship of hiodontids with the Notopteridae than with the Mormyridae. It is the swimbladder-ear connection in *Hiodon* and particularly its involvement with the perilymphatic system and the prootic bone which allows comparison with the Clupeomorpha, namely the otophysic connection of the latter (Greenwood, 1973).

Inclusion by Lauder and Liem (1983) of the Mormyridae and Gymnarchidae, with the Notopteridae and Hiodontidae, in the Notopteroidei is controversial despite clear relationship of these four taxa. In contrast, the Osteoglossidae and Pantodontidae, with the Arapaimidae, constitute a well defined lineage, the suborder Osteoglossoidei (Greenwood, 1973; Lauder and Liem, 1983) (Fig. 10.9).

Sperm relationships. To anticipate, spermatozoal ultrastructure of the few osteoglossomorphs in which it has been investigated promises to throw some light on the phylogeny of these fishes. The rare aflagellate condition of sperm seen in Gymnarchidae (*Gymnarchus niloticus*) and in Mormyridae (*Gnathonemus niger*, *G. senegalensis*, *Hyperopisus bebe*, *Mormyrus rume*, *Petrocephalus bovei*) is in agreement with the association of these families in the Mormyroidei (Nelson, 1984) or as sister-groups in a wider Notopteroidei (Lauder and Liem, 1983) (Fig. 10.9). The simple anacrosomal aquasperm of the notopteroid *Papyocranus afer*, the African Knife Fish, appears to be a plesiomorph condition, contrasting with the apomorphic, aflagellate sperm of mormyroids, but does not in itself necessitate exclusion of the mormyroids from the Notopteroidea in which Greenwood (1973) placed them. The peculiar aquasperm of the only other investigated osteoglossoid, *Pantodon bucholzi*, endorses the separate status of the Pantodontidae recognized by (Nelson, 1984) but the spermatozoa of other osteoglossoids have yet to be described ultrastructurally.

Suborder Osteoglossoidei

Diagnosis. Maxilla toothed; no intracranial penetration of swimbladder (Nelson, 1984). Lauder and Liem (1983) unite the Osteoglossoidei on the synapomorphies fusion of hypurals; septum bisecting the eye, extending between the retina and the lens; and articulation between the ventrolateral peg of the parasphenoid and the entopterygoid.

Pantodon buchholzi. The West African *Pantodon buchholzi*, the Butterfly fish (Fig. 10.10), is the only species of the Pantodontidae. Its anacrosomal spermatozoon is 80-85 µm long (van Deurs, 1973, 1974; van Deurs and Lastein, 1973) (Fig. 10.11).

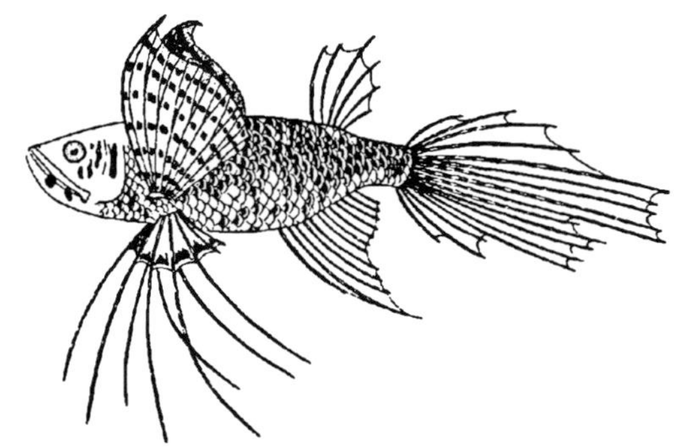

Fig. 10.10. *Pantodon bucholzi*. After *Traité de Zoologie*. (ed. P.-P. Grassé). XIII. Agnathes et Poissons. Masson et Cie, Paris. Fig. 1573.

Nucleus. The nucleus (Fig. 10.11A) is elongated, comprising a cylindrical, electron dense chromatin rod, about 7 µm long and 0.6 µm wide, tapering anteriorly. At maturity the nuclear envelope is not visible. Between the nuclear rod and the cell membrane some granular material is normally seen.

Midpiece. The midpiece (Fig. 10.11C, G) is very elongate, length about 45 µm, and consists of 9 helical mitochondrial derivatives, each describing about 20-25 turns with an inclination of 60-70° and each about 1000 Å thick. The cristae are modified as columnar derivatives. End to end fusion of the original mitochondria to form these derivatives is also seen in snakes, some birds and all mammals.

Helical dense fibres. Nine helical dense fibres (Fig. 10.11C, G) are also present in the midpiece, alternating with the mitochondria in the spermatid but between the mitochondria and the axoneme in the sperm; they are not part of the axonemal complex; each is cross striated, with a major period of 750-900 Å, and they are equated with flagellar rootlets (van Deurs, 1973).

Fenestrated sheath and cytoplasmic canal. Behind, but not as an extension of, the midpiece there is a fenestrated sheath, 6 µm long (Fig.10.11D, 10.11H). This is a 0.5 µm wide swelling of the flagellum containing a membranous structure around

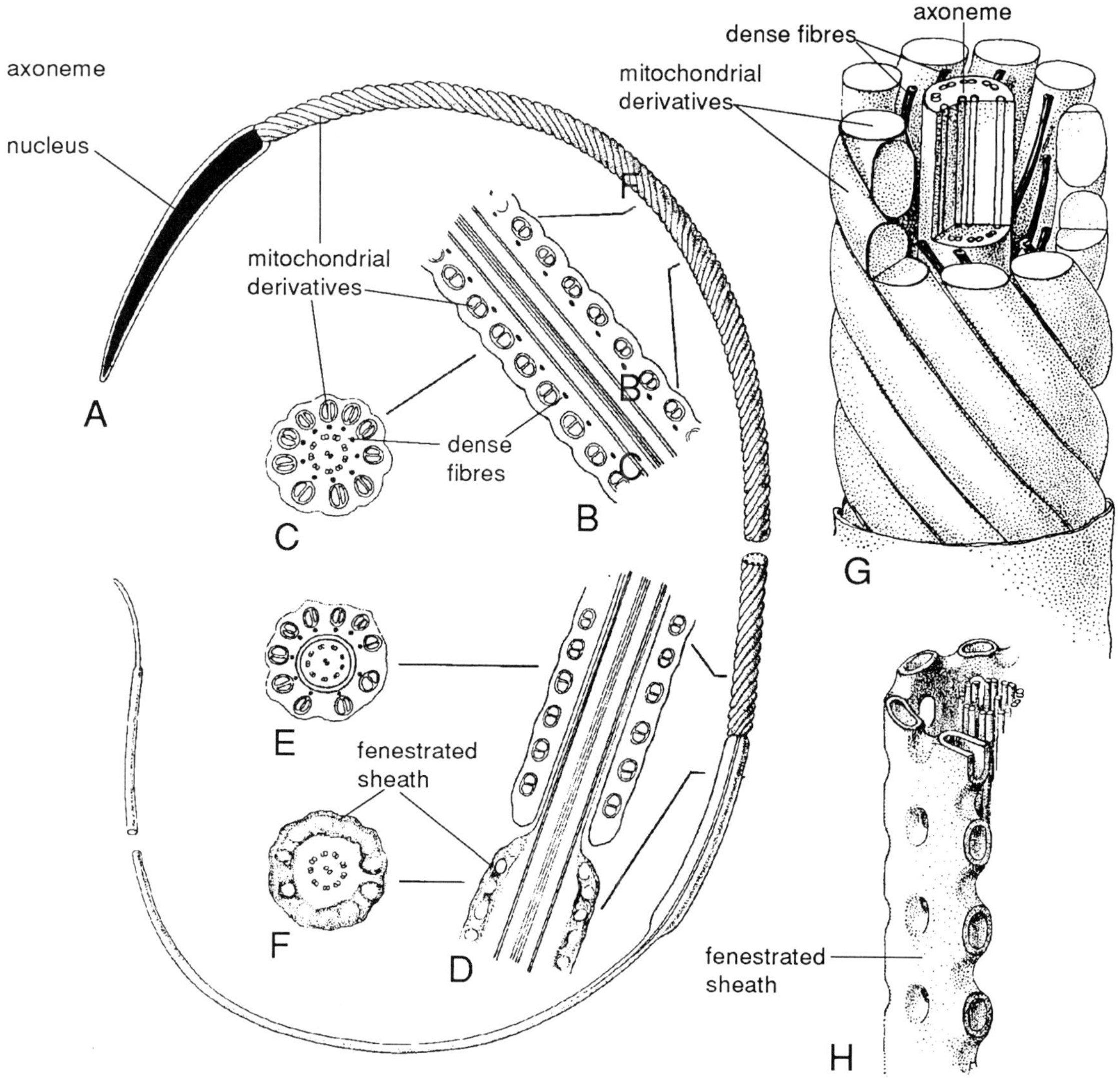

Fig. 10.11. *Pantondon buchholzi*. Schematic model of the spermatozoon. A. The nucleus is elongated. The centriolar complex is covered by the anterior part of a sheath formed by the 9 helical mitochondrial derivatives ("threads"). Behind the midpiece follows the fenestrated sheath region and the free flagellum. B. Longitudinal section of the midpiece anterior to the flagellar canal in which the flagellum runs. Axoneme, dense fibres and mitochondria are seen. C. Cross section corresponding to B. D. Longitudinal section through the junction between the midpiece and the anterior part of the fenestrated sheath region. Through the posterior part of the midpiece, the flagellum runs in a flagellar canal. Leaving the canal, the flagellum swells and a membrane structure is found beneath the flagellar membrane. E. Cross section of the posterior part of the midpiece, showing axoneme, flagellar canal, dense fibres, and mitochondria. F. Cross section of the fenestrated sheath region. G. Schematic three dimensional reconstruction of part of the midpiece, as seen in mature spermatozoon, showing the helical arrangement of the 9 mitochondrial threads and the 9 dense fibres surrounding the flagellum (cristae omitted). H. Schematic three dimensional reconstruction of part of the fenestrated sheath, showing that the sheath is composed of an outer and an inner membrane penetrated by pores. The pore walls connect the outer and inner membranes (plasma membrane omitted). After Deurs, B. van and Lastein, U. (1973). *Journal of Ultrastructure Research* **42**, 517-533. Figs. 1, 28 and 29.

the axoneme. Sections tangential to the surface show a fenestrated plate, with pores 350 Å in diameter and 850 Å apart, centre to centre; the sheath is a double membrane penetrated by the pores. Both the midpiece and the sheath are separated from the axoneme by a space (cytoplasmic canal).

Centriole and fibrous body. The centriolar complex consists of a single centriole (basal body), although an additional centriole, presumably the proximal centriole, is sometimes observed parallel to it. A lamellate body between the nucleus and the basal body is reminiscent of the fibrous body of the sperm of *Lepisosteus* and possibly of that of *Neoceratodus* and (van Deurs and Lastein, 1973) is particularly similar to the "intracentriolar lamellated body" in the spermatozoon of *Poecilia*. The lamellated body is a curved cap, about 0.5 µm wide and 1300-1500 Å thick. It is composed of four layers: (1) an electron dense lamella, 200-300 Å thick; (2) a more electron lucent layer, 450-520 Å thick, striated perpendicularly to the electron dense lamella; (3) a further electron dense lamella, 200-300 Å thick; and (4) a more electron lucent striated layer, 300-400 Å thick. From this layer some dense material appears to connect with the centriole (van Deurs and Lastein, 1973).

Flagellum. About 25 µm of the 9+2 flagellum is free; no fins are described.

***Pantodon* fertilization biology.** From sperm structure the mode of fertilization in *Pantodon* is deduced by van Deurs and Lastein (1973) to be either internal fertilization or deposition of bundles of spermatozoa near the female genital opening or on the eggs. It has been reported that the male places its large anal fin under the female so that the genital duct is in contact with that of the female and that the eggs are fertilized on extrusion (See Breder and Rosen, 1966: 144). The similarity of this form of fertilization to that of the lamprey is noteworthy and the possession in the sperm of both of 9 accessory fibres around the axoneme, although these are of a different nature, is a striking convergence, conceivably for a similar, if unknown, function. Whether the osteoglossomorphs, an ancient group, originally had internal fertilization is a matter of conjecture. The absence of fins is equivocal in this respect.

Suborder Notopteroidei

Diagnosis. Lauder and Liem (1983) include notopterids, mormyrids (with gymnarchids) and hiodontids in the Notopteroidei, typified by two synapomorphies: the "osteoglossomorph type" of otophysic connection; and ventral throat musculature containing an anteroposteriorly orientated intermandibularis posterior muscle that is not fused with the interhyoideus. Nelson (1984) retains a suborder Mormyroidei, separate from the Notopteroidei. Notwithstanding these nomenclatural differences, the relationship of mormyrids to notopterids recognized by Lauder and Liem appears acceptable although the sister-group relationship of mormyrids and gymnarchids (the "mormyroids") is not contested. The Notopteroidei (Mormyridae *s. lat*., Notopteridae and Hiodontidae) are regarded as the sister-group of the Osteoglossoidei by Lauder and Liem

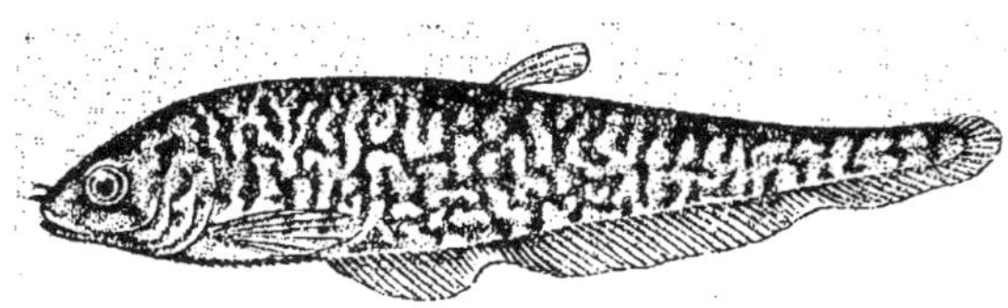

Fig. 10.12. *Papyocranus afer*. After Sterba, G. (1962). *Freshwater Fishes of the World*. Vista Books, London. Fig. 53.

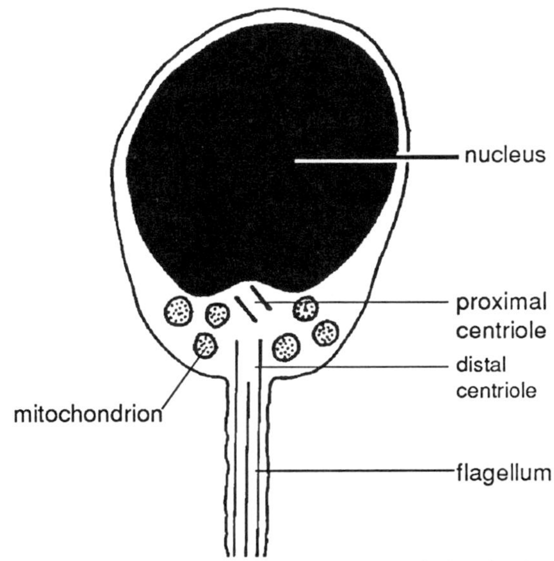

Fig. 10.13. *Papyocranus afer*. Diagrammatic longitudinal section of the spermatozoon. After Mattei, X. (1970). In *Comparative Spermatology*. (ed. B. Baccetti, pp 59-69). Academic Press, New York. Fig. 4: 21.

(1983).

Notopteridae

Diagnosis. Maxilla toothed; anterior prongs of the swimbladder lateral to the skull or intracranial.

Notopterids are elongate and laterally flattened nocturnal fish which propel themselves by undulations of the long anal fin and are capable of breathing air. *Notopterus* occurs in India and Southeast Asia; the closely related *Papyocranus* (Fig.10.12) and *Xenomystus* are tropical African (Lauder and Liem, 1983; Nelson, 1984). Only the monotypic *Papyocranus* has been examined for sperm ultrastructure.

Papyocranus afer. The freshwater African Knife fish, *Papyocranus* (=*Notopterus*)*afer*, has an anacrosomal aquasperm (Mattei, 1970) (Fig. 10.13).

The proximal centriole is partly contained in the small nuclear fossa and is in line with but tilted relative to the distal centriole and flagellum. There are several irregularly arranged small mitochondria. No cytoplasmic canal is indicated.

Hiodontidae

Diagnosis. The Hiodontidae, Mooneyes, are included in the Notopteroidei *s. lat.* of Lauder and Liem (1983) and *s. strict.* of Nelson (1984), the latter author excluding the mormyroids. They also have an elongated anal fin but this is only moderately long and is not confluent with the caudal fin. They are the only osteoglossomorphs from North American waters (Lauder and Liem, 1983; Nelson, 1984). Hiodontids have have not been investigated for sperm ultrastructure.

Although recognizing here the Notopteroidei *s. lat.* as the sister group of the Osteoglossoidei, we will distinguish a suborder Mormyroidei, in accordance with Nelson (1984) for the Mormyridae and Gymnarchidae while accepting in the phylogram (Fig. 10.9) that cladistically mormyroids have a lower hierarchical position than the notopteroids. They might alternatively be regarded as a superfamily, the Mormyroidea.

Suborder Mormyroidei

Diagnosis. Maxilla toothless. The most outstanding features of the constituent families Mormyridae and Gymnarchidae are the electrogenic organs, derived from caudal muscles, and the greatly enlarged cerebellum. The huge cerebellum results in a brain size relative to body weight rivalling that of humans. The Mormyridae, elephant fish, and the Gymnarchidae, inhabit freshwaters in tropical Africa and the Nile. Mormyrids, unlike the other osteoglossiform families, which have very few species, have undergone an evolutionary radiation which has produced over 300 extant species (Lauder and Liem, 1983; Nelson, 1984).

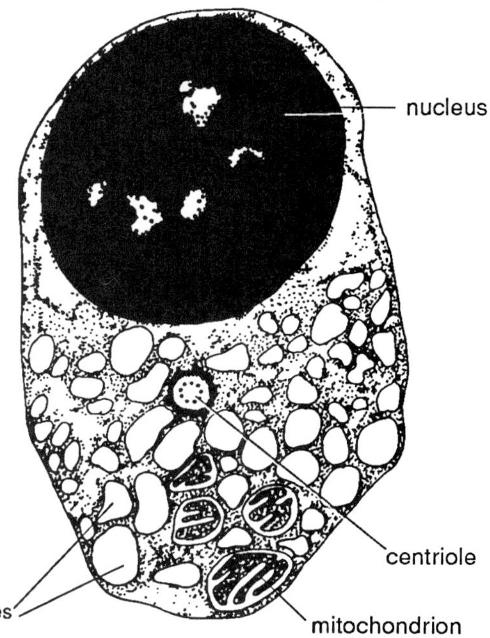

Fig. 10.14. *Petrocephalus bovei.* Longitudinal section of spermatozoon. After a micrograph of Mattei *et al.* (1972). *Journal de Microscopie*(Paris) **15**, 67-78.

Mormyridae. Sperm literature. The five examined species of Mormyridae *Gnathonemus niger*, *G. senegalensis*, *Hyperopisus bebe*, *Mormyrus rume* and *Petrocephalus bovei* (Fig. 10.14) have relatively uniform aflagellate sperm (Mattei *et al.*, 1972).

General ultrastructure. The nucleus occupies one pole of the cell, the centrioles are situated between it and the mitochondria which are grouped at the opposite pole, as in flagellate teleost sperm. These mormyrid sperm are intermediate between the typical teleost anacrosomal aquasperm and the more modified sperm of *Gymnarchus* (below).

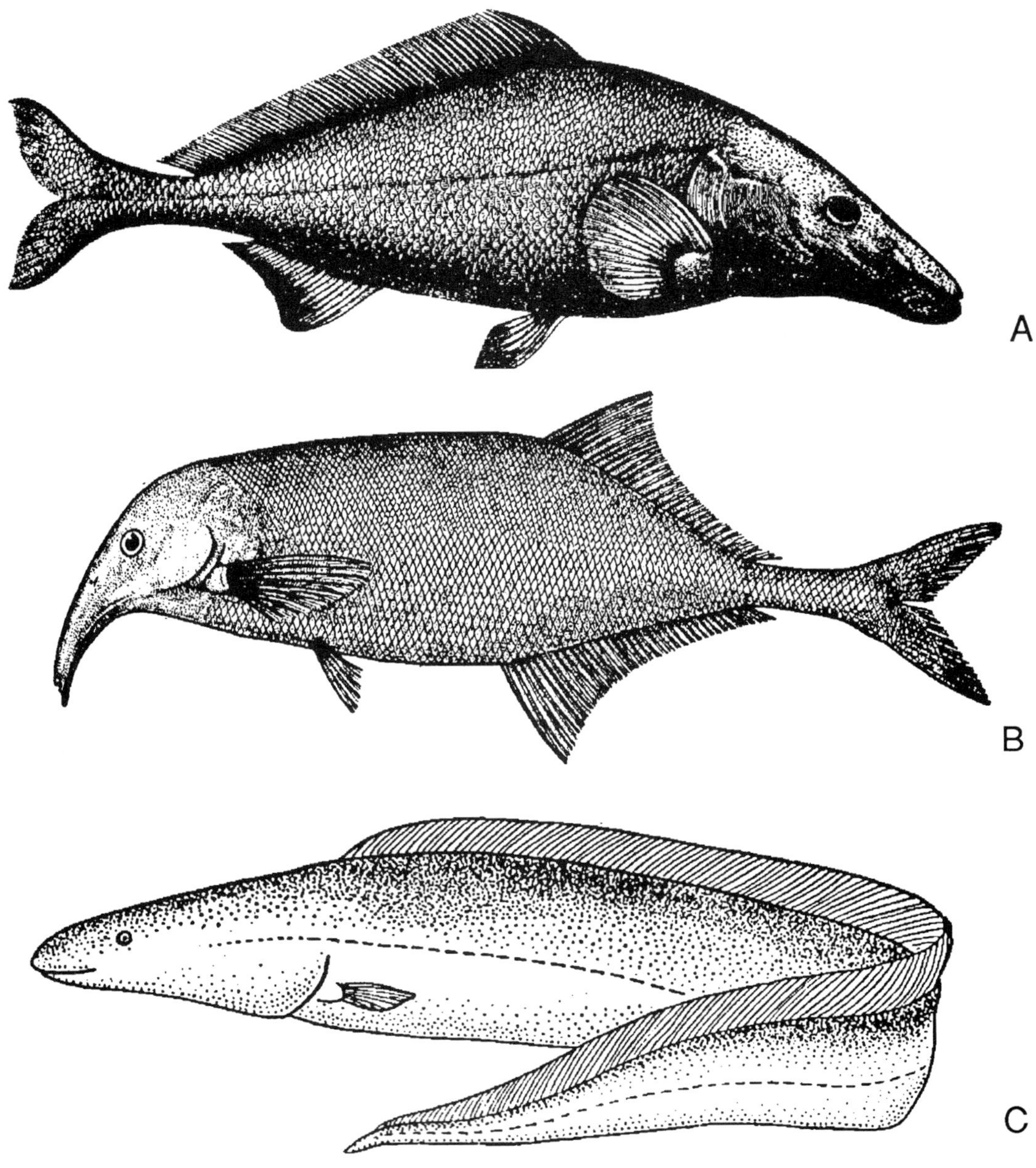

Fig. 10.15 . Mormyroids. A. *Mormyrops*. B. *Gnathonemus*. C. *Gymnarchus*. From Grassé, P.-P. (ed.) (1958). *Traite de Zoologie*. XIII. Agnathes et poissons. Masson et Cie, Paris. Fig. 1576.

Gymnarchidae. *Gymnarchus niloticus*. The sperm of *G. niloticus*, the only species of its genus, (Fig. 10.16) has a very polymorphic nucleus with a double envelope with some nuclear pores; the internal membrane is adherent to the chromatin. The cell membrane is thick and osmiophilic, lined internally by a regular arrangement of "subcuticular" 25-30 nm thick tubular fibrils. There is abundant cytoplasm consisting mostly of sinuous tubules which in some sections have the appearance of small vesicles. In the vicinity of the nucleus there are some globular mitochondria, 1 µm wide, lipid masses and myelin ves-

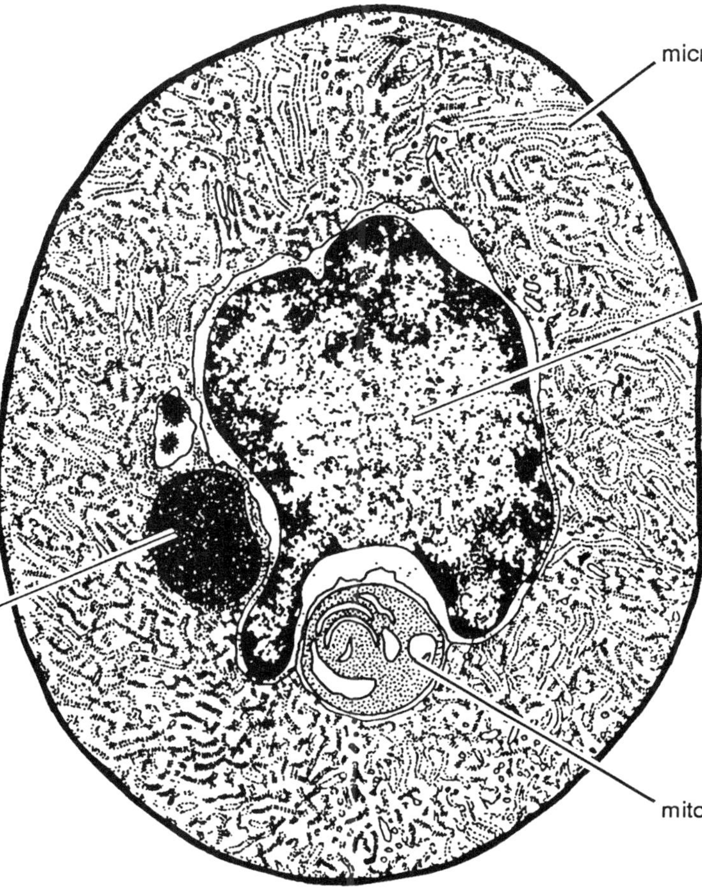

Fig. 10.16. *Gymnarchus niloticus*. Section of a spermatozoon showing the cytoplasm containing abundant microtubules and the quiescent, polymorphic nucleus pierced in places by nuclear pores. Drawn from a micrograph by Mattei*et al.* (1967a). *Comptes Rendus Hebdomadaires des Séances de l'Académie des Sciences* D **265**, 2010-2012. Pl. II, Fig. 4.

icles. The two centrioles lie parallel at an equal distance from the nucleus (Mattei *et al.*, 1967a).

Mormyroid fertilization biology. The mode of fertilization in mormyroid fishes is unknown but their lack of copulatory organs strongly suggests that fertilization is external, as is it known to be in *Notopterus* and *Papyocranus*. The correlation of aflagellarity and suspected external fertilization in mormyroids is intriguing. Amoeboid movement has been suggested as a means of motility for these aflagellate sperm during fertilization (Mattei, 1970, 1988). Relationships indicated from sperm ultrastructure are discussed above.

Chapter 11

SUPERORDERS ELOPOMORPHA AND CLUPEOMORPHA

ELOPOMORPHA

Diagnosis and relationships. The Elopomorpha are unified by three characters: presence of rostral and prenasal ossicles; initial fusion of the angular and retroarticular bones in the lower jaw; and the presence of a leptocephalus larva (Lauder and Liem, 1983; Smith, 1984). A phylogeny with sperm apomorphies superimposed is shown in Fig. 11.1.

Greenwood (1973) considered the Elopomorpha to be the plesiomorph sister-group of the Euteleostei (Fig. 10.7) whereas Lauder and Liem (1983) (Fig. 11.1) and Smith (1984) (Fig. 10.8) saw elopomorphs as the plesiomorph sister-group of a clupeomorph-euteleost assemblage. Division of elopomorphs into three orders, the Elopiformes, Anguilliformes and Notacanthiformes, is arbitrary and probably not sustainable cladistically. An alternative phylogeny in which the Megalopidae (*Megalops* and *Tarpon*) are elevated to ordinal rank as the Megalopiformes while the Notacanthiformes are subsumed in the Anguilliformes is given by Smith (1984). This trichotomy indicates failure to find synapomorphies which would indicate a sister-group relationsip of any of these three entities to one of the other two. Association of the Notacanthiformes with the Anguilliformes is, however, in agreement with Greenwood *et al.* (1966). Lauder and Liem (1983) also placed the Albuloidei (including notacanthids) with the Anguillodea in the Anguilliformes. They also consider the Elopidae, Megalopidae and Anguilliformes to form an unresolved trichotomy.

Sperm literature. Some 19 species of the Elopoidei, Albuloidei and Anguilloidei have now been investigated for sperm ultrastructure (Order Elopiformes: Suborder Elopoidei; Elopidae - *Elops lacerta*, Mattei and Mattei, 1974; Suborder Albuloidei; Albulidae - *Albula vulpes*, Mattei and Mattei, 1972, 1973, 1974; *Pterothrissus belloci*, Mattei and Mattei, 1974. Order Anguilliformes: Suborder Anguilloidei; Anguillidae - *Anguilla anguilla*, Baccetti *et al.*, 1981; Ginzburg and Billard, 1972; Gibbons *et al.* 1983; *A. australis schmidtii*, Todd, 1976; *A. dieffenbachii*, Todd, 1976; *A. japonica*, Colak and Yamamoto 1974a, b; Congridae - *Congermuraena bertini*, *Cynoponticus ferox*, *Paraconger notialis*; Ophichthyidae - *Echelus myrus*, *Ophichthus ophis*; *Pisodonophis semicinctus*, *Mystriophis rostellatus*, undetermined Congridae, Mattei and Mattei , 1974; Muraenidae - *Lycodontis afer*, *Muraena helena*, *Muraena robusta*; Heterenchelydae - *Pythonichthys microphthalmus*, Mattei, 1970, Mattei and Mattei 1974). In addition Mattei (1988) has briefly referred to the ultrastructure of the spermatozoon of unnamed notacanthiforms. The great value of spermatology for taxonomy and phylogeny is strikingly demonstrated by this assemblage.

Elopomorph sperm autapomorphies. Sperm ultrastructure in Elopomorpha shows unique features and, with some exceptions in that of the Muraenidae, is remarkably uniform (Fig. 11.2).

Two features constitute spermatozoal synapomorphies linking the Elopiformes, Anguilliformes and Notacanthiformes and therefore comprise additional autapomorphies of the Elopomorpha. They are a 9+0 flagellum, a constant feature, and division of the proximal centriole into two elongate bundles of 4 and 5 triplets, which may extend as a free pseudoflagellum. A third feature is the striated centriolar rootlet, seen in all but the Muraenidae.

General elopomorph sperm ultrastructure. Because of the relative uniformity of the elopomorph sperm, a general description, embracing all investigated species but indicating the chief variations will be given here. It is largely derived from the account of Mattei and Mattei (1974).

Nucleus. The Muraenidae are exceptional in having a rounded nucleus (Fig. 11.2A), as in most

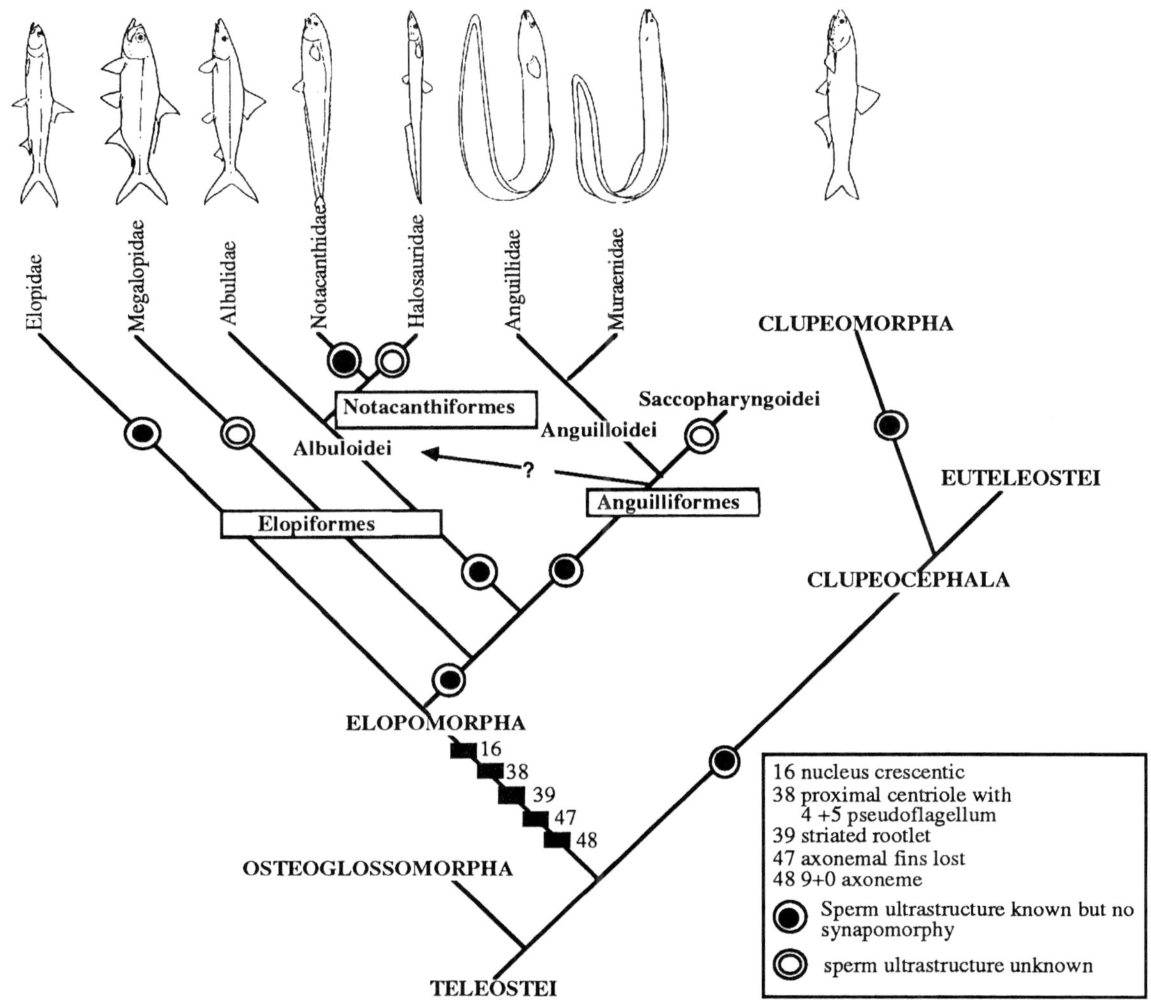

Fig. 11.1. Phylogeny of the Elopomorpha. Spermatozoal apomorphies have been superimposed on a phylogeny modified from Lauder and Liem (1983). *Bulletin of the Museum of Comparative Zoology* **150**, 95-197. Fig. 24.

teleosts, with the distal area slightly depressed. The other elopomorphs investigated have an elongate nucleus of variable shape and size (6-20 µm long). The shape is gently curved with a hook-shaped "lateral"(here termed anterior) end which is directed either anteriorly, as in *Congermuraena* (Fig. 11.2D), or posteriorly, as in *Albula vulpes* (Fig. 11.2E) (Mattei and Mattei, 1974). A length of 6.3 µm in *Anguilla japonica* accords (Çolak and Yamomoto, 1974b). As the "lateral" limit of the nucleus is furthest from the basal body (distal centriole) it is properly termed the anterior end by Todd (1976), although the head is carried at right angles to the flagellum.

Except for *Elops* and the albulid *Pterothrissus* (Fig. 11.2B), the nucleus shows a perceptibly rounded mid-region and two edges of smaller section. It has a concavity in which the base of the proximal centriole is located. At its lateral end the nucleus terminates with a slender or depressed area.

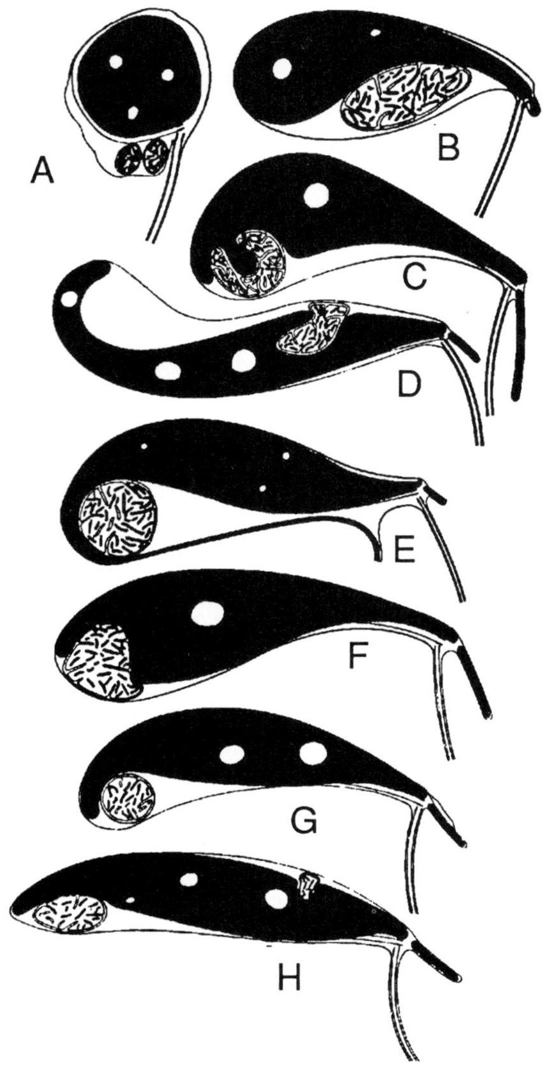

Fig. 11.2. Diagrams of various spermatozoa of Elopomorpha. A. Muraenidae. B. *Pterothrissus belloci*. C. Undetermined congrid. D. *Congermuaena bertini*. E. *Albula vulpes*. F. *Pythonichthys microphthalmus*. G. *Paraconger notialis*. H. *Pisodonophis semicinctus*. From Mattei, C. and Mattei, X. (1974). In *The Functional Anatomy of the Spermatozoon* (Ed B.A. Afzelius), pp 211-221. Pergamon Press, Oxford. Fig. 1.

Occurrence of an implantation fossa is denied for *Anguilla japonica* (Çolak and Yamomoto, 1974b). In *Elops* and *Pterothrissus*, the centriolar and mid-region of the nucleus are attenuated, whereas the lateral end is rounded. In all species the nuclear material is not compact but has areas of clear nucleoplasm. There is usually a depression in the nucleus in which the mitochondria are situated, the exceptions being *Anguilla anguilla*, *A. japonica* and *Muraena*.

Mitochondria. In *Muraena*, the mitochondria remain in the centriolar area, In the other species, excepting *Anguilla anguilla* and *A. japonica* (see below), the mitochondria are located at well defined sites along the nucleus (Mattei and Mattei, 1974) (Fig. 11.2A-H). In *Anguilla anguilla* (Fig. 11.6) there are several mitochondria throughout the length of the nucleus (Gibbons *et al.*, 1983). In this species the mitochondria are said to lie in the bowl of a nucleus shaped like a long narrow spoon but illustrated transverse sections (Fig. 11.6) suggest that the mitochondria are at the edge of the nucleus. A subterminal location of a single mitochondrion is illustrated by Billard and Ginsburg (1973) for *A. anguilla*. It seems possible that this was a different species from that similarly identified by Gibbons *et al.* (1983, above). In contrast, in the Japaneses Eel, *Anguilla japonica*, a posterior sleeve of cytoplasm surrounding the tail is said to contain the mitochondrion (Çolak and Yamomoto, 1974a) or mitochondria (Çolak and Yamomoto, 1974b), the usual teleostean condition. *Anguilla australis* and *A. dieffenbachii* differ from these in location of the mitochondrion in the terminal ("lateral") hook of the nucleus.

In *Paraconger*, *Mystriophis*, *Pythonichthys* (Fig. 11.2F) and *Echelus*, the mitochondrion reaches the anterior extremity or tip (" lateral edge") of the nucleus; in *Albula* (Fig. 11.2E), it is located in the curve of the tip of the nucleus, much as *Anguilla australis* and *A. dieffenbachii* (above). In the undetermined congrid (Fig. 11.2C) the nucleus presents a tubercle very close to the tip on which the mitochondrion is anchored. In other cases the mitochondrion has not migrated to the end of the nucleus but has stopped at its mid-region. It may rest against the nucleus, as in *Pterothrissus* (Fig. 11.2B), or penetrate into a recess in the nucleus as in *Congermuraena* (Fig.11.2D). In *Pisodonophis* (Fig. 11.2H), two mitochondria are present, one at the lateral edge of the nucleus at the level of the rear face of the sperm, the other in a recess of the mid-region of the nucleus at the anterior face (Mattei and Mattei, 1974). The arrangement in the Notacanthiformes is described below under that

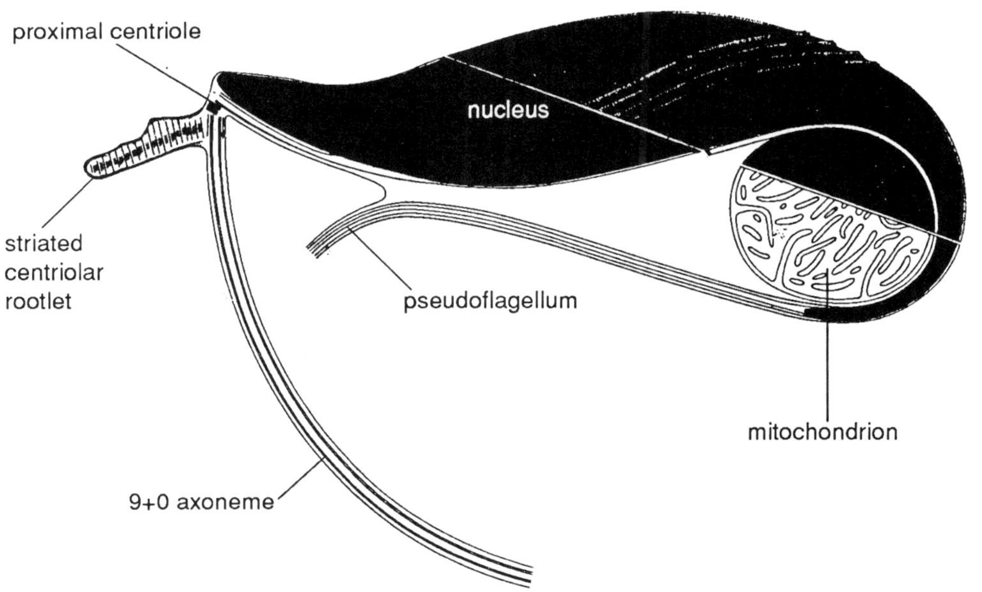

Fig. 11.3. *Albula vulpes.* Exemplifying full development of elopiform spermatozoal characteristics. After Mattei, C. and Mattei, X. (1973). *Zeitschrift für Zellforschung und Mikroskopische Anatomie* **142**, 171-192. Fig. 11.

order.

Proximal centriole and pseudoflagellum. The well developed proximal centriole persists. In all cases a diverticulum of the nuclear membrane penetrates into the centriolar cylinder during late spermiogenesis and provokes its dissociation into two bundles, one of five triplets, the other of four (Fig. 11.4) (Mattei and Mattei, 1973, 1974; Todd, 1976). This dissociation is not described by Çolak and Yamamoto (1974a,b) for *Anguilla japonica* but may be suspected in view of its occurrence in *A. anguilla*, *A. australis* and *A. dieffenbachii.*

Except for the Muraenidae, in which the structure has not been analyzed further, the four triplet bundle is transformed into doublets A-C by an opening of the B subtubule (Mattei and Mattei, 1973, 1974; Todd, 1976) (Fig. 11.5). The fate of the five triplet bundle varies with the species and even the individual.

In *Albula vulpes* (Mattei and Mattei, 1973, 1974) (Fig. 11.3, 11.5) and supposedly in *Anguilla anguilla* (Ginzburg and Billard, 1973) the five triplets retain their structure as far as the tip of the nucleus; beyond it the components may appear as doublets A-C. (However, Gibbons *et al.*, 1983, indicate that in *Anguilla anguilla* A-C doublets coexist with triplets on the nucleus). In the other species, the transformation of the five triplet bundle is achieved in the mid or distal part of the nucleus. In the undetermined congrid, the five triplets yield five doublets A-C; in *Paraconger*, *Mystriophis*, *Pisodonophis*, *Pythonichthys*, *Ophichthus* and *Elops* four triplets are altered into doublets A-C, whereas the fifth one, through loss of the C subtubule, transforms into doublet A-B, the latter being the one nearest the four triplet bundle. In *Pythonichthys* the transformation of the triplet into a doublet A-B is achieved where the proximal centriole dissociates into two bundles.

At the tip ("lateral edge") of the nucleus the two bundles tend to join. In *Albula vulpes*, both membranes of the nuclear envelope extend as a curved edge of the nucleus; the centriolar components are distributed around this membranous axis and this formation may project from the cell as a "pseudoflagellum". The free pseudoflagellum may include doublets A-C or doublets A-C and triplets (Mattei and Mattei, 1974); in *Anguilla anguilla* it contains only 9 doublets. Dynein arms are absent (Gibbons *et al.*, 1983). In *Elops*, *Cynoponticus* and *Albula* the pseudoflagellum is conspicuous. In *Pisodonophis* some elements of the proximal centriole form an intracellular cylinder surrounding what seems to be an exten-

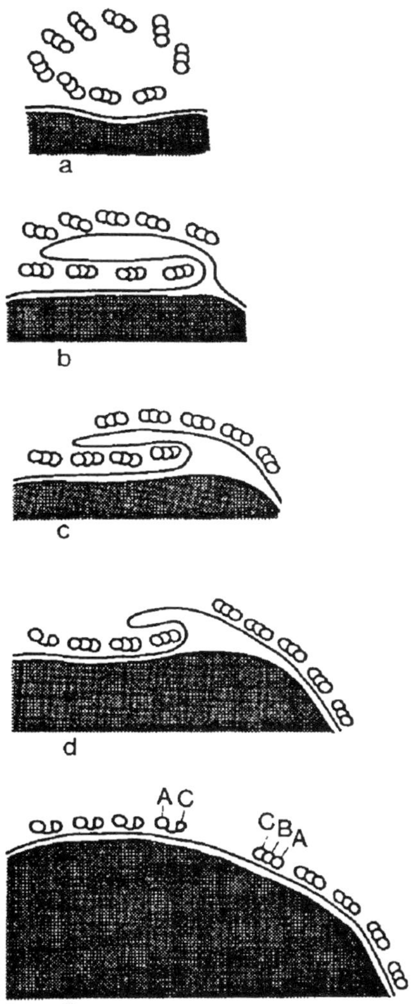

Fig. 11.4. *Albula vulpes.* Dislocation of the proximal sperm centriole by penetration of a diverticulum of the nuclear membrane. A,B and C are subtubules of the centriolar triplets. From Mattei, C. and Mattei, X. (1973). *Zeitschrift für Zellforschung und Mikroskopische Anatomie* **142**, 171-192. Fig. 19.

sion of the outer membrane of the nuclear envelope (Mattei and Mattei, 1974).

The function of the pseudoflagellum may be to stiffen and stabilize the orientation of the large head as the sperm are propelled by the true flagellum. The function of the striated rootlet is uncertain (Gibbons *et al.*, 1983). Alternatively, Billard and Ginsburg (1973) suggest that the striated centriolar rootlet (described below), and the extension of the proximal

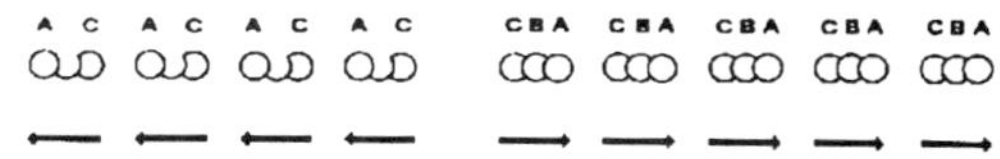

Fig. 11.5. *Albula vulpes.* Constitution of the extension of the proximal centriole in the mid-region of the nucleus. Subtubules A,B, and C are indicated on each of the elements (basically triplets) of the centriole. The arrows indicate the orientation of each of these. From Mattei, C. and Mattei, X. (1973). *Zeitschrift für Zellforschung und Mikroskopische Anatomie* **142**, 171-192. Fig. 25.

centriole, exist as temporary structures playing a consolidating and stabilizing role in joining the centriolar-flagellar complex to the sperm head during elongation. These alternative hypotheses requires evaluation from additional studies.

Distal centriole. This has a classical triplet structure and emits a motile flagellum though of the 9+0 type. Microtubules, abundant in spermatids, persist in *Congermuraena* and in the case of *Pterothrissus* make up a prominent bundle along the nucleus.

Centriolar rootlet. A rootlet exists in all examined elopomorph sperm except the Muraenidae. It is shorter than 1 µm in *Elops* and *Pterothrissus*, approximately 1 µm in *Albula*, 1-1.8 µm in *Anguilla australis*, 1.6-2.0 µm in *Anguilla dieffenbachii*, 1-2 µm in *Congermuraena*, *Paraconger* and *Ophichthus*, 2-4 µm in *Mytriophis*, *Pisodonophis* and *Pythonichthys* and longer than 4 µm in the undetermined congrid (Mattei and Mattei, 1974; Todd, 1976). It shows conspicuous periodicity (160-200 Å in *Anguilla australis* and *A. dieffenbachii*, Todd, 1976) and is supported by a central axis (Mattei and Mattei, 1974). An axis is supposedly absent in *Albula* and perhaps *Pterothrissus* but is illustrated for the former by Mattei and Mattei (1973) (Fig. 11.3).

Gibbons *et al.* (1983) suggest, at least for *A. anguilla*, that projection of the rootlet beyond the cell body may be a shrinkage artefact during fixation as it is almost never seen by light microscopy of unfixed sperm. It seems unlikely, however, from Fig. 11.2 that there is sufficient cytoplasm around the rootlet to obscure it *in vivo* and, whether exposed or not, the rootlet remains a striking feaure of all non-muraeinid elopomorph sperm.

Axoneme. The doublets of the 9+0 true axoneme run close to the plasma membrane (Mattei and Mattei, 1983; Gibbons *et al.*, 1983), except proximally where Y links clearly span the gap between tubules and membrane. The axoneme is about 35 µm long in *Anguilla anguilla*, in which it has a flexible terminal piece (Gibbons *et al.*, 1983).

The elopomorph sperm axoneme, and specifically that of *Anguilla anguilla*, lacks the central-sheath and -tubules (central singlets), the radial spokes and the outer dynein arms (Baccetti *et al.*, 1981; Gibbons *et al.*, 1983), though a short outer arm is present on doublets in the extreme proximal part of the axoneme (Baccetti *et al.*, 1981). Mattei and Mattei (1973) state that both arms are present on each doublet but illustrate only the inner one. Electrophoretic deficiencies in dynein bands have been demonstrated in various animal groups for such axonemes with doublets lacking an arm (Baccetti *et al.*, 1979a).

Axonemal deficiencies and motility. The structural and molecular deficiencies of the elopomorph centriolar and axonemal apparatus have not impaired motility; indeed, the frequency of beat of the eel flagellum, at 95Hz, is the highest known in eukaryotic flagella. This rapidity is tentatively ascribed to absence of a regulatory response to Ca^{2+} such as occurs in most cilia and flagella with a complete 9+2 structure (Gibbons *et al.*, 1983).

The sperm progress forwards at 140 µm per sec,

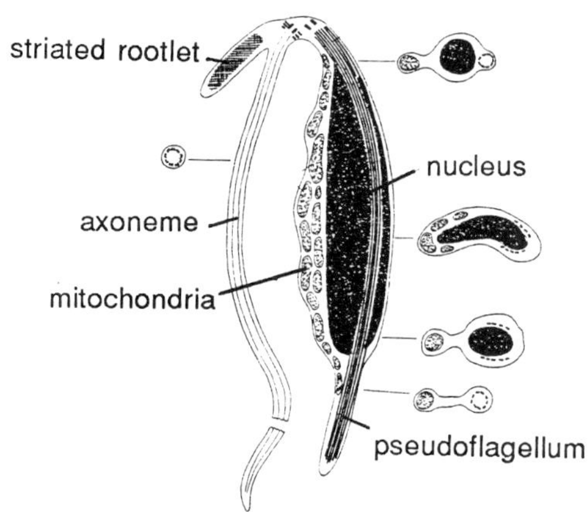

Fig. 11.6. *Anguilla anguilla*. Structure of spermatozoon, shown diagrammatically. After Gibbons *et al.* (1983). *Journal of Submicroscopic Cytology* **15**, 15-20. Fig. 10.

at about 21° C, as a result of a left-handed helical wave propagated distally along the flagellum at a true beat frequency of about 95Hz. Simultaneously the sperm rolls at a frequency of approximately 18 Hz. Thus these three dimensional waves, which have a propulsive efficiency of about 1.5 µm/beat are significantly less effective than planar waves, which typically have efficiencies of 4.5 µm/beat (Gibbon *et al.*, 1983).

Genetic control. In contrast with the sperm tails, ependymal cilia of *Anguilla* have a normal 9+2 axoneme with two dynein arms on each doublet. This indicates that the structure of ciliary and flagellar axonemes in *Anguilla* is controlled by at least partly different groups of genes (Baccetti *et al.*, 1981) both in terms of numbers of central singlets and representation of dynein on the doublets. In man 9 + 0 sperm flagella may, similarly, coexist with 9+2 cilia in the same individual (Baccetti *et al.*, 1979b), indicating control of microtubule numbers by at least partly different genomes, but, in contrast, the "armless" syndrome affects both cilia and flagella and is thought to indicate a single genetic control of human dynein (Afzelius *et al.*, 1975).

Order Elopiformes

Diagnosis. Pelvic fins abdominal; body slender, usually compressed; gill openings wide; caudal fin deeply forked. With leptocephalus larva. Marine with some extension into brackish and freshwaters (Nelson, 1984). The phylogeny in Fig. 11.1 observes the classification of the Elopomorpha into three orders by Nelson (1984) but paraphyly of the Elopiformes is indicated and the orders are not of equivalent rank.

Suborder Elopoidei

Diagnosis. Among other features, the gular plate is well developed (Nelson, 1984). The elopoids here include the Elopidae (ten pounders) and the Megalopidae (tarpons), the latter family placed in a separate order by Smith (1984).

Family Elopidae. *Elops*. The sperm of *Elops lacerta*, a species regarded by Patterson (1973, p. 235), as the most archaic of the teleosts (adult, Fig. 11.7), has been briefly referred to by Mattei and Mattei (1974) (see general account, above).

In *Elops*, as in the albulid *Pterothrissus*, the centriolar and mid-region of the nucleus are attenu-

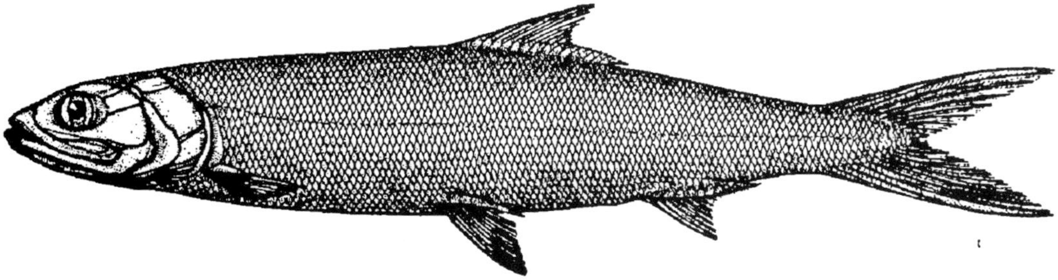

Fig. 11.7. *Elops lacerta*. After Grassé, P.-P. (1958) (ed.). *Traité de Zoologie*. XIII Agnathes et Poissons. Masson et Cie, Paris. Fig. 1565.

ated whereas the "lateral" (anterior) end is rounded. The two genera agree also in having the shortest striated rootlet at less than 1 μm.

Suborder Albuloidei

Diagnosis. Among other features, the gular plate is reduced to a thin median splint or is absent. The Bone Fish or Queen Fish and relatives from tropical seas and off Japan (Nelson, 1984).

Family Albulidae. Two albulid species have been investigated (*Albula vulpes*, Mattei and Mattei 1972, 1973, 1974) (Fig. 11.8); and *Pterothrissus belloci*, Mattei and Mattei 1974). With small variations, their sperm conform to the general elopmorph pattern described above.

Albula. This pattern finds full expression in *Albula* in which a striated rootlet is developed (albeit modestly) and (as in *Elops* and *Cynoponticus*) the pseudoflagellum is well developed and migration of the mitochondrion towards the tip (lateral edge, *sensu* Mattei and Mattei, 1974) of the nucleus is pronounced; the tip of the nucleus is unusual in *Albula* in its strong posterior hook. *Albula* may be unique in the investigated elopomorphs in persistence to the end of the nucleus of five intact triplets in one of the extensions of the proximal centriole (Mattei and Mattei 1972, 1973, 1974).

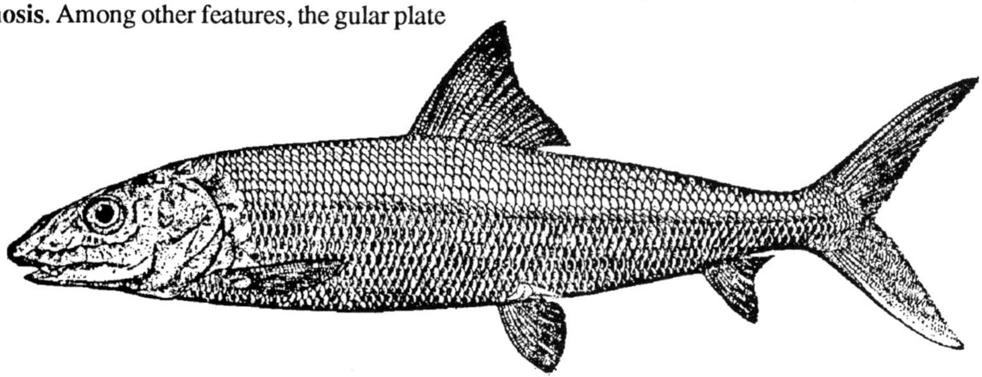

Fig. 11.8. *Albula vulpes*. After Jordan, D.S. (1907). *Fishes*. Henry Holt, New York. Fig. 203.

Pterothrissus (Fig. 11.2B) agrees with *Albula* in the modest development of the striated rootlet. In *Pterothrissus*, as in *Elops*, the centriolar and midregion of the nucleus are attenuated, whereas the lateral end is rounded. Persistence of microtubules from the spermatid as a prominent bundle along the nucleus is peculiar to *Pterothrissus* (Mattei and Mattei, 1974).

Order Anguilliformes

Diagnosis. Pelvic fins absent in living forms. Marine and freshwater eels. With leptocephalus larva. A single suborder Anguilloidei (Nelson,

1984).

Family Anguillidae. ***Anguilla*.** Sperm ultrastructure has been examined in four species of the freshwater eel genus *Anguilla* (*A. anguilla*, Baccetti *et al.*, 1979a, 1981; Billard and Ginsburg, 1973; Ginzburg and Billard, 1972 (Fig. 11.9); Gibbons *et al.*, 1983 (Fig. 11.6); *A. australis schmidtii* (Fig. 11.10) and *A. dieffenbachii*, Todd, 1976; *A. japonica*, Colak and Yamamoto 1974a, b). Todd (1976) summarizes dimensions for the spermatozoa of *Anguilla australis*, *dieffenbachii*, *anguilla* and *japonica* as follows: head length 6, 8, 8-11, and 6.3 µm; head width 2, 3, -, 1 µm; flagellum length 26-30, 28-44, 24-36, and 30.5 µm. Other data are given in the general account above.

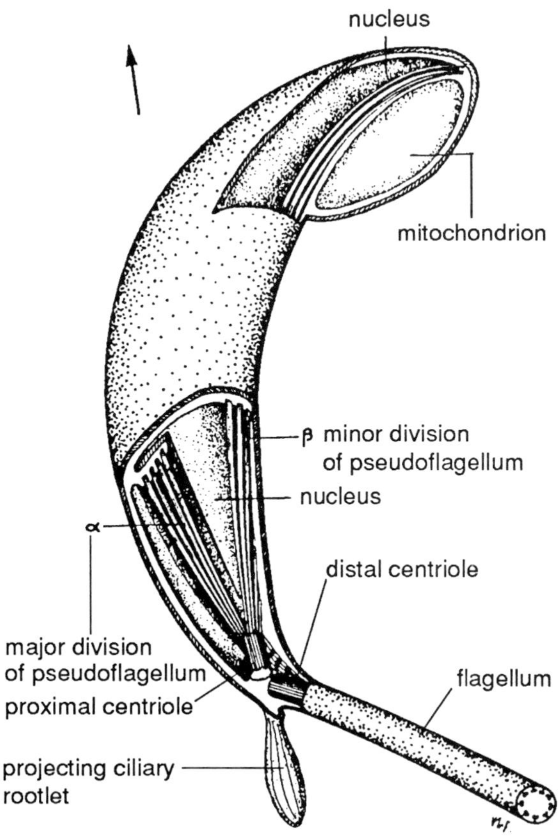

Fig. 11.9. *Anguilla anguilla*, the Eel. Schematic representation of the spermatozoon. The arrow indicates the direction of movement. After Billard, R. and Ginsburg, A.S. (1973). *Annales de Biologie Animale, Biochemie, Biophysique* **13**, 523-534. Fig. 1.

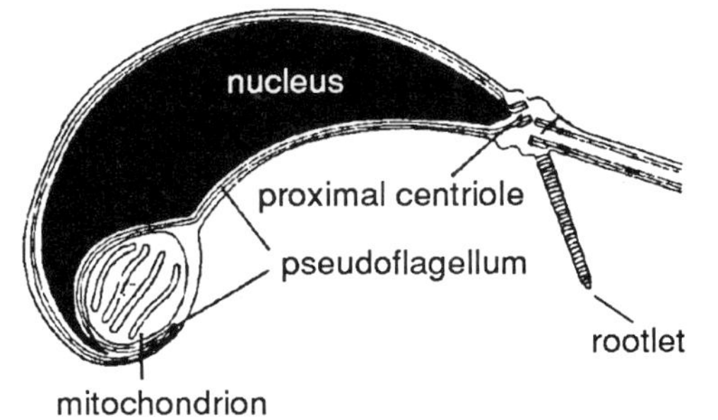

Fig. 11.10. *Anguilla australis.* Schematic reconstruction of a spermatozoon. After Todd, P.R. (1976). *Cell and Tissue Research* **171**, 221-232. Fig. 20.

The sperm of *Anguilla anguilla* is illustrated in Fig. 11.6 and 11.9 and that of *A. australis* in Fig. 11.10.

The sperm of *Anguilla japonica*, as described by Colak and Yamamoto (1974a, b), is the simplest in the Elopomorpha excepting the Muraenidae and as such is an interesting approach to that isolated family. The description requires confirmation, however, as it contains inconsistencies, such as the supposed 0.1 µm (clearly 1 µm) width of the nucleus.

Baccetti *et al.* (1979a) describe the axoneme of *Anguilla anguilla.* It is composed only of 9 doublets and lacks central central singlets as well as any system of radial connections. The sperm nevertheless move actively and the tail beats with large frequent waves propagating backwards. Each doublet possesses only the inner, straight arm which is about 20 nm long. A very short segment of outer arm in present in the proximal segment of the axoneme. Short Y-links bind doublets to the plasma membrane. The inner arms and Y-links are strongly ATPase positive with the Wachstein and Meisel test. Electrophoretically *Anguilla* axonemes have a predominant band corresponding with the B band of sea urchin sperm and clearly relating to the presence in the eel sperm of only the inner dynein arm.

The sperm of the following families are described chiefly in Mattei and Mattei (1974), summarized in the general description, above.

Family Congridae. Examined species are *Congermuraena bertini* (Fig. 11.2D), *Cynoponticus ferox*, *Paraconger notialis.* and an undetermined congrid (Fig. 11.2C) (see Mattei and Mattei, 1974,

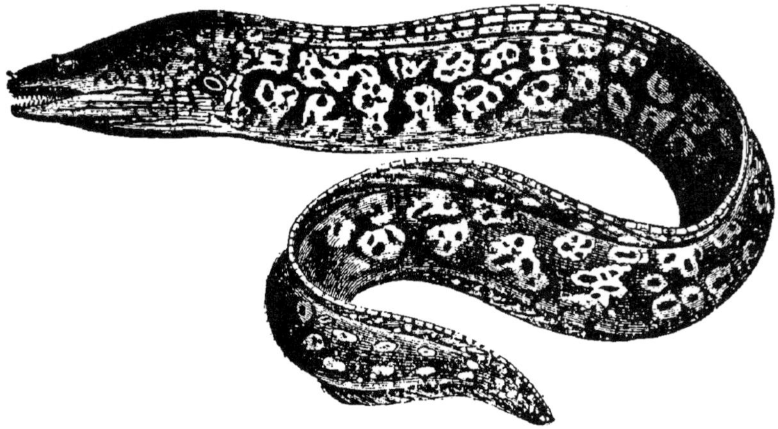

Fig. 11.11. *Muraena helena.* Norman, J.R. (1937). *Illustrated Guide to the Fish Gallery. British Museum (Natural History).* Trustees of the British Museum, London. Fig. 26.

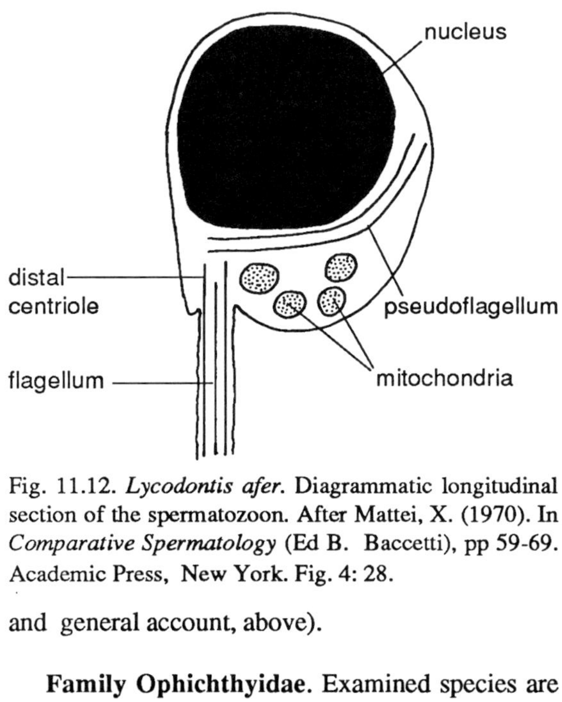

Fig. 11.12. *Lycodontis afer.* Diagrammatic longitudinal section of the spermatozoon. After Mattei, X. (1970). In *Comparative Spermatology* (Ed B. Baccetti), pp 59-69. Academic Press, New York. Fig. 4: 28.

and general account, above).

Family Ophichthyidae. Examined species are *Echelus myrus*; *Ophichthus ophis*; *Pisodonophis semicinctus*(Fig. 11.2H) and *Mystriophis rostellatus* (see Mattei and Mattei, 1974, and general account, above).

Family Muraenidae. Examined species are *Lycodontis afer* (Mattei, 1970, Mattei and Mattei, 1974); *Muraena helena* (Fig. 11.11); and *Muraena robusta* (see Mattei and Mattei, 1974).

The isolated position of the Muraenidae (e.g. *Lycodontis afer*, Fig. 11.12) is confirmed by their lacking a fundamental elopomorph spermatozoal apomorphy: development of a striated rootlet. A flagellar rootlet is probably plesiomorphic for the teleost spermatid but its retention into the elopomorph spermatozoon must be considered apomorphic. Two features normally considered plesiomorphic for teleost sperm are also distinctive of muraenids: the rounded form of the head and, possibly shared with *Anguilla japonica*, location of the mitochondria in the centriolar region.

Family Heterenchelydae. The sperm of *Pythonichthys microphthalmus* (Mattei, 1970, Mattei and Mattei 1974) is illustrated in Fig. 11.2F and is briefly discussed in the general account, above.

Order Notacanthiformes

Diagnosis. Body eel-like but pelvic fins present; deep-sea (Nelson, 1984). These are regarded as members of the Albuloidei, within the Anguilliformes, in phylogenies by Lauder and Liem (1983) and Smith (1984).

Sperm ultrastructure. In a brief reference to unnamed notacanthiforms, Mattei (1988) indicates that their sperm have the basic elopomorph apomorphies: the proximal centriole extends along the nucleus and divides into two bundles of tubules arranged in five triplets in one of them and four doublets, arising from tubules A and C of the centriole, in the other; the doublets of the 9+0 axoneme have a single arm in an internal position and are in contact with the flagellar membrane.

Elopiform-Anguilliform sperm relationships. Greenwood *et al.* (1966) advocated uniting Elopiformes and the Anguilliformes in the Elopomorpha, with the presence of a leptocephalus larva as a shared character. The two distinctive spermatozoal synapomorphies support this union (Mattei and Mattei, 1974).

Sperm morphology therefore unifies the Elopomorpha and has resolved some taxonomic and phylogenetic problems. The interrelationship of elopiforms, anguilliforms and "notacanthiforms" and inclusion of the albuloids among these is confirmed. Nybelin considered the mutual possession of a leptocephalus as a poor, possibly plesiomorphic, character for relationship of *Elops* and *Albula* which were considered unrelated on the basis of features of their somatic anatomy, the albulids being supposedly closer to clupeoids (Nybelin, 1973, and references therein). We have seen that sperm structure clearly unites the groups to which *Albula* and *Elops* belong, as upheld from somatic anatomy by Forey (1973), Nelson (1984), and others, while indicating no close affinity with clupeoids (*q.v.*).

In Chapter 5 a number of alternatives for the derivation of the unique elopmorph sperm were discussed. It is a complex anacrosomal aquasperm. It is probably derived from an aquasperm with the "primitive" morphology but some consideration must be given to the possibility that it is derived from a primary introsperm. If so, the simplified muraenid aquasperm would be regarded as a parallelism, by virtue of relationship, with later teleosts unless it were considered that higher teleosts arose from an elopomorph at the root of the Muraeinidae, a not implausible alternative. If, however, the simple aquasperm is plesiomorphic, and non-homoplasic, for fish, the muraeinids would have to be considered more primitive than usually thought in retaining the simpler sperm facies.

CLUPEOCEPHALA

The Clupeocephala are the Clupeomorpha and the entire Euteleostei (Lauder and Liem, 1983) (see also Fig. 11.1). Clupeocephalan synapomorphies are increased versatility of the feeding apparatus brought about by an important innovation. The upper pharyngeal jaws, which are supported by the first three pharyngobranchials, have a well anchored armour of teeth formed by the complete fusion of the toothplates to the endochondral pharyngobranchial elements. Similarly, in the lower pharyngeal jaw the tooth plates fuse to ceratobranchial five. Further significant characters uniting the Clupeomorpha and Euteleostei are the co-ossification of the angular and articular bones of the lower jaw; exclusion of the retroarticular bone from the quadratomandibular joint surface; and reduction or loss of the neural arch on ural centrum one (Patterson, 1977b; Lauder and Liem, 1983; Nelson, 1973, 1984).

Superorder Clupeomorpha

Diagnosis. Clupeomorphs are unique in having the otophysic connection involving a diverticulum of the swim bladder which penetrates the exoccipital and extends into the prootic within the lateral wall of the brain case (Greenwood *et al.*, 1966; Nelson, 1984). Lauder and Liem (1983) (Fig. 11.13) list three unifying synapomorphies: an autogenous hypural number one; hypural two fused with ural centrum one (two characters also found in several primitive euteleosteans); and fusion of the median and lateral extrascapulars with the supraoccipital and parietal bones. Grande (1985) defines the Clupeomorpha on two synapomorphies in addition to the otophysic connection: one or more abdominal scutes, each primitively consisting of a single (unpaired) element which crosses the ventral midline of the fish; and a supratemporal commissural sensory canal primitively passing through the parietals and supraoccipital. Living forms are confined to the order Clupeiformes.

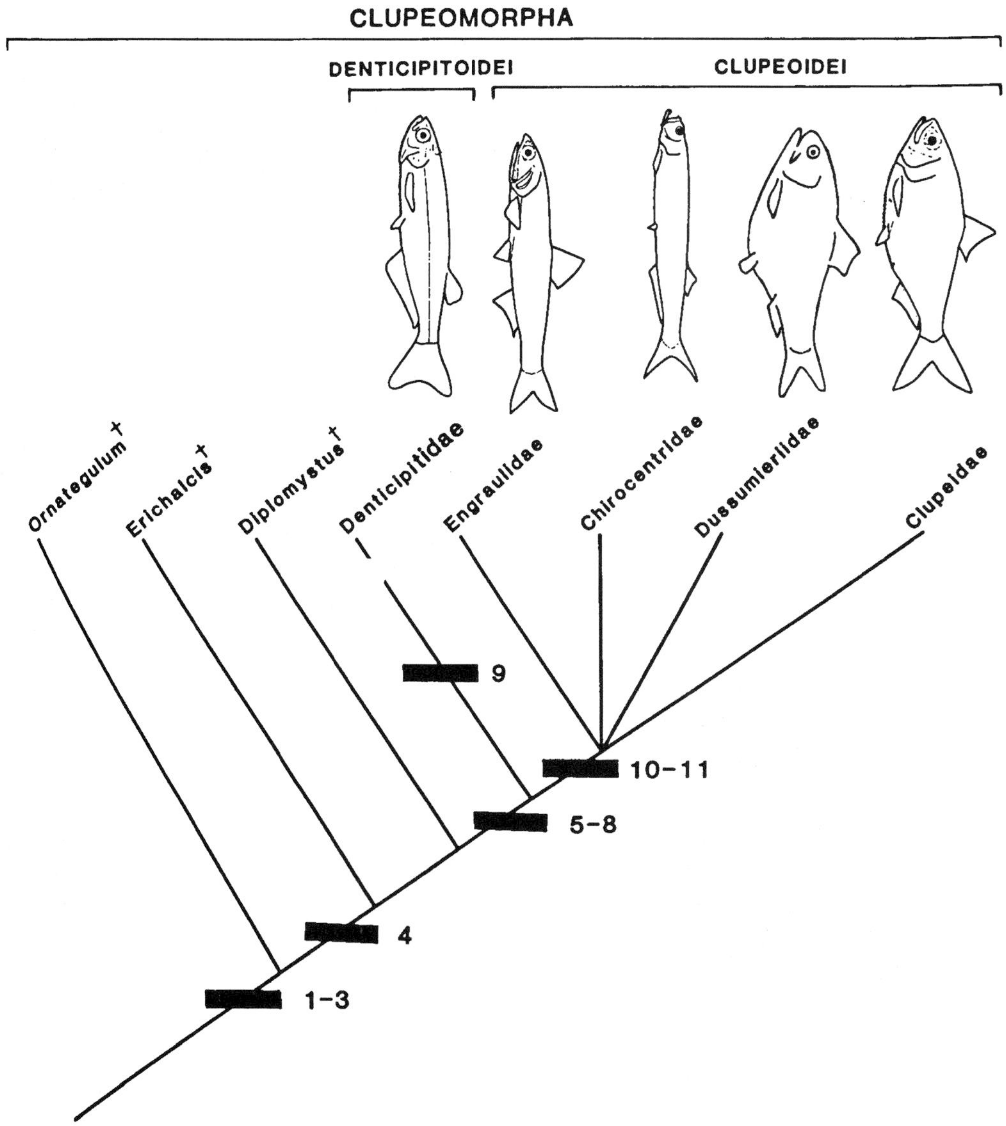

Fig. 11.13. Interrelationships of the major groups of the living Clupeomorpha. Major specializations characterizing the various lineages are: 1, autogenous hypural number one; 2, hypural two fused with ural centrum one (these two characters are also found in several primitive euteleosteans and not in *Erichalcis*; 3, fusion of the median and lateral extrascapulars with the supraoccipital and parietal bones (indicated in part by the presence of a sensory canal in the supraoccipital); 4, abdominal scutes present; 5, presence of a recessus lateralis; 6, temporal foramen present; 7, post-temporal groove present; 8, prootic and pterotic bullae present and enclosing an intracranial swimbladder diverticulum (this character is not known in fossil forms but the pterotic bulla may be a unique feature of all clupeomorphs); 9, denticles on skull bones and some trunk scales; 10, third pharyngobranchial bones with long medial processes; 11, two divisions of the levator arcus palatini muscle. The interrelationships of the clupeoids are a matter od dispute and additional families such as the Pristigasteridae and Congothrissidae may be recognized. From Lauder and Liem (1983).*Bulletin of the Museum of Comparative Zoology* **150**, 95-197. Fig. 26. (See references therein).

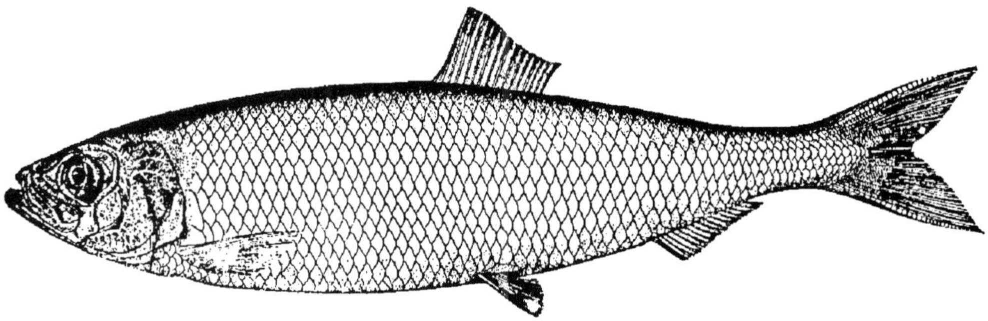

Fig. 11.14. *Clupea harengus*. The Herring. After Jordan, D.S. (1907). *Fishes*. Henry Holt, New York. Fig. 209

Order Clupeiformes

Diagnosis. Herrings (Fig. 11.14) and their relatives, mostly plankton feeders, with numerous gill rakers which serve as efficient straining devices.

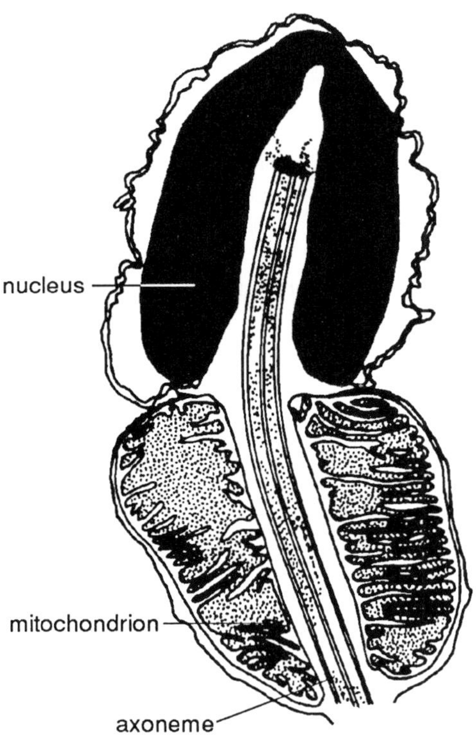

Fig. 11.15. *Anchoa guineensis*. Diagrammatic longitudinal section of the spermatozoon. After Mattei, C., Mattei, X., Marchand, B. and Billard, R. (1981). *Journal of Ultrastructure Research* **74**, 307-312. Fig. 1.

Primarily marine, many move easily into brackish and freshwater. There are about 317 Recent species but they make up about one third of the world's total commerical fishing catch (Nelson, 1984; Grande, 1985). Specialized character complexes, listed under Clupeomorpha above, attest to their monophyly (Greenwood *et al.*, 1966; Lauder and Liem, 1983; Nelson, 1984; Grande, 1985).

Relationships. The relationships of the Clupeomorpha, and clupeiforms, with other teleosts are obscure. They have traditionally been classified with, or related to, the elopoids, or occasionally (e.g. Greenwood, 1973), with the osteoglossoids, both attributions indicating their primitive status, but Nelson (1973) states that there are no shared advanced characters to support such relationships and has proposed characters of the jaws which support division of teleosts into three groups: Osteoglossomorpha, Elopomorpha and a group, not formally named, containing the clupeoids and all other teleosts (Euteleostei). This view accords with the phylogeny derived by Smith (1984) from larval morphology (Fig. 10.8) and that of Lauder and Liem (1983), for the clupeomorph-euteleost assemblage (Clupeocephala), from somatic anatomy of the adult. Synapomorphies of the Clupeocephala are listed above.

Sperm literature. The ultrastructure of the sperm of three clupeoid species has been illustrated in diagrammatic sketches by Mattei (1970), for a clupeid (*Ethmalosa fimbriata*) (Fig. 11.16A) and two engraulids (the anchovy, *Anchoa* (=*Engraulis*) *guineensis*, Fig. 11.16B; and the sardine, *Sardinella aurita*, Fig. 11.16C) and that of *A. guineensis*, particularly the axoneme and mitochondrion, has been further de-

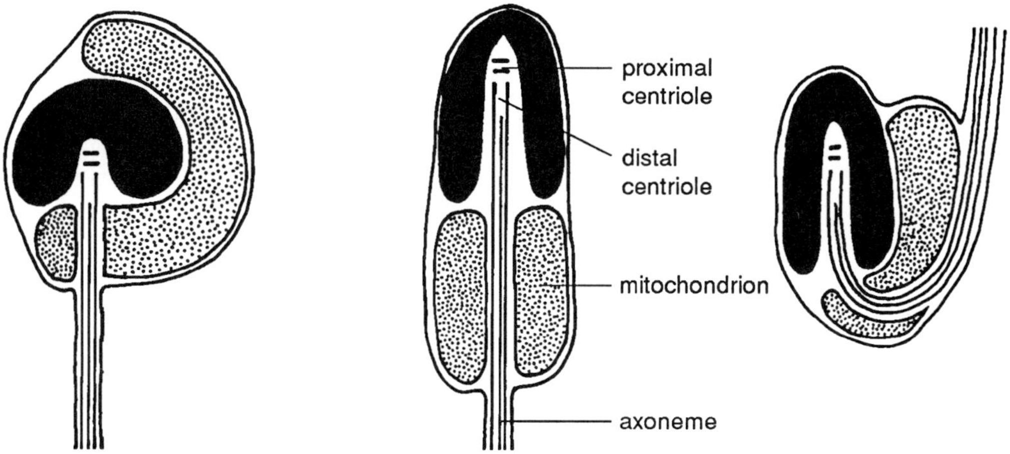

Fig. 11.16. Diagrammatic longitudinal sections of clupeid sperm. A. *Ethmalosa fimbriata.* B. *Anchoa (=Engraulis?) guineensis.* C. *Sardinella aurita.* From Diagrammatic longitudinal section of the spermatozoon. After Mattei, X. (1970). In *Comparative Spermatology*. (Ed B. Baccetti), pp 59-69. Academic Press, New York. Fig. 4: 25, 27 and 22.

scribed by Mattei *et al.* (1981) (Fig. 11.15). The sperm of *Clupea harengus*, the Atlantic Herring, are known by light microscopy (Retzius, 1905).

General sperm ultrastructure. The engraulids, at least, show a departure from the commoner teleostean aquasperm ultrastructure in penetration of the nucleus almost to its tip by the basal fossa and contained axoneme, a tendency present but less developed in *Ethmalosa* ; furthermore, in all three the single mitochondrion is ring- or C-shaped, a feature known in the holostean *Lepisosteus* and in teleosts in four families in addition to the Clupeidae, and Engraulidae, namely the salmoniform families Alepocephalidae (*Xenodermichthys* sp.), Searsidae (*Searsia* sp.) Salmonidae and Galaxiidae. Each of these has some axonemal doublets (varying with the family) with an A subtubule which appears solid owing to intrusion of a septum. Among these teleost families (*q.v.*), the affected doublets are 1, 2, 5, 6, 7; or 1, 3, 5, 6, 7, as in the three clupeomorphs; or 1, 2, 3, 5, 6, 7. Association of a ring-shaped mitochondrion and septate A subtubules is seen also in *Lepisosteus* (1, 2, 6 and 7 septate) and in the holothurian echinoderm *Cucumaria* (1, 2, 5, 6, 7 septate) but is attributed to the general asymmetry of the spermatozoon rather than to a direct correlation with the axonemal condition (Mattei *et al.*, 1981). Occluded A subtubules also occur in families in which the mitochondria are not annular, including the Carangidae and Hemiramphidae (*q.v.*, below). From micrographs (Mattei *et al.*, 1981) no definite flagellar fins appear to be present in the clupeomorphs.

Chapter 12

INFRADIVISION EUTELEOSTEI

Relationships. The Euteleostei are regarded as the apomorph sister-group of the Clupeomorpha (Fig. 11.1) but there seems to be no known unique character which is present in all species of euteleosts (Lauder and Liem, 1983; Nelson, 1984). Spermatologically they are correspondingly diverse. Euteleostean traits are, nevertheless, recognizable although they have been lost in the more derived forms: an adipose fin; nuptial tubercles; and an anterior membranous component to the first uroneural (Patterson and Rosen, 1977). Identical cladistic relationships of the basal euteleosts have been suggested by Lauder and Liem (1983) (Fig. 12.1) and by Fink (1984).

A contribution of spermatology, and departure

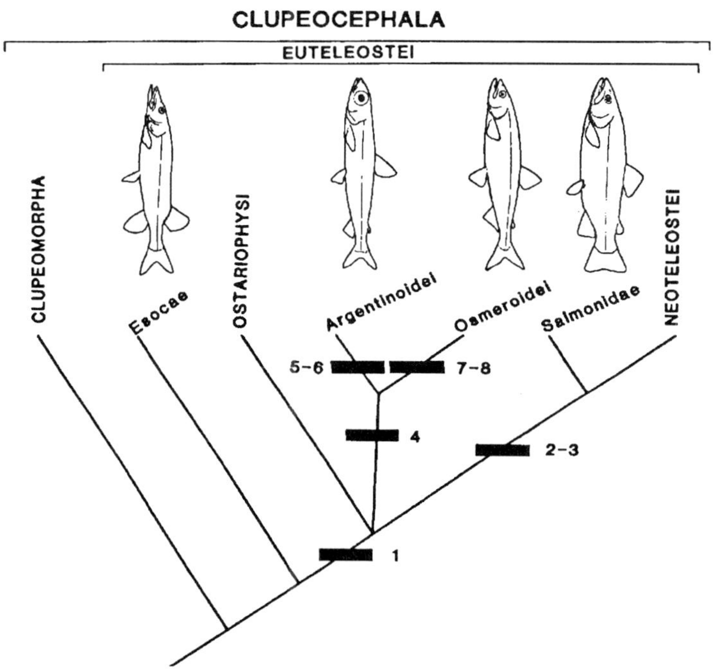

Figure 12.1. Interrelationships of primitive euteleosteans, the Clupeocephala: Clupeomorpha and Euteleostei. The characters are: 1, loss of the dentigerous toothplate on basibranchial four; 2, cartilage nodules (homologous to the rostral cartilage of euteleosts) between the ethmoid and premaxillae; 3, both the basioccipital and exoccipital articulate with the first vertebra; 4, fusion of the posterior neural arches in the caudal fin with either the uroneural or ural vertebra one, 5, specialized "tongue-bite" mechanism between basihyal and mesopterygoid teeth; 6, loss of basisphenoid (this bone is also absent in ostariophysans; 7, basibranchials toothless; 8, "crumenal organ" present and the posterior aspect of the fifth ceratobranchial associated with a complex accessory cartilage. From Lauder and Liem (1983).*Bulletin of the Museum of Comparative Zoology* **150**, 95-197. Fig. 28. (See references therein). From sperm ultrastructure salmonids are considered in the present work to be monophyletic with the Argentinoidei and Osmeroidei.

from the phylogeny of Lauder and Liem (1983) shown in Fig. 12.1, is the demonstration from mutual presence of a ring-shaped mitochondrion that the Salmonidae and Argentinoidei (with the Osmeroidei?) probably form a monophyletic clade.

SUPERORDER ESOCOMORPHA

Order Esociformes

In naming this Superorder and order we have here recognized the independent phylogenetic position of the esocids and their removal from the Salmoniformes which has been advocated by Fink and Weitzman (1982) and by Lauder and Liem (1983). These authors consider the "Esocae" to be the plesiomorphic sister-group of an ostariophysan - salmoniform - neoteleostean assemblage (Fig. 12.1). The superordinal and ordinal ranks have little meaning, or, at least, equivalence in an Hennigian analysis but are introduced for conformity with the remaining classification of fish.

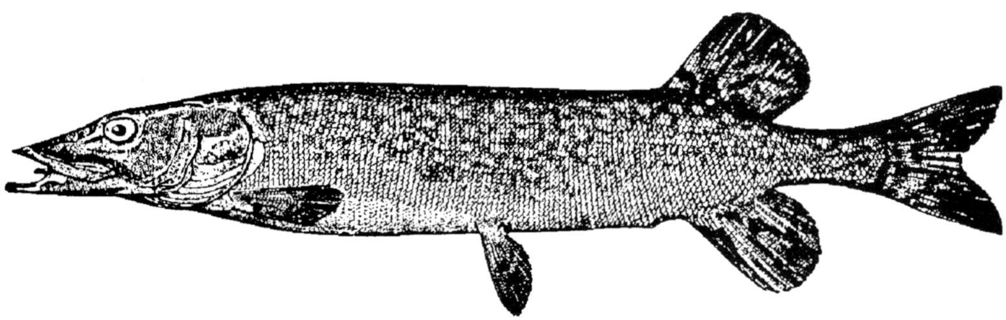

Fig. 12.2. *Esox lucius*. After Jordan, D.S. (1907). *Fishes*. Henry Holt, New York. Fig. 96.

Suborder Esocoidei

Diagnosis and relationships. Esocoids, represented in Fig. 12.2 by *Esox lucius*, the Northern Pike, were considered to be closely related to salmonoids by Rosen (1973) and were placed, with reservations noted here, as a suborder, in the Salmoniformes by Nelson (1984). The teeth on the tongue of esocids are small and uniform, as are the teeth on the basibranchial elements behind the tongue. The absence of the adipose fin may merely be related to the posterior position of the dorsal fin. The maxillae are toothless, although they do form part of the gape as in salmonds, swinging anteriorly during prey capture as in *Amia* and *Salmo*. All esocoids occur in temperate and arctic freshwaters of the Northern Hemisphere and are predatory (Lauder and Liem, 1983). The yolk-sac circulation differs from that of all salmoniform fish for which it has been described (Martin, 1984). Fink and Weitzman (1982), Fink (1984), and Lauder and Liem (1983) (Fig. 12.1) interpose the Ostariophysi between the "Esocae", which they consider the sister-group of all other teleosts, and the salmoniforms. The Esocoids may thus be the most primitive euteleostean clade. The group contains the pikes and pickerels (Esocidae) and the mudminnows (Umbridae).

Fink and Weitzman (1982) have questioned inclusion of the Western Australian *Lepidogalaxias* in the Esocae by Rosen (1974) and we have, with reservations, relegated it to the Salmoniformes.

Esocoid sperm. The spermatozoon of only one species, *Esox lucius*, has been briefly described (Billard 1970a; Stein, 1981) (Fig. 12.3). Billard (1970a) considers it to show a strong resemblance to the aquasperm of Cyprinidae (see Fig. 12.7), being placed in his group I spermatozoa which include also the Carp but which also embrace the sperm of the perciform *Tilapia*. Stein (1981), in contrast, finds it similar to the sperm of *Perca* and *Cottus*, both of which it resembles in the form of the head.

Nucleus. The nucleus (Fig. 12.3) is rounded in one plane but flattened in the other, imparting bilateral symmetry (Billard, 1970a; Stein, 1981). It is 2 µm long, 1.8 µm in maximum diameter and 1.5 µm in minimum diameter. The chromatin is in the form of dense masses (Billard, 1970a), a perciform rather than cyprinid condition. The nuclear membrane is strongly undulated. Stein states that the pike shows no relationship to the other salmoniforms which belong to his group 1 sperm in which the head is somewhat elongated and in which the midpiece is central behind the head whereas it overlaps the head in the pike as it does in cyprinids.

Midpiece. Dilatation of intermembrane spaces and fusion of mitochondria is rudimentary compared

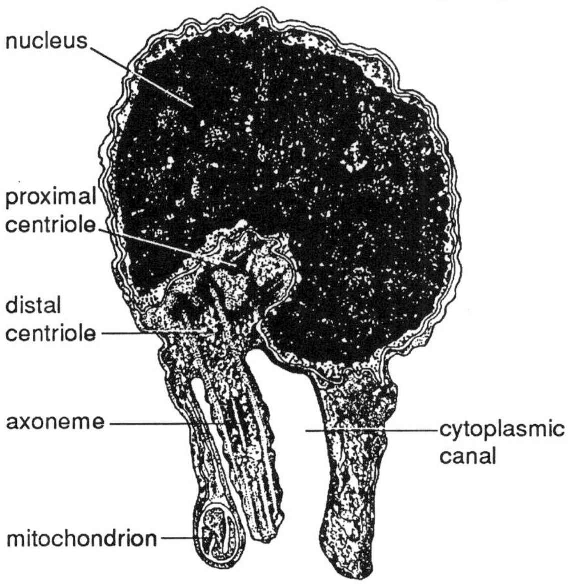

Fig. 12.3. *Esox lucius.* Longitudinal section of spermatozoon. Drawn from a micrograph of Stein (1981). *Zeitschrift für Angewandte Zoologie* **68**, 183-198. Fig. 1.

with the trout.

Flagellum. The flagellum is inserted laterally where one extremity of the proximal centriole is engaged in a nuclear fossa situated on the ventral face of the head while the other is applied to the distal centriole. Inner and outer dynein arms are present in the classical 9+2 axoneme. Anteriorly, as in the trout, the doublets are connected by an internal dense ring and a further peculiarity is the presence of an additional outer, approximately radial arm (Billard, 1970a). A cytoplasmic expansion [fin] is present on one side only (Billard, 1970a).

Remarks. *Esox* differs from cyprinids, and agrees with salmonids, in having flagellar fins (Billard, 1970a; Stein, 1981) (albeit unpaired in the pike) but this is a plesiomorphy for Osteichthyes and does not indicate esocoid-salmonoid affinity.

Similarities of the sperm of *Esox* to those of both cyprinids and perciforms, noted above, add weight to the argument that esocoids are not salmoniforms. Failure of the mitochondria to unite in a ring is plesiomorphic relative to salmoniforms. These facts are not inconsistent with, though they do not strongly confirm, placement at the base of the euteleosts.

SUPERORDER OSTARIOPHYSI

Orders Cypriniformes, Siluriformes and Characiformes

Diagnosis and relationships. The 6,000 species in the Ostariophysi constitute nearly three quarters of the freshwater fish of the world, from carps (cypriniforms) and the South American tetras (characiforms) to the poisonous marine catfish and the weakly electric gymnotids (siluriforms) (Lauder and Liem, 1983; Fuiman, 1984). The Ostariophysi *sensu strictu* (the Otophysi within a wider Ostariophysi of some workers) include all fish whose four or five anteriormost vertebrae are modified to form an otophysic connection, the Weberian apparatus (Rosen and Greenwood, 1970; Fink and Fink, 1981; Fuiman, 1984), a convincing autapomorphy. To this may be added the synapomorphies: second supraneural absent; distinctive specialization of several anterior supraneurals which form a joint with the third and fourth vertebrae; caudal skeleton with a uniquely compound terminal centrum; and hypural two fused to the compound centrum (references in Lauder and Liem, 1983). From adult characters, Fink and Fink (1981) concluded that this cypiniform-characiform-siluriform assemblage (the Otophysi in their nomenclature) is a monophyletic group; the Siluriformes included not only the Siluroidei but also the Gymnotoidei. Characiforms were seen as the plesiomorph sister-group of the siluriforms, and both formed the apomorph sister-group of the cypriniforms. These views were endorsed by Lauder and Liem (1983) (Fig. 12.4).

The sister-group of the Ostariophysi *s. strict.* (Otophysi) is the Anotophysi, consisting of the Chanoidei and the Gonorhynchoidei, neither of which has been examined for sperm ultrastructure. This more widely constituted Ostariophysi is convincingly monophyletic on the basis of at least seven synapo-

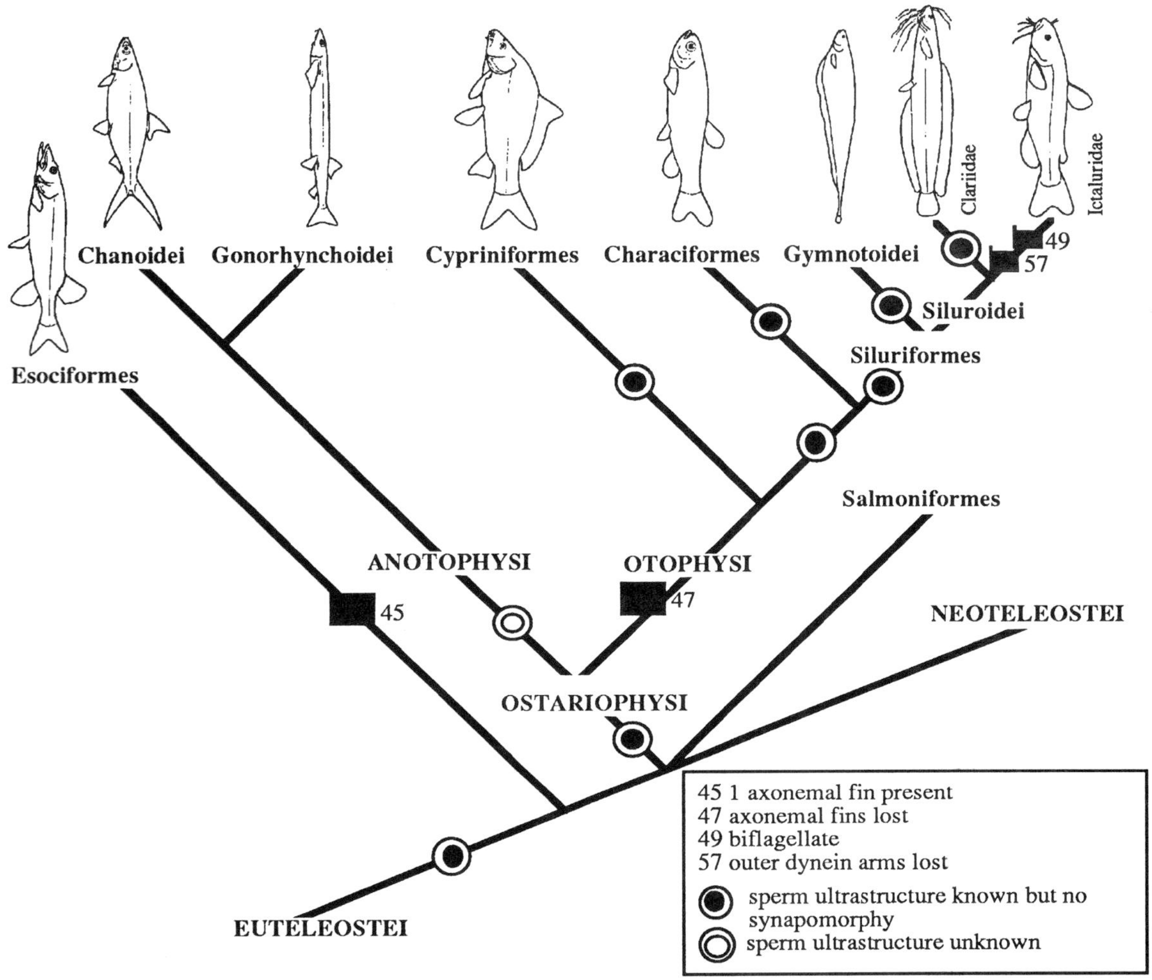

Fig. 12.4. Phylogeny of the Euteleostei. Modified after Lauder, G.V. and Liem, K.F. (1983). *Bulletin of the Museum of Comparative Zoology* **150**, 95-197. Fig. 30. With sperm apomorphies superimimposed. Fish figures from Nelson (1980).

morphies of which one is the presence of in the epidermis of cells which exude an alarm substance when wounded; this causes a fright reaction which is evoked even in non-ostariophysans (references in Lauder and Liem, 1983).

Fuiman (1984), in a study of developmental characters, endorsed collective monophyly of the three orders of the Ostariophysi *s. strict*. The sole gymnotoid considered, *Eigenmannia*, emerged as an apomorphic member of the siluriform clade, in accordance with the novel association of siluroids and gymnotoids by Fink and Fink (1981). It is important to note that some paraphyly was indicated within each of the three orders (Fig. 12.5) but the significance of this is uncertain in view of the small number of species considered.

Application of ordinal rank to each of the three main constituents of the albeit monophyletic Ostariophysi is at best arbitrary, leaving aside paraphyly within each group. Not only discerned relationships of fish groups but also rank names will be greatly altered when a full Hennigian analysis of the Pisces is made. Roberts (1973) is not illogical in recognizing monophyly of the siluroids, cyprinoids and characoids by placing them as suborders in an enlarged order Cypriniformes but this nomenclatural step will be avoided here pending the broader analysis.

Order Cypriniformes

Cyprinidae. Sperm literature. Several species of the Cyprinidae (minnows and carps) have been examined for sperm ultrastructure: *Abramis brama*,

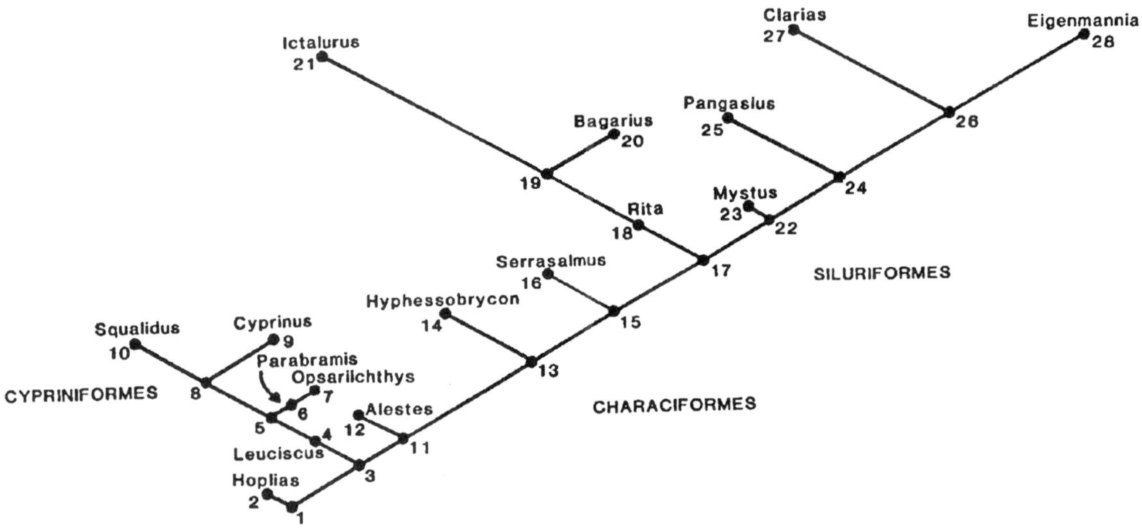

Fig. 12.5. Wagner tree of ostariophysan phylogeny based on larval characters. Stem lengths are proportional to the number of character-state changes on a given stem. From Fuiman, L.A. (1984). In *Ontogeny and Systematics of Fishes*. Special Publication Number 1, pp. 126-137. American Society of Ichthyologists and Herpetologists. Fig. 67.

Diagnosis. Cypriniforms are mainly freshwater fish which have their greatest diversity in southeast Asia but are native to all continents excepting Australia. The mouth is usually protractile and always toothless. An adipose fin is absent. The Weberian apparatus typical of the Ostariophysi (see above) is present (Nelson, 1984). On adult somatic chartacters, the group appears monophyletic, synapomorphies being the kinethmoid bone and the unique upper jaw mechanism, the dorsomedial palatal process, and the structure of the premaxilla and pharyngeal jaw complex (references in Lauder and Liem, 1983).

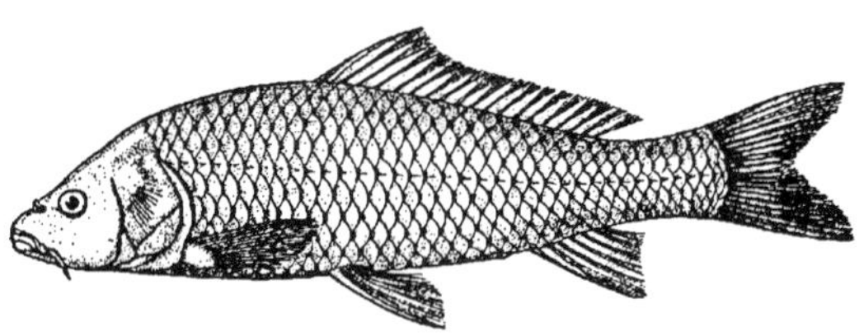

Fig. 12.6. *Cyprinus carpio*. The Carp. After Grassé, P.-P. (1958). (ed.). *Traité de Zoologie*. XIII. Agnathes et Poissons. Masson et Cie, Paris. Fig. 1638.

Stein, 1981; *Alburnus alburnus alborella* (Fig. 12.7); *Barbus barbus plebejus*, Baccetti *et al.*, 1984 (Fig. 12.7); *Blicca bjorkna*, Stein, 1981; *Brachydanio rerio*, Wolenski and Hart, 1987; *Carassius auratus*, Baccetti *et al.*, 1984; Fribourgh *et al.*, 1970; Munoz-Guerra *et al.*, 1982 (Fig. 12.7); *Chondrostoma toxostoma*, Baccetti *et al.*, 1984 (Fig. 12.7); *Cyprinus carpio*, Billard 1970a; Fujimura *et al.*, 1957; Kudo, 1980; Stein, 1981 (adult, Fig. 12.6); *Leuciscus cephalus*, Baccetti *et al.*, 1984; Stein, 1981 (Fig. 12.7); *Leuciscus souffia*, Baccetti *et al.*, 1984 (Fig. 12.7); *Leuciscus leuciscus*, *Phoxinus phoxinus*, Stein, 1981; *Rhodeus ocellatus*, Ohta and Iwamatsu, 1983; *Rutilus rubilio*, Baccetti *et al.*, 1984 (Fig. 12.7); *Rutilus rutilus* and *Scardinius erythrophthalmus*, Stein, 1981.

Cyprinid sperm-type. Stein recognized a "type 2 sperm" for cyprinids alone, characterized by an almost spherical head, on the basis of ultrastructural investigation of the sperm of the cyprinids *Abramis brama*; *Blicca bjorkna*; *Cyprinus carpio*; *Leuciscus cephalus*; *L. leuciscus*, *Phoxinus phoxinus*; *Rutilus rutilus*; and *Scardinius erythrophthalmus*. However,

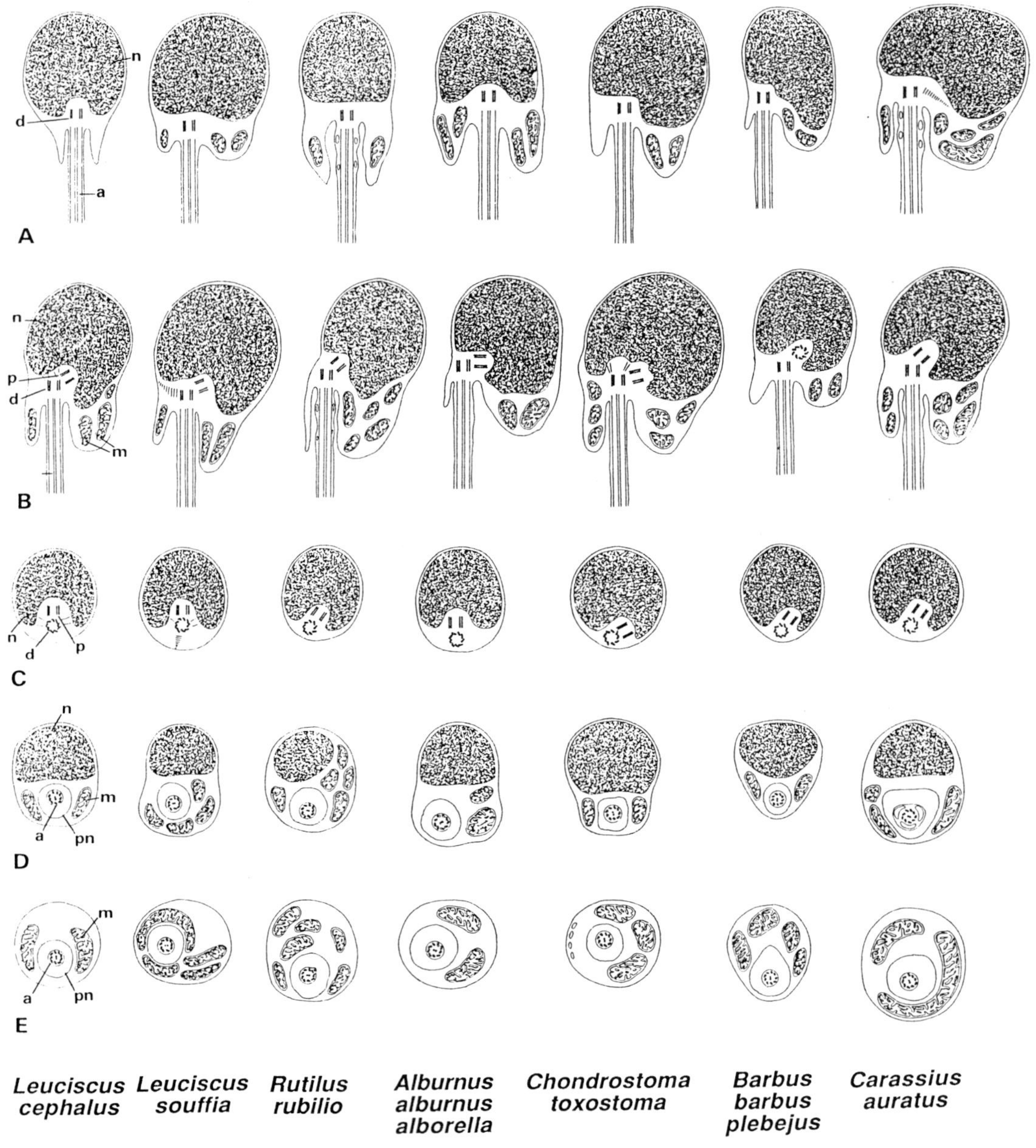

Fig. 12.7. Variations in sperm of seven cyprinid species. A. Longitudinal frontal section. B. Longitudinal sagitttal section. C. Cross section at centriolar region. D. Postnuclear cross section. E. Cross section of the midpiece region. a. axoneme. dc. distal centriole. m. mitochondrion. n. nucleus. pc. proximal centriole. pn. postnuclear (cytoplasmic) canal. From Baccetti *et al.* (1984). *Gamete Research* **10**, 373-396. Figs. 2-8.

although the nucleus was spheroidal in most species examined by Baccetti *et al.*, (1984), it was slightly ellipsoidal in *Leuciscus souffia* and *B. barbus*. (For type 1 sperm of Stein see Salmoniformes). Cyprinid sperm resemble the spermatozoon of *Esox lucius* (Fig. 12.3).

The sperm of the Goldfish, *Carassius auratus* is illustrated by Fribourgh *et al.* (1970), Munoz-Guerra *et al.* (1982) and Baccetti *et al.* (1984) (Fig. 12.7); it closely resembles that of *Cyprinus carpio* (see Fujimura *et al.*, 1957; Stein, 1981). These, like the sperm of *Brachydanio rerio* described by Wolenski and Hart (1987), conform to the general cyprinid pattern.

Cyprinid variation. The valuable study by Baccetti *et al.* (1984), on seven cyprinid species (Figs. 12.7) nicely exemplifies variation in sperm ultrastructure within a family. Cyprinids have simple anacrosomal aquasperm which show significant differences, even intragenerically. Variation occurs with regard to tail length, the position of the centrioles and of the proximal centriole relative to the distal centriole and to the nucleus, and the number of mitochondria, a number related to the depth of the cytoplasmic canal. The sperm have a spheroidal or slightly elliptical nucleus, always eccentrically placed on the tail; two variously orientated centrioles, and a postnuclear cytoplasmic region of varying size which contains the mitochondria (2 to 10) and surrounds the periaxonemal postnuclear canal (cytoplasmic canal).

Head. The head diameter is said to be uniformly 2 μm, agreeing with 1.9 μm for *Rhodeus ocellatus* (Ohta and Iwamatsu, 1983) though given as 3.2 μm by Fribourgh *et al.* (1970) for *Carassius auratus*.

Mitochondria. Asymmetry in addition to that of the centrioles is seen in the distribution of the mitochondria. Most of them are located in the area adjacent to the nucleus, only 1 (in *Leuciscus cephalus*, *L. souffia*, and *Carassius*) or none (as in all other species) being present in the opposite area. In these seven species the mitochondria never fuse to form a mitochondrial derivative (Baccetti *et al.*, 1984). The report of a single large unilaterally located mitochondrion for *Rhodeus ocellatus* by Ohta and Iwamatsu, (1983) requires confirmation.

A relationship between the number of mitochondria and axonemal length is not demonstrable but a clear correlation exists between the number of mitochondria (see below) and the length of the cytoplasmic canal, which exceeds 1.5 μm in *Rutilus* and *Carassius*, and is only 0.1 μm long in *Barbus* (Baccetti *et al.*, 1984).

Proximal centriole The proximal centriole lies in the sagittal plane of the sperm in only one case (*Leuciscus cephalus*) and is inclined at 40° to this plane in *Leuciscus souffia* and *Rutilus*, 50° in *Carassius*, 60° in *Alburnus*, 80° in *Barbus* and 90° in *Chondrostoma*. It is rarely perpendicular to the distal centriole (only in *Alburnus* and *Barbus*) but is inclined at an angle of 110° in *Leuciscus souffia*, 120° in *Leuciscus cephalus* and *Chondrostoma*, 125° in *Carassius*, and 140° in *Rutilus*. These variations in centriolar geometry are clearly correlated with the position of the nucleus with respect to the axis of the tail (Baccetti *et al.*, 1984).

Distal centriole and rootlet. The distal centriole is linked to the surface of the nuclear fossa by fibres in *Leuciscus cephalus*, *L. souffia*, *Chondrostoma toxostoma* and *Carassius auratus*. *L. souffia* is unique in the seven species in having a large striated rootlet which links the distal centriole to the adjacent plasma membrane (Fig. 12.7). Such roolets are widely reported for spermatids, for instance those of the salmonid *Salmo gairdneri* (Billard, 1983b). Membranous vesicles intervene between the plasma membrane and the axoneme in the anterior region of the flagellum in *L. cephalus* and *Barbus barbus plebejus*, for almost its whole length in *Rutilus rubilio*, or are numerous over an unspecified length in *Carassius auratus* (Baccetti *et al.*, 1984). In *C. aura-*

Fig. 12.8. *Carassius auratus*. Transverse section of axoneme, showing absence of fins in Cypriniformes. Drawn from a micrograph of Fribourgh, J.H. (1970). *Copeia* **2**, 274-279. Fig. 6.

tus, at least, they extrude from the cytoplasmic canal.

Glycogen. Glycogen granules are reported for the cytoplasm of several species and are probably general for cypriniform sperm.

Flagellum. The tail is of moderate length (from 36 to 60 μm) with a 9+2 axoneme; both dynein arms are present. The length of the tail varies from 36 μm in *Rutilus rubilio*, 42 μm in *Chondrostoma toxostoma*, 58 μm in *Carassius auratus* to 60 μm in *Barbus barbus*. Because the proximal centriole is always eccentric, the axonemal axis is tangential to the nucleus in *L. cephalus*, *Rutilus* and *Alburnus*, lateral to it in *Chondrostoma*, *Barbus* and *Carassius* (as also in *Rhodeus ocellatus*, Ohta and Iwamatsu, 1983) and almost central with respect to the nucleus in *L. souffia* (Baccetti *et al.*, 1984).

All cyprinid sperm examined to date lack the one or more fins seen on the flagellum of most teleost sperm (Fig. 12.8).

Phylogeny and sperm structure. Some correlation between phylogeny and sperm structure can be discerned. Thus in terms of asymmetry and mitochondrial number, *Leuciscus cephalus* and *L. souffia* seem the most closely related. Of all the spermatozoal characteristics, the number of mitochondria seems the only character closely linked with phylogeny. Thus it maintains *Carassius* (Cyprininae), with 10, and *Barbus* (Barbinae), with 2, in distinct positions, and to order the other species (all belonging to the Leuciscinae) according to mitochondrial number from *Leuciscus* (2-3) to *Alburnus* (2 large mitochondria), *Chondrostoma* (3-4) and *Rutilus* (5-6). Even these aquasperm are therefore seen as potentially useful for phylogenetic evaluation at any level (Baccetti *et al*, 1984).

In view of the widespread occurrence of flagellar fins throughout the Osteichthyes, absence in cyprinids (as in all other investigated Ostariophysi) appears to be an apomorphic loss and an ostariophysian synapomorphy.

Cobitidae. *Acanthophthalmus semicinctus*. The sperm of *Acanthophthalmus semicinctus*, the Coolie Loach (adult, Fig. 12.9), in a further cypriniform family, the Cobitidae, has been examined by Jamieson (unpublished).

Nucleus. The nucleus is approximately spherical, 1.7 μm wide, but has an eccentric fossa sufficiently

Fig. 12.9. *Acanthophthalmus semicinctus*. After Sterba, G. (1962). *Freshwater Fishes of the World*. Vista Books, London. Fig. 484.

large to house the entire basal body (Fig. 12.10) and, anteriolateral to this, the proximal centriole. The regularity of the spherical nucleus is enhanced by the homogeneous nature of the chromatin despite some

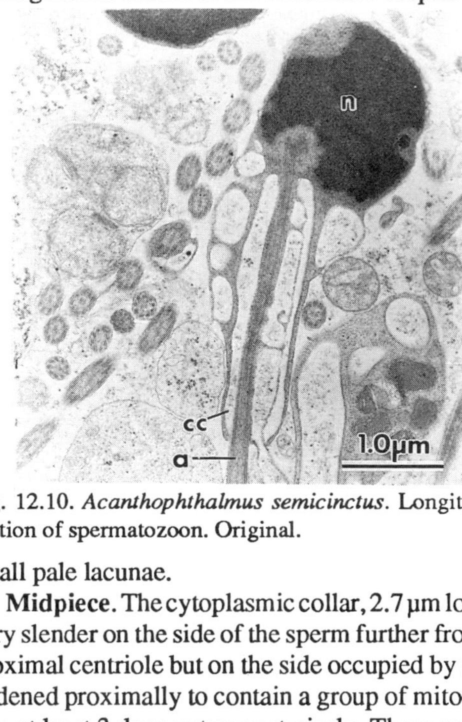

Fig. 12.10. *Acanthophthalmus semicinctus*. Longitudinal section of spermatozoon. Original.

small pale lacunae.

Midpiece. The cytoplasmic collar, 2.7 μm long, is very slender on the side of the sperm further from the proximal centriole but on the side occupied by this is widened proximally to contain a group of mitochondria at least 3 deep anteroposteriorly. These are mutually adpressed and vary from spherical to angular.

Flagellum. Because of the eccentric location of the implantation fossa and its contained centrioles, the long axis of the flagellum is almost tangential to the surface of the nucleus.

Remarks. The spherical head of this cobitid species conforms with the type 2 sperm recognized by Stein (1981) for the related cyprinids. In this and other respects, including the thin, trailing midpiece collar, and absence of axonemal fins, the cypriniform sperm resembles the characiform sperm described below.

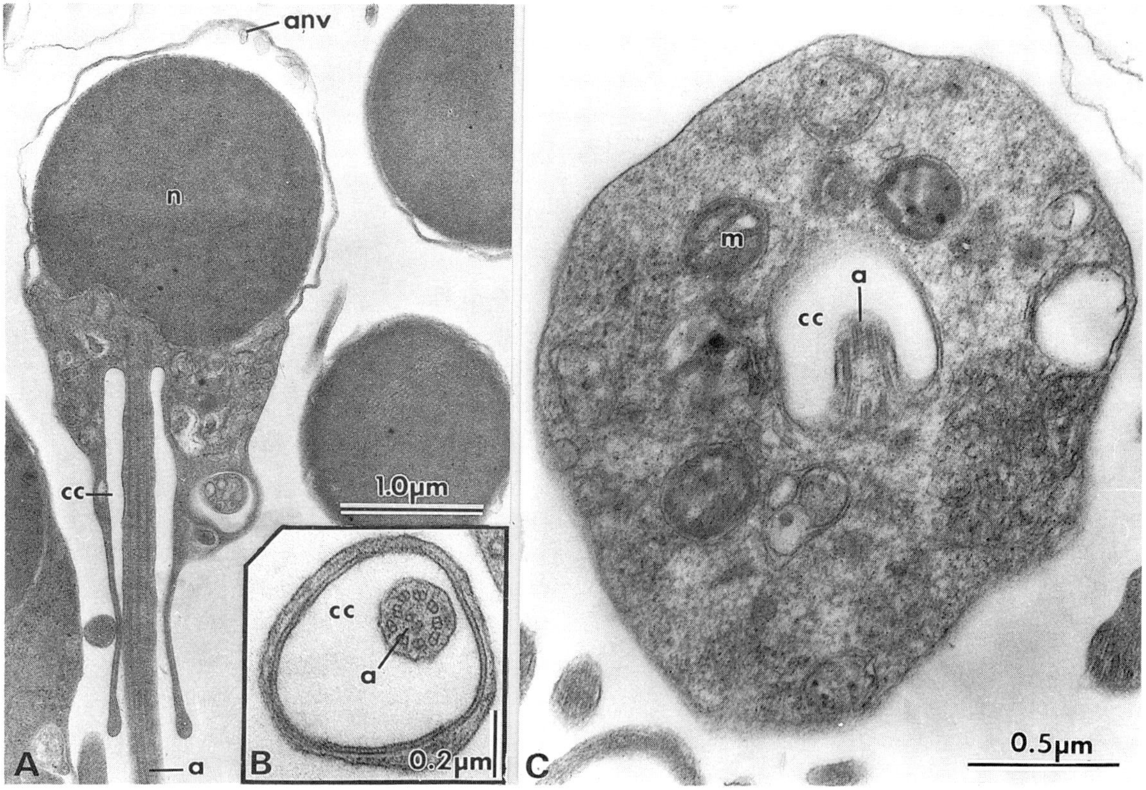

Fig. 12.11. *Paracheirodon (=Hyphessobrycon) innesi*, the Neon Tetra. A. Longitudinal section of the spermatozoon. B. Transverse section of the cytoplasmic canal and axoneme. C. Transverse section of the midpiece at beginning of the cytoplasmic canal. a. axoneme. anv. anterior vesicles. cc. cytoplasmic collar. m. mitochondrion. n. nucleus. Original.

Order Characiformes

Diagnosis and relationships. Characiforms, natives of Africa, South America and southernmost N. America, differ notably from cypriniforms in retaining teeth, which are multicuspid, and in presence, in most, of an adipose fin (Fuiman, 1984; Nelson, 1984). The Characiformes and Gymnotiformes are often allied and regarded jointly as the sister-group of the Cypriniformes, characiforms being considered the most primitive "otophysans". In contrast, as mentioned above, Fink and Fink (1981) present detailed evidence in support of their view that cypriniforms are the sister-group of characiforms, siluroids and gymnotoids. Despite their popularity as aquarium fish, the spermatozoal ultrastructure of the Characiformes is known only from *Paracheirodon (=Hyphessobrycon) innesi*, the Neon Tetra (family Characidae), described here.

***Paracheirodon innesi*. Acrosome**. Although there is no acrosome, some sperm (Fig. 12.11A) have small vesicles anterior to the nucleus which are possibly acrosomal vestiges.

Nucleus. The nucleus is an electron dense sphere, 2 µm wide, which may show some small pale lacunae, and has two contiguous small embayments, one for the proximal, the other for the distal centriole. Only the tips of the centrioles lie within these fossae (Fig.12.11A). As in *Acanthophthalmus*, the regularity of the spherical nucleus is enhanced by the homogeneous nature of the chromatin.

Midpiece. The cytoplasmic canal and corre-

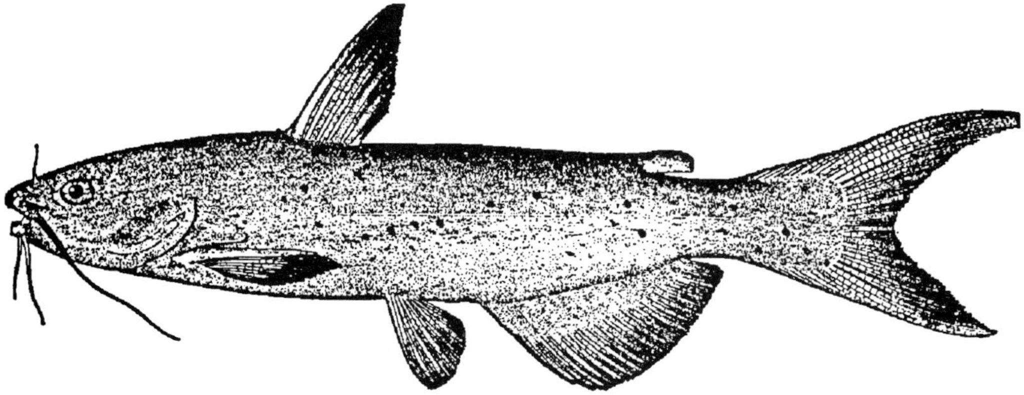

Fig. 12.12. *Ictalurus punctatus*. After Jordan, D.S. (1907). *Fishes*. Henry Holt, New York. Fig. 312.

sponding collar are long, commencing shortly behind the nucleus and extending for about 2.8 µm. Mitochondria are confined to the anterior half of the collar and the posterior half consists of a thin sleeve consisting of little more than two apposed plasma membranes (Fig. 12.11A, B). The two membranes are occasionally separated by large vacuoles containing round multivesicular bodies; these are probably cytoplasmic residues undergoing exocytosis. The thick part of the collar consists of a wide zone of cytoplasm around the cytoplasmic canal and axoneme (Fig. 12.11C) in which are embedded, in transverse section, several simple cristate mitochondria in an irregular layer. Various inclusions and multivesicular bodies are scattered amongst and posterior to them.

Centrioles and axoneme. The basal body (distal centriole) and flagellum is at a slight angle to the apparent anteroposterior axis of the nucleus. The proximal centriole is anteriolateral and almost parallel to the basal body, the long axes of the two being slightly convergent anteriad (Fig. 12.11A). The lumina of the microtubules of the 9+2 axoneme are patent (Fig. 12.11B).

Phylogeny. The absence, interpreted here as an apomorphic loss, of axonemal fins in Characiformes, Cypriniformes, and apparently siluriforms, a rare condition for Osteichthyes, provides a tentative endorsement of the relationship of these three orders which has been alluded to above.

The sperm of *Paracheirodon innesi* correspondingly further conforms to the cypriniform type in the small (2 µm wide) spherical, dense nucleus and the failure (albeit a plesiomorphy) of the mitochondria to unite as a single structure. Extension of an amitochondrial portion of the collar behind the mitochondria is a doubtfully significant divergence from the cypriniform condition.

Order Siluriformes

Diagnosis and relationships. Siluriforms have an almost worldwide distribution as siluroids (catfish) whereas the gymnotoids are a small group restricted to South and Central America. Siluriforms are morphologically the most peculiar and highly modified ostariophysans. Gymnotoid-siluroid synapomorphies include absence of the intercalar, of scleral bones, and of a supraorbital bone; presence of only one pharyngobranchial tooth plate; absence of articular process of the intercalarium; distally bifurcate Baudelot's ligament; posterior pectoral fin rays offest from the anteriormost ray; and absence of medial ossification of the dorsal and anal fin radials. Both have electroreceptive capability (references in Lauder and Liem, 1983). They exhibit great diversity in trophic mechanisms and are far more diverse than characiforms and cypriniforms in the structure of the Weberian complex, swimbladder and caudal fin skeleton (Fuiman, 1984; Nelson, 1984; Roberts, 1973).

Sperm literature. Accounts of siluroid sperm ultrastructure are restricted to *Ictalurus punctatus* (Ictaluridae) (adult, Fig. 12.12; sperm Fig. 12.13) (Jaspers *et al.*, 1976; Poirier and Nicholson 1982;

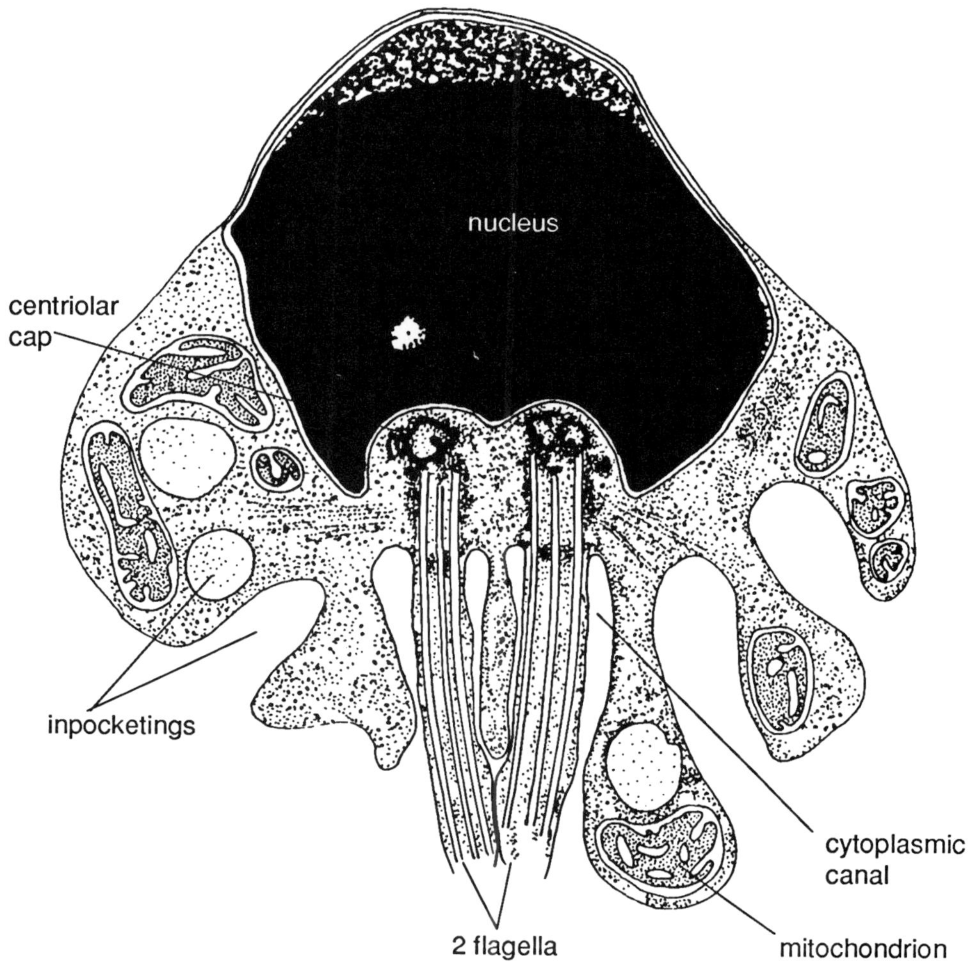

Fig. 12.13. *Ictalurus punctatus*. Longitudinal section of spermatozoon. After a micrograph by Poirier, G.R. and Nicholson, N. (1982). *Journal of Ultrastructure Research* **80**, 104-110. Fig. 2.

Yasuzumi 1971); *Clarias senegalensis* (Clariidae) (Mattei 1970); and, SEM only, *Rhamdia sapo* (Maggese *et al*., 1984). A single gymnotoid, *Sternarchus albifrons*, has been investigated (Jamieson, unpublished, see below).

Suborder Siluroidei

The catfish form a diverse group of marine to freshwater fish with about 2,000 species in 31 families. Among the outstanding siluroid specializations are a defence mechanism involving serrated spines in the dorsal and pectoral fins and devices for holding the spines erect; highly specialized barbels; increased numbers of branchiostegal rays; and an axillary gland. Catfish also share a number of losses which constitute synapomorphies: lack of true scales; absence of supraorbital, parietal, subopercular, symplectic, first and second infrapharyngobranchials, and of epipleural and epineural intermuscular bones (Roberts, 1973; Lauder and Liem, 1983; Fuiman, 1984; Nelson, 1984).

Siluroid sperm types. Siluriforms have uniflagellate and biflagellate anacrosomal aquasperm. Biflagellate spermatozoa are reported as normal for *Ictalurus punctatus*, the Channel Catfish (Fig.

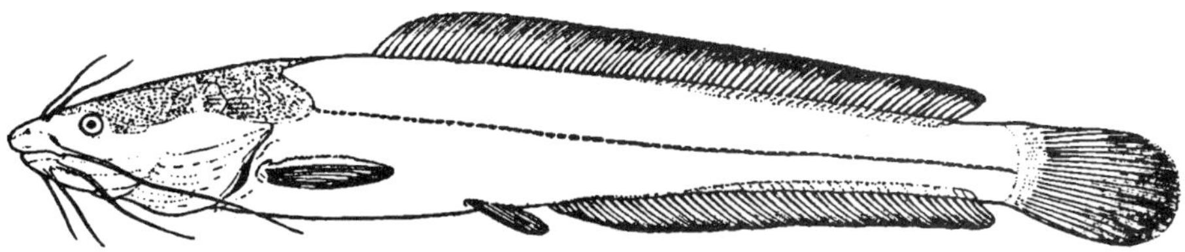

Fig. 12.14. *Clarias senegalensis*. After Grassé, P.P. (1958). Traité de Zoologie. XIII Agnathes et poissons. Masson et Cie, Paris. Fig. 1651.

12.12), of the North American freshwater catfish family Ictaluridae by Poirier and Nicholson (1982) (Fig. 12.13). They dismiss a claim by Jaspers *et al.* (1976) from an optical and TEM investigation that less than 5 per cent of the sperm are biflagellate. SEM plates for the long-whiskered catfish, *Rhamdia sapo* (Pimelodidae), unaccompanied by a description, indicate clearly that its sperm are also biflagellate and round-headed (Maggese *et al.*, 1984).

***Ictalurus* sperm.** Jaspers *et al.* (1976) give the following dimensions for 525 sperm from mixed wild and domestic stocks of this commercially important species : total length of head and midpiece 3.9 µm; head length 2.3 µm; midpiece length 1.6; head width 2.4 µm; midpiece width 3.1 µm; flagellum lnegth 94.9 µm. The standard error was very small.

Head. The head and contained nucleus are rounded and each of the two basal bodies lies in a moderate basal nuclear fossa (Fig. 12.13). The condensed chromatin is granular in appearance. The double nuclear envelope and the plasma membrane are tightly apposed over the anterior one half to two thirds of the sperm head. In this region the surface is slightly undulated and the individual membranes may be difficult to distinguish. Slightly more posteriorly in the head, the nuclear membrane appears wavy and the space between the two membranes varies. The nuclear membrane lining the nuclear fossa is distinct and constitutes a basal plate.

Midpiece. The midpiece extends anteriorly, enveloping up to half of the nucleus in a collar-like arrangement. The mitochondria, located at the periphery of the midpiece at some distance from the centrioles, are not fused and are circular to oblong in section. Inpocketings occur at irregular intervals along the surface of the midpiece and in cross section have the appearance of double-walled vacuoles. The cytoplasmic canal surrounding each flagellum also appears double-walled; its inner wall is thinner than the outer wall lining the canal. Putative beta glycogen granules, 150-300 Å in diameter, are present throughout the midpiece.

Centriolar complex. A centriolar cap, circular in section, 0.35 µm wide with a 62 Å wall, extends from the proximal end of each centriole of the two centrioles, while satellite-like structures surround the distal portion. Thin filaments connect each cap with the outer nuclear membrane in the fossa. Microtubules which extend from satellites in bundles of up to 20 radiate throughout the length and width of the midpiece.

Axonemes. The axonemes are of the 9+2 type but the outer dynein arms are absent; inner arms are tentatively recognized. Structures similar in location to Y links, interdoublet links, and radial spokes [Afzelius rays] are present in the axoneme within the midpiece. Projections from the A subtubule curve below the adjacent doublet to connect with one of the central singlets (Poirier and Nicholson, 1982). A

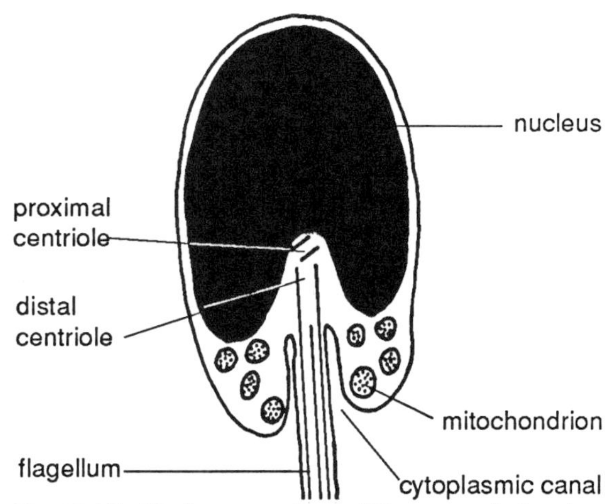

Fig. 12.15. *Clarias senegalensis*. Diagrammatic longitudinal section of the spermatozoon. After Mattei, X. (1970). In *Comparative Spermatology* (Ed B. Baccetti), pp 59-69. Academic Press: New York. Fig. 4: 10.

peripheral sheet or ridge [fin] is absent (Jaspers *et al.*, 1976).

Clarias. The sperm of the African catfish *Clarias senegalensis* (Clariidae), the ultrastructure of which is known only from a diagram by Mattei (1970), is uniflagellate.

Unlike *Ictalurus* the second centriole does not function as a basal body but forms the proximal centriole, at an angle to the basal body. As in *Ictalurus* there is a moderate basal nuclear fossa and there are numerous mitochondria in the midpiece which envelopes the proximal region of the flagellum in a cytoplasmic canal.

Phylogeny. Absenceof flagellar fins in sperm of siluroids (deduced from micrographs)and of gymnotoids (below) is a tenuous link between siluriforms and the Cypriniformes and Characiformes.

Poirier and Nicholson (1982) rightly state that no phylogenetic relationship can be detected in the possession, as in *Ictalurus*, of biflagellate sperm in other fish: *Porichthys notatus*, *Opsanus tau*, *Protopterus annectens*, *P. aethiopicus*, *Polypterus senegalus* and *Lepadogaster lepadogaster* (also in malapterurids and apogonids, Mattei and Mattei, 1988). The statement that the centriolar caps in *Ictalurus* are unique structures must be questioned as they strongly resemble those of *Lepisosteus* and, apparently, of *Neoceratodus* (*q.v.*) though no phylogenetic link is apparent from this. The uniflagellate sperm of *Clarias* relates the biflagellate sperm of *Ictalurus*, as an apomorphic form, to the more usual uniflagellate teleostean aquasperm, including that seen in the gymnotoid *Sternarchus albifrons*.

Suborder Gymnotoidei

Relationships. We have seen above that gymnotoids are now regarded as apomorphic members of the Siluriformes (Fink and Fink, 1981; Lauder and Liem, 1983; Fuiman, 1984). Grouped into four families, they are mostly nocturnal and insectivorous and are remarkable for their ability to generate and detect weak electric signals which are used in navigation and intraspecific communication and in *Electrophorus* for stunning or killing the prey.

Sternarchus albifrons. The sperm of *S. albifrons* (adult, 12.16; sperm, Fig. 12.17) (family Apteronotidae) is a simple anacrosomal aquasperm significantly less modified than the biflagellate ictalurid sperm.

Nucleus. The chromatin of the subspheroidal nucleus (Fig. 12.17A, B, C, F) consists of numerous dense masses in a paler granular matrix.

Midpiece. The mitochondria (Fig. 12.17A, B, C, F) are grouped irregularly around the long cytoplasmic canal for as much as half of its length. They are in one or two tiers longitudinally and vary from subspherical to elongate. The intra-mitochondrial matrix is strongly electron dense. This is reflected by unusual density of the general cytoplasm and of the axoneme. The cytoplasmic collar is approximately one and a half times the length of the nucleus. Its posterior half, behind the mitochondria, forms a thin trailing sheath, seen in longitudinal section in Figs. 12.17A and B and in transverse section in Fig. 12.17D.

Centrioles. The proximal centriole is approximately at right angles to, and in the same longitudinal axis as, the basal body (Fig. 12.17C and inset, 12.17F).

Axoneme. The 9+2 axoneme lacks fins. At the endpiece (Fig. 12.17E) it becomes roughly oblong in cross section where the two central singlets are lost.

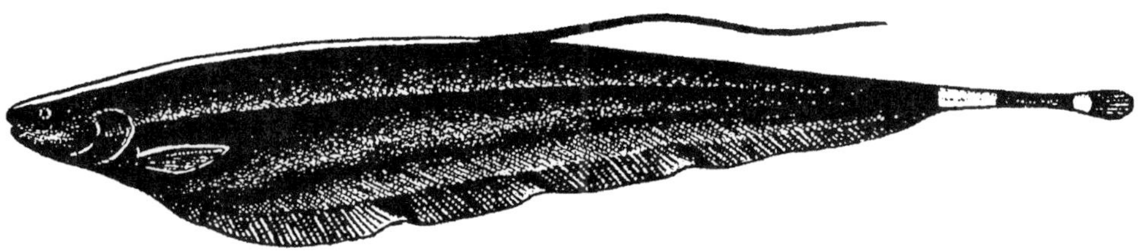

Fig. 12.16. *Sternarchus albifrons*. After Sterba, G. (1962). *Freshwater Fishes fo the World*. Vista Books, London. Fig. 514.

Sperm phylogeny. The long amitochondrial posterior sheathlike extension of the midpiece collar of the *Sternarchus* sperm is seen also in the cypriniform *Acanthophthalmus semicinctus* (Fig. 12.10) and in the characiform *Paracheirodon innesi* (Fig. 12.11A) but is seen also in some Percomorpha and Atherinimorpha. It does not appear to occur below the Euteleostei. It may be an advanced, though

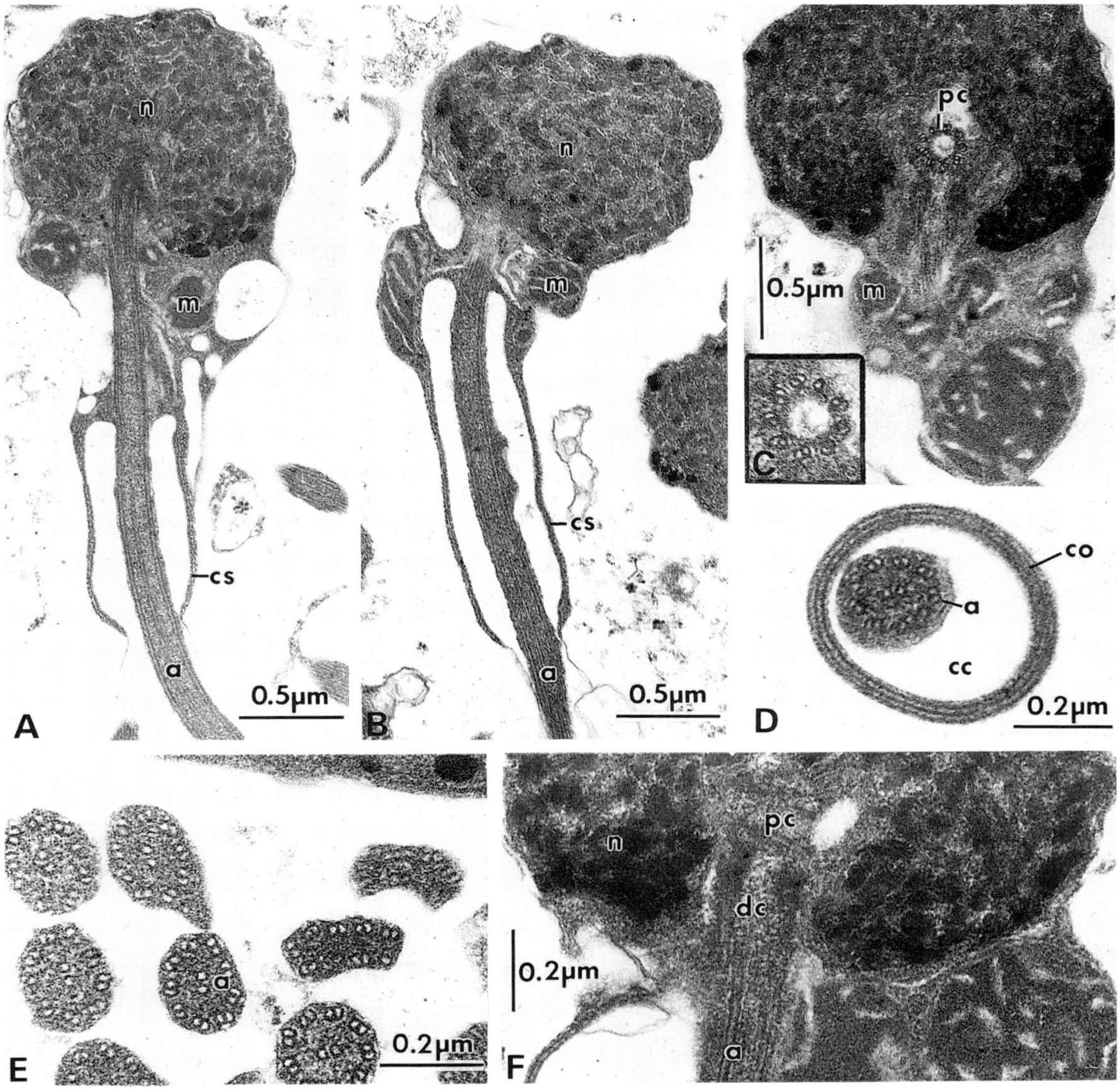

Fig. 12.17. *Sternarchus albifrons*. Spermatozoa. A and B. Longitudinal sections of two spermatozoa. C. Longitudinal section of the head and midpiece passing transversely through the proximal centriole which is shown in the inset at higher magnification. D. Transverse section of posterior amitochondrial extension of collar, enclosing cytoplasmic canal and axoneme. E. Transverse sections through the endpiece of the axoneme. F. Longitudinal section through the nucleus and midpiece at right angles to C. a. axoneme. cc. cytoplasmic canal. m. mitochondrion. n. nucleus. Original.

apparently homoplasious feature of euteleosts. The absence of axonemal fins in *Sternarchus* supports its placement in the Ostariophysi but sperm structure cannot be said to particularly endorse or refute a siluriform status for gymnotoids.

Chapter 13

ORDER SALMONIFORMES ("PROTACANTHOPTERYGII")

Diagnosis and relationships. The Salmoniformes is the only order now contained in the so-called Protacanthopterygii. This higher grouping was conceived, in a now classical work, by Greenwood *et al.* (1966) for salmoniforms, stomiatiforms, alepocephaloids, myctophids, neoscopelids and, questionably, Ostariophysi. Rosen (1973) removed all but the salmoniforms. In a cladistic classification based on gill arch and caudal skeleton anatomy and secondary sexual characters he saw salmoniforms, including esocoids, as a monophyletic entity, the reduced Protacanthopterygii, postulating that argentinoids are the plesiomorphic sister-group of salmonoids and osmeroids. Monophyly of the Protacanthopterygii has been questioned, however, especially if the Esocoidei are included. Thus Fink and Weitzman (1982) and Lauder and Liem (1983) (12.1) regard argentinoids and osmeroids as sister-groups forming a group paraphyletic relative to the salmonoids.

It will be shown below that sperm ultrastructure supports the monophyly of the argentinoid - osmeroid - salmonoid assemblage. It does not, however, endorse the doubtful union with the esocoids. Nelson (1984), who followed Rosen (1973, 1974) in including the esocoids considered that it would not be inappropriate to regard the esocoids as a separate order plesiomorphic to the Salmoniformes and possibly to all other euteleosts, as proposed by Fink and Weitzman (1982). We have formalized this by giving esocoids ordinal rank in the previous chapter.

Lauder and Liem (1983) rejected the Protacanthopterygii on the grounds that one of its chief synapomorphies, fusion of gill elements to the endochondral gill arch elements, is a general clupeocephalan, not specifically protacanthopterygian, feature and that the only contained order, the salmoniforms, even when the the "Esocae" were excluded, was paraphyletic. Although we consider at least the argentinoid-salmonoid salmoniforms to be monophyletic we follow these authors in regarding the Protacanthopterygii as a redundant, and because of its confused history, undesirable term.

The Salmoniformes are characterized by inclusion of the maxilla in the gape of the mouth. They occur almost exclusively outside the tropics in both hemispheres. All except the Argentinoidei and *Nesogalaxias* spawn in fresh water. Nelson (1984) recognizes the constituent suborders Esocoidei (here excluded), Argentinoidei, Lepidogalaxioidei and Salmonoidei. Contrary to the phylogram of Lauder and Liem (1983) (Fig. 12.1) Nelson includes the Osmeroidei as a superfamily Osmeroidea, in the Salmonoidei. As indicated above, the Esocoidei are here placed earlier in the phylogram (Fig. 13.1), as supported by Fink and Weitzman (1982), Fink (1984) (Fig. 13.2) and Lauder and Liem (1983) who regard them as the sister-group of all other euteleosts. We have modified the phylogram of Lauder and Liem (1983) (Fig. 12.1, 13.1) to show the salmonoid - argentinoid - osmeroid assemblage as a monophyletic clade.

Suborder Argentinoidei

Diagnosis and relationships. The Argentinoidei have a complex posterior branchial structure ("epibranchial organ") termed the crumenal organ (Nelson, 1984). Unique developmental characters are described by Ahlstrom *et al.* (1984). The group contains the argentinoid and alepocephaloid fishes, united in the enlarged Argentinoidei by Greenwood and Rosen (1971).

The wider relationships of the Argentinoidei are uncertain but it is suggested that they are the sister-group of the Osmeroidei on the basis of fusion of the posterior neural arches in the caudal fin with either the uroneural or ural vertebra one (references in Fink,

1984; Lauder and Liem, 1983) (Fig. 13.1).

Sperm literature. Species in two families of the suborder Argentinoidea briefly investigated are: *Xenodermichthys* sp. (Alepocephalidae) and *Searsia* sp. (Searsidae) (Mattei *et al*., 1981).

Sperm ultrastructure. Mattei *et al*. (1981) have shown that in both of these argentinoid species there (Marshall, 1989) and possibly indicates relationship of galaxioids with the Argentinoidei rather than the Salmonoidei in which, as in Fig. 13.1, they are currently placed. The unifying occlusion of doublet 3 might, alternatively, be a homoplasy.

Sperm relationships. The ultrastructure of argentinoid sperm, particularly the ring-shaped mi-

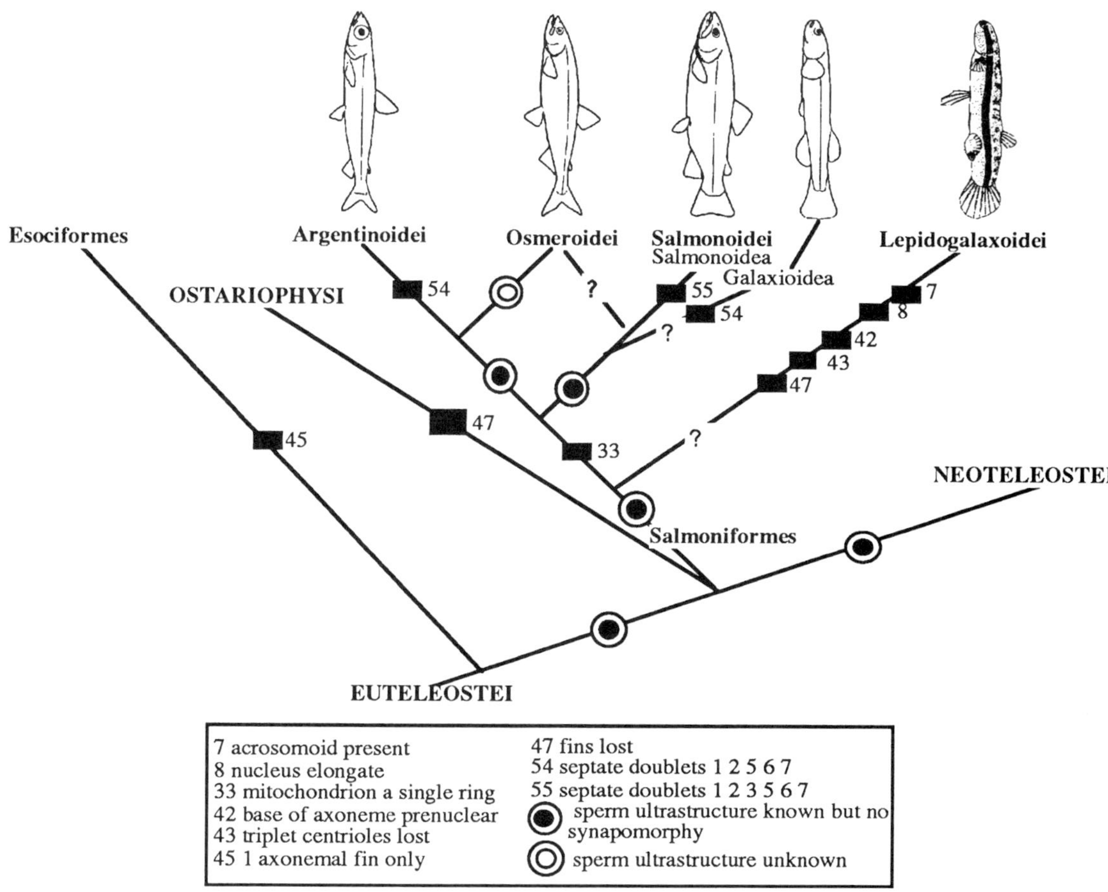

Fig. 13.1. Phylogeny of the Salmoniformes within the context of the Neoteleostei, including sperm apomorphies. Original. partly after Lauder and Liem (1983).

is a single ring-shaped mitochondrion (as in salmonoids) and that the A subtubule is septate in axonemal doublets 1, 2, 5, 6, and 7 (contrast 1, 2, 3, 5, 6, and 7 in *Salmo*). This configuration is a hitherto unrecognized synapomorphy of the Alepocephalidae and Searsidae, and therefore of the Argentinoidei *s. lat*., so far as this small sample shows. This arrangement is, however, also seen in the Galaxiidae tochondrion, leads us to link argentinoids and salmonoids in the phylogeny (Fig. 13.1) more closely than usual.

Suborder Osmeroidei

Diagnosis. The Osmeroidei usually have an adipose fin (Nelson, 1984). They are the true smelts,

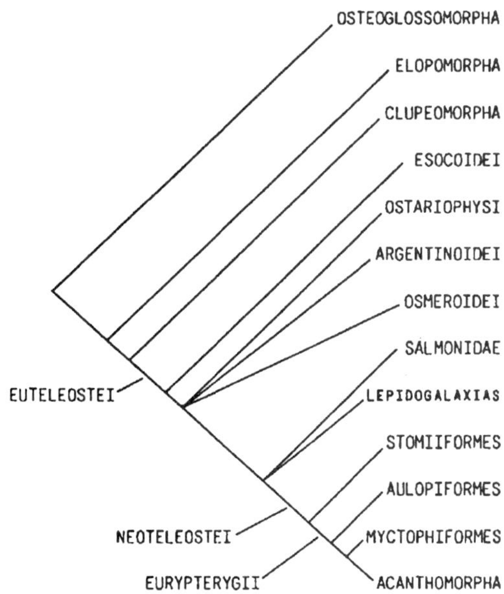

Fig. 13.2. Phylogeny of the Teleostei. From Fink, W.L. (1984). In *Ontogeny and Systematics of Fishes*. Special Publication Number 1, pp. 202-206. American Society of Ichthyologists and Herpetologists. Fig. 108.

marine anadromous or landlocked and freshwater forms in the Pacific, Arctic and Atlantic oceans and their drainages (Hearne, 1984).

Osmeroid sperm. A single species of the Osmeroidei, *Plecoglossus altivelis*, the Ayu Fish (Fig. 13.3), the only species of the family Plecoglossidae, an anadromous salmoniform of Japan, China and Korea (Nelson, 1984), has been examined for sperm ultrastructure. The very brief account forms part of a valuable investigation of penetration of the egg. The sperm has a somewhat flattened ellipsoid head about 1.2 µm wide, 1.0 µm thick, and 2.1 µm long. There is no acrosome and the flagellum is deeply inserted into the nucleus. The flagellum is 21.2 µm long and is terminally thin (Kudo, 1983). From a micrograph the shape of the nucleus, with its deep implantation fossa is typically salmoniform. A section of a mitochondrion is seen on one side of the flagellum but it is not possible to deduce whether this has the annular form usual in salmoniforms.

Suborder Salmonoidei

Diagnosis and relationships. According to

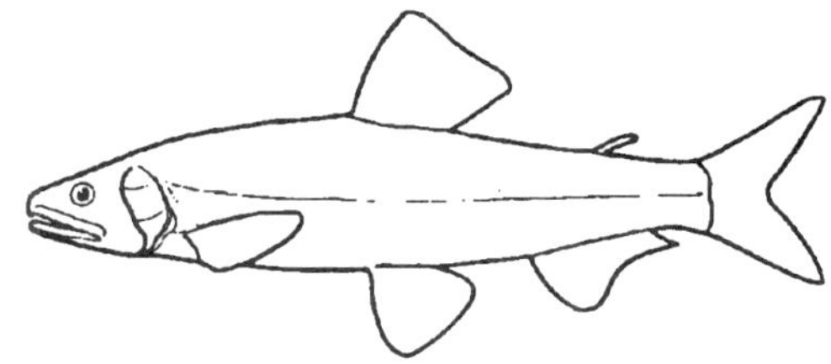

Fig. 13.3. *Plecoglossus altivelis*. From Nelson, J.S. (1984). *Fishes of the World*. 2nd edition, John Wiley, New York. p. 164.

Nelson (1984) the Salmonoidei contain the superfamilies Osmeroidea, Galaxioidea and Salmonoidea but we have seen that the Osmeroids are excluded, as the sister-group of the Argentinoidei, by Fink and Weitzman (1982) and Lauder and Liem (1983).

Galaxioids lack pyloric caeca and a mesocoracoid. They include the Galaxiidae which are diadromous salmoniforms of cool temperate southern hemisphere freshwaters, and the freshwater to partially marine Retropinnidae of New Zealand (Nelson, 1984) both of which possibly have an osmeroid derivation (MacDowall, 1984). However, Rosen (1974) associates galaxiids with the salmonids and retropinnids with the osmerids. Galaxiid sperm ultrastructure is described below and suggests argentinoid affinities.

The Salmonoidea have an adipose fin, a mesocoracoid and pyloric caeca. They contain the freshwater and anadromous family Salmonidae (Nelson, 1984).

Salmonoidea. Sperm literature. In the Salmonoidea, only the family Salmonidae has been investigated for spermatozoal ultrastructure. Species examined are: *Coregonus wartmanni*, *Hucho hucho*, Stein, 1981; *Oncorhynchus gorbuscha*, Drozdov *et al.*, 1981; *Oncorhynchus keta*, Kobayashi and Yamamoto, 1987; *Oncorhynchus kisutch*, Lowman, 1953; *Oncorhynchus tshawytscha* Zirkin, 1975 (adult, Fig. 13.4); *Salmo gairdneri*, Billard, 1970a; 1983a; Fribourgh and Soloff, 1976; Mattei *et al.*, 1981; Stein, 1981; *Salmo salar*, Nicander, 1968; *Salmo trutta* Furieri, 1962; *Salmo trutta fario*, Billard, 1983a; Stein, 1981; *Salvelinus alpinus*, Stein, 1981; *Salvelinus fontinalis*, Fribourgh, 1978; Stein, 1981 (adult, Fig. 13.5); *Thymallus thymallus*, Stein, 1981.

Thymallus and *Coregonus*, investigated by Stein (1981), who attributed each to a distinct family, are now placed in the subfamilies Thymallinae and

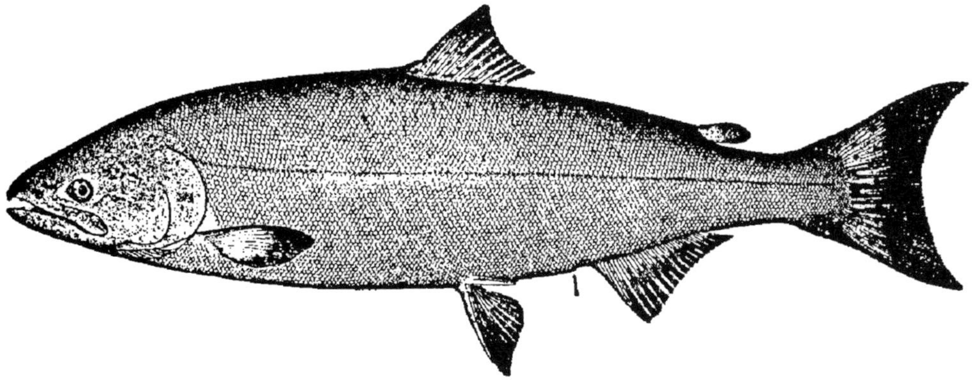

Fig. 13.3. *Oncorhynchus tschawytscha*. Quinnat Salmon. After Jordan, D.S. (1907). *Fishes*. Henry Holt and Co., New York. Fig. 222.

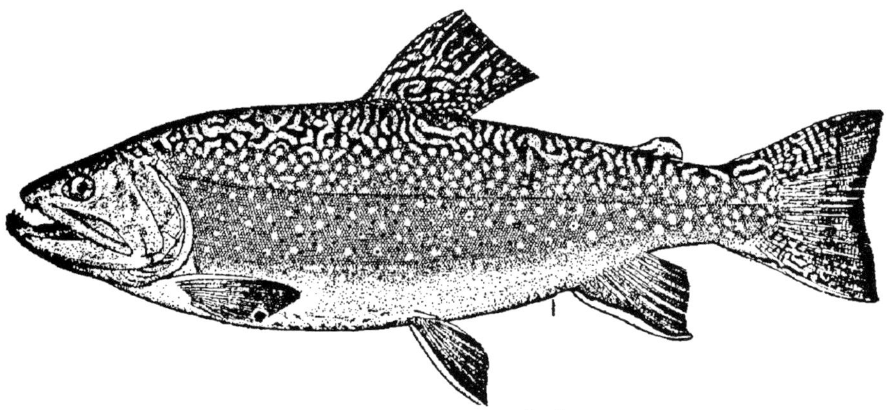

Fig. 13.5. *Salvelinus fontinalis*. Speckled Trout. After Jordan, D.S. (1907). *Fishes*. Henry Holt and Co., New York. Fig. 248.

Coregoninae in the Salmonidae (Nelson, 1984).

Salmonoid sperm type. With the Salmoninae, *Thymallus* and *Coregonus* are attributed a type 1 sperm by Stein (1981) in which the head is somewhat [but only slightly] elongated compared with the subspherical head of his type 2 sperm of the Cyprinidae and Esocidae. A type 3 sperm is recognized by Stein (1981) for *Perca* and *Cottus*, in which the head is rounded but flattened (Fig.13.6).

The most complete account of salmonid sperm ultrastructure is that for two species of trout, *Salmo gairdneri* and *S. trutta fario* (Billard, 1983a) (Fig. 13.7) on which the following account is chiefly based.

Putative acrosome. A putative acrosome has been reported in spermiogenesis of *Salmo gairdneri*, the Rainbow Trout (Billard, 1983b). (For a discussion of acrosomes in the Neopterygii see Chapter 5).

Nucleus. The head and contained nucleus is ovoid, 2.5 µm long and 1.5-2 µm wide. Some of the chromatin is compact but most is granular with electron-lucent patches. Apically, in the absence of an acrosome, a dark layer is produced by adhesion of nuclear material to its envelope. Elsewhere the nuclear material is occasionally detached from the envelope. A roughly cubical basal depression constitutes the implantation fossa (Billard, 1983a). (In contrast there is only slight development of a nuclear fossa in *Esox*.).

An osmiophilic body at the base of the nucleus in *Coregonus wartmanni* and *Salvelinus alpinus* (Stein, 1981) is possibly homologous with the round bodies seen at the anterior border of the nucleus in some Percicthyidae (see Chapter 15).

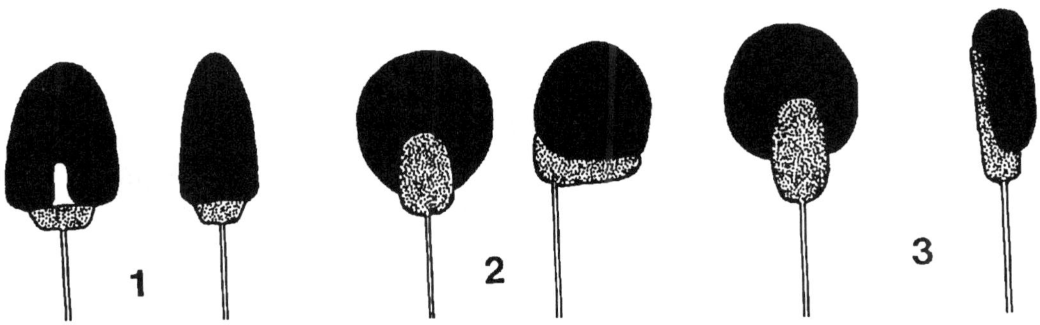

Fig. 13.6. Contrasted types of teleost sperm. A. Type 1. *Thymallus thymallus* and *Coregonus wahrmanni* (Salmonidae). B. Type 2. Cyprinidae and *Esox lucius*. C. Type 3. *Perca fluviatilis* and *Cottus gobio*. After Stein, H. (1981). *Zeitschrift für Angewandte Zoologie* **68**, 183-198. Fig. 1.

Midpiece. In *Salmo gairdneri* and *S. trutta fario*, as in other salmonids, the base of the flagellum is surrounded by a short mitochondrial collar which surrounds the cytoplasmic canal. There is a single incompletely closed ring-shaped mitochondrion, encircling the cytoplasmic canal, as also shown for *S.*

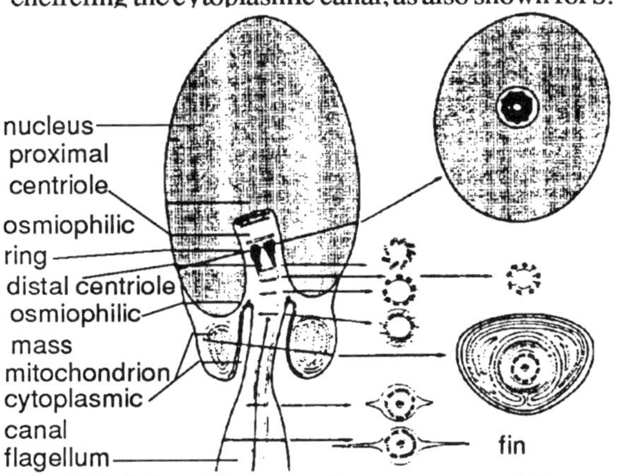

Fig. 13.7. *Salmo*. Diagrams of spermatozoon of *S. trutta fario* and *S. gairdneri*. After Billard, R. (1983a). *Cell and Tissue Research* **228**, 205-218. Fig. 1.

trutta by Furieri (1962), who claims that sometimes several mitochodria are present. The midpiece is irregularly shaped; considerable amounts of cytoplasm may be present adjacent to the mitochondrial ring. The cytoplasmic canal, bounded by the usual fold of the plasma membrane, separates the midpiece from the flagellum. At the bottom of this fold, anteriorly, the plasma membrane is surmounted by an osmiophilic mass and is more dense. A second membrane is present throughout the length of the midpiece between the mitochondrion and the plasma membrane which externally borders the cytoplasmic canal (Billard, 1983a). This membrane is seen in other orders, see, for instance, the characiform *Paracheirodon* (Fig. 12.12). Fribourgh (1978) suggests the presence of separate mitochondria in the sperm of the Brook Trout, *Salvelinus fontinalis* but Stein

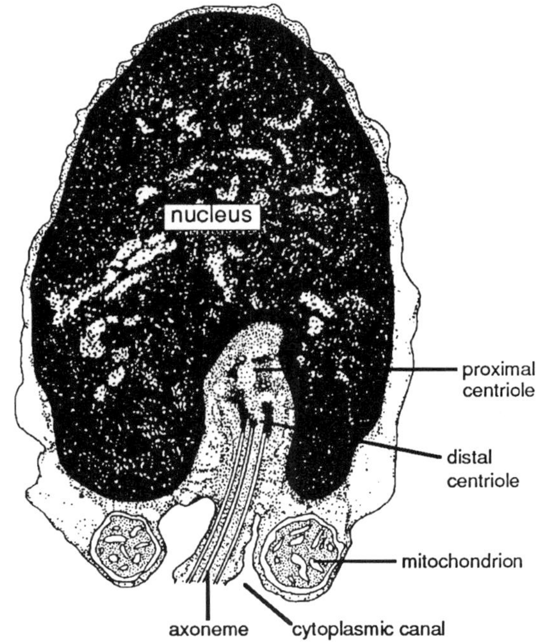

Fig. 13.8. *Oncorhynchus tshawytscha*. Longitudinal section of spermatozoon. The nucleus is extremely electron dense but clumps of material can often be seen. After a micrograph by Zirkin, B.R. (1975). *Journal of Ultrastructure Research* **50**, 174-184. Fig.13.

(1981) (Fig. 13.6) states that there is a single mitochondrial ring in this species and in *Salmo*, *Coregonus*, *Salvelinus* and *Thymallus*.

Centriolar complex. The two, mutually perpendicular centrioles, each 30 nm long by 22 nm wide, are located in the fossa. The distal is slightly displaced relative to the proximal centriole so that the two are not in the same plane, the extent of the displacement varying from sperm to sperm. The anterior part of the distal centriole, embedded in electron dense material, is attached by several filamentous bundles to a series of highly osmiophilic axes arranged parallel to the proximal centriole and apparently attached to its triplets. Twisted microtubular structures are present between the distal centriole and the lateral walls of the fossa (Billard, 1983a). The distal centriole is of the triplet type (Stein, 1981; Billard, 1983a).

Flagellum. Usual metazoan flagellar structure is described by Billard (1983a). The triplets of the distal centriole give way posteriorly to doublets. They are conjoined by a ring, and each is linked to the plasma membrane at the level of the base of the implantation fossa by an arm [Y-link equivalent] shortly in front of the osmiophilic mass. The arms disappear at the commencement of the two central singlets. The plane of the two singlets passes through the plane of the longitudinal axis of the proximal centriole. The two central tubules are interconnected by a straight line and by two semicircular lines which constitute the sheath. Nine arms, supposedly connecting the sheath with the A tubules of the doublets are described and clearly constitute Afzelius rays (Billard, 1983a). Distally the B tubules disappear first while the A tubules lose their [dynein] arms. At the tail tip only one or two tubules remain (Billard, 1983a).

At about 1 µm behind the midpiece, in *Salmo gairdneri* and *S. trutta fario* sperm (Billard, 1983a), extensions of the flagellar plasma membrane [fins], appear. In a survey of 200 transverse sections of flagella, two opposite lateral extensions (82%) and more rarely (6.4%) perpendicular ones were found. Sometimes (12%) there were three extensions. The two fins are not always equal in length; their orientation relative to the central singlets varies, from the same plane (38%), to perpendicular to it (16%), or orientated differently from these cases (46%). It is here suggested that this variation may represent fixation of dynamic variation in the individual axoneme. The A tubule of the 1, 2, 3, 5, 6, and 7 doublets appears darker and septate in the region of the fins. Two lateral fins were observed on the 9+2 flagellum in *Salmo*, *Coregonus*, *Salvelinus* and *Thymallus* by Stein (1981).

Salmonoid sperm and phylogeny. Relationship of the Salmonoidei and Argentinoidei is endorsed by the synapomorphic ring-shaped single mitochondrion and some elongation of the nucleus. The presence of flagellar fins is a symplesiomorphy but contrasts with the apomorphic loss of these in the related Ostariophysi.

Galaxioidea sperm. A single species of the superfamily Galaxioidea, *Galaxias olidus*, has been examined for sperm ultrastructure (Marshall, 1989).

Nucleus. The nucleus is moderately elongate, with a length of 2.0-2.3 µm (mean 2.2 µm) and a width of 0.7 µm (n=4). The basal fossa is deep to the extent that the longitudinal sagittal section appears inverted U-shaped and most cross sections doughnut-shaped. The chromatin consists of an electron dense matrix containing scattered darker particles, and closely conforming to the double nuclear membrane.

Midpiece. There is a single, annular mitochondrion subdivided by few large cristae. This surrounds a short, narrow cytoplasmic canal. There appears to be a short extension of the mitochondrion along one side of the basal region of the nucleus.

Centrioles and flagellum. Probably because of the moribund condition of examined sperm, no centrioles have been observed. The axoneme, inserted in the longitudinal axis of the head, has a 9+2 pattern of microtubules and, in different regions, has 1, 2 or no fins. The fins are orientated in the plane of the central

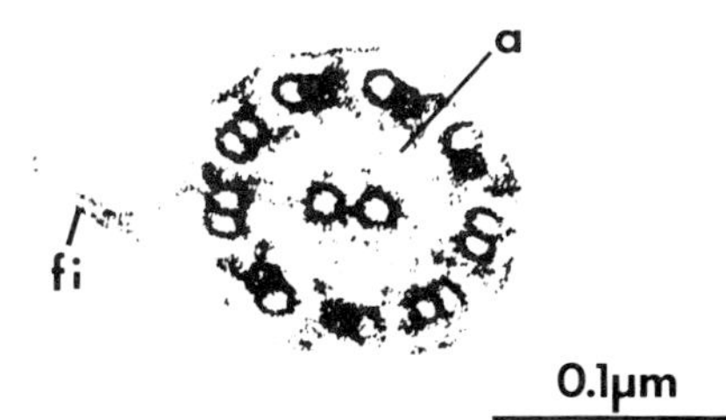

Fig. 13.9. *Galaxias olidus*. Transverse section of flagellum, showing occluded A mirotubules in doublets 1, 2, 5, 6 and 7. Courtesy of C.J. Marshall.

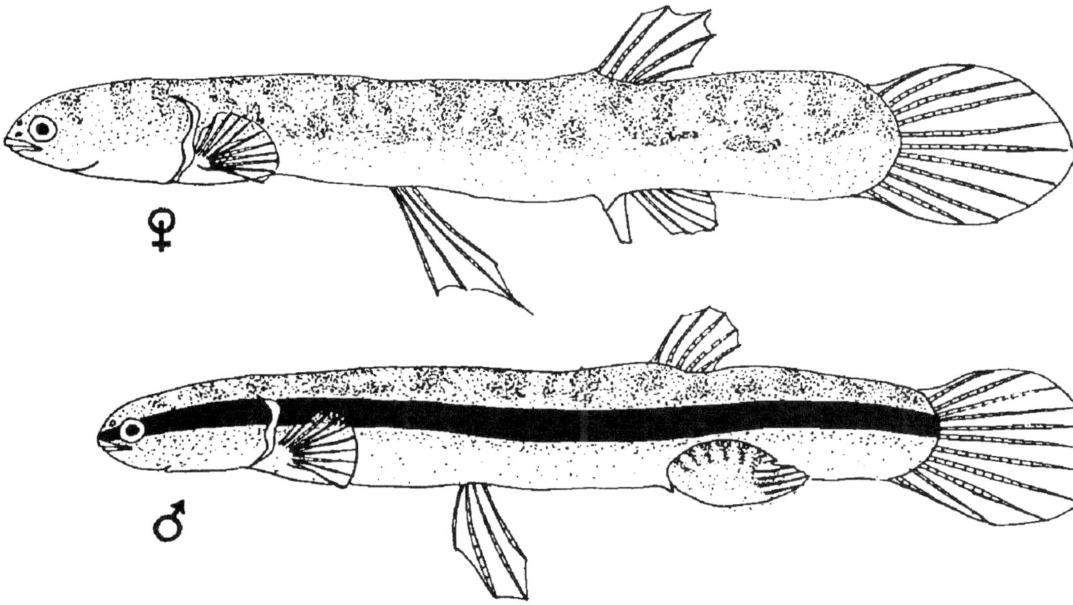

Fig. 13.10. *Lepidogalaxias salamandroides*. After Pusey, B.J. (1983). *Journal of the Australia New Guinea Fishes Association.* **1**, 9-11. Fig. 1.

singlets. The lumina of the subtubules of the doublets are complete in many sections and in all of those which are cut through the nucleus. In some sections the A subtubule of doublets 1, 2, 5, 6 and 7 appears dark and occluded because it is partitioned by a septum (Fig. 13. 9). Each doublet has two dynein arms (Marshall, 1989).

Sperm relationships. The sperm of *Galaxias olidus* thus closely resembles a salmonid sperm but the nucleus, though no longer, is more slender and therefore appears more elongate and is more deeply penetrated by the basal fossa, and the axoneme has a different pattern of occlusion of the doublets. As noted above, the galaxiid pattern of occlusion agrees with that in argentinoids. Salmonid occlusion differs only in septation of doublet 3 in addition to those occluded in *Galaxias* and it is uncertain what phylogenetic significance should be attached to this difference.

Suborder Lepidogalaxioidei

Relationships. The single species of the suborder Lepidogalaxioidei and family Lepidogalaxiidae, *Lepidogalaxias salamandroides*, the Salamander Fish (Fig. 13.10), is known only from a small creek in South-western Australia. It was first described as a galaxiid (see Pusey, 1983) but was considered by Rosen (1974) to be more closely related to the esocoids than to galaxoids and osmeroids despite similarities with these. Fink (1984) found *Lepidogalaxias* "a potpourri of contradictory and reductive characters". He paired it with salmonids but saw it as the sister-group of the Neoteleostei (Stomiiformes, Aulopiformes, Myctophiformes and Acanthomorpha) (Fig. 13.2).

Sperm literature. The sperm of *Lepidogalaxias salamandroides* has been investigated ultrastructurally by (Leung, 1988).

Lepidogalaxias. The spermatozoon (Fig. 13.11) differs from all other neopterygian sperm in having an acrosome (excepting putative homologues noted below for a few species, see also Chapter 5) with one or two perforatoria, and in lacking triplet centrioles (Leung, 1988). It is unusual but not unique in attachment of the flagellum anterior to the nucleus, a condition seen (with no phylogenetic construction) in the biflagellate sperm of *Polypterus*.

Acrosome. A flattened structure filled with a homogeneous matrix on the anterior end of the spermatozoon resembles an acrosome vesicle. It has a dense cylindrical core about 0.6 µm long (range 0.49-0.71 µm, n=6) identified as a perforatorium. In some spermatozoa two perforatoria are present.

Nucleus. The nucleus is exceptionally long for the Pisces, at about 20 µm, a length exceeded only by the sperm nuclei of sharks and *Neoceratodus*. It is 0.3 µm in maximum diameter. Its chromatin is dense and homogeneous with occasionally a single axial lacuna. In some sperm the anterior end of the nucleus bifurcates. Its tapered posterior end, bounded by the nuclear envelope and plasma membrane, projects freely parallel to the anterior region of the axoneme.

Spermatozeugmata. In the tubules of the tissue surrounding the testis spermatozeugmata occur in which the sperm are parallel and approximately in register; the anterior ends of the sperm are less compactly arrayed than the posterior ends.

The acrosome and phylogeny. The presence of an acrosome in this teleost is intriguing. An acrosome is known from the Ceratodontiformes, Lepidosireniformes, Coelacanthiformes, Polypteriformes, and Acipenseriformes but not in taxa of presumed more recent origin, the Lepisosteiformes, Amiiformes or teleostean groups with the exception that perinuclear intermembranous scales in *Lepadogaster lepadogaster* (Gobiesocidae) have tentatively been identified as a vestigial acrosome (Mattei and Mattei, 1978) and acrosome-like vesicles are present in the *Paracheirodon* and *Melanotaenia* sperm and immature *Gambusia* sperm (see also Chapter 5).

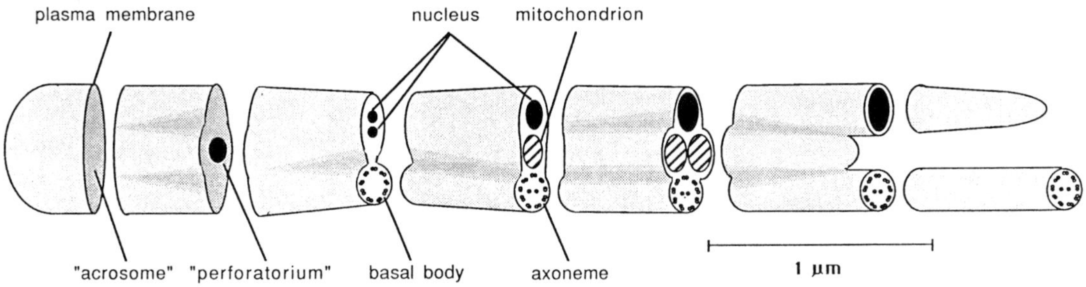

Fig. 13.11. *Lepidogalaxias salamandroides*. Schematic model of the spermatozoon. From Leung, L.K-P. (1988). *Gamete Research* **19**, 41-49. Fig. 27.

Mitochondria. Two elongate mitochondria, 10-13 µm long and 0.28 µm wide, are situated between the nucleus and the axoneme, parallel to these. The anterior ends of the mitochondria are usually at some distance (0-1.3 µm) behind the anterior end of the nucleus. The cristae are predominantly longitudinal. Two to four microtubules are occasionally associated with the mitochondria.

Centrioles. No typical triplet centrioles have been observed. A basal body, 0.46-0.54 µm long, located at the anterior end of the flagellum consists of 9 doublets. It is amorphous, lacking the circular configuration, in cross section, of the flagellum with which it is continuous. Electron dense material is associated with the tubules and radial links of the basal body. Some tubules narrower than these tubules are present on the inner side of the doublets.

Flagellum. The 9+2 axoneme is approximately 53 µm long; dynein arms, radial links, central sheath and central cross bridges are present. Towards its posterior end, as is usual in flagella, A and B subtubules separate from each other and the arrangement of the tubules becomes disorientated (Leung, 1988).

Presence of an acrosome in *Lepidogalaxias* correlates with absence of a micropyle from the egg (Pusey and Stewart, 1989), an exceptional condition for a teleost. This in turn is presumably an adaptation to some (specialized?) feature of its internal fertilization. It is possible that the acrosome also functions in the known embedment of the sperm in the wall of the crypt region of the oviduct (Pusey and Stewart, 1989).

An acrosome is here considered a basic (plesiomorphic) feature of fish sperm irrespective of whether fertilization was primitively internal or external in fish. The presence of the acrosome in *Lepidogalaxias*, a condition shared, in salmoniforms, with the spermatid of *Salmo* (Billard, 1983b), if a plesiomorphy, would contribute little to the hypothesis of relationship of their two suborders. It might, however, be considered to indicate a more basal position for the Salmoniformes within the teleosts than has been accepted.

Leung (1988) suggested that internal fertilization in *Lepidogalaxias*, whether primitive or not, is adaptive to the highly acidic waters in which it lives. The adverse effects of acidity on teleost sperm had been demonstrated by Urho *et al.* (1984). Leung considered the alternatives that internal fertilization and the complex form of the sperm (including presence of the acrosome) were retentions from an ancestral condition or were secondarily acquired in response to the acid environment and concluded that the former hypothesis was more parsimonious. Pusey and Stewart (1989) perhaps rightly consider it more parsimonious to regard internal fertilization in *Lepidogalaxias* as secondary. This view is here considered to be endorsed by the clearly secondary absence of the micropyle, a structure already present in the Chondrostei. Pusey and Stewart (1989) suggest that internal fertilization has arisen in relation to sperm competition while also suggesting that it may allow the female to confine the laying of fertilized eggs to optimal conditions for their development.

Chapter 14

NEOTELEOSTEI

SCOPELOMORPHA AND PARACANTHOPTERYGII

SUPERORDER SCOPELOMORPHA

Diagnosis. The grouping Scopelomorpha is accepted here with hesitation. Among characteristic features, the premaxilla forms the gape of the mouth and the maxilla is excluded; the upper jaw is not protrusible; an adipose fin is usually present; and the caudal fin is usually forked (Nelson, 1984). Scopelomorphs contain the orders Aulopiformes and Myctophiformes. Both are included as suborders in the Myctophiformes by some workers (for instance, Okiyama, 1984) while Fink (1984) sees the Aulopiformes as the sister group of the Myctophiformes + Acanthomorpha, (see Chapter 13, Fig. 13. 2).

Lauder and Liem (1983) depict the Aulopiformes and Myctophiformes as having independent though close origins from the main neoteleost line and therefore would consider the Scopelomorpha an artificial, paraphyletic assemblage (see also Fig. 14.3).

Order Aulopiformes

Diagnosis. Aulopiformes are marine, benthic to pelagic fish; many are synchronous hermaphrodites. Members of the order have a specialization of the gill arches which is apparently unknown in other teleosts: the second pharyngobranchial is greatly elongated posterolaterally, extending away from the third pharyngobranchial, with the uncinate process of the second epibranchial contacting the third pharyngobranchial (Rosen, 1973; Lauder and Liem, 1983; Nelson, 1984).

Trachinocephalus. The spermatozoon of the only aulopiform species studied, *Trachinocephalus myops*, the Snake Fish, is an anacrosomal aquasperm known only from a line drawing (Mattei, 1970) (Fig. 14.2). *Trachinocephalus* is a member of the Atlantic, Indian and Pacific Ocean family Synodontidae, lizard or snake fish (Fig. 14.1).

The nucleus is an asymmetrical cone with a deep

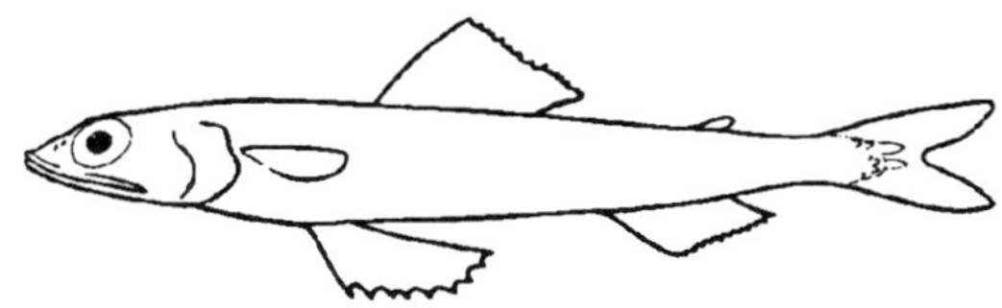

Fig. 14.1. An example of a Lizard or Snake Fish (subfamily Synodontinae). From Nelson, J.S. (1984). *Fishes of the World*. 2nd edition, John Wiley and Sons, New York. p. 181.

basal fossa which contains the proximal centriole at right angles to the basal body. The large (single?) mitochondrion is depicted on one side only of the base of the flagellum.

Sperm relationships. Little can be determined with regard to aulopid relationships from this de-

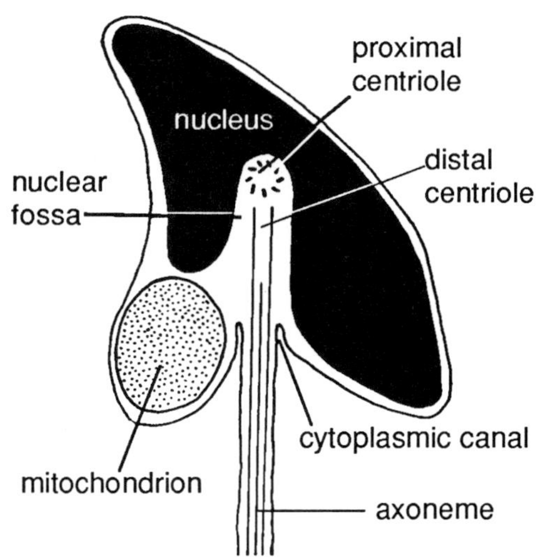

Fig. 14.2. *Trachinocephalus myops*. Diagrammatic longitudinal section of the spermatozoon. After Mattei, X. (1970). In *Comparative Spermatology* (ed. B. Baccetti), pp. 59-69. Academic Press, New York. Fig. 4:12.

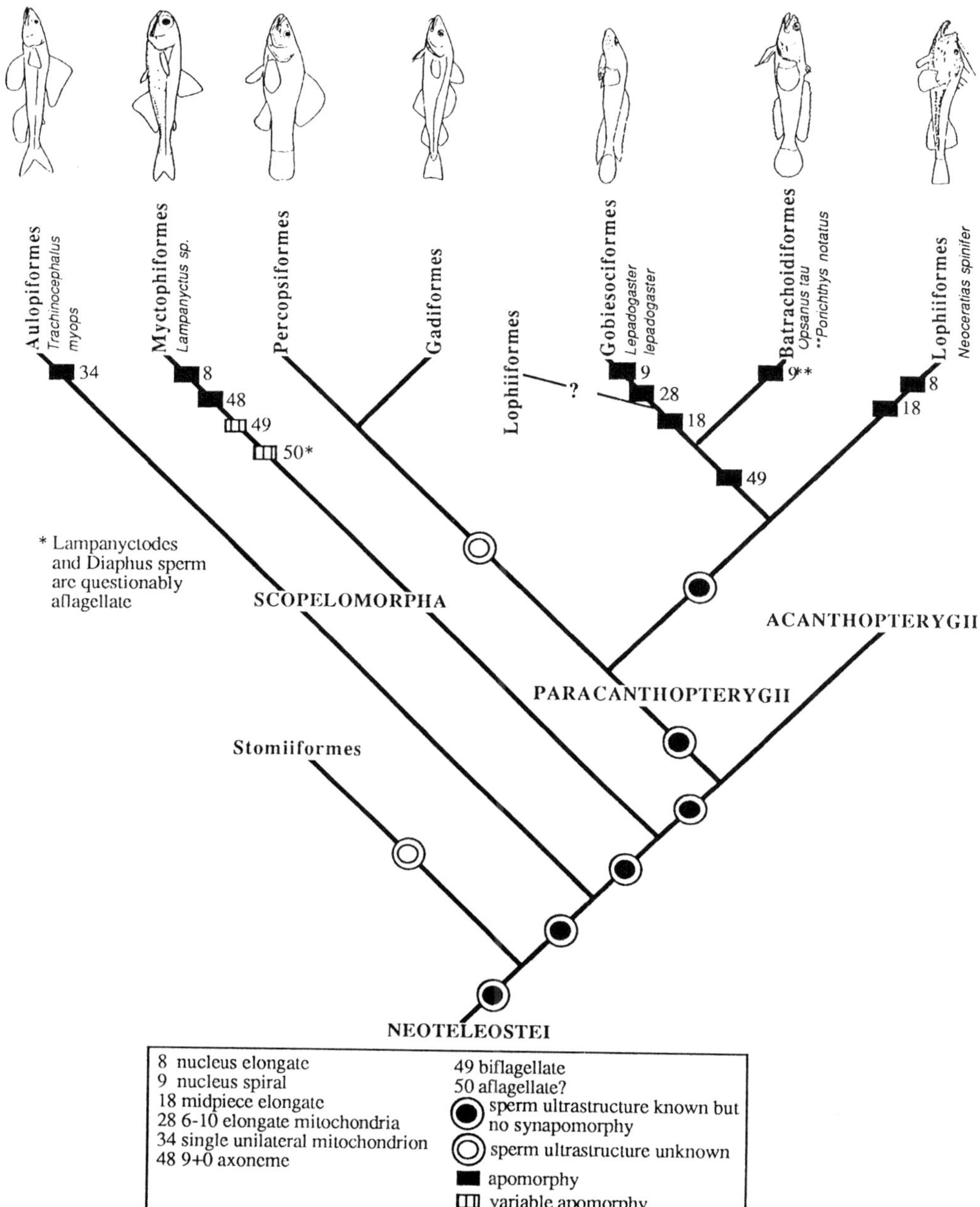

Fig. 14.3. Interrelationships of the major groups of the Paracanthopterygii. Modified after Lauder, G.V and Liem, K.F. (1983). *Bulletin of the Museum of Comparative Zoology* **150**, 95-197. Fig. 37. With spermatozoal apomorphies superimposed and Gobiesociforms regarded as the sister-group of batrachoidiforms.

scription beyond noting that the the mitochondrion, although possibly single, as in the argentinoid-salmonoid assemblage, differs in being unilateral; and that the conical form of the nucleus with its deep implantation fossa and a single mitochondrion are not ostariophysan features.

Order Myctophiformes

Diagnosis. Myctophiforms differ from the Aulopiformes in having upper pharyngobranchials and retractor muscles like those of generalized paracanthopterygians (Rosen, 1973). The head and body are compressed; the mouth is large and terminal; an adipose fin is present; there are usually 8 pelvic fin rays and 7-11 branchiostegal rays. All are deep sea pelagic and benthopelagic species (Nelson, 1984). The type-family, the Myctophidae (Fig. 14.4), exemplified here by *Lampanyctus,* is characterized by photophores arranged along the sides of the body, hence the name lantern fish; they are amongst the most common deep sea fish.

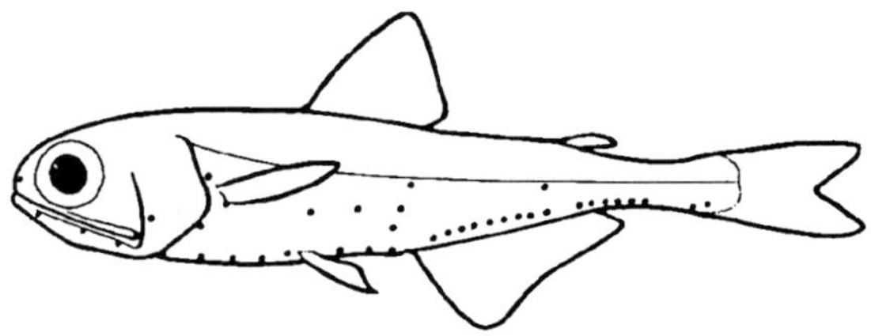

Fig. 14. 4. A lantern fish (Myctophidae). From Nelson, J.S. (1984). *Fishes of the World.* 2nd edition, John Wiley and Sons, New York. p. 185.

Fig. 14.5. *Lampanyctus* sp. Spermatozoon *in toto.* From Mattei, X. and Mattei, C. (1976a). *Journal de Microscopie et de Biologie Cellulaire* **25**, 187-188. Fig. 1e.

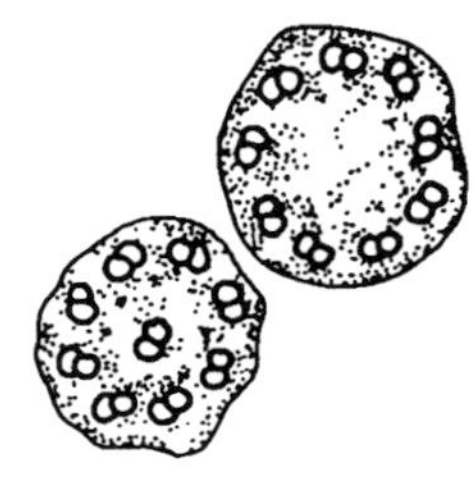

Fig. 14.6. *Lampanyctus* sp. Cross sections of axonemes, showing 9 doublets but no central singlets. One of the sections is abnormal in having one of the doublets displaced to the centre. After a micrograph of Mattei, X. and Mattei, C. (1976a). *Journal de Microscopie et de Biologie Cellulaire* **25**, 187-188. Fig. 1d.

***Lampanyctus* sperm.** The African, Senegalese, *Lampanyctus* sp., (Myctophidae, Lampanictinae) has a biflagellate spermatozoon (Fig. 14.5) with a 9+0 axonemal structure (Fig.14.6).

The nucleus is S-shaped and about 6 µm long; the mitochondria are not organized into a true midpiece; and the two flagella, inserted laterally considerably proximal to the base of the nucleus, are 55 µm long. The two centrioles (described only for the spermatid) are parallel (Mattei and Mattei, 1976a).

Sperm relationships. The asymmetry of the nucleus and of insertion of the flagella are reminiscent of the less conspicuous condition in the uniflagellate sperm of the aulopiform *Trachinocephalus* but are not sufficient to conclusively support sister-group relationship of the two orders.

Biflagellarity is known elsewhere in teleosts only in the Ictaluridae, Apogonidae, Malapteruridae, and in the Batrachoididae and the related Gobiesocidae, but only the uniflagellate elopomorph sperm share the 9+0 axoneme. No phylogenetic significance can be attached to these resemblances to *Lampanyctus* nor is their functional significance clear.

***Lampanyctodes* and *Diaphus* sperm.** Scanning electron microscopy of the sperm of the lampanyctines *Lampanyctodes hectoris* and unspecified observations on *Diaphus danae* are said by Young *et al.* (1987) to have revealed aflagellate sperm but this is not ascertainable from the micrograph which they provide.

SUPERORDER PARACANTHOPTERYGII

Diagnosis and relationships. This superorder, defined largely on features of the caudal skeleton, jaw muscles and nerve patterns (see review by Rosen, 1973) consists of forms of presumed early origin, possibly derivatives of the neoscopelid-myctophid or the ctenothrissiform lineage via the Percopsiformes (but see Fig. 14.3 based on Lauder and Liem, 1983, for percopsiform relationships). Paracanthopterygii may represent a side branch in teleost evolution, the remaining fishes (Acanthopterygii) either sharing a common ancestry within or being derivatives of a related lineage (Nelson, 1984).

Lauder and Liem (1983) support the view that the Paracanthopterygii are the plesiomorphic sister-group of the Acanthopterygii (Fig. 14.3). They list the following synapomorphies for the Paracanthopterygii: increase in the number of abdominal vertebrae; decrease in the depth of the head and trunk; suboperculum enlarged and operculum reduced; anterior vertebrae crowded and linked; a trend toward various patterns of fusions of the hypurals; in the caudal skeleton the second preural centrum possesses a complete spine which is formed by fusion of the first epural with the crest on the second preural centrum.

Paracanthopterygians are predominantly marine; only five genera of the percopsiforms, the gadiform *Lota*, a few brotulids, some batrachoidiforms and the fluviatile gobiesocids are freshwater fish (Lauder and Liem, 1983).

Spermatozoal ultrastructure has not been studied in the Gadiformes and Percopsiformes but is known for the Ophidiiformes, Batrachoidiformes, Gobiesociformes and Lophiiformes.

Order Ophidiiformes

General. In the Ophidiiformes the pelvic fins, when present, are inserted at the level of the preopercle or more anteriorly, with one or two soft rays in each. Dorsal and anal fins are long and usually joined with the caudal fin (Nelson, 1984). The Bythitidae, in the Ophidiiformes, are live-bearing. In an excellent light microscopical paper on ophidioid spermatophores, Nielsen *et al.* (1968) suggest that the development of these structures may be a device ensuring that the spermatozoa are kept alive in the female until the eggs mature.

***Ophidion* sperm.** The spermatozoon of *Ophidion* sp.(suborder Ophidioidei, Ophidiidae, Ophidiinae) has been examined ultrastructurally by Mattei *et al.* (1989).

Nucleus. The nucleus is 8 μm long and circular in cross section. The base is 0.8 μm wide and its anterior end is drawn out into a very slender process, 2.5 μm long and 0.05 μm wide.

Midpiece. The "intermediate piece" is short, measuring about 0.6 μm long and 1 μm wide. It contains two centrioles, 7 to 8 mitochondria and a vesicular cytoplasm. In places the mitochondria appear from a micrograph to be in two tiers of which the more posterior encloses a short cytoplasmic canal.

Flagellum. The flagellum is about 100 μm long. It has the 9+2 structure and two lateral fins which are generally inclined at 10° and 30° relative to the plane of the central singlets. As many as eight 10 nm wide longitudinal microtubules are present under the plasma membrane near the lateral extremity of each fin and appear to act as a cytoskeleton (Mattei *et al.*, 1989).

Sperm phylogeny. *Ophidion* shows elongation of the nucleus which is common on adoption of internal fertilization but is otherwise unmodified apart from the filamentous form of the apex of the nucleus. These modifications must be interpreted as independent developments which do not cast light on the phylogenetic affinities of ophidioids.

Order Gadiformes

General. The pelvic fins, when present, are inserted below or in front of the pectorals. and have up to 17 rays. Scales are cycloid (for other features, see Nelson, 1984). The only gadiform species for which sperm have been studied, optically, *Lota vulgaris*, has a uniflagellate anacrosomal aquasperm (Retzius, 1905). Gadiforms are seen as the sister-group of the percopsiforms by Lauder and Liem (1983) (see also Fig. 14.3).

Order Batrachoidiformes

Diagnosis. In batrachoids, among other features,

the body is usually scaleless (or in some has small cycloid scales); the head is large with eyes tending to dorsal; the large mouth is bordered by the premaxilla and maxilla; there is a short spinous dorsal fin; the pelvic fins are anterior to the pectorals, with one spine and two or three soft rays; the entopterygoid is not ossified; there are three pairs of gills; and a swim bladder is present. They are mostly marine (Lauder and Liem, 1983; Nelson, 1984). The Batrachoidiformes are envisaged by Lauder and Liem (1983) as the sister-group of the Lophiiformes (but see Fig.

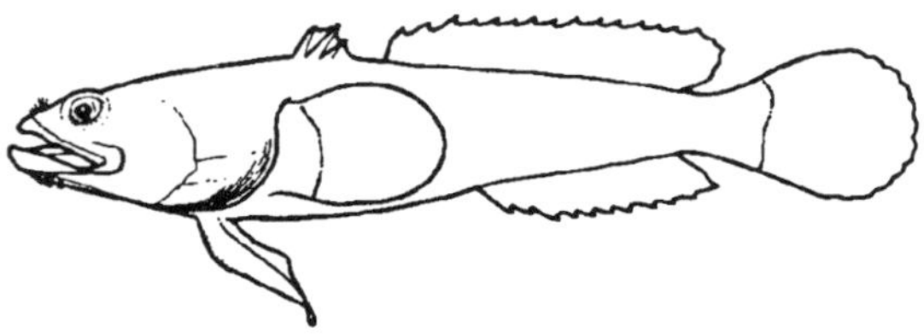

Fig. 14.7. A toad fish (subfamily Batrachoidinae). From Nelson, J.S. (1984). *Fishes of the World.* 2nd edition, John Wiley and Sons, New York. p. 199.

14.3), with the synapomorphy of elongate pectoral radials, a view which receives some support from the mutual possession of helical nuclei, a rare condition. The two orders are seen by them as the sister-group of the Gobiesociformes. Synapomorphies of the batrachoidiform - lophiiform - gobiesociform clade are: skull roof flattened; parasphenoid and frontal bones either approaching each other or sutured to each other; large sphenotics flaring forward and laterally; and progressive reduction in the ossification of the palatopterygoid (Lauder and Liem, 1983).

Biflagellarity of batrachoidiform and gobiesociform sperm, as far as they are known, is a remarkable similarity and if not a synapomorphy may be a homoplasy by virtue of relationship (symparamorphy *sensu* Jamieson, 1984). Elongation of the pectoral radials is a somewhat slender link between batrachoidiforms and lophiiforms in contrast with the more convincing reasons for allying the three orders. It may be that the biflagellarity of batrachoidiform and gobiesociform sperm is a true synapomorphy of the two orders and that lophiiforms are their sister-group, as indicated in Fig. 14.3.

Batrachoid sperm. The two investigated species of the Batrachoidiformes (adult, Fig. 14.7), are *Opsanus tau*, the Oyster Toadfish (Hoffman, 1963;

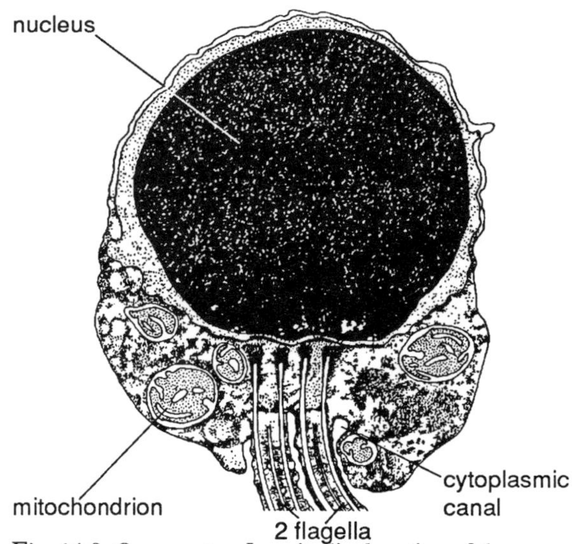

Fig. 14.8. *Opsanus tau.* Longitudinal section of the spermatozoon. Drawn from micrographs after Casas, M.T., Munoz-Guerra, S. and Subirana, J.A. (1981). *Biology of the Cell* **40**, 87-92. Fig. 2e and f.

Casas *et al.*, 1981) and *Porichthys notatus*, the Plainfin (Stanley, 1965b). Both are spawners with parental care (Breder and Rosen, 1966) and have biflagellate aquasperm, simple in the former, more complex in the latter.

Opsanus. The light microscope study of Hoffman (1963) provides little information beyond demonstrating the biflagellarity of the sperm of *Opsanus tau.* The ultrastructural account of Casas *et al.*, (1981) is directed to an investigation of the chromatin but yields additional data.

The sperm of *Opsanus tau* (Fig. 14.8) is a typical teleostean aquasperm except for its biflagellarity.

Nucleus. The nucleus is spheroidal. Upon maturation of the spermatozoon the chromatin fibres become less apparent than during spermiogenesis. The chromatin is uniformly distributed through the nucleus. It appears as granules and filaments of various sizes (50-100 Å). Thus this histone-containing nucleus does not show the high degree of compaction achieved in other species which contain protamines in their sperm nuclei. The individuality of the 200 Å fibres, seen in spermatids, is lost. *O. tau* may represent an intermediate case between similarly histone-containing sperm in which the 200 Å diameter fibres are preserved at maturity (*Holothuria* and *Limulus*) and those in which complete condensation is achieved through the effect of protamines (Casas *et al.*, 1981).

Midpiece. The midpiece contains many irregularly arranged round mitochondria and there is a short cytoplasmic canal around the base of each flagellum.

Motile apparatus. The two centrioles are mutually parallel and each is associated with a separate slight indentation of the nucleus (Casas *et al.*, 1981). It is not stated whether the 9+2 flagella possess fins. These are not present in other biflagellate sperm.

Porichthys. In the late spermatid of *Porichthys notatus* (unillustrated), in contrast with *Opsanus*, the nucleus is spiral, although again with no evident acrosome, and the mitochondria are elongate, lying close around the basal portions of the flagella to form a distinct midpiece. The parallel centrioles are each connected by a fan-like array of filaments to the nuclear membrane which is slightly indented (Stanley, 1965b). There is a suggestion of a similar connection in micrographs of the *Opsanus* sperm.

Batrachoid sperm and phylogeny. The high cellular DNA content led Hinegardner (1968) to regard *Opsanus tau* as primitive, a view which Casas *et al.* (1981) consider might be supported by the presence of histones and the low degree of condensation of the sperm nucleus. They note, however, that Mattei (1969) considers that rotation of the nucleus during spermiogenesis, seen in *O. tau*, is an advanced feature. Somatically batrachoids are in fact highly modified if basal euteleosts. The histone-rich sperm nuclei in *O. tau* perhaps indicate their primitive origins while their biflagellarity, with that of *Porichthys notatus*, is a further apomorphy shared (homplasically?) with gobiesociforms. Whether this biflagellarity is a true synapomorphy or a homoplasy, the fact that biflagellarity is, seen, independently, in basal pre-teleostean lineages (lungfish and *Polypterus*) indicates that this condition cannot be regarded *per se* as indicating an advanced position for batrachoids.

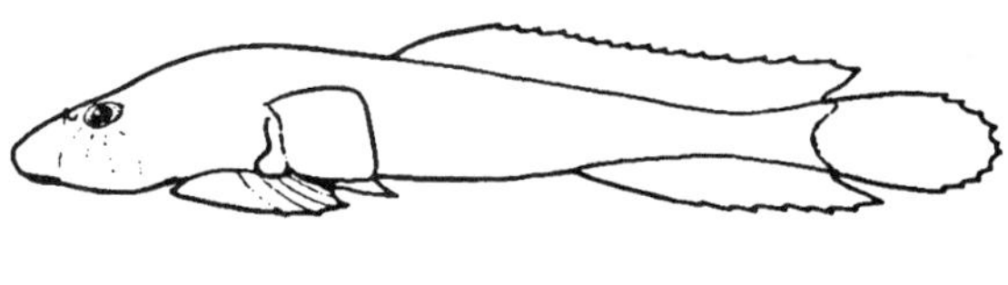

Fig. 14. 9. A cling fish (Gobiesocidae). From Nelson, J.S. (1984). *Fishes of the World.* 2nd edition, John Wiley and Sons, New York. p. 211.

Order Gobiesociformes

Diagnosis. Gobiesociforms (cling fish) (Fig. 14.9) are primarily marine bottom dwellers in shallow waters and occur worldwide in tropical and temperate seas. They have a scaleless head and body; 5-7 branchiostegal rays; pelvic fins well anterior to the pectorals; and no swim bladder (for other characters, see Nelson, 1984). Synapomorphies listed by Lauder and Liem (1983) are: extreme dorsoventral flattening; presence of a joint between cleithrum and supracleithrum; pelvic fins modified into a sucker; presence of a joint between a convex condyle of interoperculum and concave fossa of epihyal; hypertrophied sternohyoideus with a medial and lateral head; asymmetrical heart with reduced sinus venosus and large accessory common cardinal chambers; and

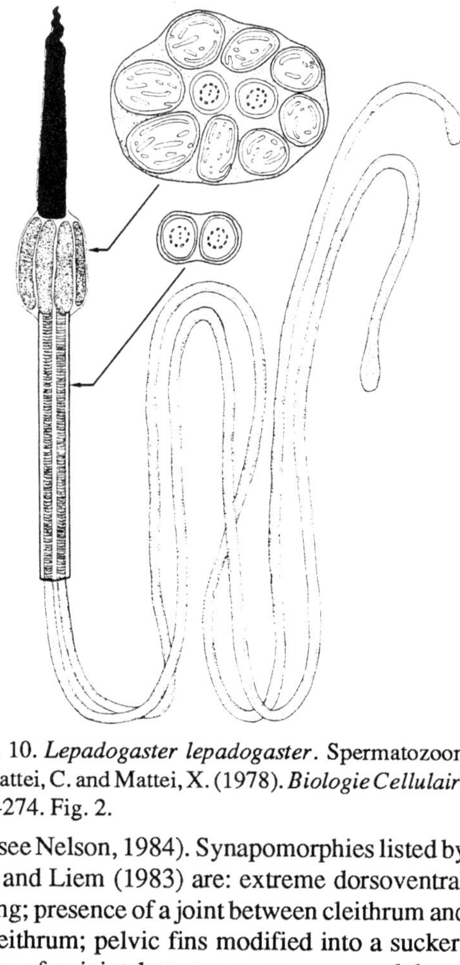

Fig. 14. 10. *Lepadogaster lepadogaster*. Spermatozoon. After Mattei, C. and Mattei, X. (1978). *Biologie Cellulaire* **32**, 267-274. Fig. 2.

a short ventral aorta with only three afferent branchial arteries.

Contrary to Lauder and Liem (1983), gobiesociforms are here (Fig. 14.3) tentatively regarded as the sister-group of the Batrachoidiformes on the basis of shared biflagellarity of sperm in conjunction with batrachoidiform - Gobiesociform - lophiiform similarities. However, Gosline (1970) considered that similarities (absence of scales, flattened head, anterior pelvic, incomplete circumorbital series) are convergent between the two and that gobiesociforms have almost all the diagnostic features of notothenioid blennioids in the Perciformes. Some current classifications therefore regard them as perciform derivatives within the superorder Acanthopterygii. Allen (1984) found little resemblance between gobiesociform and notothenioid larvae. Sperm ultrastructure (biflagellarity and/or spiral nucleus) appears to confirm the gobiesociform - batrachoid - lophiform relationship.

***Lepadogaster*.** Like the batrachoids, the Gobiesociformes (adult, Fig. 14.9) have biflagellate sperm, as seen in the clingfish *Lepadogaster lepadogaster* (Mattei and Mattei, 1978b), resembling the complex sperm of *Porichthys*. More species need to be studied to confirm if biflagellarity is general for these sister-groups.

General sperm morphology. The sperm of *L. lepadogaster* is about 90 µm long of which the nucleus constitutes 8 µm, the intermediate piece 13 µm and the free portion of the flagella 70 µm with an end piece of 1 µm (Mattei and Mattei, 1978b) (Fig. 14.10). An acrosome remnant is recognized in the spermatid (Mattei and Mattei, 1978a).

Nucleus. The elongate nucleus is straight except for a pointed anterior region which forms a helix with three turns. Nuclear pores are present basally.

Midpiece. The "intermediate piece" consists of two regions: an anterior portion in which 6-10 elongate mitochondria are parallel to the flagella from which they are separated by two cytoplasmic canals; and a longer, 10 µm long, region in which the two flagella are enclosed in a common sheath, consisting of the wall of the cytoplasmic canals, which contains fine granules.

Centrioles and flagellum. The two centrioles are mutually parallel. The outer dynein arms are absent from the doublets of the 9+2 axonemes. No fins are present in numerous micrographs of cross sections of the tails. In the distal portion of each flagellum the doublets lose their secondary fibres [B subtubules] and the 9 peripheral and two central singlets become disordered, The terminal piece is enlarged (Mattei and Mattei, 1978).

Cell penetration. The sperm of *L. lepadogaster lepadogaster* use the spiral nucleus to penetrate other cells, including spermatids, while in the testes. This was not observed in the supposed subspecies *L. lepadogaster purpurea*. This difference was considered by Mattei and Mattei (1978b) to vindicate the former separation of these entities as distinct species.

Sperm and phylogeny. Biflagellarity may have developed in a common ancestor of Gobiesociformes and Batrachoidiformes (see above). The parallel centrioles presumably represent the normal orientation of the basal body, here duplicated.

Order Lophiiformes

Diagnosis. The anglerfish. The first ray of the spinous dorsal fin, if present, is located on the head and modified as an illicium (lure), a line and bait device for attracting prey to the mouth. Further synapomorphies are modification of the pectoral girdle correlated with a "walking" mechanism; and reduction of all palatopterygoid bones. Sixteen families; all marine; most species in deep water (Lauder and Liem, 1983; Nelson, 1984). They appear to be closely related to the Batrachoidiformes (Nelson, 1984), the two being considered sister-groups by Lauder and Liem (1983), though here tentatively regarded as the sister-group of batrachoidiforms + gobiesociforms (Fig. 14.3).

***Neoceratias spinifer*.** *Neoceratias spinifer* (the only species of the Neoceratiidae) (Fig. 14.11) is a deep-sea, North Atlantic species with a dwarf parasitic male. An illicium is absent.

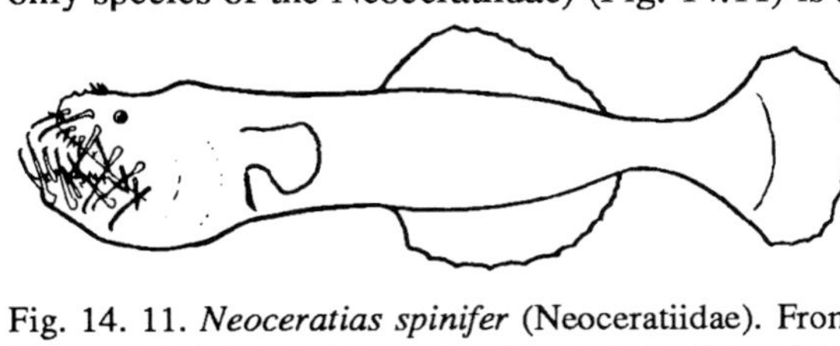

Fig. 14. 11. *Neoceratias spinifer* (Neoceratiidae). From Nelson, J.S. (1984). *Fishes of the World*. 2nd edition, John Wiley and Sons, New York. p. 207.

The complex sperm, briefly described from a recently dead fish by Jespersen (1984) differs from that of batrachoids and gobiesociforms in being uniflagellate but shows similarities to those of the batrachoid *Porichthys* and the gobiesociform *L. lepadogaster*. It differs notably from these, apart from its single flagellum, in having numerous small mitochondria.

Form. The spermatozoon is needle-shaped.

Head. The head is about 11 µm long with an helical anteriorly tapering nucleus. A small acrosome and acrosomal filament present in the spermatid are lost in spermiogenesis.

Midpiece. The midpiece, containing uniformly scattered small spherical unmodified mitochondria, forms a cylinder, 8 µm long and 2 µm wide, around the flagellum from which, throughout its length, it is

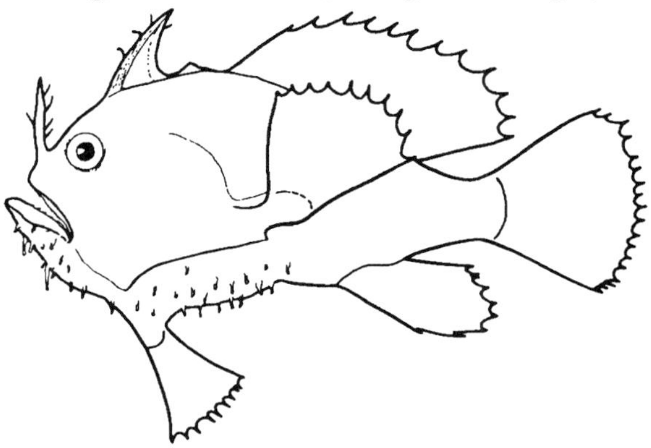

Fig. 14. 12. A frog fish (Antennariidae, subfamily Antennariinae). From Nelson, J.S. (1984). *Fishes of the World.* 2nd edition, John Wiley and Sons, New York. p. 202.

separated by a cytoplasmic canal.

Centrioles and flagellum. The two mutually perpendicular centrioles lie in a deep basal nuclear fossa, the posterior giving rise to the 9+2 flagellum (Jespersen, 1984).

Antennarius senegalensis. In contrast with *Neoceratias*, the frogfish *Antennarius senegalensis* (Antennariidae) (Fig. 14.12) has a basic teleostean ectaquasperm classifiable as type I, known only from an illustration by Mattei, 1970) (Fig. 14.13).

Sperm relationships. The correlation of complex spermatozoa with absence of copulatory organs in *Porichthys*, *Lepadogaster* and *Neoceratias*

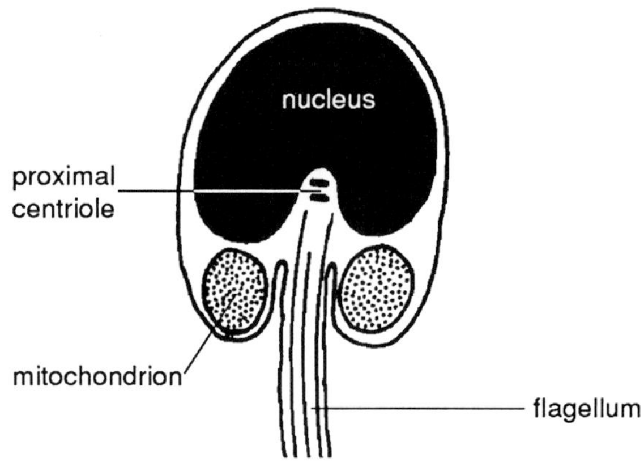

Fig. 14. 13. *Antennarius senegalensis*. Diagrammatic longitudinal section of the spermatozoon. After Mattei, X. (1970). In *Comparative Spermatology* (ed. B. Baccetti), pp. 59-69. Academic Press, New York. Fig. 4: 15.

(Briggs, 1955; Jespersen, 1984) is rare in teleosts. Nothing is known about the mode of fertilization in *Lepadogaster* and *Neoceratias* though *Porichthys* is supposedly free spawning. However, the spawning of the deep-sea angler fish *Linophryne arborifera*, a species closely related to *N. spinifer*, has been described (Bertelsen, 1980). The eggs are are suspended from the genital pore of the female in long sheets of a mucoid substance in which they are embedded. These sheets are presumed to be brought into close contact with the attached male. Fertilization of *L. lepadogaster* and *N. spinifer* is probably similar to that in *L. arborifera*. The spermatozoa of these fishes are probably adapted for penetrating the mucoid covering of the eggs.

Chapter 15

SUPERORDER ACANTHOPTERYGII

PERCOMORPHA: ZEIFORMES THROUGH MUGILOIDEI

Diagnosis. The Acanthopterygii comprise 15 orders, in two series (Percomorpha and Atherinomorpha) and no less than 246 families (Nelson, 1984). They have the following synapomorphies, shared between percomorphs and atherinomorphs: the presence of an interarcual cartilage in the dorsal gill arch skeleton between the first and second pharyngobranchial; insertion of the retractor dorsalis principally or entirely on the third pharyngobranchial; and a capability of the synophyseal and alveolar parts of the premaxilla for significant downward and forward displacement (Lauder and Liem, 1983). Rosen (1973, 1982) discusses various other features of the upper jaw mechanism but none is unique to the Acanthopterygii and this group remains poorly defined (Lauder and Liem, 1983; Nelson, 1984).

The basic sperm type is the anacrosomal aquasperm but several groups have secondarily developed similarly anacrosomal introsperm.

Series Percomorpha

Diagnosis and constitution. No definition is available for the Percomorpha as they do not apear to have any unique autapomorphies not already present in their presumed common ancestry with the atherinomorphs. Constituent orders of the Percomorpha are the Lampriformes*, Beryciformes*, Zeiformes, Gasterosteiformes*, Indostomiformes*, Pegasiformes*, Syngnathiformes, Dactylopteriformes, Synbranchiformes*, Scorpaeiniformes, Perciformes, Pleuronectiformes, and Tetraodontiformes (*sperm ultrastructure unknown). The various groups will be treated here chiefly in the sequence adopted by Nelson (1984) in which definitions are given. The phylogeny of this group is problematical (see Lauder and Liem, 1983; Nelson, 1984; Rosen, 1973). A tentative phylogeny, from Lauder and Liem (1983) is presented in Fig. 15.1.

Spermatozoal diversity and relationships. There is great uncertainty as to internal percomorph relationships and the considerable spermatozoal diversity known does not notably contribute to assessment of affinities. Spermatozoal ultrastructure of only a small fraction of species has been studied, however, with whole orders awaiting investigation, and there are indications that spermatology will contribute substantially to classification and phylogeny on fuller enquiry. Thus, peculiarities of sperm ultrastructure are at least compatible with a relationship between some orders, for instance the Dactylopteriformes and Perciformes and the Zeiformes and Tetraodontiformes and there is a widespread perciform type, with flocculent chromatin and a thin-walled sleeve trailing behind the mipiece, and with it investing the cytoplasmic canal, seen in several families exemplified by the Centropomidae, Percichthyidae, Mugiloididae, Cichlidae and others (see below). Most percomorphs have simple anacrosomal aquasperm, with or without some asymmetry, and have external fertilization.

Internal fertilization has evolved independently in a few percomorph groups (see Nelson, 1984; Thresher, 1984). In these groups, the spermatozoa are elongated, though again derivable from the anacrosomal aquasperm, and differ between groups (e.g. the scorpaeniforms *Oligocottus maculosus* and *Sebastiscus marmoratus*, and the perciform *Cymatogaster aggregata*).

Order Zeiformes

Diagnosis and relationships. The Zeiformes (dories and boarfish) are a heterogeneous and probably paraphyletic assemblage of marine fish. Features include a pelvic fin with one spine and 5-9 soft rays; caudal fin usually with 11 branched rays; dorsal fins spines 5-10; anal fin spines 0-4; soft rays of dorsal,

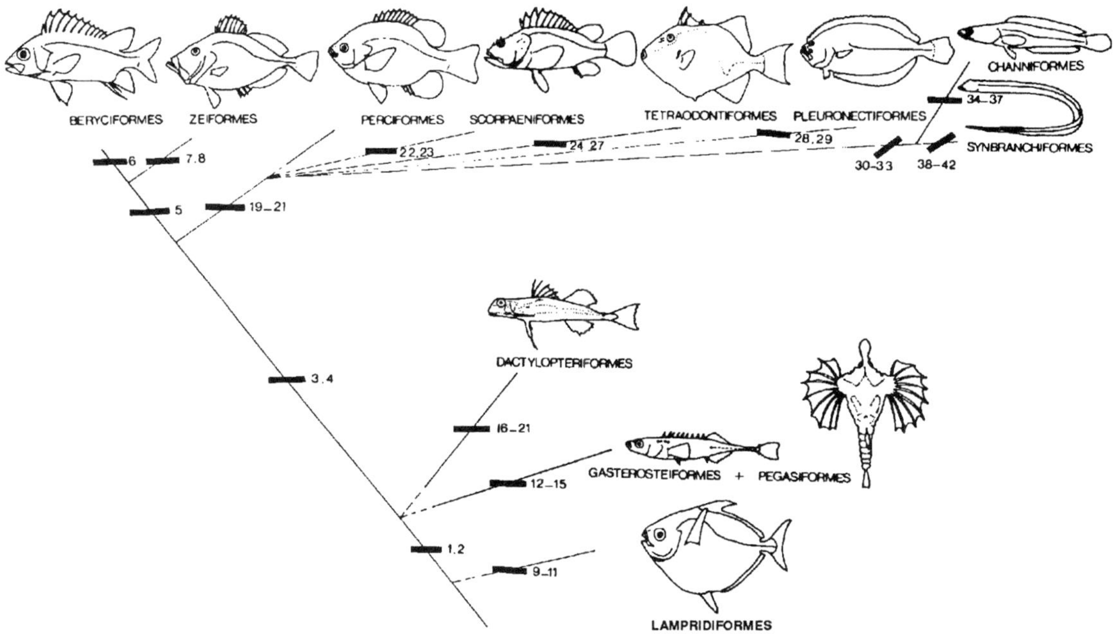

Fig. 15.1. Interrelationships of the major groups of the percomorph Acanthopterygii. Major specializations characterizing the various lineages are: 1, separate soft and spinous dorsal fins; 2, pelvic girdle attached to cleithrum; 3, second circumorbital bone with subocular shelf; 4, pelvic fin with spine and five rays; 5, sacculith (otolith) morphology specialized; 6, presence of specialized procurrent caudal spines; 7, decrease in number of vertebrae; 8, deepening of the body; 9, maxilla slides out with premaxilla during jaw protrusion; 10, no true spines in fins; 11, pelvic girdle attached to a large specialized hypocoracoid; 12, body encased in bony armor; 13, small mouth at end of tubular snout; 14, number of branchiostegals reduced (1-5); 15, trend to either reduce or modify gills; 16, no lateral line; 17, enlarged pectoral; 18, body covered with scutelike scales; 19, no free second ural centrum; 20, 17 principal caudal rays; 21, caudal skeleton with 5 hypurals; 22, hypurals fused into two distinct, large plates; 23, third circumorbital bone with an extension attached to the preoperculum; 24, entire branchiostegal region covered by a thick layer of scaleless or scaled skin, 25, no suborbitals, parietals, nasals, sensory canals in the skull, and anal spines; 26, gill opening restricted to a very small slit just below the base of the pectoral fin; 27, specialized dermal protective devices; 28, median fins extend along much of body profile; 29, bilateral asymmetry affecting topography of the eyes and coloration; 30, hemispheres of forebrain coalesced; 31, adductor mandibulae complex with specialized A1 and A2 divisions; 32, fourth branchial arteries modified; 33, fins without spines; 34, otic bullae for sacculith, utriculith and lagenolith are contained in the prootic bone; 35, metapterygoid with prominent anterodorsally directed uncinate process; 36, two ventral aortae emerge separately from the bulbus arteriosus; 37, gas exchange with air in the suprabranchial and buccopharyngeal cavities, which remain in open communication throughout the breathing cycle; 38, frontals turned down and sutured to basisphenoid; 39, large parietals meet in midline; 40, suspensorium articulates with basisphenoid, frontal, vomer and lateral ethmoids; 41, interarcual cartilage ossified; 42, the elongate heart is located far posteriorly in the body cavity. (Many of these characters, e.g. 1-3, 10, and 17, are poor and considerable future work will be necessary to clarify the relationships of the Acanthopterygii). From Lauder and Liem (1983). Fig. 50.

Fig. 15.2. *Zeus faber*, the European John Dory. After Grassé, P.-P. (1958) (ed.). *Traité de Zoologie.* XII. Agnathes et Poissons. Masson et Cie, Paris. Fig. 1717.

anal and pectoral fins not branched; body usually thin and deep; jaws usually greatly distensible; swim bladder present. Some members (the caproids) are considered better placed in the Perciformes (Nelson, 1984). Lauder and Liem (1983) recognize a decrease in the number of vertebrae; and deepening of the body as apomorphies relative to an ancestor shared with perciforms. Patterson (1964) regards zeiforms as the sister-group of the Beryciformes on the basis of shared specializations in the structure of the otoliths. However, Rosen (1984) has revealed seven synapomorphies shared between zeiforms and the Tetraodontiformes and advocates their inclusion in the latter. The caproids are seen by him as the sister-group of the enlarged Tetraodontiformes.

Zeus faber. In the Zeiformes, *Zeus faber*, the European John Dory, has a broadly conical asymmerical sperm (but nevertheless type I) with very deep basal nuclear invagination in which the proximal centriole is perpendicular to the basal body.

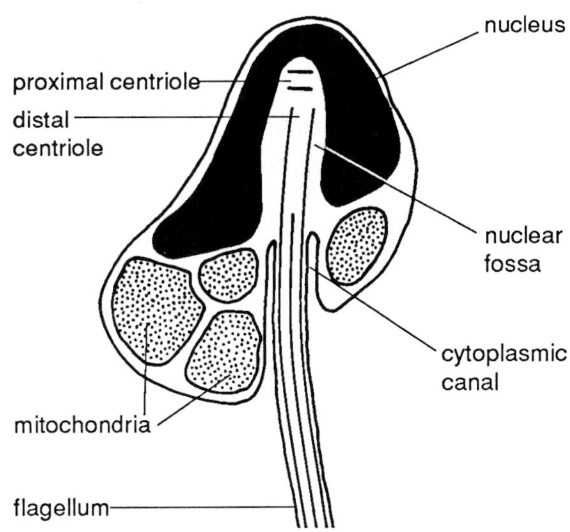

Fig. 15.3. *Zeus faber*. Diagrammatic longitudinal section of the spermatozoon. After Mattei, X. (1970). In *Comparative Spermatology* (ed. B. Baccetti), pp. 59-69. Academic Press, New York. Fig. 4: 18.

There appear to be several mitochondria disposed asymmetrically relative to the axoneme (illustration by Mattei, 1970) (Fig. 15.3).

Sperm relationships. The deep invagination of the nucleus in *Zeus* and in Tetraodontiformes, is a synapomorphy which, although not restricted to these forms, adds some support to placement by Rosen (1984) of Zeiformes in the Tetraodontiformes.

Order Syngnathiformes

Diagnosis and relationships. Syngnathiforms have a small mouth at the end of a tube-shaped snout, excepting *Enchelyocampus* which lacks even a short tubiform snout (Nelson, 1984). They include the sea horses, pipe fish and cornet fish, among others. Union with the Gasterosteiformes has been advocated (see Lauder and Liem, 1983; Fritzsche, 1984) and they are subsumed in this order in Fig. 15.1.

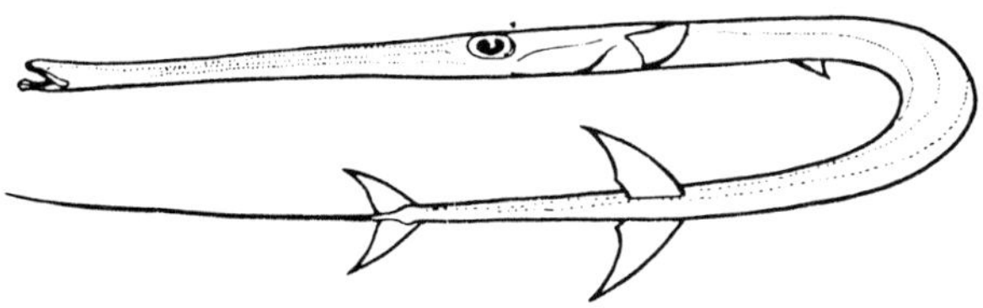

Fig. 15.4. *Fistularia*, a cornet fish. From Nelson, J.S. (1984). *Fishes of the World.* 2nd edition, John Wiley and Sons, New York. p. 250.

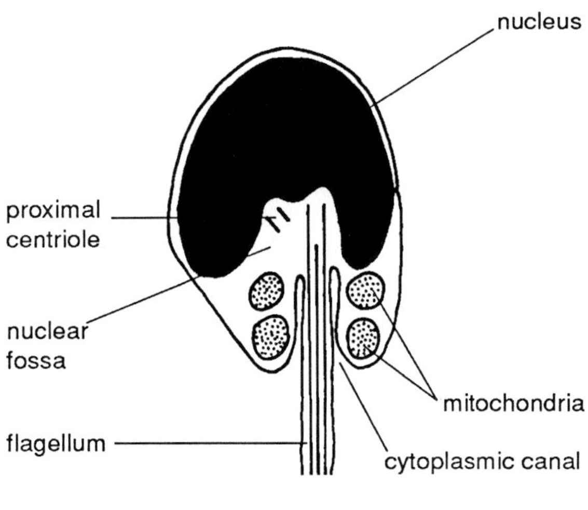

Fig. 15.5. *Fistularia tabacaria.* Diagrammatic longitudinal section of the spermatozoon. After Mattei, X. (1970). In *Comparative Spermatology* (ed. B. Baccetti), pp. 59-69.

Fistularia tabacaria. In the syngnathiform family Fistulariidae, *Fistularia tabacaria*, the Blue-spotted Cornet Fish, has a basic anacrosomal aquasperm but the proximal centriole is oblique and to the side of the basal body, each in its own small embayment of a wide, deep basal nuclear fossa. The mitochondria, symmetrically disposed, appear to be in two tiers (illustration by Mattei, 1970) (Fig. 15.5).

Order Dactylopteriformes

Diagnosis and relationships. These are the flying gurnards, benthic fish which produce sounds by stridulation, using the hyomandibular bone and which "walk" on the sea bed by alternately moving the pelvic fins. Affinity with the pegasids and syngnathiforms is suspected (Nelson, 1984). They appear to be among the more basal groups of the Percomorpha (Fig. 15.1). Autapomorphies include absence of a lateral line; enlarged pectoral fins; scutelike scales covering the body; absence of a free second ural centrum; presence of 17 principal caudal rays and of 5 hypurals in the caudal skeleton (Lauder and Liem, 1983).

Dactylopterus volitans. The only species of the Dactylopteriformes examined, *Dactylopterus* (=*Cephalacanthus*) *volitans*, the Flying Gurnard (Fig. 15.6) (family Dactylopteridae), has a uniflagellate anacrosomal aquasperm, described by Boisson *et al.* (1968a) and Mattei, 1970 (Fig. 15.7). It is strongly modified but is clearly derivable from the usual teleostean type. The modĕ of reproduction is unknown but is presumably by external fertilization.

Nucleus. The moderately elongate nucleus is penetrated so deeply that only a thin layer of chromatin covers the anterior end of the expanded fossa. The nucleus is grossly asymmetrical about the fossa.

Midpiece. Three or four large mitochondria are situated at the base of the large posterior prolongation of the nucleus on the same side of the axoneme. A cytoplasmic collar, surrounding the base of the flagellum, is thin on one side but houses the mitochondria on the other.

Centrioles and flagellum. The nuclear fossa houses two mutually perpendicular triplet centrioles (type I arrangement) of which one, the distal centriole, forms the basal body. From an osmiophilic mass associated with the tip of the distal centriole, arise

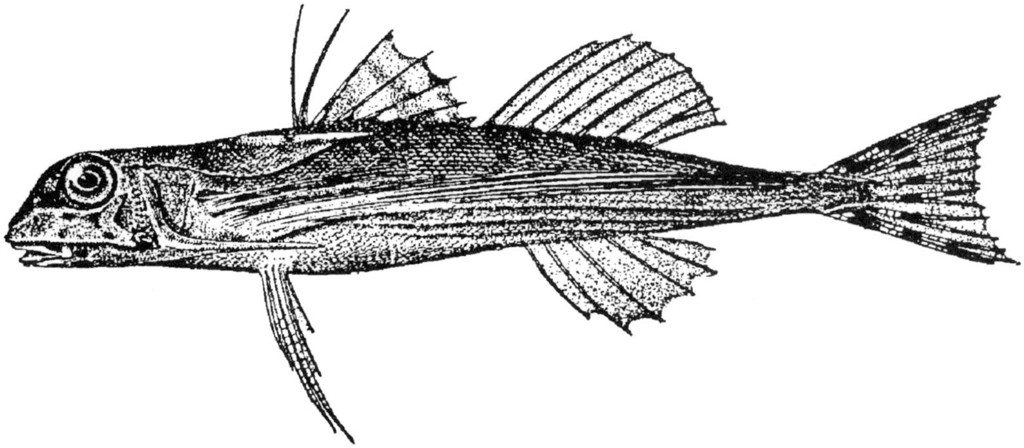

Fig. 15.6. *Dactylopterus* (=*Cephalacanthus*) *volitans*, the Flying Gurnard. After Jordan, D.S. (1907). *Fishes*. Henry Holt, New York. Fig. 579.

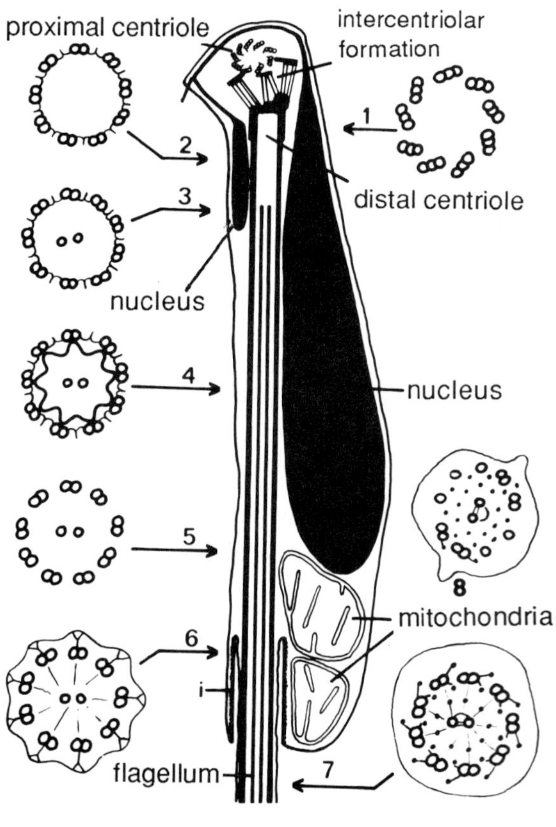

Fig. 15.7. *Dactylopterus* (=*Cephalacanthus*) *volitans*. Diagrammatic sections of the spermatozoon. After Boissin *et al.* (1968a). *Comptes Rendus des Séances de la Société de Biologie de l'ouest Africain* 162, 820-823. Fig. 1.

bundles of fibrils which reach the proximal centriole where each is capped by a dense lamina.

The 9+2 flagellum is exceptional in having 9 very thin accessory fibres peripheral to the doublets and two circlets, each of 9, internal to these. Presence of rudimentary fins is indicated in the illustration (for further details see Boissin *et al.*, 1968a).

Sperm relationships. Sperm ultrastructure is at least compatible with a relationship between Dactylopteriiformes, e.g. *Dactylopterus* (=*Cephalacanthus*) *volitans* (Fig. 15.7), and the Eleotridae in the Perciformes, as exemplified by *Hypseleotris galii* (Jamieson, unpublished) (Fig. 16.13) though it is possible that resemblances are homoplasies by virtue of relationship (symparamorphies of Jamieson, 1984). In both of these species the nucleus is so deeply invaded by the basal fossa as to retain only a very thin layer of chromatin (in *Hypseleotris* subapically no chromatin) between the outer and inner nuclear envelope over the head of the flagellum; two centrioles are present; and the mitochondria are restricted to one side of the axoneme. Asymmetry of the head is more pronounced in *Dactylopterus volitans* of the two species. In the perciform *Ophioblennius atlanticus* (Fig. 16.10D) asymmetry is extreme. The mode of fertilization of these three species, in which sperm morphology approaches the complex condition, is unknown but there is no reason to doubt that it is external. Axonemal fins, apparently small in *D. volitans*, are, in contrast, very well developed in *Hypseleotris galii*.

Order Scorpaeniformes

Diagnosis and relationships. Scorpaeniforms, the "mail-cheeked fish" constitute the fourth largest order of fish, with over 1000 species.They are benthic or epibenthic in tropical to polar waters with some freshwater representatives. The suborbital stay is the sole defining character (autapomorphy) but is possibly not indicative of monophyly (see Washington *et al.*, 1984a,b); in addition, the hypurals are fused into two distinct, large plates (Lauder and Liem, 1983). Relationships recognized by Matsubara (1943) with families currently placed in other orders and derived from a common percoid ancestor are shown in Fig. 15.8.

Modes of reproduction. These vary widely. Many families spawn individual pelagic eggs while others spawn demersal clusters of adhesive eggs but most produce pelagic egg masses enclosed in a gelatinous matrix. There are strong trends towards internal fertilization and the genus *Sebastes* and the comephorids of Lake Baikal are live-bearing (Washington *et al.*, 1984b).

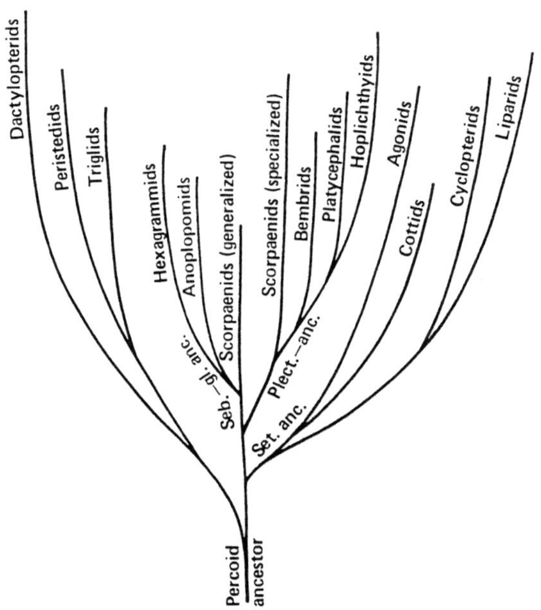

Fig. 15.8. Internal relationships of the Scorpaeniformes. After Matsubara, K. (1943). *Transactions of the Sigenkagaku Kenyusho.* (*Fide* Washington *et al.*, 1984a). Fig. 240.

Cottus gobio. In *Cottus gobio*, the Miller's Thumb or River Bullhead, (family Cottidae) (Fig. 15.9), the sperm nucleus, rounded in one longitudinal plane is depressed in the perpendicular plane where it overlaps a midpiece consisting of several rounded mitochondria.

Fig. 15.9. *Cottus gobio*, the Miller's Thumb or River Bullhead. After Buckland, F. (1891). *Natural History of British Fishes*. S.P. C. K., London. p. 24.

The mitochondria lie on each side of a long cytoplasmic canal and as many as three are seen in a longitudinal row. The midpiece in this species is unusually long, 1.7 µm compared with 2.8 µm for the head; the length of the midpiece varied from 0.6-1.2 (mean 0.8) µm in 9 non-cottids examined. There are two very long flagellar fins (Stein, 1981). The sperm is diagrammatically illustrated in Fig. 13.5C. "Spawning" in *Cottus* occurs with the ventral surfaces of the partners in close contact and internal fertilization is suspected (Breder and Rosen, 1966).

Oligocottus maculosus. The relationship with *Cottus* of *Oligocottus maculosus*, the Tide Pool Sculpin (Fig. 15.10), of which the sperm is described by Stanley (1966,1969) (Fig. 15.11), is seen in the

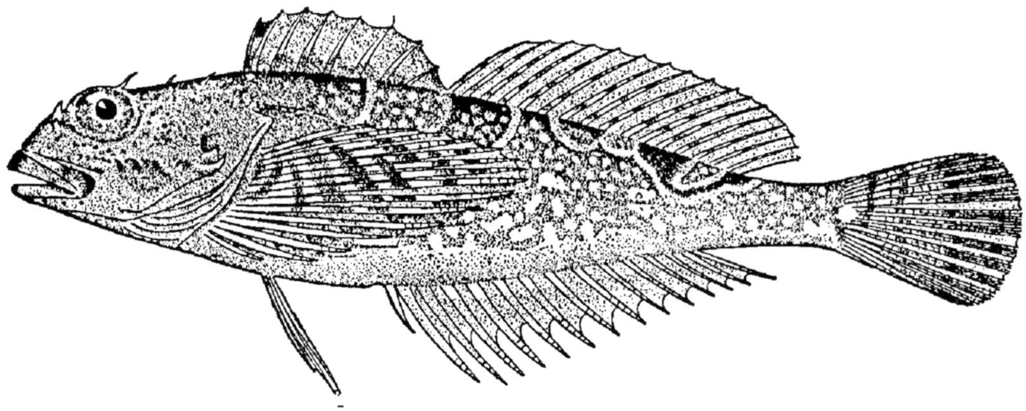

Fig. 15.10. *Oligocottus maculosus*. After Jordan, D.S. (1907). *Fishes*. Henry Holt, New York. Fig. 566.

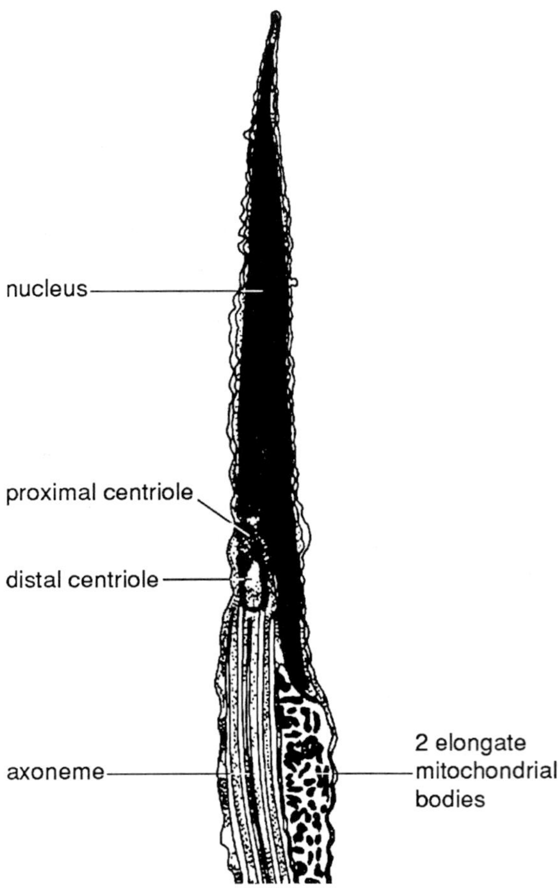

Fig. 15.11. *Oligocottus maculosus*. Longitudinal section through the "sagittal" plane of a late spermatid. After a micrograph by Stanley, H.P. (1969). *Journal of Ultrastructure Research* **27**, 230-243. Fig. 12.

spade-like form of the nucleus with the flagellar basal body and proximal centriole inserted into a groove along of the flattened sides of the nucleus, parallel with its long axis.

The nucleus is more elongate in this species, however. A further difference is transformation of the mitochondria of the spermatid into two elongate bodies surrounding the base of the tail in the cytoplasmic sleeve common to both species.

The elongation of the nucleus and development of elongate mitochondrial derivatives are features frequently seen in internally fertilizing animal sperm. It is therefore noteworthy that this species has a penis and that the existence of internal fertilization has been established (Breder and Rosen, 1966).

***Scorpaena angolensis*.** The sperm of *Scorpaena angolensis* (Scorpaenidae, Scorpaeninae) sketched

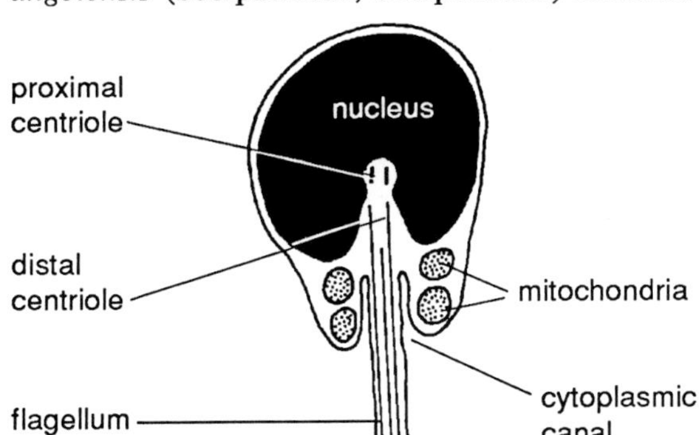

Fig. 15.12. *Scorpaena angolensis*. Diagrammatic longitudinal section of the spermatozoon. After Mattei, X. (1970). In *Comparative Spermatology* (ed. B. Baccetti), pp. 59-69. Academic Press, New York. Fig. 4: 3.

by Mattei (1970) (Fig. 15.12) appears somewhat similar to that of *Cottus* (above) but it is not known whether the nucleus is depressed in one plane nor is the mode of fertilization discussed.

Sebastiscus marmoratus (Sebastinae) is certainly ovoviviparous but the mature spermatozoon is not described. The late spermatid (Mizue, 1968) has a nucleus only about twice as long as wide, depressed in one plane. There is a mitochondrial sheath with about 8 separate mitochondria arranged in a circle around a cytoplasmic canal and it seems unlikely that these fuse at maturity. The 9+2 flagellum has two fins.

Order Perciformes

General. The Perciformes (perch-like fish) is the most diverse of all fish orders and, indeed, is the largest vertebrate order. Perciforms dominate in vertebrate ocean life and among fish of many tropical and subtropical freshwaters. These and other statements about the group have little meaning in view of the doubtful monophyly of the group for which uniquely diagnostic features are not known (Lauder and Liem, 1983; Nelson, 1984). Of the 22 suborders, we have listed only those in which sperm ultrastructure has been examined (Table 5. 1).

Suborder Percoidei

Diagnosis and habitat. Seventy years of research in systematic ichthyology have failed to find a meaningful definition of or boundaries for the Percoidei (Johnson, 1984). Percoids are best represented in the nearshore marine environment and form a significant component of the reef-associated fish fauna of tropical and subtropical seas. Association with brackish water occurs in many such families and some have one or more exclusively freshwater members, but some families are restricted to freshwaters: the north temperate Percidae and Centrarchidae, the south temperate Percichthyidae and the tropical Nandidae (Johnson, 1984). We follow Kaufman and Liem (1982), Lauder and Liem (1983) and Johnson (1984) in excluding the Pomacentridae, Cichlidae and Embiotocidae placing these, with the Labridae, in the Labroidei (Fig. 15.1).

Chief sperm types. Within the percoids but supposedly generalizable to all teleosts, Mattei (1970) recognizes two types of spermiogenesis and therefore two types of sperm. These are exemplified by the Mullidae and Haemulidae, respectively.

Mullidae. Spermiogenesis in the mullid *Upeneus prayensis* is illustrated as an example of what Mattei designates "type I" spermiogenesis (Fig. 5.2). In this four stages are recognized: (1) the young spermatid, with central nucleus, basal centrioles, and mitochondria scattered in the cytoplasm; (2) migration of the centrioles in the direction of the nucleus, drawing with them the flagellum and the cell membrane to give a cytoplasmic canal; (3) formation of the basal nuclear fossa and a 90° rotation [of the nucleus into the same axis as the flagellum] bringing the centrioles into the nuclear fossa; and (4) migration of the mitochondria to the base of the nucleus. The resultant anacrosomal aquasperm is approximately symmetrical (Fig. 5.2G).

A specific feature of the *Upeneus* sperm which is not general for type I sperm is the very deep penetration of the nucleus by its basal fossa so that in longitudinal section the nucleus has the form of an inverted U (Boisson *et al.*, 1969; Mattei, 1970). Unusually, also, the proximal centriole, which is connected to the nuclear membrane at the summit of the fossa, has its central axis in the same longitudinal axis as that of the basal body.

Both of these features, deep penetration of the nucleus and serial coaxial centrioles, are also reported by Boissin *et al.* (1969) in *Pegusa triophthalmus* (Soleidae) (see also Mattei, 1970) (Fig. 16.18), *Balistes forcipatus* (Balistidae), *Aluterus punctatus* (Monacanthidae), *Chilomycturus reticulatus* (Diodontidae), and *Scorpaena angolensis* (Scorpaenidae) and by Jamieson (Chapter 16) for *Pseudobalistes fuscus* (Balistidae) (Fig. 16.20A, B). In the cyprinodontid *Fundulus heteroclitus* the nucleus is also U-shaped but the centrioles appear to be mutually perpendicular (Yasuzumi, 1971). In longitudinal section of the *Upeneus* sperm a separate mitochondrion is seen on each side in contact with the corresponding rim of the nucleus peripheral to an anteriorly widening but fairly short cytoplasmic canal.

Haemulidae. The type I spermiogenesis of *Upeneus* contrasts with "type II" spermiogenesis, exemplified by the haemulid (=Pomadasyid) *Parapristipoma octolineatum* (Fig. 5.3) (see also Mattei, 1970).

In type II spermiogenesis, at stage (2) the centrioles behave similarly but the axis of the flagellum becomes tangential relative to the nucleus. At (3) a nuclear fossa again forms but no rotation of the nucleus occurs and the mitochondria surround the centrioles which do not enter the fossa (Mattei, 1970). In the axoneme of this type the A tubules of the doublets 1, 2, 3 and 6 have intratubular differentiation. This type is restricted to highly evolved teleosts and supposedly only the perciforms, where 25 families out of the 39 studied exhibited a spermatozoon of this model (Mattei, 1988).

Operationally this classification has some utility for mature spermatozoa, even if the development stages are not known, as insertion of the centrioles into the nucleus (type I) or their isolation from this (type II) are observable. It is does not appear, however, that all variants of even simple anacrosomal aquasperm (particularly with regard to the orientation of the axoneme) can be described by only two categories. Mattei (1970) recognizes exceptions, citing the existence of an intermediate condition in *Galeoides decadactylus* and the inapplicability of this classification to the aflagellate sperm of *Gymnarchus niloticus*. Furthermore, in erecting the classification into two sperm types, he does not attempt to place more than 70 teleost sperm investigated into either category.

Percidae. The sperm of *Perca fluviatilis* is of the same type as that of *Cottus gobio* (Type 3) in the system of Stein (1981) and is illustrated in Fig. 13.6.

Centropomidae. The Australian centropomid *Lates calcarifer*, the Barramundi (Fig. 15.13), has an anacrosomal aquasperm (Fig. 15.14) with an elliptical nucleus with a narrow, moderately deep basal fossa.

The chromatin consists of coarse masses narrowly separated by clear material, giving it a coarsely flocculent appearance. Round bodies (see Percichthyidae) are absent.

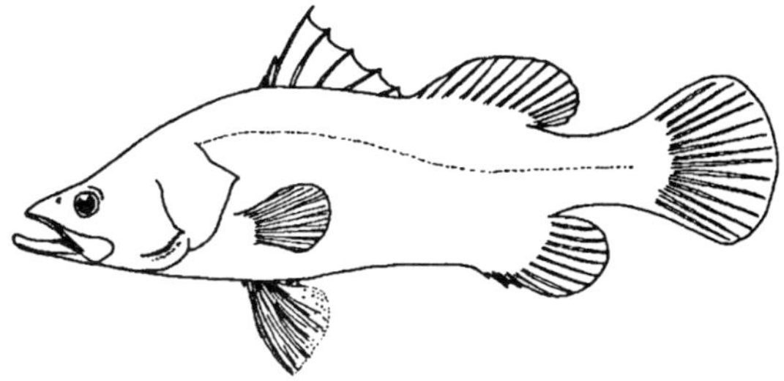

Fig. 15.13. *Lates calcarifer*, the Barramundi. Original.

There are 4 to 6 mitochondria distributed asymmetrically around a short cytoplasmic canal. The long axis of the proximal centriole is at 110° to that of the

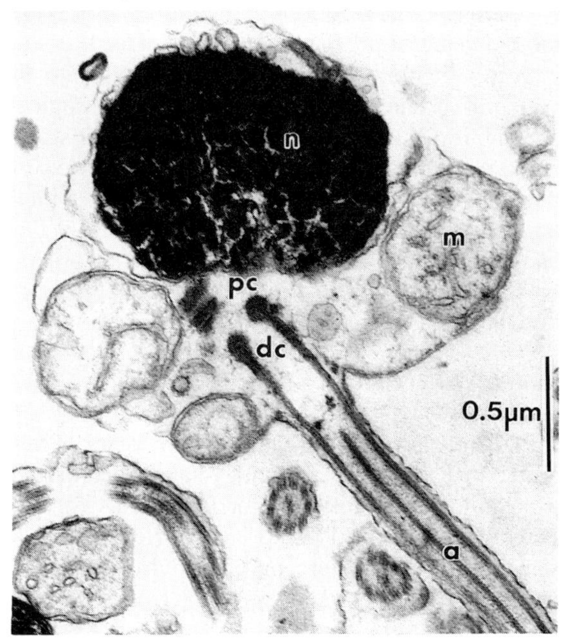

Fig. 15.14. *Lates calcarifer*, the Barramundi. Longitudinal section of a spermatozoon (thawed after cryopreservation with 20% glycerol. From Leung, L. K.-P. (unpublished). Abbreviations as Fig. 15.15.

Fig. 15.15. Sperm of Australian Percichthyidae. A-C. *Macquaria australasica*. A. Scanning electron micrograph (SEM) of spermatozoa. B. SEM of head, showing cytoplasmic canal. C. Longitudinal section (LS) of spermatozoon close to the sagittal plane. D-H. *Macquaria ambigua*. D. LS sperm. E. Transverse section (TS) of two midpieces, showing variable shape of mitochondria. F. LS through centrioles. G. TS flagellum through fins. H. TS flagellum behind fins. I-L. *Maccullochella macquariensis*. I. SEM, showing nuclear fossa. J. LS sperm, showing apical round body. K. TS head through initial, 9+0 region of axoneme and single mitochodrion. L. TS flagellum, showing single fin. M. *Maccullochella peeli*. M. LS sperm. N. TS tail and lateral fin. a. axoneme. cc. cytoplasmic canal. dc. distal centriole. fi, fin. fo. nuclear fossa. m. mitochondrion. n. nucleus. pc. proximal centriole. r. round body. Courtesy of L. K.-P. Leung.

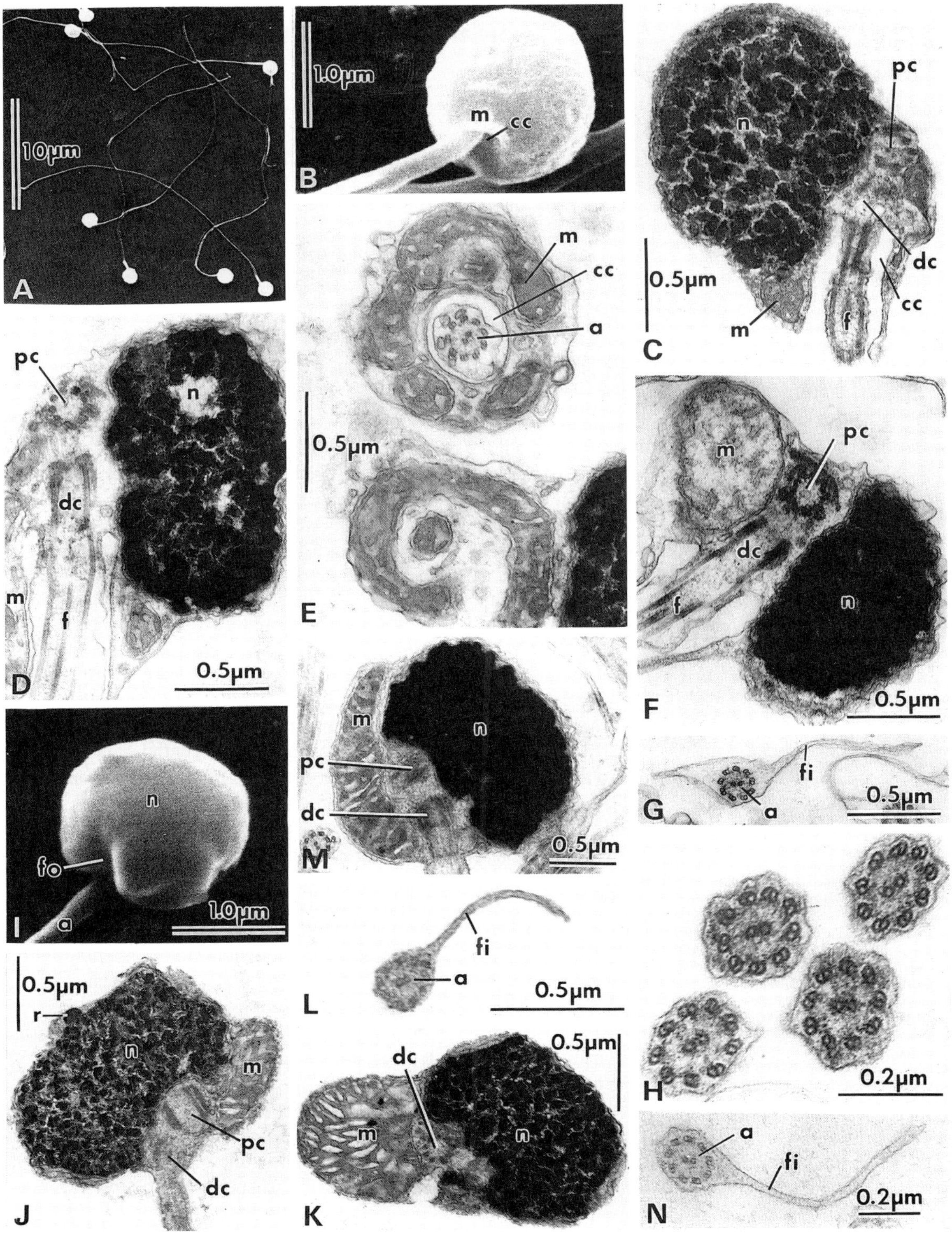
10μm
A
1.0μm
m
cc
B
n
pc
dc
cc
0.5μm
m
f
C
pc
n
dc
m
f
0.5μm
D
m
cc
a
0.5μm
E
m
pc
dc
f
n
0.5μm
F
fi
a
0.5μm
G
n
fo
1.0μm
I
a
m
n
pc
dc
0.5μm
M
fi
a
0.5μm
L
0.2μm
H
0.5μm
r
n
m
pc
dc
J
dc
0.5μm
m
n
K
a
fi
0.2μm
N

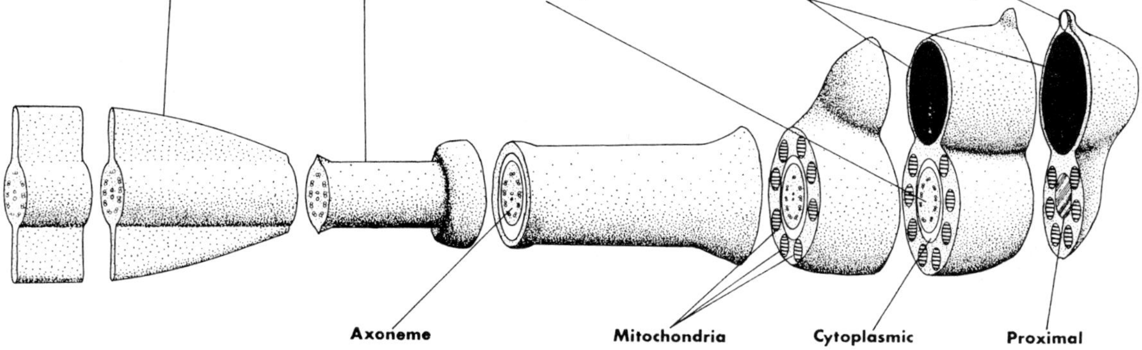

Fig. 15.16. *Macquaria ambigua*. Schematic diagram of spermatozoon. From Marshall, A. (unpublished).

axoneme and about 45° to the sagittal plane; in some sections this centriole lies in a moderately deep nuclear fossa (not visible in Fig. 15.14) but the distal centriole is wholly outside it. Each centriole is of the triplet type; the proximal has 9 dense peripheral accessory densities; an undetermined number of fibres accompanies the distal centriole. There is a long cytoplasmic canal, surrounded for most of its length by a narrow cytoplasmic sleeve. The axoneme is of the 9+2 type, with a pair of lateral fins, but lacks central singlets for a short region below the basal body (Leung, unpublished). This spermatozoon appears to be intermediate between types I and II of Mattei (1970).

Percichthyidae. Percichthyids are the so-called temperate basses, occurring in marine, brackish and freshwater habitats in tropical and temperate regions of the world (Nelson, 1984). Sperm of the Australian taxa *Maccullochella* (*Macc. macquariensis*, Fig. 15.15I-L; *Macc. peeli*, Fig. 15.15M, N) and *Macquaria* (*Macq. ambigua*, Fig. 15.15D-H, 15.16; *Macq. australasica*, Fig. 15.15A-C; *Macq. novemaculeata*) so far investigated correspond generally to the above description for *Lates* but show some notable exceptions and demonstrate what appear to be genus-specific characters with relationship to numbers of mitochondria, and of axonemal fins, and presence or absence of nuclear "round bodies" and of a cyoplasmic canal (Fig. 15.15). *Maccullochella* differs in the presence of only a single, large mitochondrion (Fig. 15.15K) (occasionally two mitochondria in *Macc. peeli*) and the presence of "round bodies" within the nuclear envelope but discrete from the general chromatin. There is a single round body in *Macc. macquariensis* (Fig. 15.15J), and there are three or four in *Macc. peeli*. The nuclear fossa is deep in *Macc. peeli*, is a moderate concavity in *Macc. macquariensis*, and is weakly developed in *Macquaria*. *Macquaria ambigua* has several mitochondria, some of which are sometimes fused as an incomplete ring (Fig. 15.15E); in *Macq. australasica* there are 3 to 6 mitochondria as some may fuse; in *Macq. novemaculata* there are approximately 5. In both genera the basal body and 9+2 axoneme are at 90° or more to the proximal centriole and tangential or nearly so to the nucleus (Fig. 15.15C, D, F, J, M). Various densities occur around the centrioles. *Maccullochella* sperm have only one lateral fin (Fig. 15.5L, N) and have no cytoplasmic canal while those of *Macquaria*, like *Lates*, have two fins (Fig. 15.15G) and a long cytoplasmic canal most of which is bounded by a thin, trailing cytoplasmic sleeve which only anteriorly expands as the mitochondrion-containing midpiece (Fig. 15.15D). Terminally, the axoneme lacks fins (Fig. 15.15H) (Leung, unpublished).

Serranidae. These are the sea basses; marine, in tropical and temperate seas; a few freshwater (Nelson, 1984).

Plectropomus leopardus. In the sperm of the serranid *Plectropomus leopardus*, the Coral Trout, the nucleus is approximately 1.4 µm long and has the form of a thick C-shape owing to presence of a deep basal fossa (Fig. 15.17) (Jamieson, unpublished). The chromatin is poorly condensed, consisting of many irregular flocculent dense masses separated by pale matrix.

The midpiece is of simple construction. Five subspherical cristate mitochondria are regularly arranged around the base of the axoneme occupying the entirety of the short collar. There is no trailing sleeve. Densification occurs between contiguous outer membranes of adjacent mitochondria. Small beadlike masses are present in the vicinity of some mitochondria.

The two centrioles are mutually at right angles, with the proximal displaced to one side of the longitudinal axis of the distal centriole. The 9+2 flagellum has two lateral fins relative to which the two central singlets are tilted. It is difficult to reconcile the symmetrical distribution of mitochondria with the tangential arrangement of the axoneme, and unilateral mitochondrion in some longitudinal sections; possi-

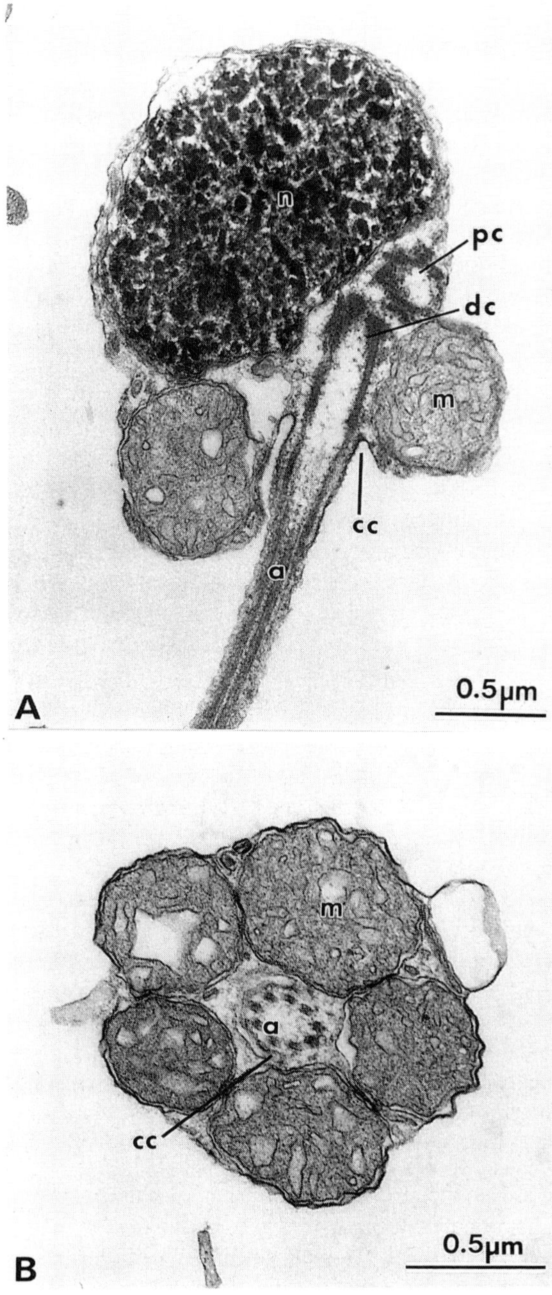

Fig. 15.17. *Plectropomus leopardus*, Coral Trout. A. Longitudinal section of spermatozoon. B. Transverse section of midpiece. a. axoneme. cc. cytoplasmic canal. dc. distal centriole. fi, fin. fo. nuclear fossa. m. mitochondrion. n. nucleus. pc. proximal centriole. Original.

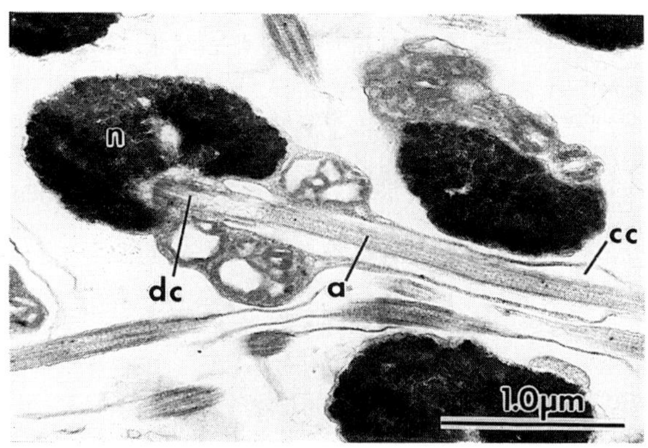

Fig. 15.18. *Nannoperca oxleyana*. Longitudinal section of the spermatozoon. From Marshall, C. J. (1989). Abbreviations as Fig. 15.17.

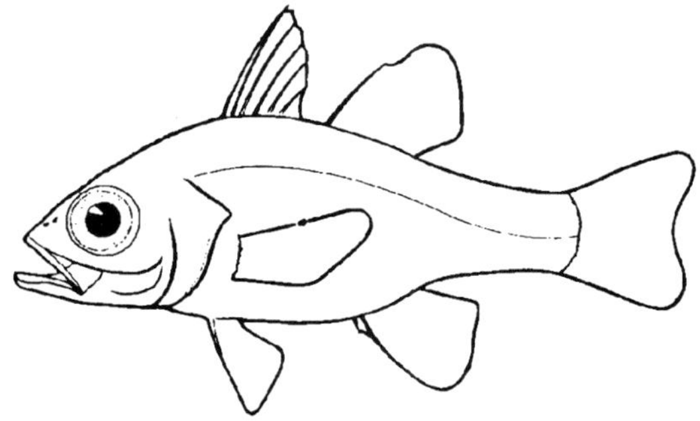

Fig. 15.19. A cardinal fish (Apogonidae). From Nelson, J.S. (1984). *Fishes of the World*. 2nd edition, John Wiley and Sons, New York. p. 289.

bly they represent a submature stage before movement of the centrioles into the nuclear fossa as decribed by Mattei (1970) for type I spermiogenesis (above).

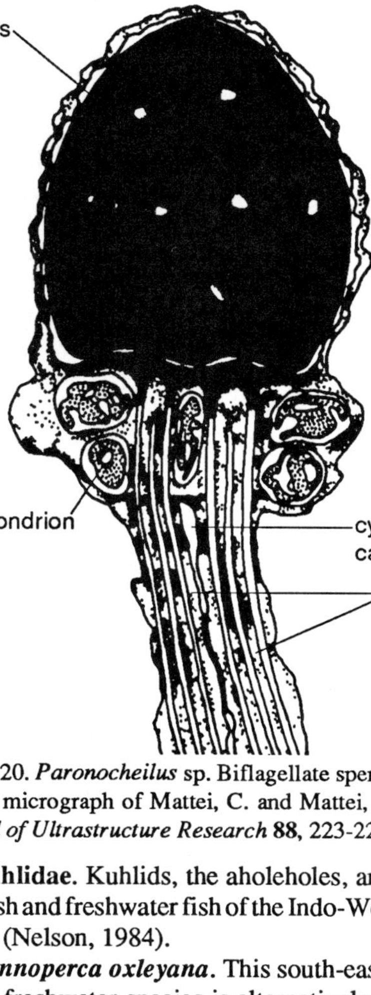

Fig. 15.20. ***Paronocheilus*** sp. Biflagellate spermatozoon. After a micrograph of Mattei, C. and Mattei, X. (1984). *Journal of Ultrastructure Research* **88**, 223-228. Fig. 7.

Kuhlidae. Kuhlids, the aholeholes, are marine, brackish and freshwater fish of the Indo-West Pacific region (Nelson, 1984).

***Nannoperca oxleyana*.** This south-eastern Australian freshwater species is alternatively placed in the family Nannopercidae, between the Tetraponidae and Centrarchidae while the Kuhlidae are placed after the Percichthyidae (Johnson, 1975).

The sperm nucleus (Fig. 15.18) is 1.3 µm long and 0.6 µm wide and is mostly located laterally to the centrioles and anterior portion of the axoneme. The centrioles nevertheless are deeply embedded in it giving a form intermediate between the type I and II spermatozoon *sensu* Mattei. The chromatin appears flocculent, consisting of large dark particles and scattered pale lacunae, and irregularly conforms to the double nuclear membrane. Three to four irregularly shaped distinct, cristate mitochondria form a ring around the long (2.4 µm) cytoplasmic canal. The two triplet centrioles are mutually perpendicular. There is, as in *Macquaria* and *Lates*, a long trailing cytoplasmic sleeve behind the midpiece. The 9+2 axoneme initially has a 9+0 configuration. The A subtubules of doublets 1 and 6 are septate. Both dynein arms are present. Two long fins, at least twice the width of the flagellum, lie approximately in the plane of the central singlets (modified from Marshall, 1989).

Apogonidae. These are the cardinal fish (Fig. 15.19); predominantly marine, in the great oceans.

The sperm of *Paronocheilus* sp. is biflagellate (Fig. 15.20).

The nucleus is ovoid, measuring 1.5 x 1 µm. Its base has two shallow fossae containing dense material except at their centres which house the anterior ends of the centrioles. The nuclear material is not compact but contains enclaves of clear nucleoplasm. The nuclear envelope is detached from the nuclear mass and retains two distinct membranes. The intermediate piece contains about 15 mitochondria, in two tiers, located around the two axonemes. It has a posterior region where the mitochondria are separated by a common cytoplasmic canal from the two flagella. This contrasts with the biflagellate sperm of gobiesociforms and siluriforms in which each flagellum has its own cytoplasmic canal. Mitochondrial cristae are sparse and the matrix is dense. The two

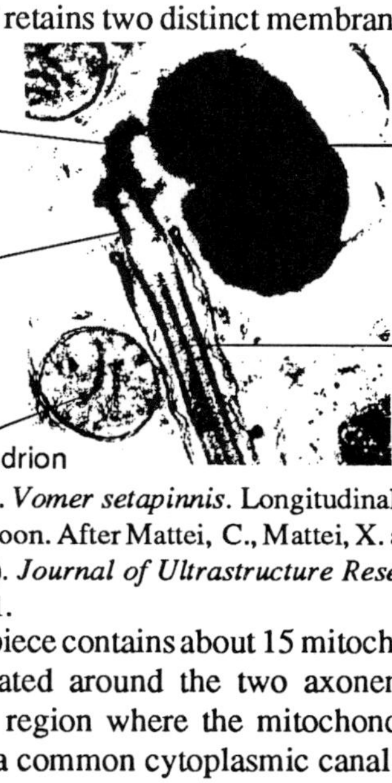

Fig. 15.21. *Vomer setapinnis*. Longitudinal section of the spermatozoon. After Mattei, C., Mattei, X. and Marchand, B. (1979). *Journal of Ultrastructure Research* **69**, 371-377. Fig. 1.

parallel centrioles are each composed of 9 triplets. Each centriole is continued as a transition region, consisting of armless doublets lacking bridges and unaccompanied by central singlets; in each axoneme a satellite is associated with one of the doublets; the two affected doublets are diametrically opposite within the intermediate piece. This region is followed by a normal 9+2 axoneme in which each

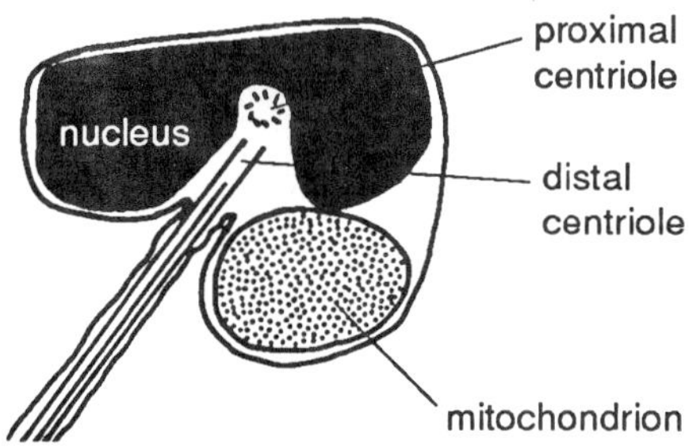

15.22. *Boops boops*. Diagrammatic longitudinal section of the spermatozoon. After Mattei, X. (1970). In *Comparative Spermatology* (ed. B. Baccetti), pp. 59-69. Academic Press, New York. Fig. 4: 14.

doublet has two dynein arms. A saccule formed from the external membrane of the nuclear envelope passes through the transition zone of each axoneme, sometimes accompanied by a mitochondrion. Usually the saccule penetrates at the level of the doublet provided with a satellite so that each axoneme appears in cross section to be divided into two groups of 3 and 6 elements. Somewhat similar divison of the axoneme by the outer nuclear membrane is seen in Acanthocephala (Mattei and Mattei, 1984).

Carangidae. *Vomer setapinnis*. In the sperm of the carangid *Vomer setapinnis*, the Atlantic Moonfish (Mattei *et al.*, 1979) (Fig. 15.21), the nucleus is lateral relative to the flagellar axis (type II sperm *sensu* Mattei). A moderately long cytoplasmic canal is present and separate mitochondria lie on each side of the axoneme in a line at right angles to the nuclear axis.

Centracanthidae. Centracanthids are marine perciforms of the eastern Atlantic and South Africa (Nelson, 1984).

***Spicara chryselis*.** *Spicara chryselis* appears to have a type I sperm *sensu* Mattei. In a micrograph of late spermatids the nucleus is ovoid but is penetrated to above its equator by the basal fossa which contains

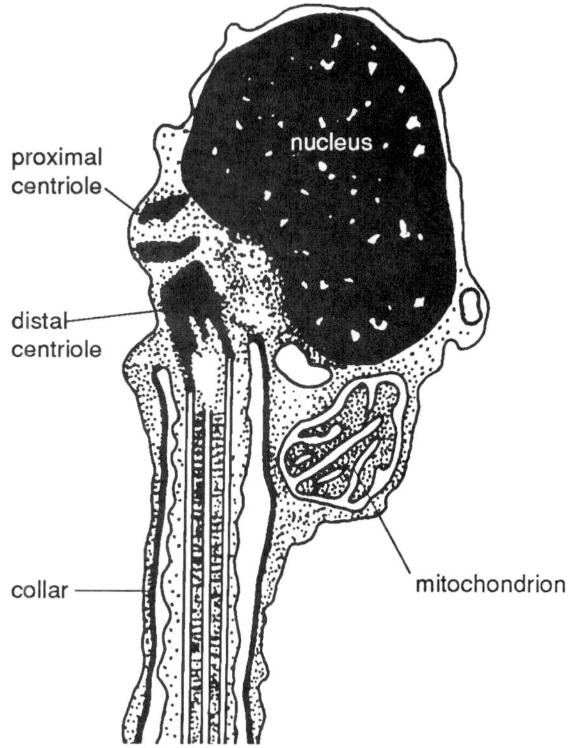

Fig. 15.23. *Liza aurata*. Longitudinal section of spermatozoon. After a micrograph by Brusle, S. (1981). *Cell and Tissue Research* **217**, 415-424. Fig. 13.

the two centrioles; three or four mitochondria in two tiers are seen at this stage (Carrillo and Zanuy, 1977).

Sparidae. Sparids are the porgies; predominantly marine, in the Atlantic, Indian and Pacific oceans (Nelson, 1984).

***Boops boops*.** From a diagram, the sparid *Boops boops* has an asymmetrical but apparently type I sperm (Fig. 15.22), resembling that of the gobiid *Periophthalmus* (below) in its chief features (Mattei, 1970).

Suborder Mugiloidei

Diagnosis. These are the mullets, with thick, streamlined bodies, deeply forked caudal fin, large cycloid or weakly ctenoid scales and no lateral line. They occur in oceans, bays, estuaries and freshwater in all except polar regions. Mugiloids (not to be confused with the Mugiloididae) appear to be related

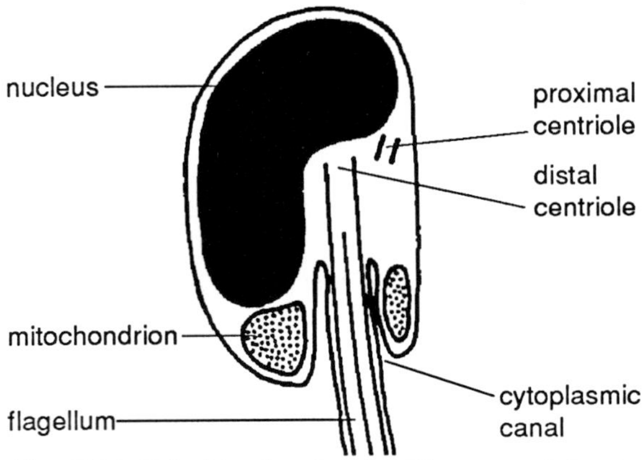

Fig. 15.24. *Galeoides decadactylus.* Diagrammatic longitudinal section of the spermatozoon. After Mattei, X. (1970). In *Comparative Spermatology* (ed. B. Baccetti), pp. 59-69. Academic Press, New York. Fig. 4: 24.

to polynemoids and sphyraenoids. A previously held relationship with atherinids has been refuted (see de Sylva, 1984).

Liza and Galeoides. The sperm of the three examined mugiloids (*Liza aurata*, Brusle, 1981, Fig. 15.23; *L. dumerilii*, van der Horst, 1976; van der Horst and Cross, 1978, both Mugilidae, and *Galeoides decadactylus*, Mattei, 1970, Polynemidae) (Fig. 15.24) are similar.

Nucleus. The nucleus, bilobed or kidney-shaped in longitudinal section, is tilted relative to the axoneme. The chromatin is very coarsely granular. In *L. dumerilii* a dense body (round body) has been observed lying in contact with the nucleus in some sperm; the nuclear and overlying plasma membrane covering the "ventral" or "dorsal" tip of the nucleus forms bulbous evaginations containing chromatin.

Mitochondria. Four subspherical mitochondria, in *L. aurata* and *L. dumerilii*, are arranged in a ring but eccentrically around the base of the axoneme from which they are separated by a long cytoplasmic canal. In *L. dumerilii* mitochondria cristae are sparse and are plate-like and tubular, the former sometimes arranged in a circular pattern as in brown adipose tissue.

Centrioles. Two fully developed centrioles are present. Despite the eccentric emergence of the flagellum, the proximal centriole, and in one plane the distal centriole, lies in the nuclear fossa, at least in *L. dumerilii*; during spermiogenesis nuclear rotation occurs; a cross striated structure above the proximal centriole is reminiscent of the axial body of *Poecilia reticulata* (van der Horst and Cross, 1978).

Axoneme. In *L. dumerilii* two classes of flagella were observed: 80 percent varied between 39 and 43 μm in length (mean 41 μm) while the remainder, considered to indicate abnormal sperm, were much shorter and varied more in length (21-30 μm, mean 26 μm) (van der Horst and Cross, 1978). No fins are described or illustrated for the 9+2 axoneme.

Remarks. Mattei (1970) recognizes the spermatozoon of *Galeoides decadactylus* as being intermediate between his type I and II categories and this appear to be true of that of *Liza* although it is classified as type I by van der Horst and Cross (1978).

The presence of a round body in *L. dumerilii* is an interesting resemblance to the percichthyid *Maccullochella* (see above).

Chapter 16

SUPERORDER ACANTHOPTERYGII

PERCOMORPHA: LABROIDEI THROUGH TETRAODONTIFORMES

Suborder Labroidei

Constitution and relationships.The constitution of the Labroidei, in terms of families, accepted here differs from that given by Nelson (1984) and accords with Greenwood and Liem (1981) and Kaufman and Liem (1982) in including the Pomacentridae (Damsel fish), Cichlidae, and Embiotocidae (surf perches) (all three regarded as percoids by Nelson), together with the Labridae (wrasses) Odacidae and Scaridae (parrot fish). Kaufman and Liem (1982) and Lauder and Liem (1983) (Fig. 16.1) consider the Pomacentridae to be the plesiomorph sister-group of all other labroids (*s. lat.*), (as earlier suggested by Stiassny, 1980); the cichlids to be the plesiomorph sister group of embiotocids and labrids; and embiotocids that of labrids. This precise sequence of families is not supported by the shared occurrence of demersal eggs and parental care of hatched young in cichlids and some pomacentrids (Richards and Leis, 1984) unless, as seems unlikely, these features are plesiomorphic for labroids.

As indicated in Fig. 16.1, shared features considered synapomorphies of the Labroidei *s. lat.* are: united or fused fifth ceratobranchials, resulting in the formation of one lower pharyngeal jaw; true diarthrosis between upper pharyngeal jaws and the basicranium; and an undivided sphincter oesophagi muscle forming a continuous sheet. Those of the cichlid - embiotocid - labrid clade are: occurrence of levator externus 4 as a continuous muscle joining the prootic region to a muscular process on lower jaw; and a predisposition for insertion of the levator posterior muscle on the lower pharyngeal jaw. Embiotocids and labrids are united by the synapomorphies loss of second pharyngobranchial toothplates; first three branchial adductor muscles covering the anterodorsal faces of the epibranchials; presence of a ligament connecting the postmaxillary process of the maxilla with the anterior border of the palatine and ectopterygoid; arrangement of tooth rows radially across the lower pharyngeal jaw, location of teeth directly over the symphysis between left and right fifth ceratobranchials, and a dominant mode of tooth replacement from the posterior margin of the toothplate. Grounds for including the former odacids and scarids within the family Labridae are a suite of synapomorphies: the levator posterior is the dominant muscle to the lower pharyngeal jaw, forming a force couple with the pharyngocleithralis muscle (Liem and Greenwood, 1981); toothplates of the fourth pharyngobranchials are absent or reduced; the fourth epibranchials are highly modified, articulating with upper pharyngeal jaws; and there is true pharyngo-cleithral articulation, functioning as sliding and hinge joint (Kaufman and Liem, 1982; Lauder and Liem, 1983).

While sperm of all three of the families Pomacentridae, Cichlidae and Embiotocidae have been examined it is remarkable that no sperm of the labroids *sensu* Nelson (1984) have been investigated ultrastructurally. Light microscopy surprisingly reveals a simple anacrosomal aquasperm for the zoarcid *Zoarces viviparus*, the Viviparous Eelpout (Retzius, 1905) which is alternatively placed in the Zoarcoidei or sometimes in the Gadiformes (see discussion in Nelson, 1984).

Sperm relationships. To anticipate, it cannot be said that our very incomplete knowledge of spermatozoal ultrastructure in labroids contributes to resolution of the problems of classification and phylogeny outlined above but further ultrastructural investigations may bring useful insights.

Pomacentridae. These are the damsel fish; marine, rarely brackish, in all tropical seas but pri-

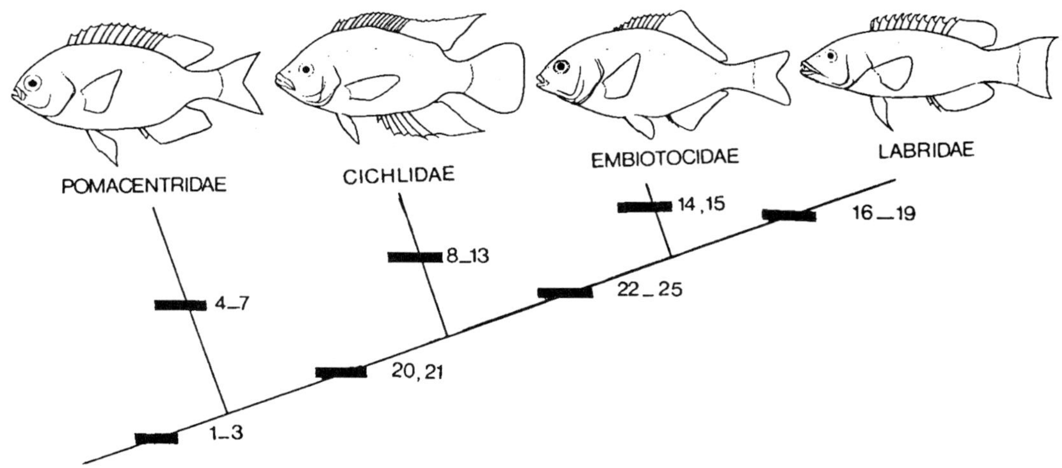

Fig. 16.1. Interrelationsips of the major labroid lineages based on the following specialized characters: 1, United or fused fifth ceratobranchials; 2, true diarthrosis between upper pharyngeal jaws and basicranium; 3, undivided sphincter oesophagi muscle; 4, strong sheet of connective tissue joining lower jaw with a ligament, which inserts on the ceratohyal bone; 5, nipple-like bony process on ventral surface of lower pharyngeal jaw; 6, pharyngo-cleithral articulation of characteristic form; 7, obliquus posterior dominant muscle to lower pharyngeal jaw; levator externus 4 and obliquus posterior vertically aligned on fourth epibranchial, separated by oblique aponeurosis or tendon; 8, transversus dorsalis muscle subdivided into four parts; 9, premaxillae and maxillae functionally decoupled; 10, cartilagenous cap on anterior border of epibranchial 2; 11, microbranchiospinae of characteristic form present on outer faces of second, third, and fourth gill arches; 12, A2 and Aw portions of adductor mandibulae complex separated completely; insertion of large ventral division of A2 onto angulo-articular; 13, head of epibranchial 4 distinctly expanded; 14, intra-uterine development of young with highly modified vascularized median fins; 15, muscular sheet joining A_1 and A_w portions of adductor mandibulae; 16, levator posterior dominant muscle to the lower pharyngeal jaw, forming a force couple with the pharyngocleithralis muscle; 17, toothplates of fourth pharyngobranchials absent or reduced; 18, fourth epibranchials highly modified, articulating with upper pharyngeal jaws; 19, true pharyngo-cleithral articulation functioning as sliding and hinge joint; 20, levator externus 4 is a continuous muscle joining prootic region to muscular process on lower jaw; 21, predisposition for insertion of levator posterior muscle on lower pharyngeal jaw; 22, loss of second pharyngobranchial toothplates; 23, first three branchial adductor muscles cover anterodorsal faces of the epibranchials; 24, ligament connecting postmaxillary process of maxilla with anterior border of palatine and ectopterygoid; 25, tooth rows arranged radially across the lower pharyngeal jaw, teeth located directly over the symphysis between left and right fifth ceratobranchials, dominant mode of tooth replacement from posterior margin of toothplate. From Lauder and Liem (1983) (After Kaufman and Liem, 1982). ***Bulletin of the Museum of Comparative Zoology*** **150**, 95-197. Fig. 52.

marily Indo-Pacific (Nelson, 1984).

Pomacentrus leucostictus. In the sperm of the pomacentrid *Pomacentrus leucostictus* a stack of membranes (not equivalent to the Golgi-derived structure in *Protopterus*) applied to the nuclear envelope of the spermatid and originating at the end of meiosis is said to persist into the spermatozoon (Mattei and Mattei, 1976) but no mature spermatozoon is illustrated.

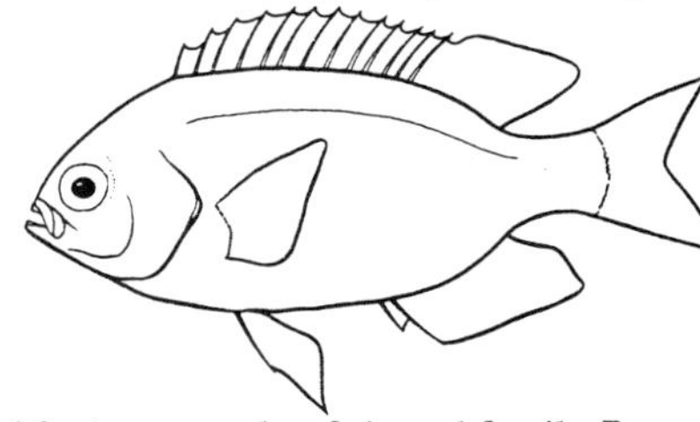

Fig. 16.2. An example of the subfamily Pomacentrinae. From Nelson, J.S. (1984). *Fishes of the World.* 2nd edition, John Wiley and Sons, New York. p. 319.

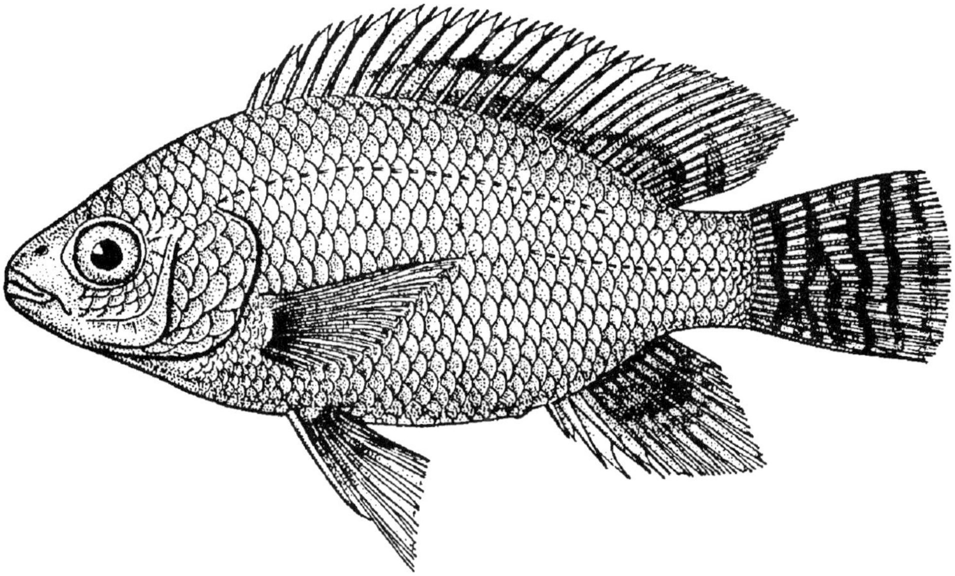

Fig. 16.3. *Tilapia nilotica*. From Grassé, P.-P. (1958). *Traité de Zoologie*. XIII Agnathes et poissons. Masson et Cie. Fig. 1732.

Cichlidae. Cichlids are freshwater and brackish water fish of Central and South America, West Indies, Africa, Madagascar, Syria and coastal India (Nelson, 1984). Species flocks characteristic of the African lakes would be an interesting subject for the study of species-specificity of spermatozoal ultrastructure but few cichlids have been examined in this respect.

Hemichromis and Tilapia. In the cichlids, *Hemichromis fasciatus* (Fig. 16.4A) and *Tilapia nilotica* (adult, Fig. 16.3; sperm, Fig. 16.4B) are shown diagrammatically to have symmetrical aquasperm the rounded nucleus of which has a moderate basal fossa which contains the centrioles (type I arrangement), the proximal centriole being perpendicular to the

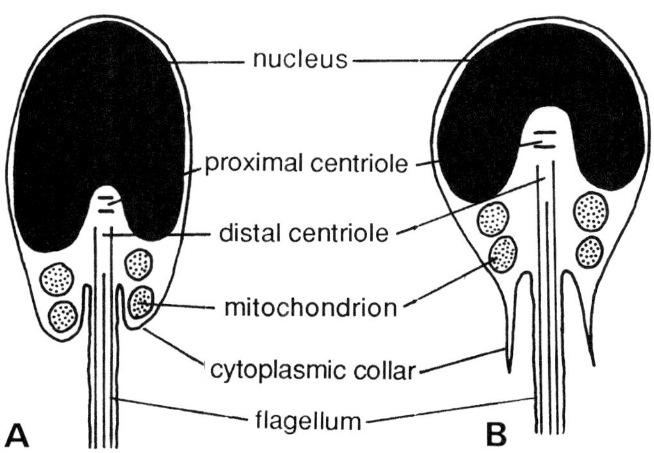

16.4. A. *Hemichromis fasciatus*. B. *Tilapia nilotica*. Diagrammatic longitudinal section of the spermatozoa. After Mattei, X. (1970). In *Comparative Spermatology* (ed. B. Baccetti), pp. 59-69. Academic Press, New York. Fig. 4: 11 and 4.29.

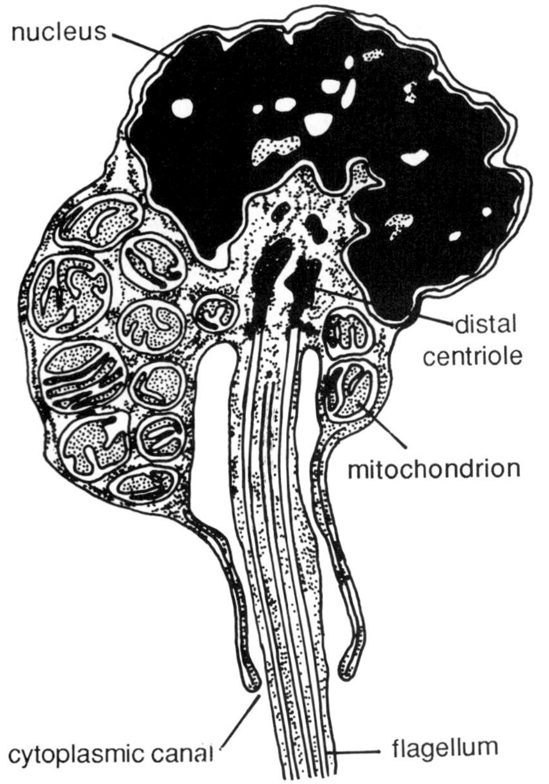

Fig. 16.5. *Oreochromis niloticus*. Longitudinal section of spermatozoon. Showing asymmetry of the midpiece around the tail axis. After a micrograph by Guha *et al.* (1988). *Proceedings of the 46th Annual Meeting of the Electron Microscopy Society of America.* San Francisco Press, San Francisco. pp. 278-279. Fig. 1.

basal body. Two tiers of separate mitochondria are shown. In both there is a short cytoplasmic collar which in *Tilapia* extends further as a short sleeve (Mattei, 1970).

Oreochromis niloticus (Fig. 16.5) has several tiers of mitochondria more abundant on one side of the axoneme than the other; a slender cytoplasmic sleeve extends for an equal length, approximately, behind the mitochondria; the implantation fossa, containing two mutually perpendicular centrioles, reaches approximately to the equator of the nucleus, the anterior surface of which is somewhat lobulated; the 9+2 axoneme has outer dynein arms but the inner arms are not conspicuous; two fins of moderate length are present, in line, as usual, with the singlets (Guha *et al.*, 1988).

Embiotocidae. The Embiotocidae is a family of coastal marine, rarely freshwater, North Pacific sea perch which are fully viviparous, delivering large, well developed young. The male has a small intromittent organ which represents a modified anterior end of the anal fin. Monophyly of the embiotocids is indicated by a suite of specialized mechansims for viviparity, including modified and vascularized median fins used in prenatal young for exchange with the convuluted and vascularized ovarian lining (Webb and Brett, 1972; Lauder and Liem, 1983).

Cymatogaster aggregata, the White Surf-fish or Shiner Surfperch (Fig. 16.6), has been shown to produce bundles of sperm which have an extracellular capsule and are therefore true spermatophores.

Each spermatophore contains some 600 parallel spermatozoa and thus differs notably from the spermatozeugmata of poeciliids, in which internal fertilization is clearly an independent development, which have the sperm heads located peripherally to a core of flagella. In the female they release the spermatozoa within an hour of insemination.

The spermatozoa (Fig. 16.7) are approximately 50 µm long. An acrosome is absent. The head (chiefly nucleus) is elongate (4 µm long) with condensed chromatin and is strongly depressed in one longitudinal plane (1 µm wide and 0.4 µm deep). Both centrioles occupy depressions in the nucleus, anterior to its midlength, which are interconnected by a thin isthmus of cytoplasm. The distal centriole is capped by an electron dense cone-shaped body. The elongate midpiece (3.5 µm long) contains 6 mitochondria, arranged three on each side; an organelle-free cytoplasmic sleeve continues 1 mm behind them. From its origin in the head and continuing through the midpiece, the anterior 4.5 µm of the 9+2 flagellum runs in a long cytoplasmic canal. Lateral protuberances of the flagella [fins] are irregularly arranged but mostly paired, in the plane of the two central singlets (Gardiner, 1978a,b).

Suborder Trachinoidei

General. Trachinoids, often included in the Blennioidei (e.g. Lauder and Liem, 1983), are generally small, primarily shallow-living temperate and tropical marine demersal or burrowing fish, including the jaw fishes, eel blennies, sand perches, weever

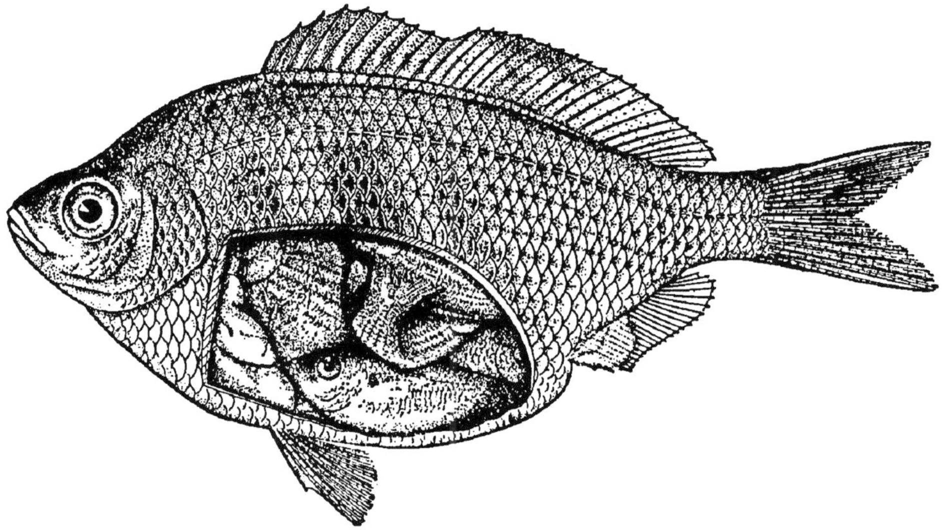

Fig. 16.6. *Cymatogaster aggegata,* the White Surf-fish or Shiner Surfperch. Viviparous female dissected to show enclosed young. After Jordan, D.S. (1907). *Fishes.* Henry Holt, New York. Fig. 22.

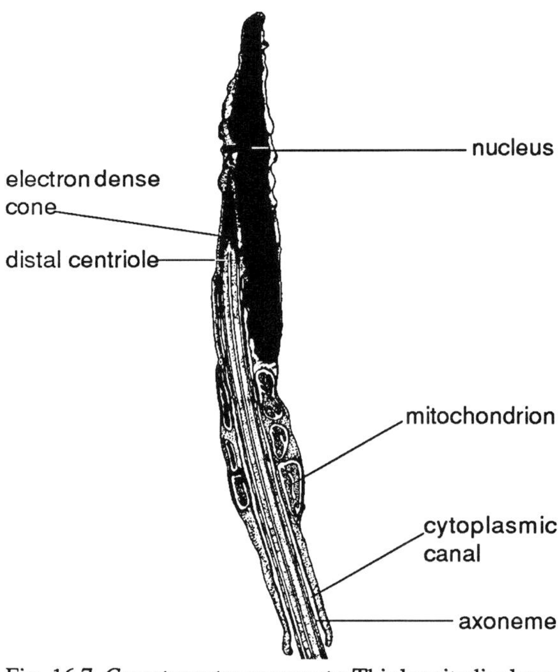

Fig. 16.7. *Cymatogaster aggregata.* This longitudinal section near the midline passes through the proximal centriole within the depression of the upper surface of the head and through the electron dense cap on the anterior end of the flagellum After a micrograph by Gardiner, D.M. (1978b). *Journal of Fish Biology* **13**, 435-438. Fig. 1B.

fish and others, some of which have affinities with the Blennioidei or Percoidei. They are probably a monophyletic assemblage (Gosline, 1968; Lauder and Liem, 1983). The Mugiloididae, examined here, appear to be the most percoid-like members of a superfamily Trachinoidea alternatively placed within the Blennioidei (see Watson *et al.*, 1984).

Mugiloididae. ***Parapercis*** sp. The sperm of the sandperch *Parapercis* sp. (adult, Fig. 16.8; sperm, Fig. 16.9) (Family Mugiloididae) is of the type II (*Parapristipoma*) kind of Mattei (1970): the proximal centriole is slightly displaced relative to the basal nuclear fossa, and the distal centriole is lateral to and at an angle, here about 45°, to the other; and there is a moderately long cytoplasmic canal. The small, ovoid mitochondria form two or more tiers in

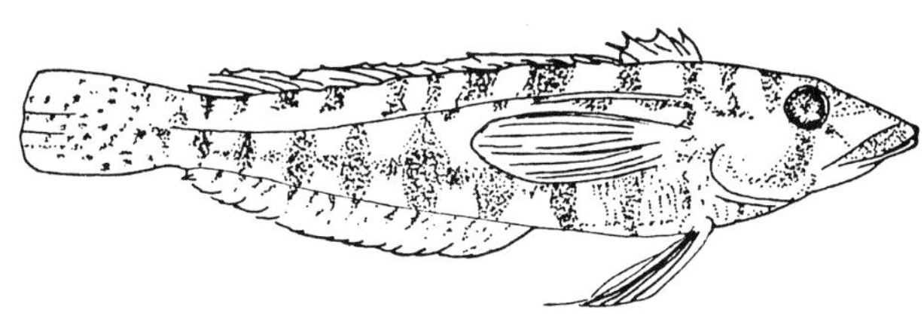

Fig. 16.8. *Parapercis* sp. Specimen from which sperm are described. From Heron Island, Great Barrier Reef, Australia. Original.

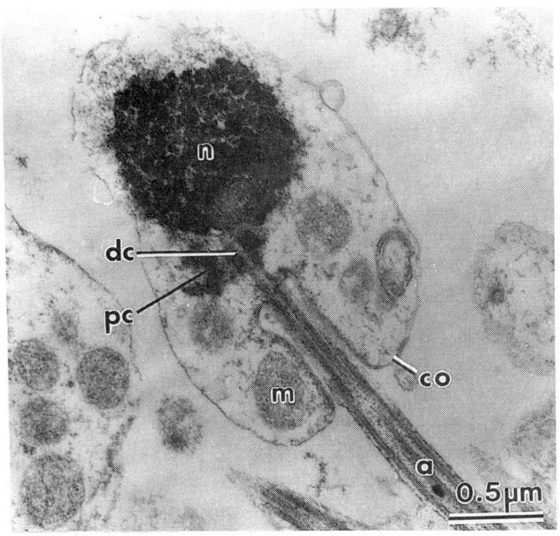

Fig. 16.9. *Parapercis* sp. Longitudinal section of spermatozoon. a. axoneme.co. cytoplasmic collar. dc. distal centriole. m. mitochondrion. n. nucleus. pc. proximal centriole. pc. proximal centriole. Original.

the broad cytoplasmic collar. An unusual feature for perciforms is that are no fins on the 9+ 2 axoneme.

Suborder Blennioidei

Diagnosis. Blennioids, worldwide in marine habitats, have less deep bodies than percoids, with a short trunk and a relatively attenuated caudal region. The dorsal and anal fins are long and low, terminating near the caudal fin, and the pectoral and usally the caudal fins are rounded. There is an exact correspondence in number between dorsal and posterior anal soft rays and the vertebrae supporting them (Rosenblatt, 1984). They are divisible into two, northern and tropical, predominantly marine groups. It is doubtful that the suborder is monophyletic. Many are characterized by demersal eggs, parental care and an advanced state of the newly hatched larvae (Matarese *et al.*, 1984).

Clinus, Blennius and Ophioblennius. Common features of the anacrosomal aquasperm of the clinid, *Clinus nuchipinnis* and bleniids, *Blennius cristatus, Blennius vandervekeni* and *Ophioblennius atlanticus,* investigated (Mattei, 1970) (Fig. 16.10A, B, C) are tilting of the axonemal relative to the nuclear axis, location of the mitochondrial material on only one side of the base of the axoneme (at least as seen in one plane), and presence of a proximal centriole in addition to the basal body. In *Ophioblennius atlanticus* (Fig. 16.10D) the curved nucleus is elongated in a direction virtually at right angles to the base of the flagellum. These sperm appear intermediate between types I and II.

The sperm of *Blennius pholis* is highly unusual in that the mitochondria migrate to the apical end of the nucleus while the centrioles and flagellum remain basal, much as in elopomorphs. Five or six mitochondria are arranged in a circle, almost contiguous centrally, in indentations at the tip of the nucleus. The head is 3.7-4.8 µm long and the nucleus is 0.6 µm wide anteriorly and 0.9 µm wide posteriorly and is therefore considerably elongated. It has electron transparent areas (vacuoles). The proximal centriole lies in a small nuclear fossa. The 9+2 flagellum is 29 µm long. The testis in *O. pholis* is supposedly both of

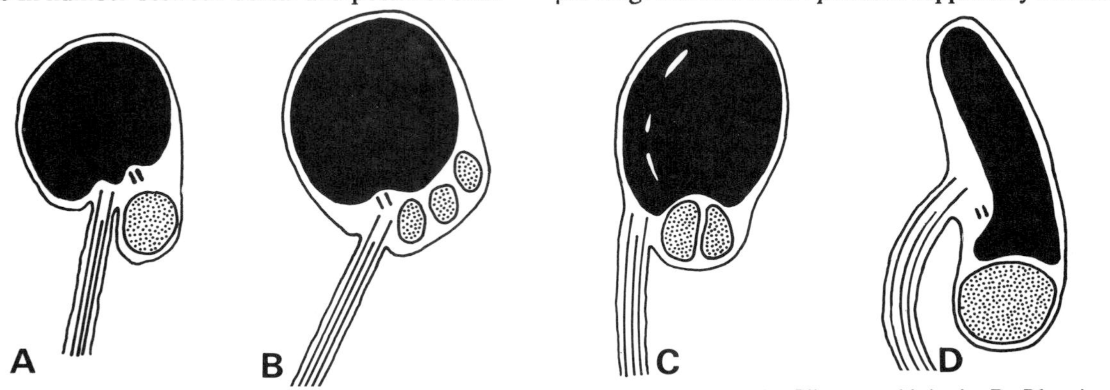

Fig. 16.10. Blenniodei. Diagrammatic longitudinal sections of the spermatozoa. A. *Clinus nuchipinnis*. B. *Blennius cristatus*. C. *Blennius vandervekeni*. D. *Ophioblennius atlanticus*. After Mattei, X. (1970). In *Comparative Spermatology* (ed. B. Baccetti), pp. 59-69. Academic Press, New York. Fig. Fig. 4: 2, 20, 26, and 16.

the "unrestricted spermatogonia testis type" of Grier (1981) and the lobular type of Billard (1986) (Silveira *et al.*, 1990). It is not entirely clear whether it is the mitochondria or the flagellar apparatus which

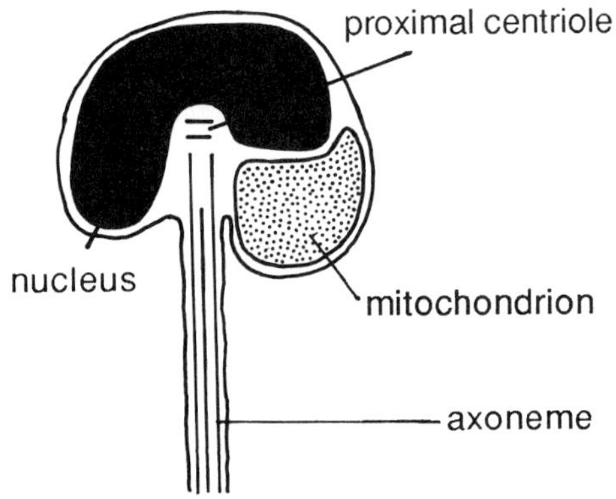

Fig. 16.11. *Periophthalmus papilio.* Diagrammatic longitudinal section of the spermatozoon. After Mattei, X. (1970). In *Comparative Spermatology* (ed. B. Baccetti), pp. 59-69. Academic Press, New York. Fig. 4: 23.

have migrated to the tip of the nucleus.

Suborder Gobioidei

Relationships. Gobioids are one of the most speciose group of fishes, with approximately 2,000 species or 10% of all teleosts (Rupple, 1984). They appear to form a monophyletic group with the Rhyacichthyidae and Eleotridae (examined here) as their most primitive representatives. The primary innovative characters defining gobiids is the development of a pelvic cup-shaped disc (Hoese, 1984).

Periophthalmus papilio. In the sperm of the gobiid *Periophthalmus papilio*, the Mudskipper, (Fig. 16.11) (family Gobiidae), the mitochondrial material, forming a single mass, again lies to one side only of the axoneme but tilting of the head relative to the axoneme is not suggested.

The nucleus is asymmetrical with a considerable basal fossa containing the proximal centriole (Mattei, 1970). It can be considered a modified type I sperm.

Hypseleotris galii. The fossa is deeper, with extreme attenuation of the chromatin in the eleotrid *Hypseleotris galii*, the Fire-tailed Gudgeon (adult, Fig. 16.12; sperm, Fig. 16.13) (family Eleotridae) (Jamieson unpublished).

Nucleus. The nucleus is elongate pyriform and is totally impaled by the axoneme and centriolar apparatus, with persistence of only a very thin layer of chromatin and investing nuclear envelope over the apical portion of the implantation fossa (Fig. 16.13E). The centriolar fossa is slightly eccentric (Fig. 16.13B), and correspondingly the nuclear layer is thicker on one side than elsewhere. A unique feature is interruption of the chromatin in one to three loci apically and subapically. As a result only apposed nuclear envelopes intervene between the implantation fossa and the plasma membrane (Fig. 16.13, see h). The largest hiatus (0.4 µm long) in the chromatin is subapical and on the face of the nucleus to which the longitudinal axis of the proximal centriole is directed.

Midpiece. The midpiece is highly asymmetrical. A single large, transversely cristate mitochondrion is situated in a unilateral collar, abutting the wide side of the nucleus, on one side of the axoneme.(Fig. 16.13C). As in *Periophthalmus* (Fig. 13.10), the nucleus is straight-edged where it abuts the mitochondrion.

Centrioles and flagellum. The basal body and 9+2 flagellum are in the long axis of the nuclear fossa. The proximal centriole is perpendicular though slightly upturned relative to the basal body. The usual pair of flagellar fins is present.

Sperm relationships. Sperm ultrastructure suggests relationship between the Dactylopteriformes, e.g. *Dactylopterus* (=*Cephalacanthus*) *volitans* (Mattei, 1970) and the Eleotridae, in the Perciformes, as exemplified here by *Hypseleotris galii*. In both species the nucleus is so deeply invaded by the basal fossa as to retain only a very thin layer of chromatin (in *Hypseleotris* subapically no chromatin) between the outer and inner nuclear envelope over the head of

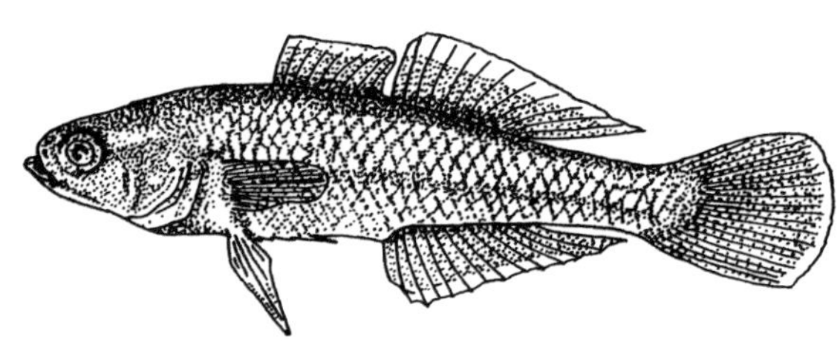

Fig. 16.12. *Hypseleotris galii*, the Fire-tailed Gudgeon. Original.

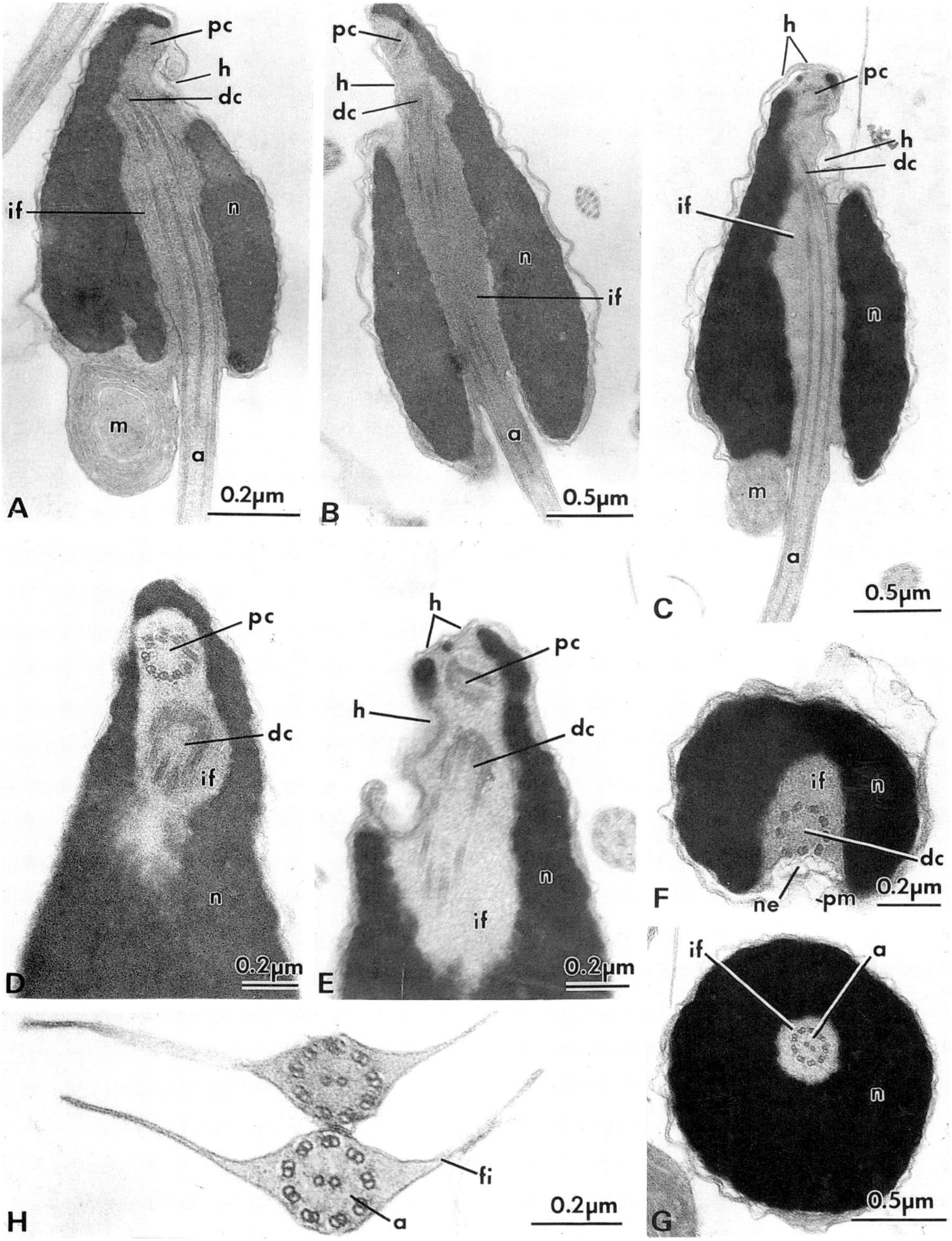
pc
h
dc
if
n
m
a
0.2μm
A
pc
h
dc
n
if
a
0.5μm
B
h
pc
h
dc
if
n
m
a
0.5μm
C
pc
dc
if
n
0.2μm
D
h
pc
h
dc
n
if
0.2μm
E
if
n
dc
ne
pm
0.2μm
F
if
a
n
0.5μm
G
fi
a
0.2μm
H

the flagellum; two centrioles are present; and the mitochondria are restricted to one side of the axoneme. Asymmetry of the head is more pronounced in *D. volitans* of the two species. A functional explanation of the deep penetration of the fossa and attenuation of chromatin is not at present apparent. A single

Fig. 16.14. *Trichiurus lepturus*, the Atlantic Cutlassfish. From Nelson, J.S. (1984). *Fishes of the World*. 2nd edition, John Wiley and Sons, New York. p. 363.

mitochondrion, as in *H. galii*, is seen in the percomorphs *Maccullochella* (Percicthyidae) (Leung, 1988b) and *Periophthalmus papilio* (Mattei, 1970).

Suborder Scombroidei

Diagnosis. Scombroids are perciform fish with a fixed premaxilla; the premaxillae are not only united with each other but also with the maxillae, forming a riigid non-protrusible upper jaw which can become elongate to form a rostrum. They include the fastest swimming fish, sailfish, swordfish and bluefin tuna (Lauder and Liem, 1983; Nelson, 1984).

Trichiurus lepturus. Like the clinid, bleniids, and eleotrids, the sperm of the trichiurid, *Trichiurus lepturus*, the Atlantic Cutlassfish (adult, Fig. 16.14), is asymmetrical.

The nucleus is on one side of the basal portion of the flagellum which lies in a long, strongly curved cytoplasmic canal. Small rounded mitochondria are shown in two tiers, on both sides of the base of the axoneme, but there is some suggestion of loss of mitochondria on one side in the plane at right angles. The persistence of the cytoplasmic canal in the spermatozoon is shown to be due to the presence of an anchoring system between the distal centriole and the bottom of the membrane invagination constituting

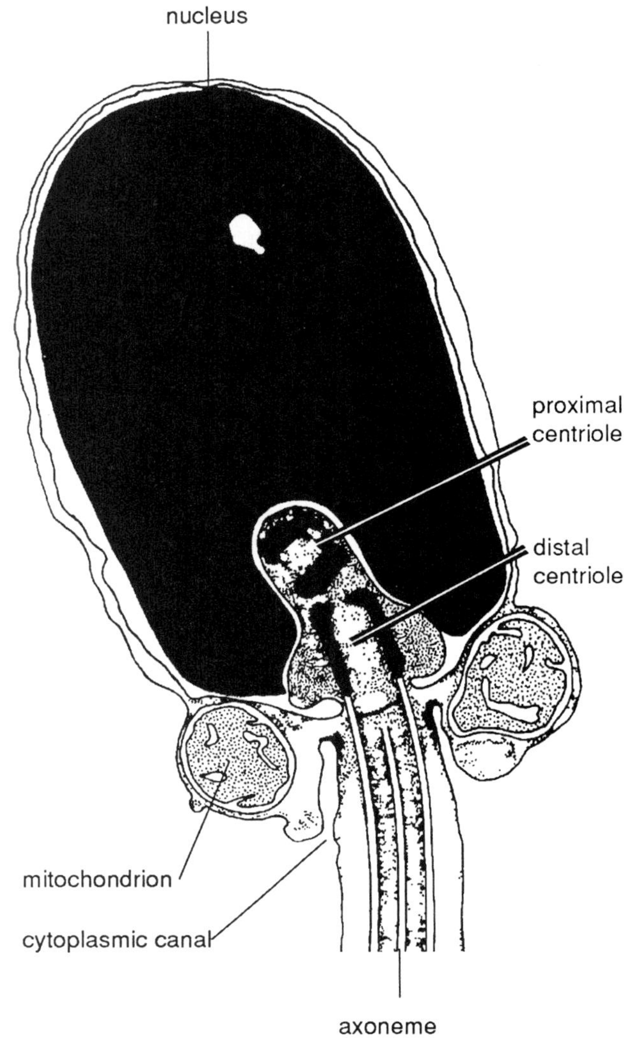

Fig. 16.15. *Platichthys flesus*. Longitudinal section of the spermatozoon. a. axoneme. co. cytoplasmic collar. dc. distal centriole. m. mitochondrion. n. nucleus. pc. proximal centriole. pc. proximal centriole. After micrographs by Jones, P.R. and Butler, R.D. (1988). *Journal of Ultrastructure Research* **98**, 71-82. Figs. 8 and 12.

Fig. 16.13. *Hypseleotris galii*. A-C. Longitudinal sections of spermatozoa, showing subapical interruption (hiatus) of the chromatin of the nucleus. A and C. "Lateral" view showing single, unilateral mitochondrion. B. Approximately "frontal view. D and E. Frontal and lateral views, respectively, of centrioles (only doublets have been seen) in nuclear fossa. F. Transverse section (TS) through distal centriole, showing its proximity to the collapsed nucelar envelope and plasma membane. G. TS of nucleus basally through the fossa and 9+2 axoneme. H. TS two axonemes, showing fins. a. axoneme. dc. distal centriole. fi. fin. h. hiatus in chromatin. if. implantation fossa. m. mitochondrion. n. nucleus. ne. nuclear envelope. pc. proximal centriole. pc. proximal centriole. pm. plasma membrane. Original.

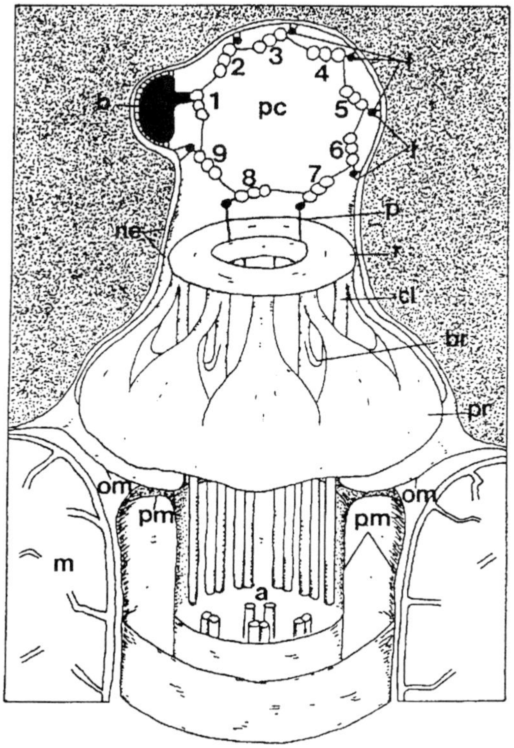

Fig. 16.16. *Platichthys flesus*. Schematic reconstruction of midpiece. The proximal centriole (pc), in the anterior region of the implantation fossa, has an associated mushroom-shaped body (b) which is the initial reference point for numbering the triplets (1-9) of the centriole. Triplets 2-9 have associated fibres (f). Fibres 2-6 and 9 are linked by short projections (I) to the nuclear envelope (ne). Fibres 7 and 8 are linked by longer projections (p) to the anterior ring (r) of the collar. This gives rise to 9 branched columns (cl) associated with the triplets of the distal centriole. Posteriorly a bridge (br) links the two branches, leaving an electron lucent window, and the outer branches [satellite rays] thicken to form a continuous posterior ring (pr). The posterior surface of the collar is bounded by a loop of the outer membrane (om) of the nuclear envelope. This separates the collar from the apex of a similar loop of the plasma membrane (pm) which itself separates the axoneme (a) and the mitochondria (m). From Jones, P.R. and Butler, R.D. (1988). *Journal of Ultrastructure Research* **98**, 71-82. Chart 1.

the canal. A helix of three granulations occurs at the base of the axoneme and is considered equivalent to the collar described in oligochaete and prosobranch sperm (Mattei and Mattei, 1976b).

Order Pleuronectiformes

Diagnosis. These include over 500 species of flatfish, flounders and soles, diagnosed as monophyletic by absence of bilateral symmetry in the adults owing to location of both eyes on one side. The body is highly compressed and they lie on the eyeless side (Lauder and Liem, 1983; Nelson, 1984). Insufficient is known of their spermatozoal ultrastructure to warrant inclusion of a phylogram.

Platichthys flesus, the flounder, has an anacrosomal aquasperm (Retzius, 1905[lm]; Jones and Butler, 1988) (Fig. 16.15, 16.16).

The bullet-shaped, homogeneously electron dense nucleus has a deep basal fossa. The anterior surfaces of 8 spherical cristate mitochondria lie within shallow depressions of the caudal surface of the nucleus. They form a ring around the opening of the nuclear fossa and the proximal region of the flagellum. The two centrioles (Fig. 16.15, 16.16) are located within the basal fossa (type I arrangement). Both are embedded in and linked by pericentriolar material which is intimately associated with the nuclear envelope. Around the proximal centriole the material is organized into a dense ring which bears nine fibres associated with the triplets, and a large fibrous body, all connected to the nuclear envelope. The material around the distal centriole is organized into a complex collar. The 9+2 flagellum is separated from the midpiece by a short cytoplasmic canal (Jones and Butler, 1988).

Pegusa triophthalmus, a soleid, has a very deep nuclear fossa (Fig. 16.18), resembling that of the mullid *Upeneus prayensis* (Fig. 5.2) in this respect and in location of the proximal centriole in the same longitudinal axis and the basal body. It is a classical Type I sperm *sensu* Mattei (1970).

Order Tetraodontiformes

Diagnosis and relationships. The Tetraodontiformes contain about 320 species of mostly shallow water, circumtropical and subtropi-

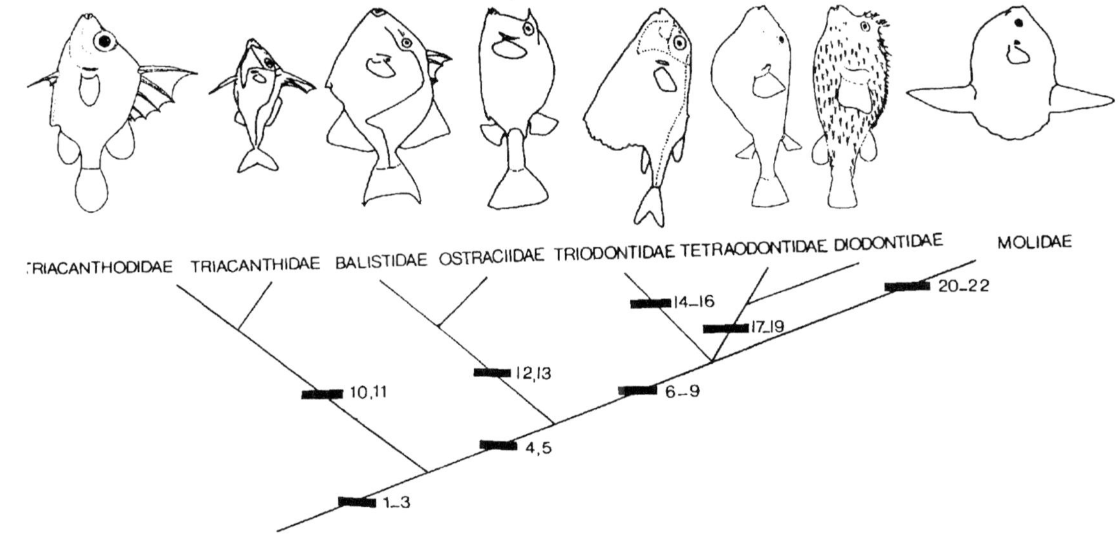

Fig. 16.17. Interrelationships of the tetraodontiform fish on the basis of the following specialized features: 1, the entire branchiostegal region is covered by a thick layer of skin; 2, the greatly restricted gill opening does not extend far below the base of the pectoral fin; 3, suborbital bones, parietals, nasals, sensory canals in the skull bones and anal fin spines are absent; 4, only a single levator operculi is present; 5, a distinct medial subdivision of the sternohyoideus is present; 6, the intermandibularis and sternobranchialis muscles are lost; 7, A_1 muscle acquires an attachment to the prefrontal region of the skull; 8, A_2 muscle is expanded dorsally and medially above A_3 to include the parasphenoid and prootic as the sites of origin; 9, teeth are either small rounded units or long rodlike structures; 10, pelvis with large pelvic spines, which can be locked in position; 11, hypertrophy of the arrector dorsalis pelvicus, arrector ventralis pelvicus, and adductor superficialis pelvicus; 12, a distinct retractor arcus palatini muscle is present; 13, differentiation of a deep A_1 muscle in the adductor mandibulae complex; 14, cleithrum greatly elongate reaching forward underneath the lower jaw; 15, a huge expansible dewlap of skin is present between the end of the pelvis and anus; 16, presence in the spiny dorsal fin of one or two small spines borne on two basal pterygiophores; 17, an inflatable diverticulum of the gut is present; 18, fourth gill arch is lost and there is no gill slit between it and the fifth arch; 19, muscle fibers from the dilatator operculi insert on the interoperculum, preoperculum and suboperculum; 20, only a single ovary is present; 21, otoliths are presumably lost; 22, caudal region with numerous structural and functional specializations creating a unique mode of locomotion. From Lauder and Liem (1983). *Bulletin of the Museum of Comparative Zoology* **150**, 95-197. Fig. 62.

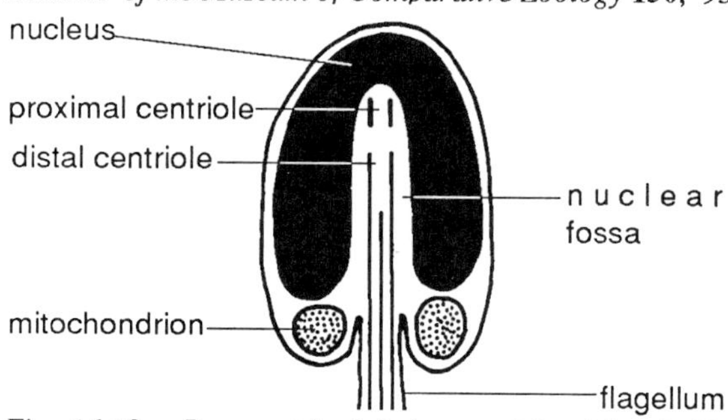

Fig. 16.18. *Pegusa triophthalmus*. After Mattei, X. (1970). In *Comparative Spermatology* (ed. B. Baccetti), pp. 59-69. Academic Press, New York. Fig. Fig. 4: 13.

cal marine forms.

They show striking examples of extreme reductive evolution and represent one of the main end lines of teleost radiation. Monophyly (Fig. 16.17) is suggested by a suite of synapomorphies: the entire branchiostegal region is covered by a thick layer of skin with or without scales; the gill opening is greatly restricted and does not extend far below the base of the pectoral fin; and all members have lost the suborbital bones, parietals, nasals, sensory canals in the skull bones and anal fin spines. Their sister-group is possibly the Acanthuroidei among the Perciformes from which they may have been derived by paedomorphosis (Patterson, 1964; Tyler, 1968; Winterbottom, 1974; Lauder and Liem, 1983).

The group contains the Suborder Balistoidei (in the classification of Nelson, 1984), including the Balistidae (leatherjackets) which contains, among others, the Balistinae (triggerfish) and Monacanthinae (file fish), and the Suborder Tetraodontoidei of which the Tetraodontidae (puffers) are one family. We will consider balistid and tetraodontid sperm.

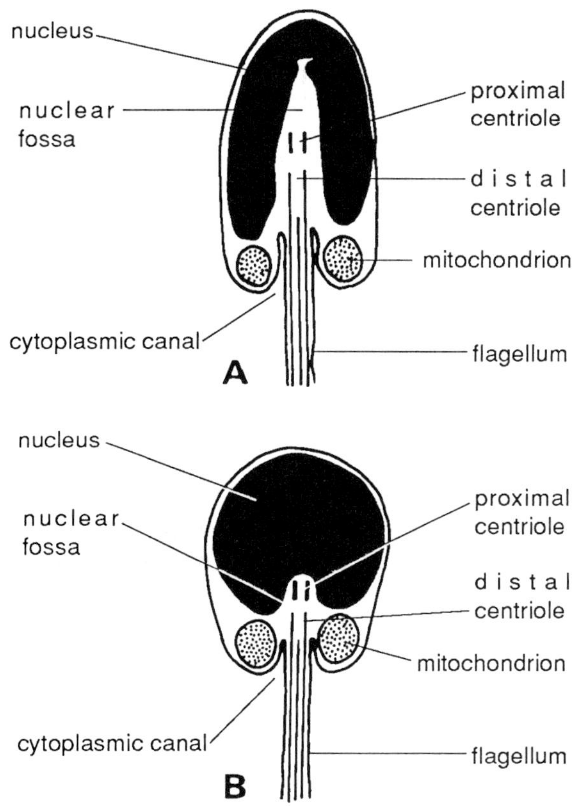

Fig. 16.19. Tetraodontiformes. A. *Balistes forcipatus* (Balistidae). B. *Chilomycterurus antennatus* (Tetraodontidae). Diagrammatic longitudinal sections of the spermatozoa. After Mattei, X. (1970). In *Comparative Spermatology* (ed. B. Baccetti), pp. 59-69. Academic Press, New York. Fig. 4: 6 and 4.

Balistid sperm. The three investigated balistid sperm (the balistine *Balistes forcipatus*, Mattei, 1970, and *Pseudobalistes fuscus* (Fig. 16.20A), and the monocanthine *Chilomycterurus antennatus* (Fig. 16.19B), Mattei, 1970) differ notably from the spermatozoon of the tetraodontid *Gastrophysus hamiltoni* (Fig. 16.20C-G) in the deep penetration of the nuclear fossa, to give an inverted U-shaped nucleus. They accord with *Gastrophysus* in alignment of the proximal centriole in the same axis as the distal centriole and symmetrical arrangement of simple mitochondria in a ring around the cytoplasmic canal but differ in orientation of the two centrioles in the same longitudinal axis whereas they are mutually perpendicular in *Gastrophysus*.

***Pseudobalistes fuscus*.** The sperm of this species are illustrated in Fig. 16.20A, B.

Nucleus. The nucleus, deeply penetrated by the implantation fossa, is approximately horseshoe-shaped in longitudinal section (Fig. 16.20A, B). It is 1.3-1.6, mean 1.5 µm long and 1.0-1.3, mean 1.1 µm, wide (n=5). The chromatin is electron dense and compact though with some indication of approximation of large masses between which several lacunae are usually present. The profile of the nucleus at its envelope has corresponding indentations but is otherwise smooth.

Midpiece. A small number of large, sparsely cristate, irregular, mutually adpressed mitochondria is grouped in a single layer around the cytoplasmic canal. In longitudinal section, the cytoplasmic collar continues as a short spurlike prolongation behind the mitochondria on each side (Fig. 16.20B). The cytoplasmic canal is approximately 0.6-0.8 µm long.

Centrioles and flagellum. The two centrioles lie within the anterior half of the deep implantation fossa (Fig. 16.20B). The proximal centriole is unusual not

Fig. 16.20. (Opposite). Tetraodontid sperm. A and B. *Pseudobalistes fuscus*. A. Spermatozoa in the testis, showing heads orientated towards the testicular epithelium. The flagella show paired lateral fins. B. Detail showing deepnuclear (implantation) fossa and serially coaxial centrioles. Inset, transverse section (TS) of a centriole, showing triplets. C-G. *Gastrophysus hamiltoni*. C. Longitudinal section of spermatozoon, showing mutually perpendicular centrioles in moderately deep nuclear fossa. D. TS of the midpiece showing 9+0 pattern of microtubules at transition from centriole to axoneme. E. TS of the midpiecethrough 9+2 region of axoneme. F. Longitudinal section of spermatozoon at right angles to C., showing mutually perpendicular centrioles in moderately deep nuclear fossa. G. TS of flagellum showing pair of lateral fins. a. axoneme. cc. cytoplasmic canal. co. collar. dc. distal centriole. m. mitochondrion. n. nucleus. pa. proximal, 9+0 region of axoneme. A and B. Original. C-G. Courtesy of Shane Hansford.

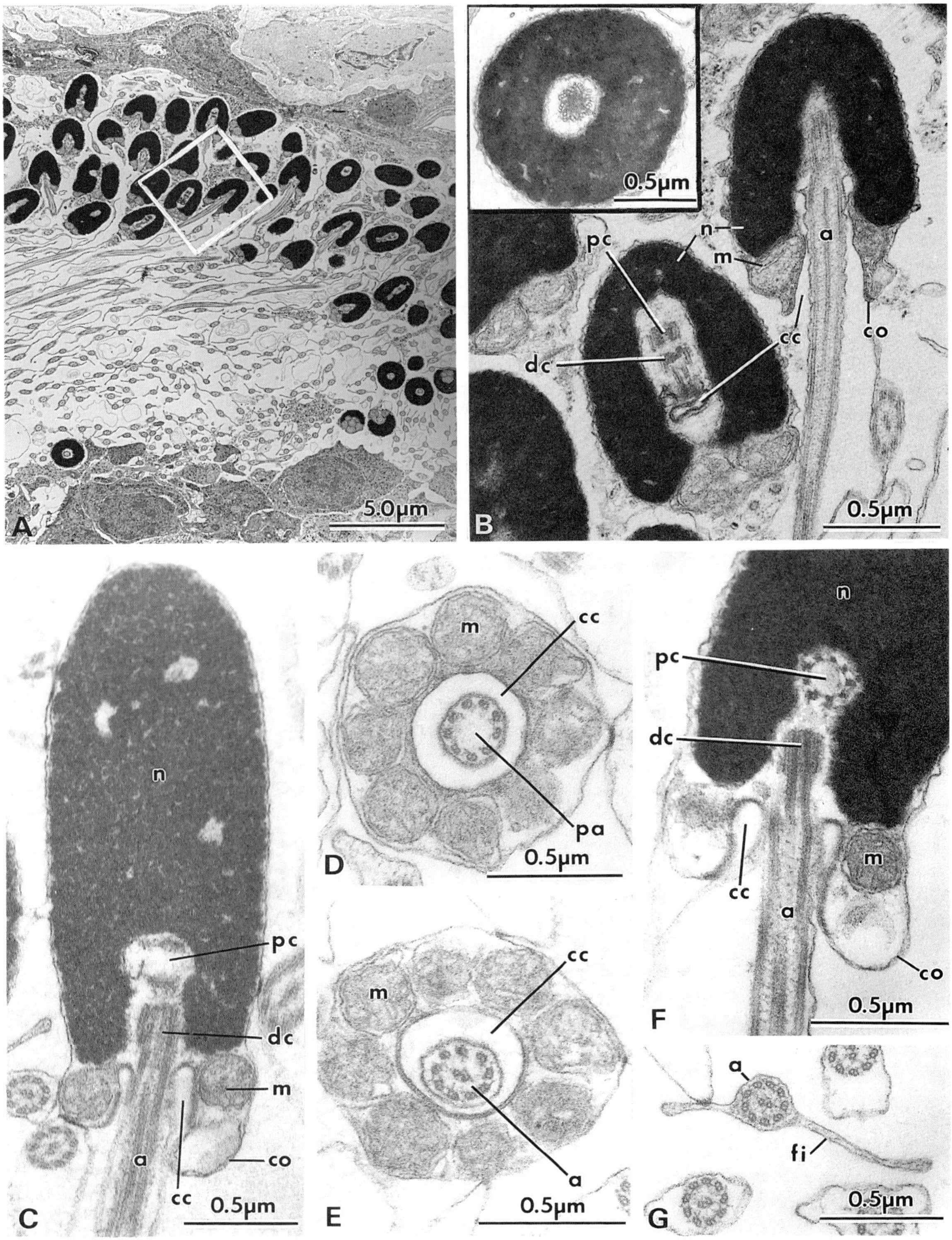
A
5.0µm
B
0.5µm
0.5µm
pc
n
m
a
cc
co
dc
C
n
pc
dc
m
a
cc
co
0.5µm
D
m
cc
pa
0.5µm
E
m
cc
a
0.5µm
F
n
pc
dc
cc
a
m
co
0.5µm
G
a
fi
0.5µm

only in being in the same axis as the basal body but also in having its longitudinal axis similarly orientated; it is near but not at the anterior limit of the nuclear fossa. At least one, apparently the distal, centriole consists of 9 triplets (Fig. 16.20B, inset). The 9+2 flagellum has long lateral fins (Fig. 16.20A) the plane of which is slightly tilted relative to that of the two central singlets.

Gastrophysus hamiltoni. (Fig. 16.20C-G). **Nucleus.** The nucleus of the sperm of *Gastrophysus hamiltoni*, a toadfish, is unusually elongate for that of an externally fertilizing spermatozoon (Fig. 16.20C). It is approximately cylindrical, 2.4 µm long and 1.0 µm wide at its greatest (midlength) diameter, with slightly convex sides and a rounded tip. Basally it is indented by a clubshaped implantation fossa, 0.6 µm long and maximally 0.3 µm wide. The chromatin is coarsely granular with some pale lacunae.

Midpiece. The eight, spherical cristate mitochondria are symmetrically arranged in a ring around the proximal region of the flagellum, shortly behind the distal centriole (Fig. 16.20D). They are housed in a cytoplasmic collar maximally 0.6 µm long, which is separated from the flagellum by a periaxonemal space bounded by the invaginated plasma membrane, the so-called cytoplasmic canal. Immediately within the plasma membrane lining the canal, in transverse section, there is a narrow, dense layer with some indication of transverse striation resembling that of a septate junction. This layer may be the equivalent of the submitochondrial net of poeciliid sperm.

Centrioles. The two mutually perpendicular centrioles lie within the implantation fossa in the same longitudinal axis (Fig. 16.20C, F). Only doublets surrounding and contiguous with a delicate ring have been demonstrated for the proximal centriole (Fig. 16.20F) which is housed in the apical expansion of the fossa. The distal centriole (basal body) consists of 9 triplets, which distally become doublets.

Axoneme. The proximal part of the axoneme within the midpiece also consists of 9 doublets (Fig. 16.20D) behind which two central singlets are added (Fig. 16.20E). The axoneme has two lateral fins relative to which the plane of the two central singlets is slightly tilted (Fig.16.20G). The lumina of the A and B subtubules are patent; weakly developed inner and outer dynein arms are visible in some sections (Hansford and Jamieson, unpublished).

Chapter 17

ATHERINOMORPHA

Diagnosis and relationships. Atherinomorphs differ from acanthopterygians, *inter alia*, in lacking a ball and socket joint between the palatine and maxilla. Their monophyly is convincingly indicated by three chief synapomorphies: the specialized oral jaw mechanism in which the rostral cartilage is not attached to the premaxilla and the protrusible upper jaw has crossed palatomaxillary ligaments with a maxillary ligament to the cranium; a large demersal egg with many oil droplets and adhesive filaments; and the third, fourth, and fifth infraorbitals and the fourth pharyngobranchial are absent (Rosen and Parenti, 1981; Lauder and Liem, 1983). The phylogram (Fig. 17.1) indicates their sister-group relationship with the Percomorpha.

Most species are surface-feeding and about 75% live in fresh or brackish water (Nelson, 1984). Atherinomorphs include the killifish (Cyprinodontidae), live-bearing top minnows (Poeciliidae), silversides (Atheriniidae), four-eyed fish (Anablepidae), rice fish (Adrianichthyidae), half beaks (Hemiramphidae), needle fish (Belonidae) and marine flying fish (Exocoetidae). These families may be placed in three orders, the Atheriniformes, Cyprinodontiformes and Beloniformes (Rosen and Parenti, 1981). The Beloniformes contain the suborders Adrianichthyoidei and Exocoetoidei, both subsumed in the Cyprinodontiformes by Nelson (1984).

In all three atherinomorph orders spermatogonia are restricted to the distal end of the testicular tubules. This is a notable contrast with the Salmoniformes, Perciformes, and Cypriniformes in which spermatogonia are distributed along the entire length of the tubules (Grier *et al.*, 1980; Grier, 1981; Grier and Collette, 1987). This dichotomy underlines the phylogenetic unity and discreteness of the Atherinomorpha indicated, from other characters, above. The telogonic atherinomorph condition is presumably apomorphic relative to the hologonic condition general in teleosts, as the conditions are termed here. Billard (1986), in a review of spermatogenesis and spermatology of some teleost species, refers to the two types of testis as lobular [hologonic] and tubular [telogonic].

Reproductive modes. Reproductive modes in the Atherinomorpha range from egg scattering to, in some Cyprinodontiformes and Beloniformes, varying degrees of viviparity (Grier, 1976). Internal fertilization is considered to have evolved from free-spawning independently among the cyprinodontiforms in several families and is attended by production of live young or rarely (*Zenarchopterus*) laying of fertilized eggs. It has been directly observed or has been inferred from the possession of spermatozeugmata, of spermatophores, or of internal embryos, in the aplocheilids *Rivulus marmoratus* and *Cynolebias*; in poeciliids; in goodeids; in *Jenynsia lineata*; in *Anableps anableps* and *A. dowi*; in *Horaichthys setnai* (all Cyprinodontiformes) and in some exocoetoid hemiramphids (Beloniformes) *viz.* the viviparous *Dermogenys pusillus*, *Hemirhamphodon pogonognathus*, and *Nomorhamphus hageni* and at least 11 species of the egg-laying genus *Zenarchopterus* (references in Parenti, 1981; Grier, 1981; Grier and Collette, 1987). Most exocoetoids are, however, externally fertilizing.

Order Atheriniformes

Diagnosis. The Order Atheriniformes was first named by Rosen (1964) for exocoetoids and cyprinodontoids and these included all taxa here placed in the Atherinomorpha. The order is defined more narrowly here but remains the least satisfactorily defined of the three atherinomorph orders. Lauder and Liem (1983) did not formally name the order, terming it merely the "atherinoids" or Atherinoidei. Members of the Atheriniformes usually have two dorsal fins,

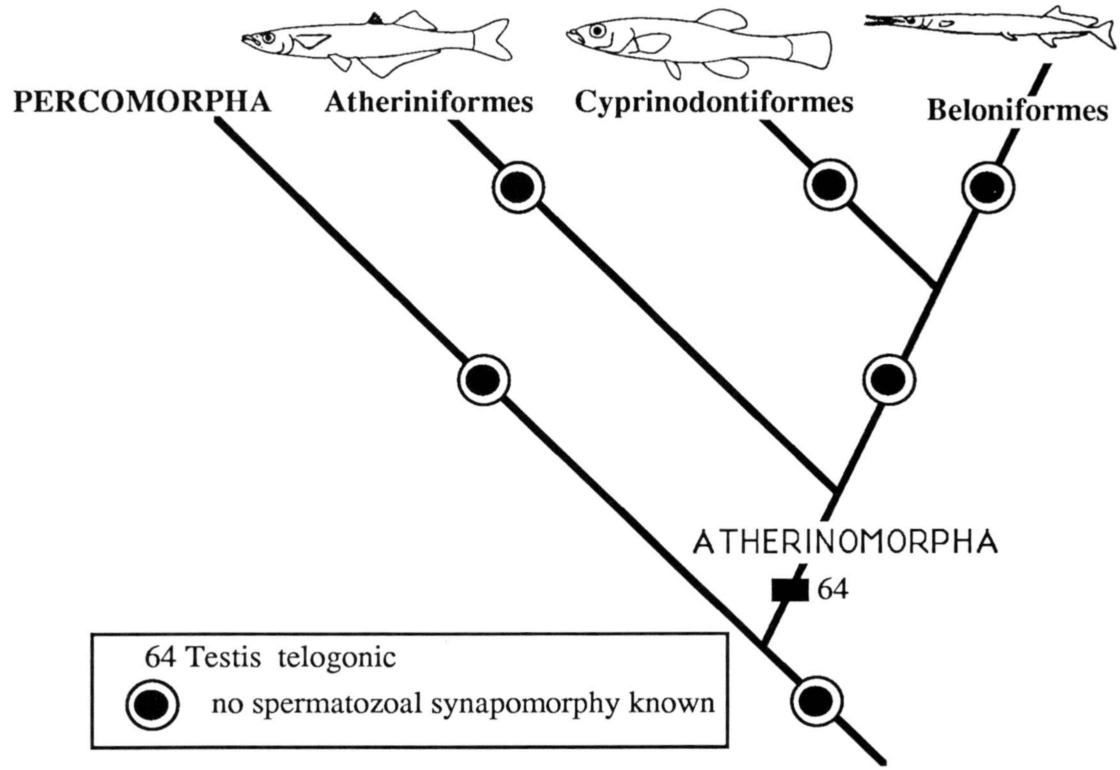

Fig. 17.1. Phylogenetic relationships of the Atherinomorpha. Modified after Lauder and Liem (1983). *Bulletin of the Museum of Comparative Zoology* **150**, 95-197. Fig. 47. Fish drawings from Nelson.

the first, if present, with flexible dorsal spines; the anal fin is usually preceded by a spine; the lateral line is absent or weak; there are 5-7 branchiostegal rays; the narial openings are paired; and parietals are present (Nelson, 1984). These characters are regarded as primitive [plesiomorphic] and therefore the Atheriniformes cannot be defined cladistically (Rosen and Parenti, 1981). Gosline (1970) considers the Atheriniformes to be related to his perciform order Mugiloidei (a relationship rejected above) but we have seen that all atherinomorphs differ in development of the telogonic testis. Despite the lack of firm apomorphies for the Atheriniformes, the telogonic condition is a unique synapomorphy (autapomorphy) of the Atheriniformes - Cyprinodontiformes - Beloniformes assemblage, setting them apart from other teleosts.

Sperm literature. Eleven species of the Atheriniformes have been examined for sperm ultrastructure: *Craterocephalus stercusmuscarum* (Fig. 17.2D-F), *C. helenae*, *C. marjoriae* (Fig. 17.2A-C), *Querichthys stramineus* (Atherinidae), *Cairnsichthys rhombosmoides*, *Iriatherina werneri* (Fig. 17.2G, H), *Melanotaenia duboulayi* (Fig. 17.2I, J), *M.*

Fig. 17.2. (Opposite). Atheriniform spermatozoa. A-C. *Craterocephalus marjoriae.* A. Longitudinal section (LS) of spermatozoon showing transverse proximal centriole with satellite rays. B. Transverse section (TS) of the proximal, 9+0 region of the axoneme with satellite rays and ring of mitochondria. C. TS of 9+2 region of flagellum with lateral fins. D-F. *Craterocephalus stercusmuscarum.* D. LS showing distal centriole partly enclosed in nuclear fossa. E. TS of proximal, 9+0, region of axoneme with satellite rays. F. TS of 9+2 region of flagellum with lateral fins. G, H. *Iriatherina werneri.* G. LS spermatozoon. H. TS of 9+2 region of flagellum with lateral fins. I, J. *Melanotaenia duboulayi.* I. LS sagittal plane of spermatozoon. J. TS flagella with lateral fins. a. axoneme. ca. transition from centriole to axoneme. cc. cytoplasmic canal. dc. distal centriole. fi. fin. m. mitochondrion. n. nucleus. s. satellite ray. Courtesy of C. J. Marshall.

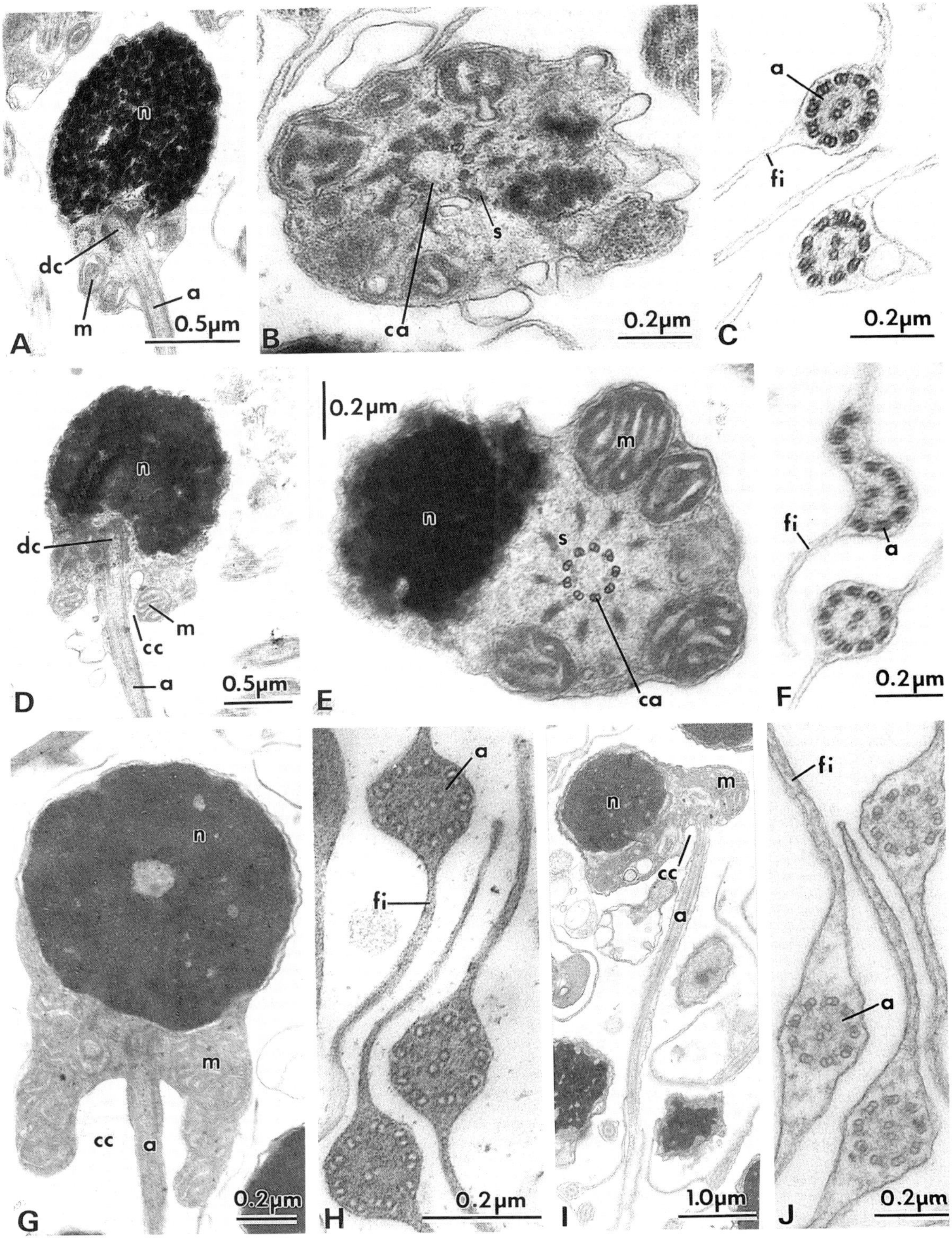
n
dc
m
a
0.5µm
A
ca
s
0.2µm
B
a
fi
0.2µm
C
n
dc
m
cc
a
0.5µm
D
0.2µm
n
m
s
ca
E
fi
a
0.2µm
F
n
m
cc
a
0.2µm
G
a
fi
0.2µm
H
n
m
cc
a
1.0µm
I
fi
a
0.2µm
J

maccullochi (Fig. 17.3A-C), *Pseudomugil mellis* (Fig. 17.3D-F), *P. signifer* (Fig. 17.3G-I) and *P. tenellus* (Fig. 17.3J-M) (Melanotaeniidae) (Marshall, 1989).

Acrosomoid. Present only in the two *Melanotaenia* species, anterior to the nucleus in longitudinal sections of many sperm, there is a single large, empty vesicle (Fig. 17.3A), here termed the acrosomoid (see Remarks). This is visible as a distinct protrusion in live, actively motile sperm, under the light microscope.

Nucleus. The nucleus is approximately isodiametric, with a diameter of 1.1 µm in *Craterocephalus marjoriae* and *Querichthys stramineus*, 1.2 µm in *Melanotaenia maccullochi* and 1.5 µm in *Iriatherina werneri* and *Melanotaenia duboulayi*. It is longer (1.3 µm) than wide (1.1 µm) in *Craterocephalus marjoriae* but is wider than long in *Craterocephalus helenae* (0.8 µm long and 1.0 µm wide) and *Cairnsichthys rhombosmoides* (0.7 µm long and 0.9 µm wide). In *Pseudomugil* it is crescentic, being 1.8 µm wide in *P. mellis* and 1.4 µm wide in*P. signifer*, while in *P. tenellus* it is hemispheroidal with a diameter of 1.1 µm.

The chromatin is usually flocculent or at least has numerous lacunae, and is sufficiently coarse grained to conform only irregularly with the double nuclear envelope; in *Iriatherina werneri*, alone, it is finely granular and conforms to the envelope. Only in this species, in *Craterocephalus helenae* and in the two *Melanotaenia* species is a nuclear fossa absent. The fossa includes the distal centriole but only in *Craterocephalus marjoriae* and the three *Pseudomugil* species is type I morphology attained by inclusion of the proximal centriole.

Midpiece. The mitochondria form a ring around the axoneme. There is a single layer of several mitochondria in *Melanotaenia duboulayi* and *M. maccullochi* (Fig. 17.3A) but in the 9 other examined atheriniforms there is more than one tier. In cross section there are 4 in *Querichthys stramineus*; 4 or 5 in *Craterocephalus stercusmuscarum* and *Iriatherina werneri*; 5 or 6 in *Craterocephalus helenae* and *C. marjoriae*; 5-7 in *Pseudomugil tenellus*; 6 or 7 in *Pseudomugil mellis*; 7 in *Cairnsichthys rhombosmoides*; and 6-10 in *P. signifer*. In longitudinal section there are 1 or 2 in *Craterocephalus helenae, C. marjoriae* and *Cairnsichthys rhombosmoides*; 2 or 3 in *Iriatherina werneri*; 3 in *Pseudomugil signifer* and *P. tenellus*; and 4 or 5 in *Pseudomugil mellis*.

There is no collar or cytoplasmic canal in *Melanotaenia duboulayi* and *M. maccullochi* but this is present in the other species. It is only 0.2 µm long in *Craterocephalus stercusmuscarum* and *Cairnsichthys rhombosmoides*; 0.3 µm in *Craterocephalus marjoriae*; 0.7 µm in *Iriatherina werneri*, and *Pseudomugil tenellus*; 0.9 µm in *Pseudomugil signifer*; 1.0 µm in *Craterocephalus helenae*; and 1.5 µm long in *Pseudomugil mellis*.

Centrioles. The proximal and distal centrioles are mutually perpendicular but not in the same longitudinal axis in *Craterocephalus marjoriae*, and in the same axis in *Pseudomugil mellis, P. signifer* and *P. tenellus*. Their long axes are parallel to each other in *Craterocephalus stercusmuscarum, C. helenae, Querichthys stramineus, Iriatherina werneri* and *Melanotaenia maccullochi* but only in *M. duboulayi* are they also serial, in the same longitudinal axis. The proximal is at about 30° to the distal centriole in *Cairnsichthys rhombosmoides*.

Satellite apparatus. Nine well developed satellite rays surround the distal centriole in *Craterocephalus helenae, C. marjoriae* (Fig. 17.2B) and *C. stercusmuscarum* (Fig. 17.2E); *Querichthys stramineus, Cairnsichthys rhombosmoides*, and *Melanotaenia maccullochi* but they appear to be weakly if at all developed in all three *Pseudomugil* species, in *Iriatherina werneri* and in *Melanotaenia*

Fig. 17.3. (Opposite). Atheriniform spermatozoa. A-C. *Melanotaenia maccullochi*. A. Longitudinal section (LS) of spermatozoon showing anterior vesicle (acrosomoid). B. Transverse section of distal centriole with satellite rays and ring of mitochondria. C. TS of 9+2 region of flagellum with lateral fins. D-F. *Pseudomugil mellis*. D. LS spermatozoon. E. TS of 9+2 region of axoneme through mitochondria and cytoplasmic canal. F. TS of 9+2 region of flagellum with lateral fins. G-I. *Pseudomugil signifer*. G. LS spermatozoon. H. TS of 9+2 region of axoneme through mitochondria and cytoplasmic canal. I. TS of 9+2 region of flagellum with lateral fins. J-M. *Pseudomugil tenellus*. J. TS of distal centriole. K. LS spermatozoon showing transverse proximal centriole and longitudinal distal centriole. L. TS of 9+2 region of axoneme through mitochondria and cytoplasmic canal. M. TS flagella with lateral fins. a. axoneme. cc. cytoplasmic canal. f. fin. m. mitochondrion. n. nucleus. s. satellite ray. v. acrosome-like vesicle. Courtesy of C. J. Marshall.

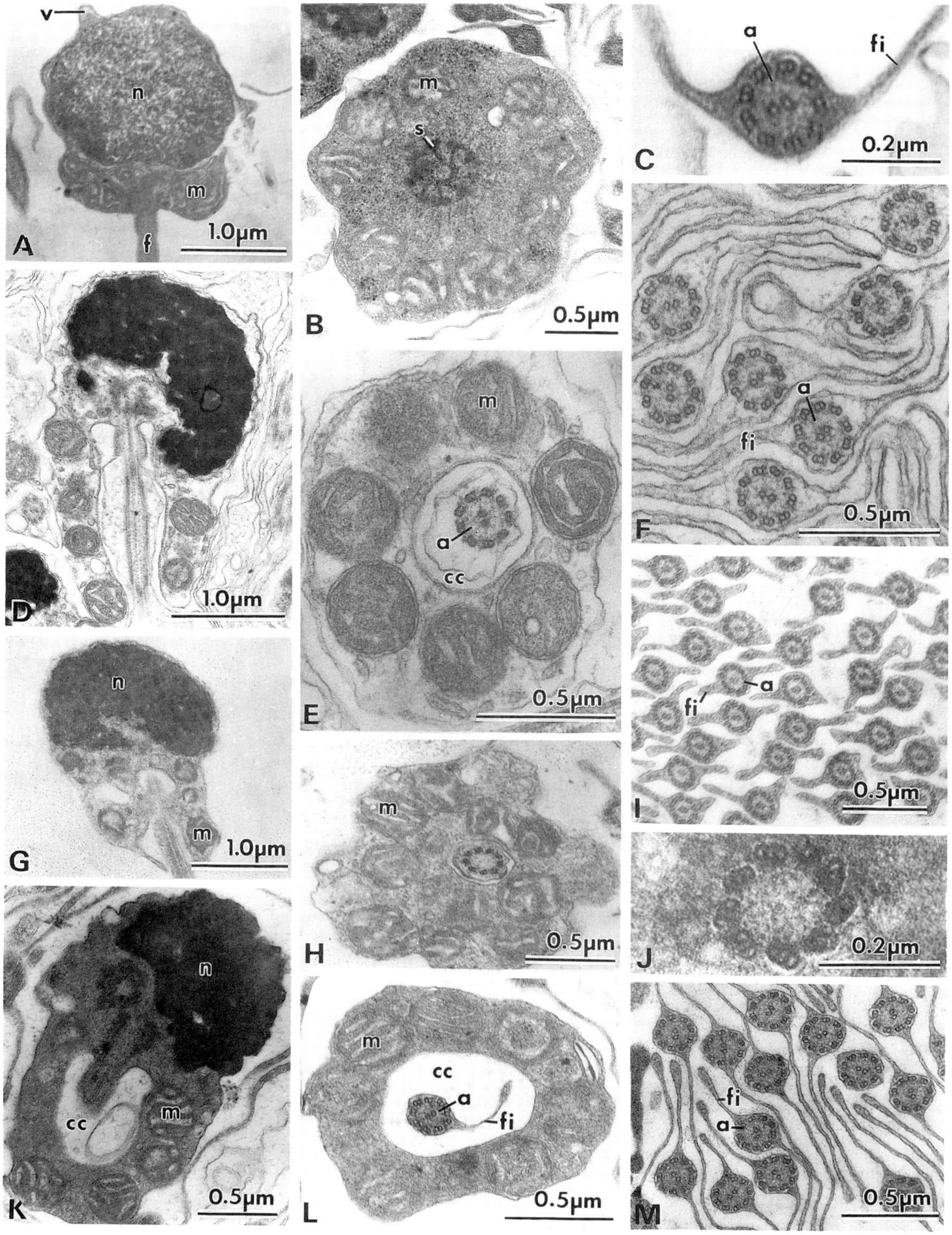

A
v
n
m
f
1.0μm
B
m
s
0.5μm
C
a
fi
0.2μm
D
1.0μm
E
m
a
cc
0.5μm
F
a
fi
0.5μm
G
n
m
1.0μm
H
m
0.5μm
I
a
fi
0.5μm
J
0.2μm
K
n
cc
m
0.5μm
L
m
cc
a
fi
0.5μm
M
fi
a
0.5μm

duboulayi.

Flagellum. In all species the flagellum bears two fins and is of the 9+2 pattern, apparently always with inner and outer dynein arms present. In *Iriatherina werneri* the flagellum is in the longitudinal axis of the nucleus but in all other species it subtends a considerable angle to the head, varying from about 40° in *Craterocephalus stercusmuscarum* to 90° in the two *Melanotaenia* species.

Remarks. The structure seen in both *Melanotaenia* species and here termed an acrosomoid is in the position occupied by the acrosome in most animal sperm but, with the exception of the internally fertilizing salmoniform *Lepidogalaxias* (Leung, 1988), an acrosome is absent from mature teleost sperm. An acrosome is present in the sperm of agnathans through chondrosteans. Evidence that the telostean aquasperm has undergone simplification is seen in the transient existence of an acrosome vestige in the spermatid of *Salmo gairdneri* (Billard, 1983b) and *Neoceratias spinifer* (Jespersen, 1984) (for further examples see Chapter 5), suggesting the occurrence of acrosome-bearing (but not necessarily internally fertilizing) sperm in the immediate ancestry of the Neopterygii. It would appear that a plesiomorphic genetic propensity to produce an acrosome has been retained in these species although its expression, which is fullest in the mature lepidogalaxiid sperm, is presumably secondary and an adaptation to a secondary internal fertilization. That the acrosomoid in *Melanotaenia* is a true homologue of an acrosome is, nevertheless, open to question and requires further investigation.

Sperm relationships. Several features seen in some or all of the investigated atheriniforms are deduced to be plesiomorphic for the order and, indeed, for the Acanthopterygii. All of these, except the flocculent chromatin and presence of axonemal fins, appear to be plesiomorphies for the Actinopterygii as a whole. These are the absence of an acrosome; the flocculent externally irregular nature of the chromatin; probably the presence of a basal nuclear fossa enclosing both centrioles (type I condition); a perpendicular arrangement of the proximal centriole relative to the distal centriole (basal body), this being the plesiomorphic condition for the Metazoa; presence of satellite rays around the basal body, as in many lower Metazoa; a moderately long cytoplasmic canal; perhaps 4 or 5 mitochondria; 2 dynein arms to each axonemal doublet; presence of two flagellar fins (at the same time an autapomorphy for the Actinopterygii); and insertion of the flagellum in or near the longitudinal axis of the head.

***Craterocephalus*.** On this basis, perhaps the most plesiomorphic atheriniform sperm are seen in the atherinid *Craterocephalus marjoriae*, with perpendicular centrioles, and perhaps in *Querichthys stramineus*, though material of the latter species was too poorly preserved for certain evaluation. *Querichthys* has been placed in the Melanotaeiniidae or in the Atheriniidae (see Allen, 1980). The apomorphic location of the proximal centriole parallel to and lateral to the distal centriole in *Querichthys stramineus*, *Craterocephalus helenae* and *C. stercusmuscarum* may be a synapomorphy of the two genera (or of *Querichthys* with the more evolved members of *Craterocephalus*) indicating that *Querichthys* is correctly placed in the Melanotaeniidae.

Marshall (1989) points out that taxonomic studies have indicated that there are two groups of hardyheads within *Craterocephalus*, one containing *C. marjoriae* and the other *C. stercusmuscarum*, and that sperm structure supports this and indicates that *C. helenae* should be placed in the *stercusmuscarum* group.

***Cairnsichthys*.** Several resemblances between the sperm of *Cairnsichthys*, conventionally placed in the Melanotaeiniidae, and sperm of *Craterocephalus*, in the Atheriniidae, appear to be symplesiomorphies (presence of satellites and of a cytoplasmic canal; location of the distal centriole in the nuclear fossa) and do not therefore justify a change in attribution.

***Melanotaenia*.** The most modified spermatozoa in the order are seen in *Melanotaenia* where the cytoplasmic canal has been lost; the mitochondria have apparently secondarily been reduced to a single layer; the centrioles are mutually parallel; the satellite apparatus is reduced in *M. duboulayi*, though not in *M. maccullochi*; and extreme displacement of the nucleus relative to the centrioles has occurred so that the axoneme is tangential to the nucleus, the complete type II condition; the anterior vesicle (acrosomoid) is a unique apomorphy possibly representing "neotenous" retention of an acrosome. The relatively modified condition of *Melanotaenia* sperm is consistent with taxonomic placement of the genus in the most advanced section of the Melanotaeniidae (Allen, 1980).

Pseudomugil. In some ways as modified are the sperm of the three *Pseudomugil* (Blue-eye) species. Here the nucleus has become crescentic but plesiomorphic location of both centrioles in the fossa is retained; satellites are apomorphically absent; the large number of mitochondria is probably apomorphic, as is elongation of the cytoplasmic canal, with 3 to 5 tiers of mitochondria along the axoneme. The blue eyes have been shuffled between the Atheriniidae and the Melanotaeniidae but are regarded as the most primitive melanotaeniids by Allen (1980). The perpendicular arrangement of centrioles in *Pseudomugil* is inconsistent with an origin from other examined melanotaeniids all of which have parallel or (*Cairnsichthys*) angled centrioles but is consistent with a basal position in the order occupied also by the atherinid *Craterocephalus marjoriae*. In other respects, as indicated, *Pseudomugil* sperm are highly modified.

Iriatherina. This has some advanced features (parallel centrioles, weak or no satellite apparatus) but is distinctive in the finely granular, compact chromatin and absence of a nuclear fossa. Absence of the fossa may be a synapomorphy with *Melanotaenia*, thus supporting current placement in the Melanotaeniidae. Generally, however, sperm structure in atheriniforms suggests that familial classification of the atheriniid-melanotaeiniid section of the order may require revision.

Order Cyprinodontiformes

Systematics. The Exocoetoidei and Adrianichthyoidei included with the Cyprinodontoidei in the Cyprinodontiformes by Nelson (1984) are here placed in the order Beloniformes, as advocated by Rosen and Parenti (1981). The Cyprinodontiformes are regarded as the sister-group of the Beloniformes, sharing with them the following apomorphies: the first epibranchial has an expanded base; the second and third epibranchials are reduced; and the 1st and 2nd infraorbital is lacking. Internal synapomorphies of the Cyprinodontiformes are the symmetrical internal skeleton of the caudal fin; absence of lobing of the caudal fin; and low-set pectoral fins associated with a large, scalelike postcleithrum. These features and the unusually long development time indicate that the order is monophyletic (Lauder and Liem, 1983) (see also Fig. 17.1).

The restricted Cypriniformes consists of the killifish and their relatives, small to medium-sized fish which live in shallow fresh and brackish water.

Two suborders, the Aplocheiloidei and Cyprinodontoidei are recognized (Parenti, 1981).

Reproductive modes. Reproduction in the group is exceptionally varied with oviparity, ovoviparity, viviparity and, in *Rivulus marmoratus*, functional hermaphroditism (see also Aulopiformes). In *R. marmoratus*, the individual has simultaneously functional ovary and testis; internal self fertilization occurs and then eggs are laid (Harrington, 1961). Viviparity may have evolved at least four times in the order. Among the viviparous forms there is a vast array of schedules and morphological modifications for internal development such as the trophotaeniae of the goodeids and the intra- and extra-follicular gestation and superfetation of some poeciliids (Able, 1984; Parenti, 1981).

Sperm literature. Sperm ultrastructure has been examined in many species of the Cyprinodontiformes (Asai, 1971; Billard, 1970b; Brummett and Dumont, 1979; Dadone and Narbaitz, 1967; Grier, 1973a,b, 1975; Grier *et al.*, 1978; Jamieson (herein), Jonas-Davies *et al.*, 1983; Mattei, 1970; Mattei and Boissin, 1966; Mattei *et al.*, 1967b; Mizue, 1969; Nicander, 1968; Porte and Follenius, 1960; Russo and Pisanó, 1973; Selman and Wallace, 1986; Thiaw *et al.*, 1986; Yasuzumi, 1971).

Suborder Cyprinodontoidei

Diagnosis. Only two basibranchials in the ventral gill arch skeleteon. Distinguished from the few other actinopterygians with this feature in having no fusion of the first basibranchial to the basihyal or to the second basibranchial. Dorsal hypohyal absent. In fresh, brackish and saltwater; pantropical and in temperate Laurasia from N. America as far east as Iran (Parenti, 1981).

The Cyprinodontoidei contain the families Cyprinodontidae, Goodeidae, Anablepidae, Poeciliidae and Jenynsiidae. Individual species are mentioned only in the context of a review of comparative sperm morphology of the investigated families.

Cyprinodontidae. The cyprinodontids (killifish) *Fundulus heteroclitus*, the Mummichog, and *Cyprinodon variegatus*, the Sheepshead Minnow, have been briefly examined for spermiogenesis by Yasuzumi (1971).

***Fundulus heteroclitus* sperm.** Yasuzumi (1971) describes some features of the mature sperm of *F. heteroclitus.* Brief SEM and TEM data on the sperm of this species are provided by Brummett and Dumont (1979). Gametogenesis and the mature spermatozoon have been described by Selman and Wallace (1986) (Fig. 17.4).

The nucleus is uniformly dense. It is horseshoe-shaped in longitudinal section in one plane, the deep fossa so formed being open along one surface and containing the proximal and distal centrioles and the base of the 9+2 flagellum (Yasuzumi, 1971; Selman and Wallace, 1986). From SEM and TEM observations the head is 1.3 μm wide, 1.0 μm thick and 2.25 μm long; the median groove extends from the posterior end almost to the anterior end; the flagellum inserts into the groove approximately 0.5 μm from the posterior end and at the point of insertion is surrounded by a low collar of a few very large mitochondria; the 9+2 pattern flagellum is 30-50 μm long and has a pair of lateral fins (Brummett and Dumont, 1979). Fins are not described or illustrated by other workers. The "central doublet" [2 singlets] terminates shortly behind the distal centriole (Selman and Wallace, 1986). This externally fertilizing cyprinodont sperm is thus similar to the internally fertilizing goodeid sperm (below).

Flagellum in other cyprinodontoids. Flagellar structure in the cyprinodontoids *Fundulosoma thierryi*; *Epiplatys ansorgei*, *E. bifasciatus*, *E. chaperi* and *E. fasciolatus* has been described by Thiaw *et al.* (1986; see also aplocheilids, below). The testis is typical of the Atheriniformes, with spermatogonia restricted to the most distal, upper ends of the tubules (Grier, 1981; Selman and Wallace, 1986).

Goodeidae. Goodeidae for which sperm ultrastructure is briefly summarized are *Ameca splendens*, *Ataenobius toweri*, *Characoden lateralis*, and *Xenotoca eiseni* (Grier *et al.*, 1978). Although live-bearing, all have sperm with the basic anacrosomal aquasperm structure. As in the perciform *Zoarces viviparus*, this is here considered to suggest relatively recent acquisition of internal fertilization. Testis structure, ovarian structure, embryonic adaptations for internal gestation, and the form of the spermatozeugmata and of the spermatozoa all indicate that goodeids have evolved internal fertilization independently of the poeciliids. Goodeid sperm have a slightly elongated nucleus, hollowed "ventrally" to form a fossa in which the centriolar complex, basal body and associated satellites and the base of the flagellum reside. The midpiece consists of a single ring of discrete mitochondria. The central flagellar doublets are said to terminate a short distance behind the basal body. The 9+2 flagellum has two lateral fins (Grier *et al.*, 1978). This sperm structure is remarkably like that of internally fertilizing cottid scorpaenids (Chapter 15).

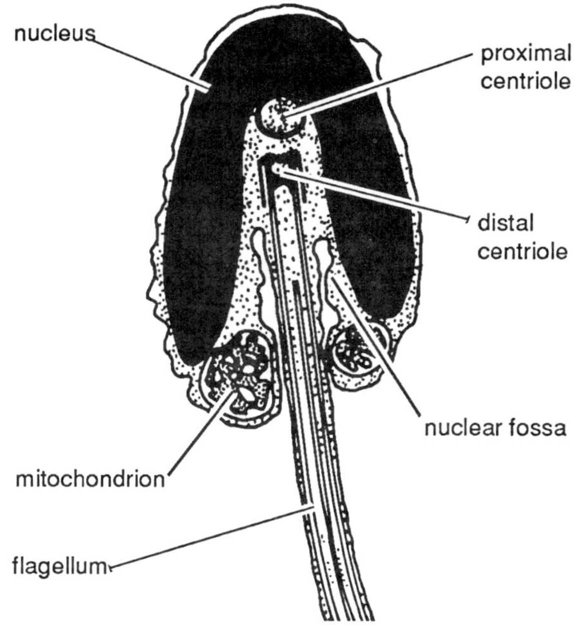

Fig. 17.4. *Fundulus heteroclitus,* the Mummichog. Diagrammatic longitudinal section of the spermatozoon. Drawn from a micrograph of Selman, K. and Wallace, R.B. (1986). *American Zoologist* **26**, 173-192. Fig. 12.d.

Jenynsiidae and Poeciliidae. Viviparous poeciliids (*Gambusia affinis, Poecilia latipinna, Xiphophorus helleri, X. maculatus, P.* (=*Lebistes*) *reticulata*) and the sole jenynsiid genus (*Jenynsia lineata*) differ from goodeids and cyprinodontids in having complex sperm with elongated nucleus and midpiece.

Jenynsia lineata. In *Jenynsia lineata*, viviparous

according to Breder and Rosen (1966), the nucleus again has a deep fossa apparently open along one surface and containing the two centrioles. The mitochondria of the spermatid fuse to form a mitochondrial sheath containing at maturity a single, cylindrical mitochondrial derivative. This is narrowly open along one side and has longitudinal cristae. Between the mitochondrial derivative and the cytoplasmic canal around the axoneme there is a "submitochondrial net" which consists of filaments enclosed between two membranes. The filaments are separated by a constant distance of about 3 Å and form a regular three dimensional pattern (Dadone and Narbaitz, 1967). In micrographs the 9+2 flagellum shows short lateral fins in only a few sections.

Poeciliidae. Five species of the live-bearing family Poeciliidae have been examined for sperm ultrastructure: *Gambusia affinis*, the Mosquito Fish, consists almost entirely of females which use males of other species to stimulate development of the egg without any genetic contribution (Lauder and Liem, 1983).

All of the poeciliids listed above have been studied principally for spermiogenesis. Spermatozoal detail is sparse. The present comparative account partly rectifies this deficiency by inclusion of the writer's observations on the sperm of *Poecilia reticulata*, *Gambusia affinis*, *Xiphophorus helleri* and *P. latipinna* (the Black Mollie variety), all obtained from aquaria in Australia.

Acrosome. An acrosome is absent in poeciliids, as in other atherinomorphs. However, in the spermatozoa or very late spermatids of *Gambusia affinis*, when their heads are still embedded in the Sertoli cells, there is an elongate vesicle on the apex of the nucleus (Fig. 17.6A, B), separated from the latter by

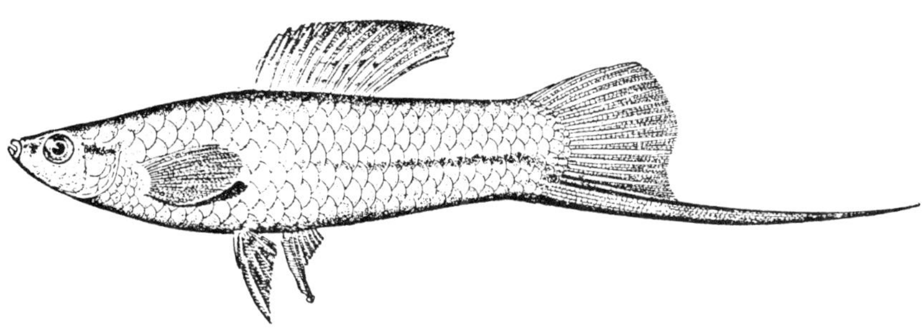

Fig. 17.5. *Xiphophorus helleri*, the Swordtail. The anal fin is modified as an intromittent organ. After Jordan, D.S. (1907). *Fishes*. Henry Holt. Fig. 329.

Grier, 1975 (Fig. 17.6A-E); *Poecilia latipinna*, the Sailfin Mollie, Mizue, 1969 (Black Mollie variety, Fig. 17.6J-L); Grier, 1973a,b (Wild form) ; *Poecilia reticulata*, the Guppy, Asai, 1971; Billard, 1970b; Mattei, 1970; Mattei and Boissin, 1966; Porte and Follenius, 1960 (Fig. 17.6F, G); *Xiphophorus helleri*, the Swordtail, Jonas-Davies *et al.*, 1983 (adult, Fig. 17.5; sperm, Fig. 17.6H, I); *Xiphophorus* (=*Poecilia*) *maculatus*, the Southern Platy, Russo and Pisano, 1973.

The evolution of viviparity in poeciliids has involved testicular modification in which spermatozeugmata are transferred to the female reproductive tract. The sperm heads are characteristically external with the tails forming the centre of the spermatozeugma (Grier, 1976), as shown for *Xiphophorus helleri* in Fig. 17.6H. *Poecilia formosa* the nuclear envelope and covered by the general plasma membrane of the sperm cell. Although this vesicle does not persist in the free spermatozoon it is possibly homologous with an acrosome and/or with the anterior vesicle of sperm of the atheriniform *Melanotaenia*. It is not known whether it occurs in other poeciliids.

The nucleus in all species is an elongate cone, apex anteriorly, strongly compressed in one plane which conventionally has been termed the lateral plane. In *P. reticulata*, at its greatest width, at the commencement of basal narrowing, its width is 0.74 μm "laterally" and 1.2 μm "dorsoventrally". Although the nucleus appears pointed when viewed dorsally, it is blunt and rounded when viewed laterally, as shown for *P. reticulata* in Fig. 17.6G. The chromatin is so strongly condensed as to reveal no substructure and is electron opaque. Basally the nucleus is rounded and slightly narrower. The length of the nucleus is relatively uniform among the different species: 3.5 µm in *G. affinis* and 3.7 µm in *P. reticulata* and *X. helleri*. The nucleus is thus in fact shaped like a parallel-sided spatula or thick, apically rounded blade. Nuclear pores are absent at maturity (Grier, 1975).

The base of the nucleus is penetrated by a deep implantation fossa which occupies approximately its posterior third and houses the basal body and anterior region of the axoneme (*G. affinis*, Fig. 17.6A-C; *P. reticulata*, Fig. 17.6G; *P. latipinna*, Fig. 17.6J). Anteriorly the fossa is circular in cross section and is enclosed on all sides by nuclear material but in its posterior region it is open "ventrally" through a posteriorly widening slit (Fig. 17.6C, inset).

There is some doubt as to whether the ventral opening of the fossa persists at maturity, though the advanced and apparently mature state of the sperm examined here suggest that it does. A similar, Sertoli embedded stage is considered mature for *P. latipinna* by Grier (1973b). A deep fossa open "ventrally" has also been described for *Gambusia affinis* by Grier (1975) and for *P. latipinna* by Grier (1973) and Mizue (1969). In *Xiphophorus helleri*, it is described as "a deep notch in the base of the nucleus" (Jonas-Davies *et al.*, 1983). However, in *Poecilia reticulata*, the groove is shown as closing off by maturity (Asai, 1971; Billard, 1970b) or, in contrast, is drawn as opening dorsally and ventrally, a doubtful observation (Porte and Follenius, 1960). It is agreed that in all species the fossa is much deeper and the surrounding nuclear material much thinner in the spermatid than in the spermatozoon. There is also a deep posteroventral nuclear fossa in the sperm of *Jenynsia*, and, indeed, in goodeids.

Mitochondria. In *P. latipinna*, the long mitochondrial sheath appears to contain many separate mitochondria in longitudinal rows (Mizue, 1969). This longitudinal array of mitochondria, with longitudinal cristae, has been clearly demonstrated for *X. helleri* (Jonas-Davies *et al.*, 1983) and is shown here for all examined species, including *G. affinis* (Fig. 17.6A, D, E); *P. latipinna* (Fig. 17.6J) and *P. reticulata* (Fig. 17.6G). This region of the sperm constitutes the midpiece. The mitochondria in the species here examined lie in a very long cytoplasmic sleeve which invests the axoneme but is separated from it by the so-called cytoplasmic canal ubquitous in osteichthyan sperm. The length of the sleeve is about 8 µm in *P. reticulata* and of the same order, though not precisely determined, in the other species. The inner and outer (plasma) membranes of the canal are continuous at its anterior limit. In the mature sperm this limit is shortly behind the nucleus (*G. affinis*, Fig. 17.6D; *P. reticulata*, Fig. 17.6G) or just within the fossa (*P. latipinna*, Fig. 17.6J; *X. helleri*, see also Jonas-Davies et al, 1983). A local dilatation of the cytoplasmic canal occurs at its anterior commencement, as noted by Jonas-Davies *et al.* (1983).

In cross section of the mitochondrial sleeve, two to four mitochondria embrace the flagellum, and are predominantly C-shaped, or at least shortly crescentic, in section (Fig. 17.6C, F, K). Each has several regularly disposed cristae, all parallel with the inner and outer mitochondrial membranes, joining the wall of the mitochondion at the tips of the C-shape, and seen in longitudinal section of the sperm to run longitudinally (Fig. 17.6A, D, E, G, J).

In longitudinal section of the axoneme, the mitochondria appear as a single row on each side, arranged end to end, there being about 7 or 8, with some variation, in a longitudinal row. Because of variations in the length and circumferential extent of individual mitochondria, the lines of apposition of adjacent mitchondria both longitudinally and around the axoneme are somewhat irregular as, therefore are

Fig. 17.6. (Opposite). Poeciliidae. Spermatozoa. A-E. *Gambusia affinis*. A, D and E. Longitudinal sections of spermatozoa. B. Detail of the acrosome-like vesicles on the apices of the nuclei embedded in Sertoli cells. C. Transverse sections (TS) of the nuclear fossa and of the midpiece. F, G. *Poecilia reticulata*. F. TS of midpiece and axonemes, showing alteration of doublets and loss of central singlets at endpiece. G. LS of two spermatozoa. That on the left shows the greater width of the nucleus, cut in the "lateral" plane, with rounded tip. That on the right shows the nucleus cut in the "dorso-ventral" plane (i.e. viewed "laterally") and appearing narrower and with pointed tip. H, I. *Xiphophorus helleri*. H. Section of part of a spermatozeugma of late spermatids, showing centrifugal orientation of the nuclei and central mass of axonemes. I. TS axonemes showing the two short lateral fins. J-L. *Poecilia latipinna*. J. Head and anterior midpiece. K. TS midpiece. L. Posterior end of midpiece. a. axoneme. dc. distal centriole. m. mitochondrion. n. nucleus. sc. Sertoli cell. v. acrosome-like vesicle. Original.

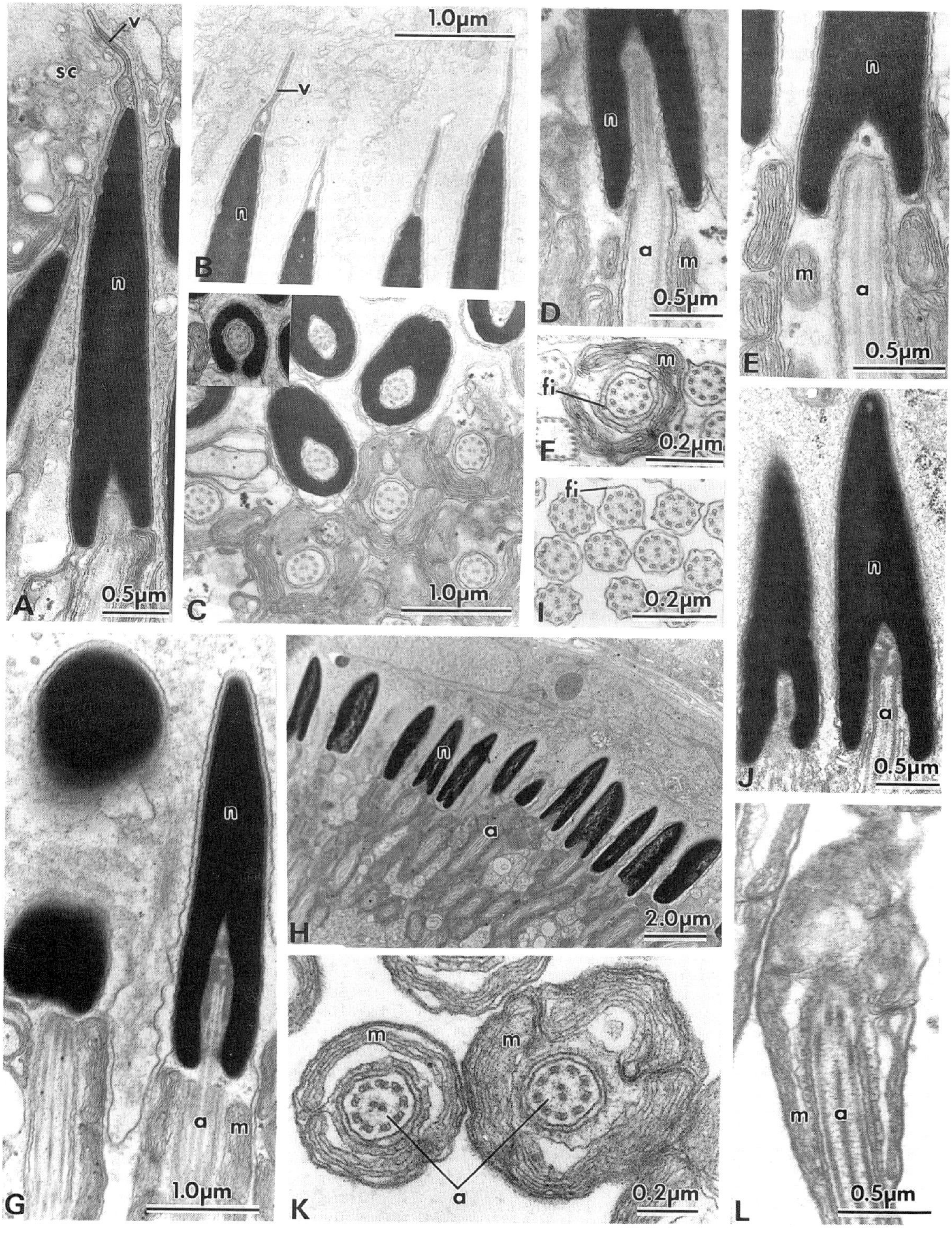
A
sc
v
n
0.5μm
B
v
n
1.0μm
C
1.0μm
D
n
a
m
0.5μm
E
n
m
a
0.5μm
F
fi
m
0.2μm
I
fi
0.2μm
J
n
a
0.5μm
G
n
a
m
1.0μm
H
n
a
2.0μm
K
m
m
a
0.2μm
L
m
a
0.5μm

the longitudinal mitochondrial columns. Although in longitudinal section most mitochondria appear several times longer than wide, rarely being isodiametric, each entire mitochondrion, in three-dimensions, is typically a C-shaped cylinder of greater diameter than length. Variation in numbers of mitochondrial columns reported in the literature may partly be ascribed to variation in the number of circumaxonemal mitochondria along the length of the midpiece. In *P. reticulata*, as noted by (Billard, 1970b), there are usually four longitudinal columns of mitochondria and at the distal extremity of the "intermediate piece" only two, contiguous crescentic columns. Five columns of mitochondria are shown for this species by Porte and Follenius (1960). The sperm of *X. maculatus* is similar but some cross sections, at least, show only three mitochondria, representing the longitudinal columns of separate mitochondria (Russo and Pisanó, 1973). Present observations confirm that dense granules present in the mitochondria early in spermiogenesis do not persist [or are infrequent] in the mature spermatozoon (Billard, 1970b).

Submitochondrial net. Densification of the plasma membrane lining the inner surface of the mitochondria, closest to the axoneme, has been recognized by some workers as a submitochondrial net in poeciliid sperm. It is considered the equivalent of the submitochondrial net of *Jenynsia*. The mesh of the net appears as a regular diamond lattice. In sagittal sections the vertices appear as electron dense profiles, about 7.5 nm in diameter, which are spaced approximately 42 nm apart. The electron dense material at the vertices appears to attach both to the outer surface of the mitochondria and to the cytoplasmic surface of the overlying plasma membrane (Jonas-

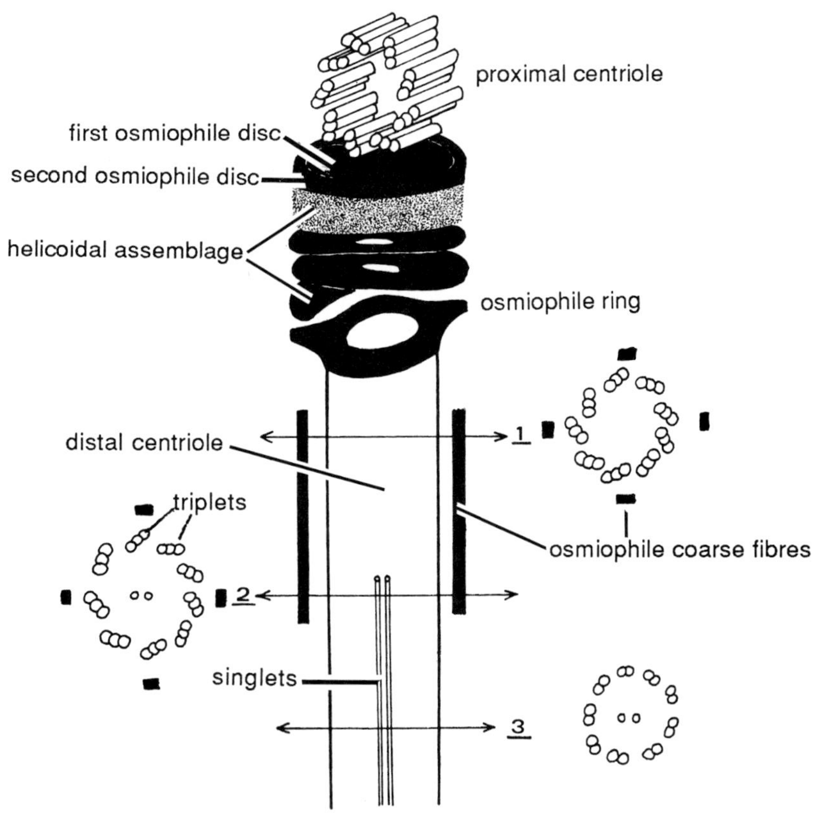

Fig. 17.7. Schematic reconstruction of the centriolar complex of *Poecilia reticulata.* 1. Transverse section (TS) of the distal centriole in the midregion. 2. TS of the base of the distal centriole. 3. TS of the flagellum. From Mattei, X. and Boissin, C. (1966). *Comptes Rendus Hebdomadaires des Sánces de l'Académie des Sciences* D **262**, 2620-2622. Fig. 10.

Davies *et al.*, 1983).

Centrioles. Of the two spermatid centrioles, only the distal centriole, forming the basal body for the flagellum persists intact in the mature spermatozoon in the four species here examined and this appears general for the examined other poecilids. The complex structure of the centriolar apparatus in *Poecilia reticulata* has been described by Mattei and Boissin (1966) and is shown in Fig. 17.7.

In late spermiogenesis the triplets of the proximal centriole become occluded and it is reduced to a remnant by maturity. In *P. reticulata,* as in other poeciliids, an intercentriolar lamellated body disappears after having contributed to an electron dense cap on the proximal end of the basal body. In *G. affinis*, as in other poeciliids, rodlike structures appear on opposite sides of the basal body during spermiogenesis. In *G. affinis* and *P. latipinna* (Grier, 1973a,b) and probably *P. reticulata* (Billard, 1970b) these appear to have a dual origin: there is an apparent fusion of two striated satellites, on one side of the basal body, to produce a long striated satellite, and in addition an amorphous electron dense structure is present on the opposite side of the basal body, in the spermatid and spermatozoon (Grier, 1975).

Flagellum. Within the nuclear fossa, the flagellum has what Grier (1973b), for *P. latipinna*, has termed a specific flagellar-nuclear geometric orientation. When viewed in cross section, a line drawn perpendicular to the plane of the two central singlets, and passing between them, will always approximately bisect the nucleus in the wide, dorsoventral plane. A considerable inclination of this plane nevertheless occurs, as shown for *G. affinis* in Fig. 17.6C.

The 9+2 flagellum has two lateral fins in the four species examined here: *Poecilia reticulata* (Fig. 17.6F), as also shown by Mattei *et al.* (1967b), *Gambusia affinis*, *Xiphophorus helleri* (Fig. 17.6I) and the Black Mollie, and in *Xiphophorus (=Poecilia) maculatus* as demonstrated by Russo and Pisanó (1973). They are probably present in all poeciliids. They are very short, rarely (as in some profiles of *X. helleri* sperm) each equalling the width of the axoneme, but their presence is taken here to attest the ect-aquasperm ancestry of the poeciliid sperm, long fins being typical of externally fertilizing species. Their orientation varies in neighbouring testicular sperm from the plane of the central singlets to about 45° to this (Fig. 17.6I), probably reflecting a dynamic situation.

At the posterior end of the flagellum, as shown in *P. reticulata* and confirmed here (Fig. 17.6F), the lateral fins are lost, the central singlets disappear, and the doublets approach the plasma membrane, reducing their distance from this to 100 Å. More posteriorly the doublets disrupt into 18 singlets; ultimately the flagellum loses its circular cross section and the number of microtubules is progressively reduced (Mattei *et al.*, 1967b).

Remarks. The elongate nuclei and, as Nicander (1968) has noted, the long mitochondrial sheaths of these poeciliids are characteristic of [though not constant in] internally fertilizing sperm.

Suborder Aplocheiloidei

Diagnosis. Pelvic fins inserted close together; metapterygoid present; three basibranchials; a dorsal ray on each of the first two dorsal radials. This definition, for the family Aplocheilidae, including all rivulines, of Nelson (1984) is here applied to the suborder *sensu* Parenti (1981). Rivulines are freshwater fish of Africa, southern Asia, southern N. America, and of S. America.

Flagellar ultrastructure. A large number of cyprinodontiforms, mostly aplocheiloids, has been investigated by Thiaw *et al.* (1986) for flagellar structure in the spermatozoon. They include the aplocheiloids *Aphyosemion guignardi*, *A. herzogi*, *A. nigrifluvi*, *A. riggenbachi*, *A. splendopleure*; *Aplocheilichthys lamberti*, *A. normani*; *Aplocheilus lineatus*; *Cynolebias wittei*; and *Notobranchius steinforti*. A similar spermatozoal morphology was reported, although not described, but a wide diversity of flagellar structure was encountered. The flagellar membrane has one, two or three lateral expansions [fins] depending on the species. Peripheral doublets of the axonemes show only the external arm, excepting two species (*Aphyosemion guignardi* and *Aplocheilus lineatus*) that completely lack the arms though, like all the other species, retaining motility. Intratubular differentiations (ITD) are present in A or B tubules of the doublets, as in central tubules of some species, whereas others are totally devoid of such differentiations. The ITD can affect all doublets or preferentially doublets 1, 5 and 6, rarely with intraspecific variation. It is suggested that these variations

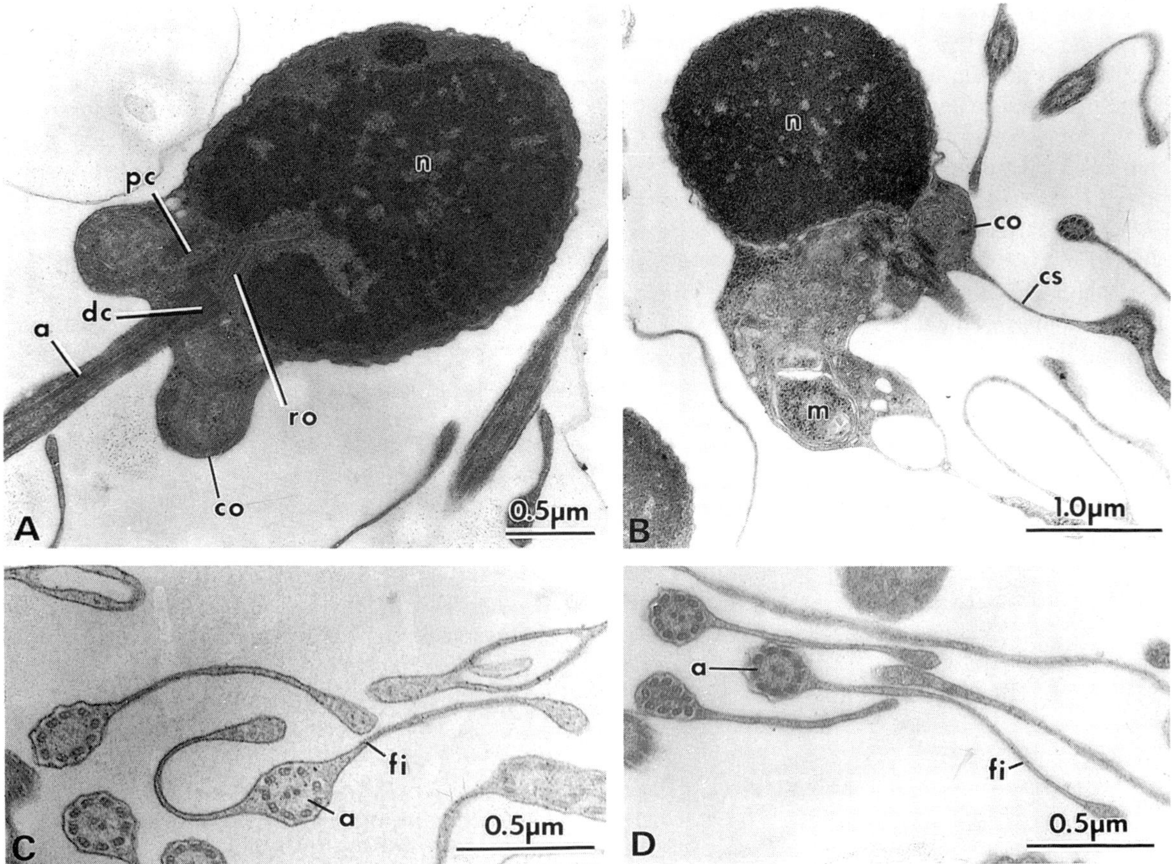

Fig. 17. 8. *Aphyosemion gardneri*. A and B. Longitudinal sections of spermatozoa. A. Showing proximal centriole parallel to distal centriole (basal body) and rootlet-like structure projecting from the basal body into the nuclear fossa. B. LS showing apparent shedding of the posterior part of the collar, seen in many sperm. C and D. Transverse sections of the flagellum. There are two lateral fins throughout most of its length but only one fin near its posterior extremity where the 9+2 pattern is disrupted. a. axoneme.co. cytoplasmic collar. cs. cytoplasmic sleeve. dc. distal centriole. fi. fin. m. mitochondrion. n. nucleus. pc. proximal centriole. pc. proximal centriole. ro. rootlet-like structure. Original.

may be due to neutral mutations (Thiaw *et al.*, 1986).

Aphyosemion gardneri. (Fig. 17.8). **Nucleus**. The nucleus of the sperm of *Aphyosemion gardneri* is spherical to slightly ellipsoid with long axis antero-posterior, 2.2-2.5, mean 2.4, µm long (n=4). It has a short basal fossa, narrower than the flagellum. The chromatin is electron dense and moderately condensed, there being many lacunae; its outline is smooth with only slight flocculence.

Midpiece. The midpiece is irregular in form, differing between individual spermatozoa. A broad anterior portion contains the centrioles and several irregularly arranged mitochondria. The cytoplasmic canal, with a maximum recorded length of 2.3 µm, commences behind the distal centriole. The cytoplasm enclosing it contains subspheroidal or elongate bodies, with concentric cristae, which appear to be modified mitochondria but consists mostly of a long trailing sleeve.

Centrioles. The longitudinal axes of the two centrioles are parallel, with the proximal located immediately anterior to the distal centriole. A rootlet-like structure, with longitudinal not transverse striae, extends obliquely from the anterior end of the distal centriole to the nuclear fossa (Fig. 17.8A).

Flagellum. The 9+2 flagellum has, as in other aplocheiloids, unusually long lateral fins approximately in the plane of the central singlets, giving the entire flagellum a width up to 6 µm. Each fin is slightly expanded terminally. Posteriorly there is only as single, unilateral fin (Fig. 17.8C, D). (Jamieson unpublished).

Order Beloniformes

Diagnosis. The Beloniformes (Synentognathi), include the needle fish, rice fish, half beaks, and flying fish. They differ from the Cyprinodontiformes, in which they are subsumed by Nelson, in having a lateral line on the body (sometimes absent); in the single, not paired, narial opening; in having 6-15 branchiostegal rays; in elongation of the lower jaw at least in some stage of the life history (Nelson, 1984); in presence of a large ventral flange on the fifth ceratobranchial; having the second and third epibranchials distinctly smaller than other epibranchials; and the second pharyngobranchial vertical reoriented

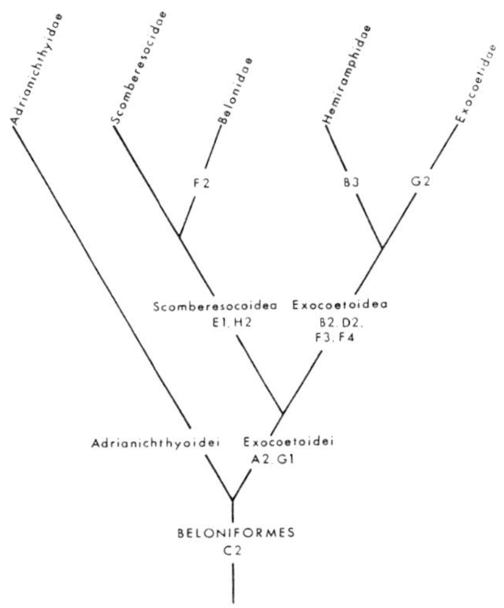

Fig. 17. 9. Cladogram of the Beloniformes. After Collette *et al.* (1984). In *Ontogeny and Systematics of Fishes.* Special Publication Number 1, pp. 335-354. American Society of Ichthyologists and Herpetologists. Fig. 184.

(Lauder and Liem, 1983).

Species of the suborder Adrianichthyoidei, family Adrianichthyidae, inhabit fresh and/or brackish waters. Most species of the other four families, comprising the suborder Exocoetoidei, are epipelagic marine fish but, in these, several genera of the Belonidae and Hemiramphidae are restricted to freshwaters and a few other genera contain estuarine and

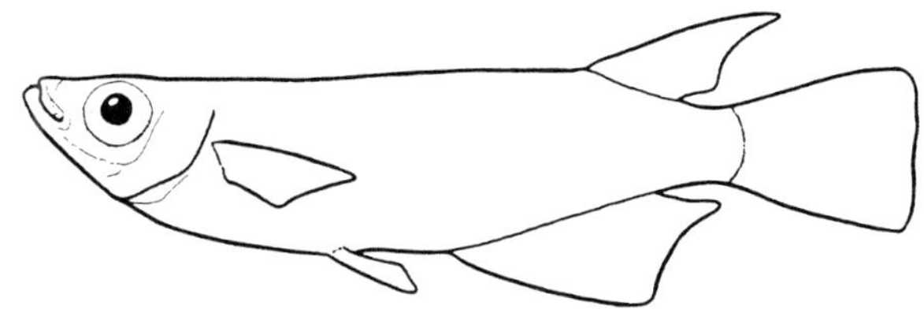

Fig. 17.10. *Oryzias latipes*, the Japanese Medaka. After Nelson, J.S. (1984). *Fishes of the World.* 2nd edition. Wiley, New York. p. 216.

freshwater as well as marine species (Collette *et al.*, 1984). The sister-group relationship of the Beloniformes with the Cyprinodontiformes is shown in Fig. 17.1.

A recent evaluation of the phylogenetic affinities of the constituent families by Collette *et al.* (1984) is shown in Fig. 17. 9.

Beloniform sperm literature. Sperm ultrastructure has been described for the adrianichthyoid *Oryzias latipes* (see Grier, 1976; Sakai, 1976; and, SEM, Iwamatsu and Ohta, 1981), the exocoetids *Fodiator acutus* (Mattei 1970) (Exocoetidae), the live-bearing *Hemirhamphodon pogonognathus* (Hemiramphidae) (Jamieson, 1989) and an externally fertilizing hemiramphid, *Arramphus sclerolepis* (Jamieson, unpublished).

Suborder Adrianichthyoidei

Diagnosis. Rosen and Parenti (1981) defined the adrianichthyoids on five characters, including great expansion of the articular surface of the fourth epibranchial and possession of a reduced autopalatine with posterior articular cartilage. They enlarged the Adrianichthyidae to include the Horaichthyidae and Oryziidae. Collette *et al.* (1984) added a further, larval, character distinguishing them from exocoetoids. The four genera, *Adrianichthys*, *Horaichthys*, *Oryzias* (Fig. 17.10) and *Xenopoecilus*, totalling only 11 species, inhabit fresh and/or brackish waters from India and Japan to the Indo-Australian Archipelago (Collette *et al.*, 1984).

***Oryzias latipes*.** The spermatozoon (Fig. 17.11) of the only adrianichthyoid species studied, the

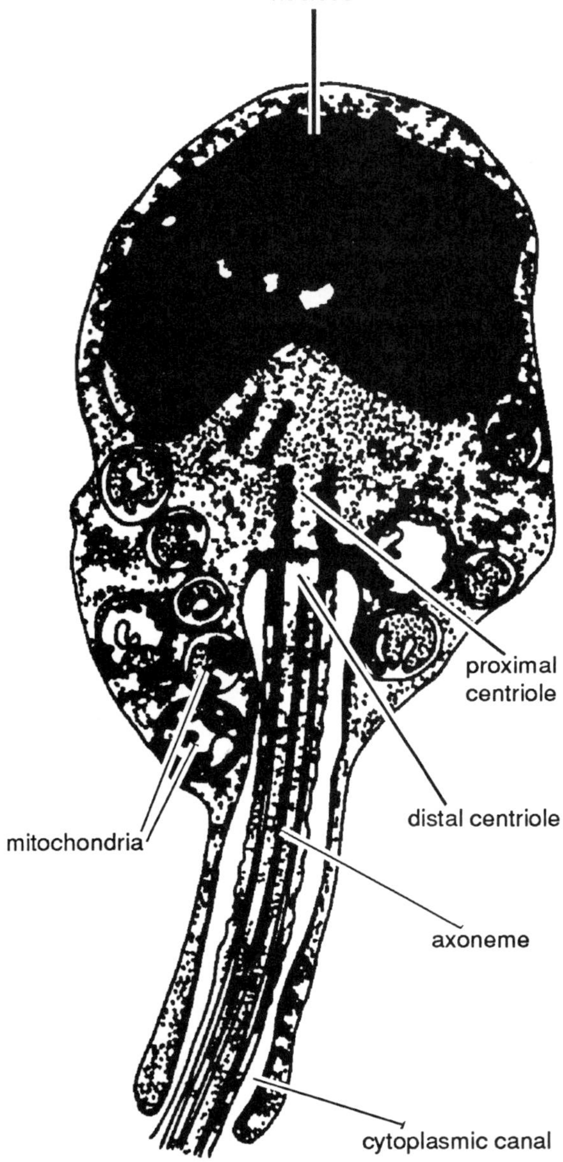

Fig. 17.11 . *Oryzias latipes.* Longitudinal section of mature spermatozoon. After a micrograph of Grier, H.J. (1976). *Cell and Tissue Research* **168**, 419-431. Fig. 14.

oryziid, *Oryzias latipes*, the Japanese Medaka, investigated by Grier (1976), Sakai (1976) and Iwamatsu and Ohta (1981), is nearer to the basic teleostean aquasperm than that of the poeciliids as befits its external fertilization. Thus the nucleus is rounded, the basal nuclear fossa is less deep and contains a persistent proximal centriole of triplet construction, and although the midpiece continues as a long "sleeve" around the base of the flagellum, the mitochondria are scattered and unmodified; the intercentriolar body differs significantly from that of poeciliids, being both smaller and disappearing earlier. Two flagellar fins are again present (Grier, 1976). Although the rounded nucleus and basal location of the fossa give some support to exclusion of *O. latipes* from the Cyprinodontidae in which it is sometimes placed, sperm structure does not yet allow a decision between placement of *Oryzias* and other adrianichthyids in the Beloniformes, as indicated in the phylogram (Fig. 17.9) and advocated by Rosen and Parenti (1981) and

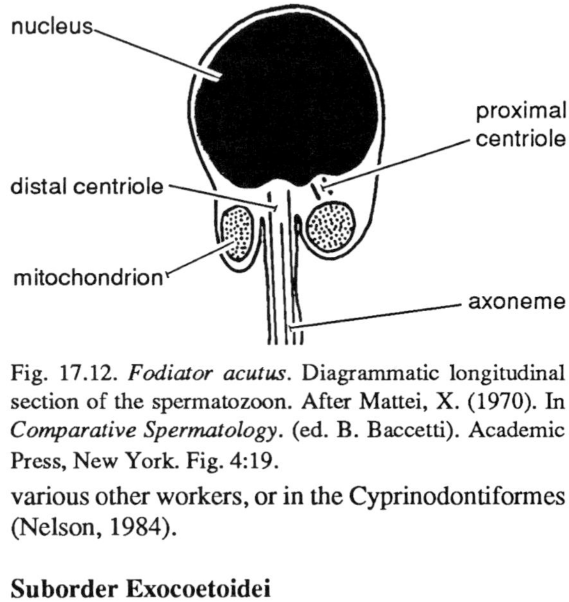

Fig. 17.12. *Fodiator acutus.* Diagrammatic longitudinal section of the spermatozoon. After Mattei, X. (1970). In *Comparative Spermatology.* (ed. B. Baccetti). Academic Press, New York. Fig. 4:19.

various other workers, or in the Cyprinodontiformes (Nelson, 1984).

Suborder Exocoetoidei

Diagnosis. Among other features, scales large, usually 38-60 in the lateral line; mouth opening small; no isolated finlets; dorsal and anal fins usually with 8-16 rays each; teeth small. This is the superfamily Exocoetoidea, within the Exocoetoidei (equivalent to the Beloniformes here) placed in the Cyprinodontiformes by Nelson (1984).

General sperm ultrastructure. The spermatozoon of the flying fish *Fodiator acutus* (family Exocoetidae) is a simple teleostean anacrosomal aq-

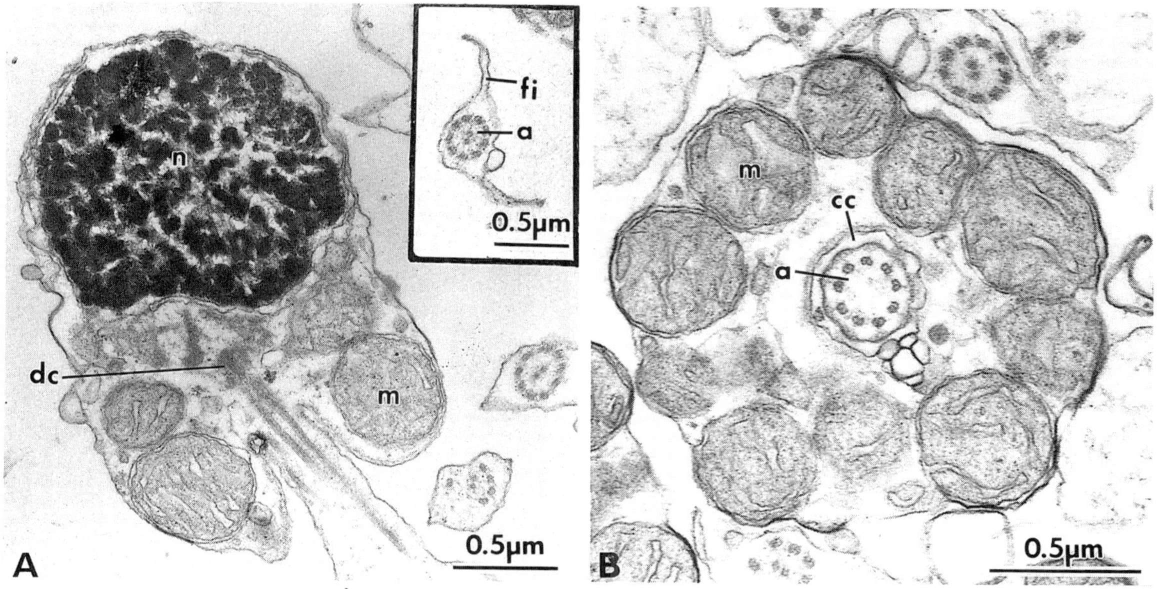

Fig. 17.13. *Arramphus sclerolepis*. A. Longitudinal section of spermatozoon. Inset, showing axoneme with two lateral fins. B. Transverse section through initial 9+0 region(ca) of axoneme showing cytoplasmic canal and 10 encircling mitochondria. a. axoneme. ca. transition from centrile to axoneme. cc. cytoplasmic canal. dc. distal centriole. fi. fin. m. mitochondrion. n. nucleus. pc. proximal centriole. Original.

uasperm (Fig. 17.12) (Mattei 1970).

The spermatozoon of the halfbeak*Hemirhamphodon pogonognathus* (adult, Fig. 17.14) shows modifications (Jamieson, 1989) (Fig. 17.15, 17.16) which are frequent though not obligate in internally fertilizing sperm, notably elongation of the nucleus and extension of the mitochondria of the midpiece as an elongate sheath around the proximal region of the axoneme. These similarities to poeciliid and jenynsiid sperm are considered homoplasic. As in the mature sperm of all but one investigated teleost, an acrosome is absent.

A comparison of the sperm of *Arrhamphus sclerolepis*, as an externally fertilizing species, with that of the internally fertilizing *Hemirhamphodon pogonognathus*, in the same family Hemiramphidae, is informative. The spermatids of the two species are relatively similar and divergence occurs in later spermiogenesis in *Hemirhamphodon*, a phenomenon in keeping with both the "biogenetic law" and von Baer's principle

***Arramphus sclerolepis*. Nucleus.** The nucleus (Fig. 17.13A) is subspheroidal, 1.6 µm long, and slightly compressed posteriorly. Basally it is indented as a poorly defined fossa, sufficient to house only part of the proximal centriole. The chromatin consists of numerous large, separate, electron dense, flocculent masses in a pale matrix. The two nuclear membranes remain separated by a considerable perinuclear cisterna.

Midpiece. The mitochondria are spherical and cristate and are arranged in two tiers longitudinally (Fig. 17.13A). Ten mitochondria are seen in transverse section of a tier, symmetrically arranged around

Fig. 17.14. *Hemirhamphodon pogonognathus*. After Weber, M. and De Beaufort, L.F. (1922). *The Fishes of the Indo-Australian Archipelago*. Brill, Liden. Vol. 4. Fig. 54.

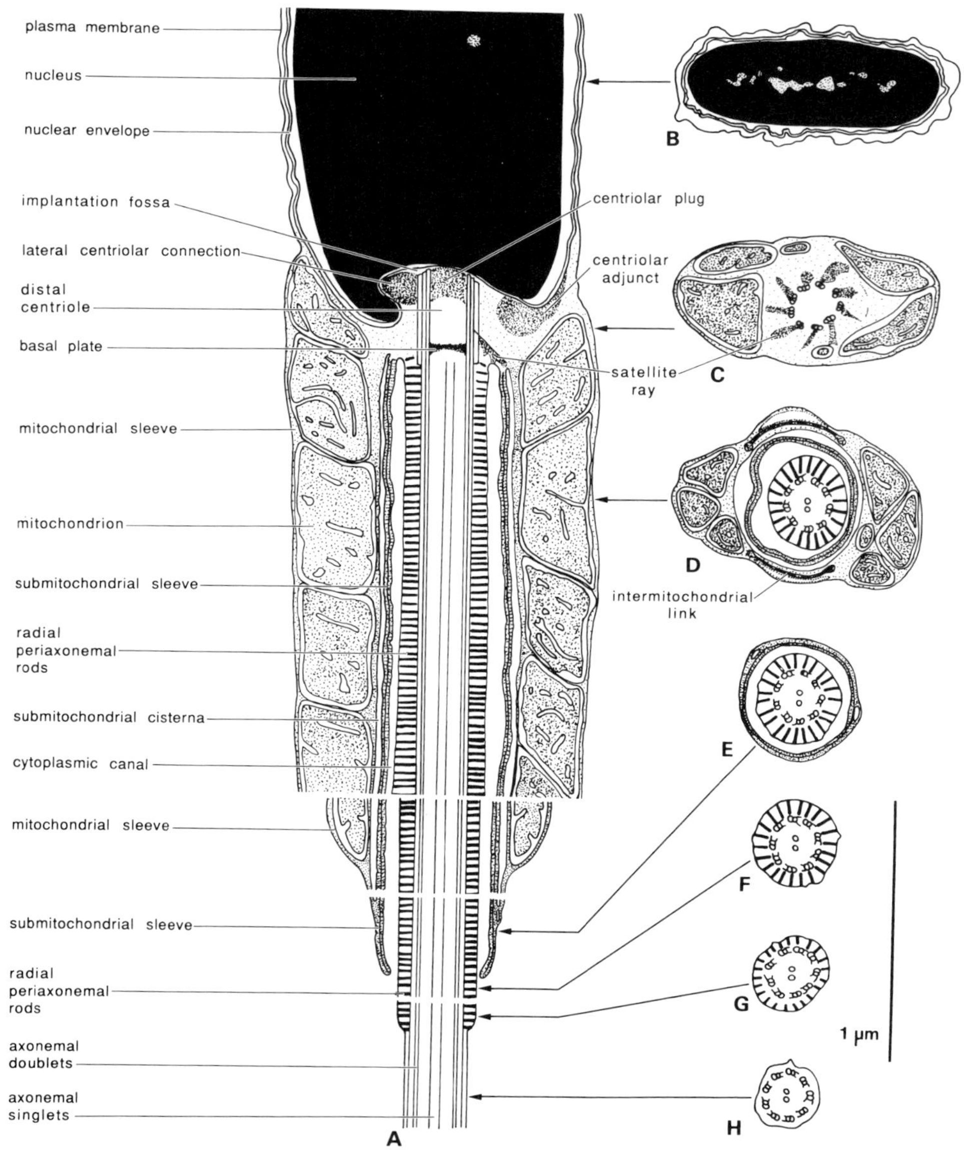

Fig. 17.15. *Hemirhamphodon pogonognathus*. Diagrammatic longitudinal and transverse sections of the spermatozoon. A. Frontal longitudinal section, showing the bilateral mitochondria. The posterior two portions are deduced from transverse sections. B-H. Transverse sections through the regions indicated in A. B. Nucleus. C. Centriole. D. Midpiece. E. Region of united mitochondrial and submitochondrial sleeves. F, G. Successive posterior regions, behind the sleeve, with periaxonemal rods. H. The simple axoneme. From Jamieson (1989). *Gamete Research* **24**, 247-259. Fig. 1.

the central axis (Fig. 17.13B). Those of the posterior tier lie in the short (0.8 µm long) mitochondrial collar, around the periaxonemal invagination (cytoplasmic canal) but those of the anterior tier surround the distal centriole (basal body) anterior to the collar (Fig. 17.13A). There is some thickening of the wall of the collar surrounding the canal (subplasmalemmal densification) which may be the equivalent of a submitochondrial net.

Centrioles. The proximal centriole (Fig. 17.13A)

is located anteriorly to and slightly to one side of the distal centriole and is tilted at an angle of approximately 45° to its long axis and, therefore, to that of the axoneme. The greater diameter of the nucleus is at right angles to the longitudinal axis of the proximal centriole; the nucleus therefore appears tilted relative to the longitudinal axis of basal body and axoneme.

Flagellum. The 9+2 flagellum has the usual lateral fins, here relatively short, at a slight angle to the plane of the central singlets (Fig.17.13A, inset). A region with 9 doublets but no central singlets intervenes, in the collar region, between the basal body and the axoneme proper, as in the tetraodontid *Gastrophysus*.

Hemirhamphodon pogonognathus. The sperm of *H. pogonognathus* (adult, Fig. 17. 14) has been described by Jamieson (1989). Some pertinent morphology is shown in Figs. 17.15 and 17.16 and is described below.

Modifications. The *Hemirhamphodon* sperm shows modifications which are frequent though not obligate in internally fertilizing sperm, and which can reasonably be considered to be apomorphic departures from a basic *Arrhamphus*-like condition. These include elongation of the nucleus and great extension of the mitochondrial collar as a long 'sleeve' around the proximal region of the axoneme.

As a further difference, the basal implantation fossa, seen also in *Arrhamphus*, houses not the proximal centriole, but the anterior half of the distal and only centriole (of triplet construction with satellite rays), and various appurtenances. A putative proximal centriole is present in the spermatid. Internal to the mitochondrial sleeve is a submitochondrial sleeve which developmentally appears to arise by separation off of the inner electron dense layer seen in the *Hemirhamphodon* spermatid and in the mature *Arrhamphus* sperm. The discrete, elongate, cristate mitochondria, in their sleeve, are unique in investigated atherinomorph sperm in being bilateral, grouped on only two, opposing sides of the axoneme.

Again, in the *Hemirhamphodon* spermatid, as in

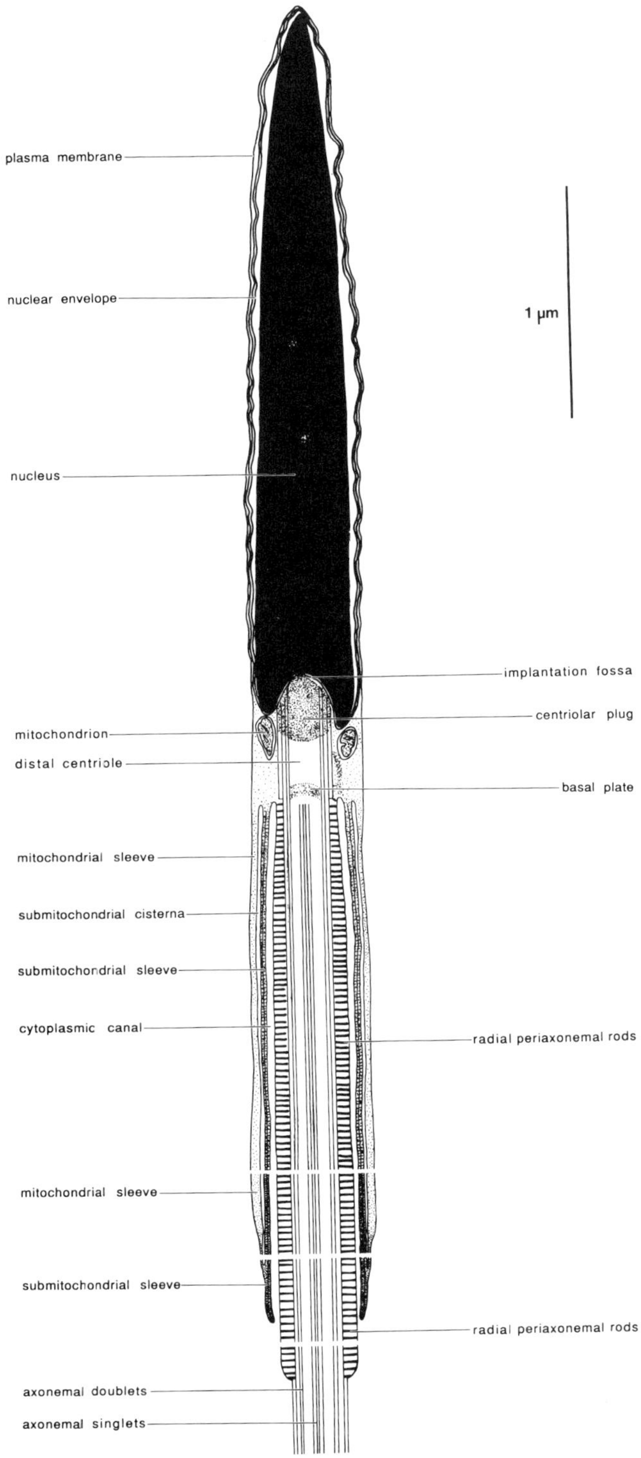

Fig. 17. 16. *Hemirhamphodon pogonognathus*. Diagrammatic longitudinal of the spermatozoon, at right angles to that shown in the previous figure. Note the absence of mitochondria, excepting rudiments in the vicinity of the centriole, in this plane. The variably delineated intermitochondrial link is not shown. From Jamieson (1989). *Gamete Research* **24**, 247-259. Fig. 2.

the mature *Arrhamphus* sperm, the mitochondrial collar is initially short, with rounded mitochondria distributed around the central axis. An arc-shaped 'intermitochondrial link' on each side of the axoneme in the mature *Hemirhamphodon* sperm is absent from the other two cell types.

The 9+2 flagellum is unique for the Animalia in having 23 radial subplasmalemmal rods, repeated longitudinally (periodicity 0.025 µm) in quasicrystalline array but in the spermatid they are absent, as in the *Arrhamphus* sperm. A negative apomorphy of *Hemirhamphodon* spermatids and spermatozoa is failure to develop the lateral axonemal fins seen in *Arrhamphus* and most other teleost sperm. These are present, although reduced, in internally fertilizing sperm of the Poeciliidae. The only major teleost group lacking fins is the Ostariophysi (the characiform - cypriniform - siluriform clade). Absence in ostariophysians appears to be an apomorphic loss as they are also present in Chondrostei, Dipnoi, and Holostei. The presence of axonemal fins in *Arrhamphus* is an important datum, indicating that the externally fertilizing hemiramphid sperm is not derived by simplification of an internally fertilizing sperm, a possibility explored but rejected by Jamieson (1989).

Chapter 18

SUMMARY AND CONCLUSIONS

by B.G.M. Jamieson and L. K.-P. Leung

Fish sperm types and the plesiomorphic mode of fertilization

For details on which this chapter is based the reader is referred to chapters 5-17.

Fish have sperm which conform to the "primitive" and "advanced" models in conventional terminology. Either type may be externally or internally fertilizing. The more primitive (plesiomorphic) or, to be more objective, the non-neopterygian fish groups (Agnatha, Chondrichthyes, Cladistia, Coelacanthimorpha, Dipnoi) have the more advanced sperm and at least two of these groups have internal fertilization. The "primitive" sperm type is restricted to, and predominant in, the Neopterygii. This paradox, of advanced sperm with or without internal fertilization in the more primitive groups, has led us to dispense with the epithets 'primitive' and 'advanced' sperm in favour of the phylogenetically, and morphologically, neutral terms introsperm (first proposed by Rouse and Jamieson, 1987), for internally fertilizing sperm, and aquasperm (Jamieson, 1986a), for sperm with an aquatic, free-swimming phase. It strongly suggests, as parsimony demands, that the simple aquasperm of the Neopterygii have developed secondarily from more complex sperm. It has further prompted us to tentatively consider the possibility that internal fertilization is plesiomorphic for fish although the ancestor of fish may be presumed to have had aquasperm as do the Cephalochordata and many invertebrate groups. Because even aquasperm may have either the "primitive" or the more complex structure, aquasperm with the "primitive" form are distinguished as "plesiosperm" (Jamieson, 1986b) or simple teleostean aquasperm. In short, while secondary development of the simple teleostean aquasperm from more complex sperm appears probable, much greater uncertainty attaches to the view that fertilization was plesiomorphically internal in the jawless ancestors of fish which themselves form a putative sister-taxon of a cephalochordate like form.

It is stressed, however, that we cannot yet entirely reject the alternative hypothesis: that some modification of aquasperm with occasional development of internal fertilization occurred in primitive fish groups but that aquasperm and external fertilization were retained in unbroken succession from the ancestors of fish, to flourish in the great radiation of the Neopterygii. In order to accommodate the fact that no extant fish group below the neopterygians has the "primitive" sperm (plesiosperm) facies, even where the sperm are aquasperm, this alternative requires the mutliple independent evolution of modified sperm in the non-neopterygians: in both aganthan groups; in chondrichthyans; in basal Actinopterygii and probably, by inheritance, early Actinopteri; and in the sarcopterygian ancestors of coelacanths and other Crossopterygii, Dipnoi and tetrapods. It also requires the independent evolution of internal fertilization in at least Chondrichthyes and coelacanths.

What seems unequivocal is that in the *higher* neopterygians, internal fertilization and specific modifications of sperm associated with this (for instance in poeciliids) are secondary. Only in the lowest neopterygians, the externally fertiizing Elopomorpha, and an exceptional and internally fertilizing salmonid (*Lepidogalaxias*), are there modified sperm which might be a survival from a plesiomorphic complex sperm and even from internal fertilization of the pre-neopterygians, though it is here considered more likely that they represent independent specializations from an externally fertilizing plesiosperm.

It has been found convenient to divide fish introsperm (internally fertilizing sperm) into acrosomal and anacrosomal types, according as a morphologically recognizable acrosome is present or absent. The anacrosomal introsperm has been subdivided into two types, simple, like the externally fertilizing aquasperm (plesiosperm), or complex, a term denoting significant departure from the ect-aquasperm ("primitive sperm" or plesiosperm) morphology while recognizing that whether this in fact involves greater complexity is sometimes an arbitrary decision. Aquasperm in fish are always externally fertilizing (ect-aquasperm) whereas in, for instance, some polychaetes they include sperm which are initially aquatic but which have modified, often internal, fertilization and which usually show corresponding morphological modification. In urochordates (Chapter 3) sperm may be ect- or ent-aquasperm without apparent alteration in their somewhat modified aquasperm (ascidiosperm). Fish aquasperm may be acrosomal, and often then of relatively complex morphology, or anacrosomal. Anacrosomal aquasperm have a simple, ectaquasperm-like (plesiosperm) morphology, though differing from the ect-aquasperm of most protochordates and invertebrates in lacking the acrosome, or they may be biflagellate, with or without significant further modification, or they may be aflagellate.

Morphological summary

At least 70 apomorphic conditions (Table 5.2) are discernible in the spermatozoa of fish, taken collectively, compared with a hypothetical ancestral ectaquasperm of the Chordata. A discussion of the chief of these provides the framework for a summary of the comparative morphology of fish sperm.

Acrosome

Absence of an acrosome in the Neopterygii is clearly apomorphic and its presence in agnathans (myxinids and lampreys); *Polypterus senegalus*; Dipnoi; Chondrichthyes; *Acipenser*; and *Latimeria* is here regarded as a plesiomorphic retention. The acrosome retains an apparently primitive, caplike form in the myxinid *Eptatretus*. A genetic potentiality for production of an acrosome probably remains in the Neopterygii, despite its absence from the mature sperm, as witnessed by its transitory appearance in the spermatid of *Salmo gairdneri* and of the lophiiform *Neoceratias spinifer*; a possible vestige in the mature sperm of the gobiesociform *Lepadogaster lepadogaster*; an acrosome-like structure in that of the salmoniform *Lepidogalaxias salamandroides*. An anterior vesicle in the atheriniform genus *Melanotaenia*, and vesicles in the gymnotoid *Sternarchus albifrons* and the percicthyid *Lates calcarifer* are possibly also homologues of an acrosome as may be an anterior Sertoli cell embedded vesicle in *Gambusia affinis*. The acrosome is exceptional in the chondrichthyan *Squalus* in having a spiral ridge. Perforatoria and their continuations as endonuclear canals are present in non-neopterygian acrosomes with the exception of hagfish, numbering one in *Lampetra*, *Polypterus*, 2 in *Neoceratodus*, and 3 in *Latimeria* and *Acipenser*. A postacrosomal ring occurs in *Lampetra*. Although perforatoria occur in some invertebrate aquasperm, notably the chelicerate *Limulus* and some polychaetes, and in nereids are also accompanied by endonuclear canals (Jamieson, 1987; Jamieson and Rouse, 1989), they are presumably apomorphic relative to a simple (monolayered) acrosome consisting of a caplike vesicle only. As hagfish sperm indicate, presence of a micropyle in the egg correlates not with absence of an acrosome but with absence of a perforatorium.

Nucleus

The nucleus in the ect-aquasperm of the Neopterygii is usually short (1-3 µm long), as in cephalochordates, and this may have been the plesiomorphic length for the Pisces. Its greatest elongation occurs in the lungfish *Neoceratodus*, at 70 µm; it is also elongate in, to take some examples, the myctophiform *Lampanyctus* sp. (6 µm); the osteoglossoid *Pantodon buchholzi* (7 µm); the gobiesociform *Lepadogaster lepadogaster* and the ophidiiform *Ophidion* (8 µm), in which it is apically helical; the lophiiform *Neoceratias spinifer* (11 µm); the agnathans *Eptatretus* (12 µm) and *Lampetra* (13-48 µm); the salmoniform *Lepidogalaxias salamandroides* (20 µm); and in *Latimeria* (~25 µm). In poeciliids and the exocoetid *Hemirhamphodon pogonognathus* the nucleus appears elongate and bladelike although only fractionally over the plesiomorph 3 µm length.

The nucleus has an apical peg, which supports the acrosome, in hagfish and *Acipenser* and in the former has a subterminal basal fossa. In lampreys chromatin is limited to a relatively short 7-14 µm long anterior region. It is spiral in the Chondrichthyes and in lophiiform *Neoceratias* and, apically, the related gobiesociform *Lepadogaster*. A basal fossa deeply penetrating the nucleus occurs in the zeiform *Zeus*; the dactylopteriform *Dactylopterus* (=*Cephalocanthus*); the Clupeomorpha; the pleuronectiform *Pegusa triophthalmus*; the tetraodontiforms *Balistes forcipatus* and *Pseudobalistes fuscus* (Balistidae), *Aluterus punctatus* (Monacanthidae) and *Chilomycturus antennatus* (Diodontidae); the scorpaeniform *Scorpaena angolensis*; the perciform *Upeneus prayensis*; and the goodeid cyprinodontiforms. The chromatin layer is particularly thin in one or more regions of the nucleus in *Dactylopterus* and in the gobioid eleotrid *Hypseleotris galii*.

The nucleus is pointed, or blade-like, with a "ventral" fossa in the scorpaeniform *Oligocottus maculosus*; the percoid embiotocid *Cymatogaster aggregata*; *Jenynsia lineata*; and poeciliids; all of which are internally fertilizing though the functional significance of the location of the fossa in these diverse forms is unknown. Sperm of elopomorphs, excepting muraeinids, have a highly distinctive crescentic nucleus.

Midpiece

A short midpiece with contained separate mitochondria is present in *Acipenser*, most Neopterygii, and the Dipnoi, probably as a reversal to the plesiomorphic condition. Although still compact it differs markedly in fusion of the mitochondria to form a single ring- or C-shaped body in disparate neopterygians: *Lepisosteus*; salmoniforms of the Salmonoidei, Argentinoidei and Galaxioidei; engraulid clupeiforms; and the percoid *Macquaria ambigua*. There is a single but elongate mitochondrial derivative open along one side in the cyprinodontiform *Jenynsia*. A single unilateral compact mitochondrion occurs in the aulopiform *Trachinocephalus*; the perciform *Maccullochella* (occasionally 2 in *M. peeli*); and in the gobioid *Hypseleotris galii*. In the elopomorphs, albulid and congrid sperm are distinctive in having a single mitchondrion in a nonbasal nuclear concavity.

The midpiece perhaps reaches its greatest elongation in the osteoglossoid *Pantodon*, at 45 µm. Some examples of other long midpieces are 20 µm in the hagfish *Eptatretus stoutii*; 12 µm in the chondrichthyan *Squalus*; 8 µm in the lophiiform *Neoceratias spinifer*; and ~5 µm in *Latimeria*.

The midpiece is highly distinctive, lying anterior to the basal body and in having an axial rod of rhizoplast origin in the Chondrichthyes including the Elasmobranchii and Holocephali. In both of these groups it is very elongate (though in Holocephali not of great absolute length) and consists of numerous mitochondria in spiral array. The midpiece is separated from the axoneme by the so-called cytoplasmic canal in chondrichthyans, *Acipenser*, Dipnoi and most neopterygians. Chondrichthyans differ in that the resultant sleeve is sloughed off at maturity. That in the Holocephali is further distinctive in being unilaterally united with the axoneme by persistent ladder-like cross connections. Salmonids are characterized by an osmiophilic body at the top of the cytoplasmic canal. In elopomorphs the mitochondrion (usually single) is displaced along the nucleus as far as its tip. Several mitochondria are also located at the tip in the percomorph *Blennius pholis*, unlike its congeners which have postnuclear mitochondria.

The arrangement of the mitochondria in elongate midpieces is highly variable taxonomically. Contrasting with the open cylinder in *Jenynsia*, those of poeciliids are in 2 to 5 columns around the axoneme, arranged end to end. In the exocoetid *Hemirhamphodon* they lie end to end bilaterally. In *Pantodon* they form 9 helical derivatives. In the gobiesociform *Lepadogaster* there are 6-10 elongate mitochondria around the bases of the two flagella. The greatest number recorded appears to be approximately 70 mitochondria in the midpiece of *Hydrolagus colliei*, despite its modest (3.5 µm) length. In the Petromyzontidae, *Mordacia* is unique in fish sperm in having the mitochondria restricted to the nuclear region, a homoplasy with non-appendicularian tunicates (Ascidiospermia). In *Lampetra planeri* and *Latimeria* the mitochondria extend from the axonemal into the nuclear region.

In the cyprinodontiform *Jenynsia* a submitochondrial sheath lies between the mitochondrial derivative and the cytoplasmic canal; a mesh equivalent to this sheath occurs in poeciliids. A differentiation of the inner layer of the cytoplasmic collar occurs in other neopterygians, for instance the gymnotoid *Sternarchus albifrons*. In the osteoglossoid *Pantodon* there is a curious 6 µm long fenestrated sheath behind the midpiece but not forming an extension of it and there are 9 non-axonemal dense fibres.

The few, globular mitochondria of the aflagellate

sperm of mormyroids lie between the centrioles and one pole of the cell in mormyrids, as in other teleosts, while in gymnarchids they are irregularly arranged.

Centrioles

Plesiomorphically, in neopterygians, if a basal nuclear fossa is present the centrioles lie within this (Type I sperm of Mattei, 1970). Location of the centrioles external to the fossa or eccentric to nucleus (Type II sperm) is a widespread percoid condition and is exemplified by the haemulid *Upeneus prayensis*; the centropomid *Lates calcarifer*; and the mugiloidid *Parapercis*. An intermediate condition occurs in the mugilid *Liza* , the polynemid *Galeoides decadactylus*, and the blennioids *Clinus nuchipinnis*, *Blennius cristatus*, and *Ophioblennius atlanticus*. These two conditions, defined for perciforms, have applicability in other orders.

A mutually perpendicular arrangement of the proximal and distal centriole (basal body) is reminiscent of the arrangement of centrioles in somatic cells and is interpreted as plesiomorphic for sperm cells. It appears to be the basic arrangement in the Neopterygii as it is seen in *Lepisosteus osseus* and is scattered through the remaining neopterygians being seen, for instance, in the cypriniforms *Alburnus alburnus* and *Barbus barbus*; in the aulopiform *Trachinocephalus myops*; the lophiiform *Neoceratias spinifer*; the scorpaeniform *Oligocottus maculosus*; in the percoids *Plectropomus lepidorus*, *Vomer setapinnis*; and the cichlids *Hemichromis fasciatus*, *Tilapia nilotica* and *Oreochromis niloticus*; the tetraodontiform *Gastrophysus hamiltoni*; the cyprinodontid *Fundulus heteroclitus* and the atheriniforms *Pseudomugil* and *Cairnsichthys* (contrasting with reduction of the proximal centriole in the related *Melanotaenia*).

The two centrioles are at an angle relative to each other exceeding 90° in many neopterygians, including the notopteroid *Papyocranus afer*; cypriniforms such as *Leuciscus leuciscus*; the siluriform *Clarius senegalensis*; and the characiform *Paracheirodon innesi*.

Rarely the proximal and distal centrioles are parallel, a condition seen in *Lampetra*; in *Pantodon* when a proximal centriole is visible; and in the mormyroid *Gymnarchus*.

Sometimes the proximal and distal centrioles are serially coaxial, i.e. in the same line. This condition is often accompanied by deep penetration of the nucleus by its basal fossa as in *Scorpaena angolensis*; the soleid *Pegusa triophthalmus*; in the tetraodontiforms *Balistes forcipatus*, *Pseudobalistes fuscus*, *Aluterus punctatus*, and *Chilomycturus reticulatus*; the scorpaeniform *Scorpaena angolensis*; and the percoid *Upeneus prayensis*.

Division of the proximal centriole into two bundles, of 4 and 5 doublets as a pseudoflagellum is diagnostic of the Elopomorpha: Elopiformes, Anguilliformes, and Notacanthiformes, as is a striated centriolar rootlet, though the rootlet is absent from muraenids.

The proximal centriole may be reduced, as in the atherinimorphs *Melanotaenia duboulayi*, pociliids, and *Hemirhamphodon pogonognathus*, or lost, as in the hagfish *Eptatretus stoutii*.

Very exceptionally the basal body is prenuclear, as in *Polypterus*; and, independently, *Lepidogalaxias*. The latter genus may be unique in investigated fish in lacking triplet centrioles, though only doublets have been seen in *Hypseliotris galii*.

The retronuclear body, between nucleus and centrioles at least during development, which unifies the dipnoans *Protopterus* and *Neoceratodus*, persists in lissamphibians, and is strikingly like the striated columns of mammalian sperm. It is uncertain whether a similarly located, lamellated body in *Pantodon* and the intracentriolar lamellated body of poeciliids is homologous with this.

Nine satellite rays, radiating from the distal centriole, which are typical of invertebrate ectaquasperm are rarely seen in fish sperm. They are recorded for the dipnoan *Protopterus* and for *Latimeria*. Surprisingly they are well developed in members of the supposedly advanced neopterygian order Atheriniformes, being well developed in *Craterocephalus*, *Querichthys* and *Cairnsichthys*, but are weakly if at all developed in some other atheriniforms, while both conditions occur in *Melanotaenia* and are sporadically present in other groups.

Axoneme

The piscine sperm has a one or, rarely, two flagella the axoneme of which is overwhelmingly of the 9+2 type. In elopomorphs and the myctophiform *Lampanyctus*, however, the axoneme is of the 9+0 pattern as the two central singlets are absent. The flagellum has been lost only in osteoglossomorphs of the Mormyridae and Gymnarchidae (with development in *Gymnarchus* of many cytoplasmic tubules) but there is an unconfirmed report of loss in the myctophyforms *Lampanyctodes hectoris* and *Diaphus danae*. Biflagellarity has developed several times: in the cladistian *Polypterus*; the dipnoan *Protopterus*; in two siluriforms, *Ictalurus punctatus* (with, here at least, loss of the outer dynein arms) and a malapterurid; in the myctophiform *Lampanyctus*; in the batrachoids *Opsanus tau* and *Porichthys notatus*; in the gobiesocid *Lepadogaster lepadogaster*; and in a percoid apogonid; as in some Amphibia.

Various arrangements of accessory fibres of different origins are developed, though rarely. In chondrichthyans two large accessory longitudinal columns are present at doublets 3 and 8. These are equal in size and oval in cross section in sharks; or equal and round in rays. In the Holocephali the column at 8 is greatly reduced. That the two fibres in *Latimeria* are homologues of the chondrichthyan columns is doubtful. Two columns also occur basally in sperm axoneme of the dipnoan *Neoceratodus*. The occurrence in *Lampetra* (but apparently not in *Mordacia*) of 9 accessory axonemal fibres, characteristic of internally fertilizing vertebrate sperm above the fish, is consistent with the argument that lampreys formerly had internal fertilization although this view is here considered equivocal. Nine weakly developed fibres spiral with the mitochondrial derivatives in the internally fertilizing osteoglossomorph *Pantodon buchholzi*.

Fins are present on the flagellum, as extensions of the plasma membrane, in most neopterygians and their sister-group, the Chondrostei (*Acipenser*). As fins are also present in the Dipnoi their acquisition, if a monophyletic event, may be attributed to the basal Osteichthyes and Teleostomi. There is a single fin on the flagellum of the sperm of *Esox* and the percicthyid genus *Maccullochella*, and in some regions of the tail in some species with two fins. Two is the usual number though three are sometimes present in aplocheilids. Fins are absent, clearly by apomorphic loss, in the Ostariophysi (cypriniforms, siluriforms, and characiforms). Fins are reduced (poeciliids) or absent (*Hemirhamphodon*, chondrichthyans, *Latimeria*, *Lepidogalaxias*) on acquisition of internal fertilization which must be a secondary mode of feritlization in the two atherinomorphs and probably is in all five taxa.

Partial occlusion (septation) of the microtubules of the doublets occurs in those fish sperm in which the mitochondrion forms a single ring or C-shape but the significance of the correlation is elusive. Septation involves doublets 1 2 6 7 in *Lepisosteus*; 1 2 5 6 7 in salmoniforms of the Argentinoidei and Galaxioidea but 1 2 3 5 6 7 in salmonids; and 1 3 5 6 7 in engraulids.

Glycogen is not characteristic of fish sperm tails but occurs in large amounts in the flagellum of *Hydrolagus colliei*; the amount in shark flagella is small.

Systematic and phylogenetic summary

An overview of sperm types in the Gnathostomata is given in Fig. 18.1. Major phylogenetic trends in the sperm of the Vertebrata are summarized in Fig. 18.7. The sperm of many fish groups remain to be investigated. A survey of known sperm types by systematic taxa has revealed the following distribution of sperm types and permits some phylogenetic conclusions.

Vertebrata. The vertebrates as a whole show no unique apomorphies in their sperm. If it be accepted that they descend from an ancestor, possibly shared with the Cephalochordata, with "primitive" sperm, vertebrates are, nevertheless, unified by two undistinguished synapomorphies, elongation of the nucleus, first seen in the Myxini but reaching its zenith in the sturgeon, and elongation of the midpiece, a trend especially prominent in the hagfish.

Agnatha. Lampreys and, it is believed, hagfish have external fertilization yet their sperm show features not normally associated in the Metazoa with externally fertilizing sperm (ect-aquasperm) (Fig. 18.1): long endonuclear perforatorium in *Lampetra* (but also in ect-aquasperm in polychaetes); elongate nucleus; an elongate mitochondrial sheath around the axoneme, in hag fishes especially, but excepting the lamprey *Mordacia* in which the mitochondria lie around the nucleus in groups; and, in *Lampetra*, nine accessory fibres around the axoneme. None of these similarities demands recognition of relationship between the two groups. This supports the view that the Cyclostomata is an artifical group unified by the symplesiomorphy jawlessness.The presence of a penial tube in lampreys has been tentatively interpreted as evidence of their former internal fertilization but possibly represents only an advanced, unusually intimate form of external fertilization. As one or more endonuclear canals and enclosed perforatoria occur in the Petromyzontiformes, and some members of both the Actinopterygii and Sarcopterygii, their presence is here regarded as plesiomorphic for this entire assemblage, i.e. vertebrates ex-

centrioles in parallel (Fig. 18.7). Great elongation of the nucleus is a further synapomorphy. The postacrosomal ring, restriction of chromatin to the anterior region of the nucleus, and presence of 9 accessory axonemal fibres are questionable petromyzontiform apomorphies as, though seen in *Lampetra*, they have yet to be confirmed for *Mordacia*.

Gnathostomes. Whereas the acrosome vesicle in

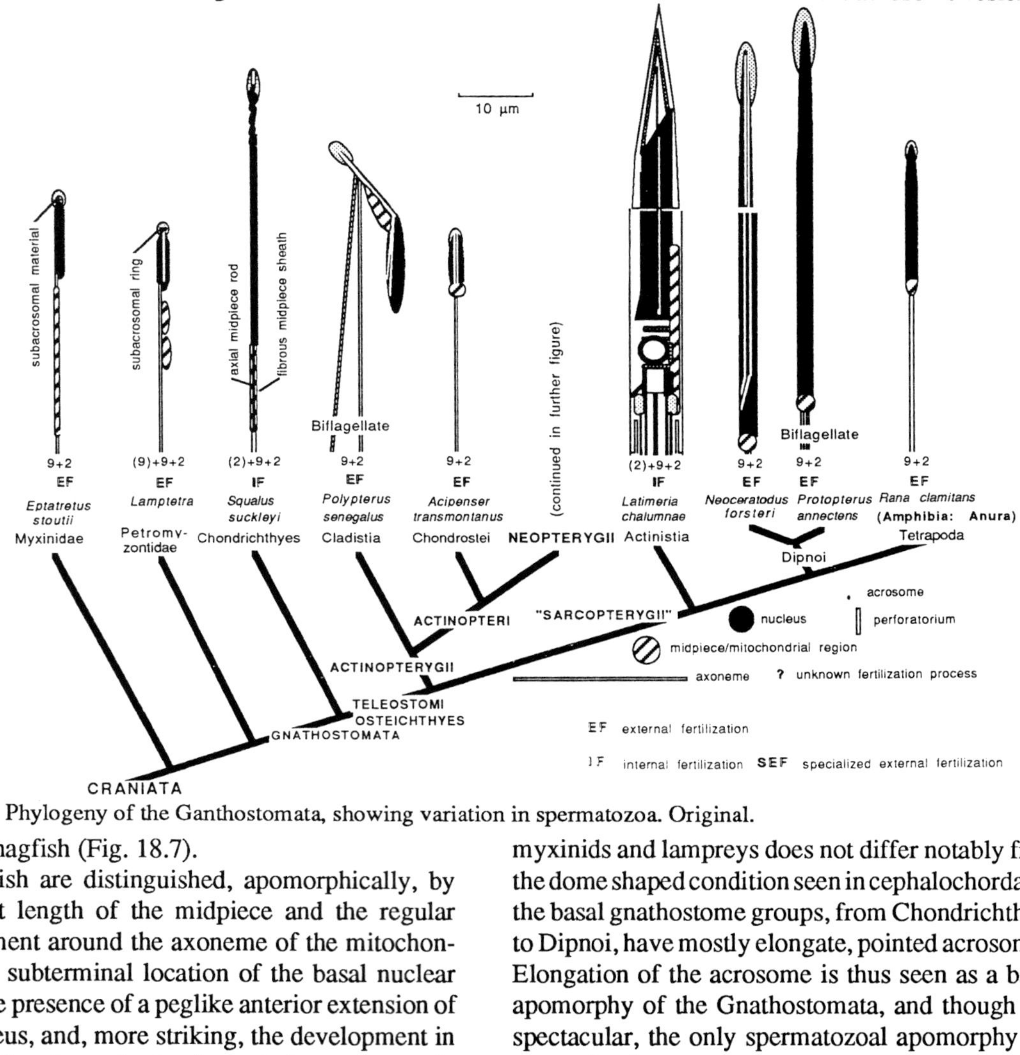

Fig. 18.1. Phylogeny of the Ganthostomata, showing variation in spermatozoa. Original.

cepting hagfish (Fig. 18.7).

Hagfish are distinguished, apomorphically, by the great length of the midpiece and the regular arrangement around the axoneme of the mitochondria, the subterminal location of the basal nuclear fossa, the presence of a peglike anterior extension of the nucleus, and, more striking, the development in the egg of a micropyle. A micropyle is not seen again until its clearly independent development in the Actinopteri.

The two lamprey genera, *Lampetra* and *Mordacia*, are united and the Petromyzontiformes are apparently distinguished, by arrangement of the two

myxinids and lampreys does not differ notably from the dome shaped condition seen in cephalochordates, the basal gnathostome groups, from Chondrichthyes to Dipnoi, have mostly elongate, pointed acrosomes. Elongation of the acrosome is thus seen as a basal apomorphy of the Gnathostomata, and though unspectacular, the only spermatozoal apomorphy apparent for the group.

Chondrichthyes. In the Gnathostomata, the Holocephali and Elasmobranchii are unified, as the Chondrichthyes (Fig. 18.1, 18.7), by a suite of apomorphies: moderately elongate, conical apical acrosome; long nucleus and midpiece, both of which are helical; midpiece composed of many subspherical mitochondria, with fibrous axial core or rod; a fibrous sheath around the midpiece; location of the basal body behind the midpiece; and location of two longitudinal columns in the axoneme outside doublets 3 and 8. It seems probable that the extreme asymmetry of the columns in the Holocephali is apomorphic (and therefore an autapomorphy) relative to their equal development in the Elasmobranchii. An apomorphy linking the Selachimorpha and the Batidoidimorpha is not yet apparent but, whereas the accessory axonemal columns in the latter retain the plesiomorphic circular form seen also in Holocephali, those of sharks are distinguished by an elliptical cross section. Presence of a spiral ridge on the acrosome is a further, though doubtful, apomorphy of sharks. Holocephali are apomorphic, in addition to asymmetry of their axonemal columns, in possession of scalariform bridges from the cytoplasmic collar to the axoneme and in the great amount of glycogen in the flagellum.

Osteichthyes. Two axonemal fins are present in the Actinopterygii excepting the Cladistia (*Polypterus*), and in the Dipnoi. They are possibly independent apomorphies of each of these two groups but are here tentatively regarded as a spermatozoal autapomorphy, the only one known, for the bony fish. It is uncertain whether the undulating membrane of Lissamphibia is homologous with a dipnoan axonemal fin.

Actinoptergii. The bizarre biflagellate sperm of *Polypterus* somewhat obscures initial trends in the ray-finned fish but one, constituting an apomorphy for the group (Fig. 18.7), is shortening of the midpiece, already clear in sturgeons and typical of the secondarily simplified sperm of neopterygians.

Cladistia. The Cladistia (*Polypterus*) have biflagellate acrosomal aquasperm (Fig. 18.1). Biflagellarity has clearly evolved more than once in fish sperm (also occurring in ictalurids, myctophids, apogonids, batrachoids, gobiesocids, and Dipnoi, as in some amphibians), but sharing of this condition constitutes weak support for the hypothesis of a special (sister-group) relationship between cladistians and the Dipnoi. The prenuclear implantation of the flagella in *Polypterus* is a notable difference from the plesiomorphic, and usual, postnuclear implantation in the dipnoan *Protopterus* but such a variation (albeit of a single flagellum) can occur in unified subgroups, as between ascothoracican and cirripedian maxillopod Crustacea. On the other hand, the occurrence of a single flagellum in the sperm of the dipnoan *Neoceratodus* may indicate that biflagellarity in Dipnoi (*Protopterus*) is an independent development from that in *Polypterus*. Biflagellarity and prenuclear basal bodies are here regarded as weak autapomorphies of the Cladistia (Fig. 18.7).

Actinopteri. The Actinopteri (Actinopterygii, commencing here with the Chondrostei, above the Cladistia) have developed egg micropyles, as their synapomorphy, independently evolved in myxinids. Two axonemal fins, regarded as an osteichthyan autapomorphy (Fig. 18.7), could, alternatively have evolved in Actinopteri independently of the Dipnoi.

Chondrostei, forming the sister-group of the Neopterygii and with them constituting the Actinopteri, probably represent a transitional stage towards the neopterygian condition of a single micropyle and no acrosome. The sturgeons have not only numerous egg micropyles but also a sperm acrosome. The nucleus is elongate and the sperm is not referable to the simple aquasperm (plesiosperm) model but more closely resembles the ent-aquasperm of some polychaetes and other invertebrates (see Jamieson and Rouse, 1989). The possibility that it is derived from a primarily internally fertilizing sperm has been mooted in Chapter 5. However, the modified form conceivably has developed for external fertilization in relation to features of the egg. In molluscs, for instance, elongation of the nucleus correlates with the diameter of the egg. In sturgeons, although the micropyles give easy access of sperm through the thick egg envelope, the plasma membrane is relatively impermeable and may necessiate the retention of the acrosome and possibly the elongate, somewhat lanceolate form of the nucleus. The presence of lateral fins on the sperm flagellum which we consider in the sperm of "higher" fish to indicate their origin from externally fertilizing precursors here, because it is its first appearance, does not unequivocally indicate an externally fertilizing ancestry.

No distinguishing apomorphies are apparent for sturgeons beyond the presence of 3 endonuclear canal and the multiple micropyles contrasting with the single micropyle of Neopterygii. The Neopterygii (see below) are distinguished from other

Actinopterygii (Cladistia and Chondrostei) in apomorphic loss of the acrosome and with it the endonucler canals and perforatoria which persisted as cladistian-chondrostean plesiomorphies.

Crossopterygii. Coelacanthimorpha. *Latimeria chalumnae* (Fig. 18.1) has an acrosomal introsperm and viviparity. The matrophagous or adelphophagous development of the young might be considered evidence for relationship to the Chondrichthes but relationship with tetrapods is better supported by the evidence. The sperm is complex but most of its features are plesiomorphic for gnathostomes. The two lateral elements in the sperm tail are probably more similar to those of the dipnoan *Neoceratodus* than to those of Chondrichthyes and with the retronuclear body are seen not only in Dipnoi but also (Mattei, 1988) in Lissamphibia. The two columns and the retronuclear body are therefore here seen as synapomorphies of the Sarcopterygii (Actinistia - Dipnoi - Tetrapoda). The actinistian sperm is then distiguished from the basal hypothetical sarcopterygian sperm in location of the mitochondrial material as an incomplete ring around the base of the nucleus (confirmation required from better fixed sperm) and loss of the axonemal fins.

Dipnoi. Sperm ultrastructure reinforces separation of *Neoceratodus* from *Protopterus* in separate families of the Dipnoi (Fig. 18.1) while offering an apomorphy (retronuclear body or its derivative) linking the two families. Both have acrosomal aquasperm. In *Neoceratodus forsteri* the sperm head is about 70 µm long, an exceptional length for any spermatozoon other than those of some Insecta. It remains to be demonstrated that this is an adaptation to exceptional features of the egg. Sperm do not actually support monophyly of the Dipnoi, though possession of the retronuclear body and accessory axonemal columns confirms their sarcopterygian status and relationship to *Latimeria* and lissamphibians. *Neoceratodus* is distingushed by emergence of the perforatorium posteriorly from the nucleus while *Protopterus* shows apomorphic biflagellarity and loss of the endonuclear canal and perforatorium (Fig. 18.7).

Neopterygii. Irrespective of whether internal fertilization is primitive or derived in fish as a whole, neopterygians are diagnosed from the Agnatha through Dipnoi in having lost the acrosome, giving an anacrosomal aquasperm. The neopterygian egg has developed a micropyle, which is single presumably as an adaptation preventing polyspermy. Divergence from the simple anacrosomal (plesiosperm) type is found only in the few neopterygians which have external fertilization of a presumably specialized type (e.g. mormyrids and the lophiiform *Neoceratias*) or have secondary internal fertilization (notably in the Atherinomorpha, including poeciliids, jenynsiids, goodeids and the exocoetid *Hemirhamphodon*, of which the goodeids, alone, retain a near-plesiosperm morphology; and also in some scorpaenids and embiotocids). The single occurrence of an acrosome at maturity, in *Lepidogalaxias*, is interpreted as a redevelopment from a basic propensity of teleost spermatids and is unlikely to be a direct retention of an ancestral condition.

Ginglymodi and Halecomorphi. The Ginglymodi (*Lepisosteus*) and Halecomorphi (*Amia*) (Fig. 18.2) have anacrosomal aquasperm, slightly elongate and with a ring-shaped mitochondrion in *Lepisosteus*, simple in *Amia*.

Osteoglossomorpha. In the Osteoglossomorpha (Fig. 18.2), the aflagellate condition seen in sperm of the Gymnarchidae and Mormyridae is in agreement with the association of these families in the Mormyroidei. The simple anacrosomal aquasperm of the notopteroid *Papyocranus afer* supports exclusion of the mormyroids from the Notopteroidea. The peculiar aquasperm of the only other investigated osteoglossoid, the internally fertilizing *Pantodon bucholzi*, justifies the separate status of the Pantodontidae within the Osteoglossoidei; nine helical non-axonemal dense fibres are present in the midpiece; behind the midpiece there is a peculiar fenestrated sheath.

Elopomorpha. The Elopomorpha (Fig. 18.3) provide a striking example of the utility of spermatozoal ultrastructure for determining phylogenetic relationship. It unequivocally supports relationship of Elopiformes, Notacanthiformes and Anguilliformes, as members of the Elopomorpha. Autapomorphies of the Elopomorpha (known for Elopiformes, Notacanthiformes and Anguilliformes) are a 9+0 flagellum and division of the proximal centriole into two elongate bundles of 4 and 5 triplets, together with the presence of a leptocephalus larva. In all except the Muraeinidae the nucleus is large and crescentic and

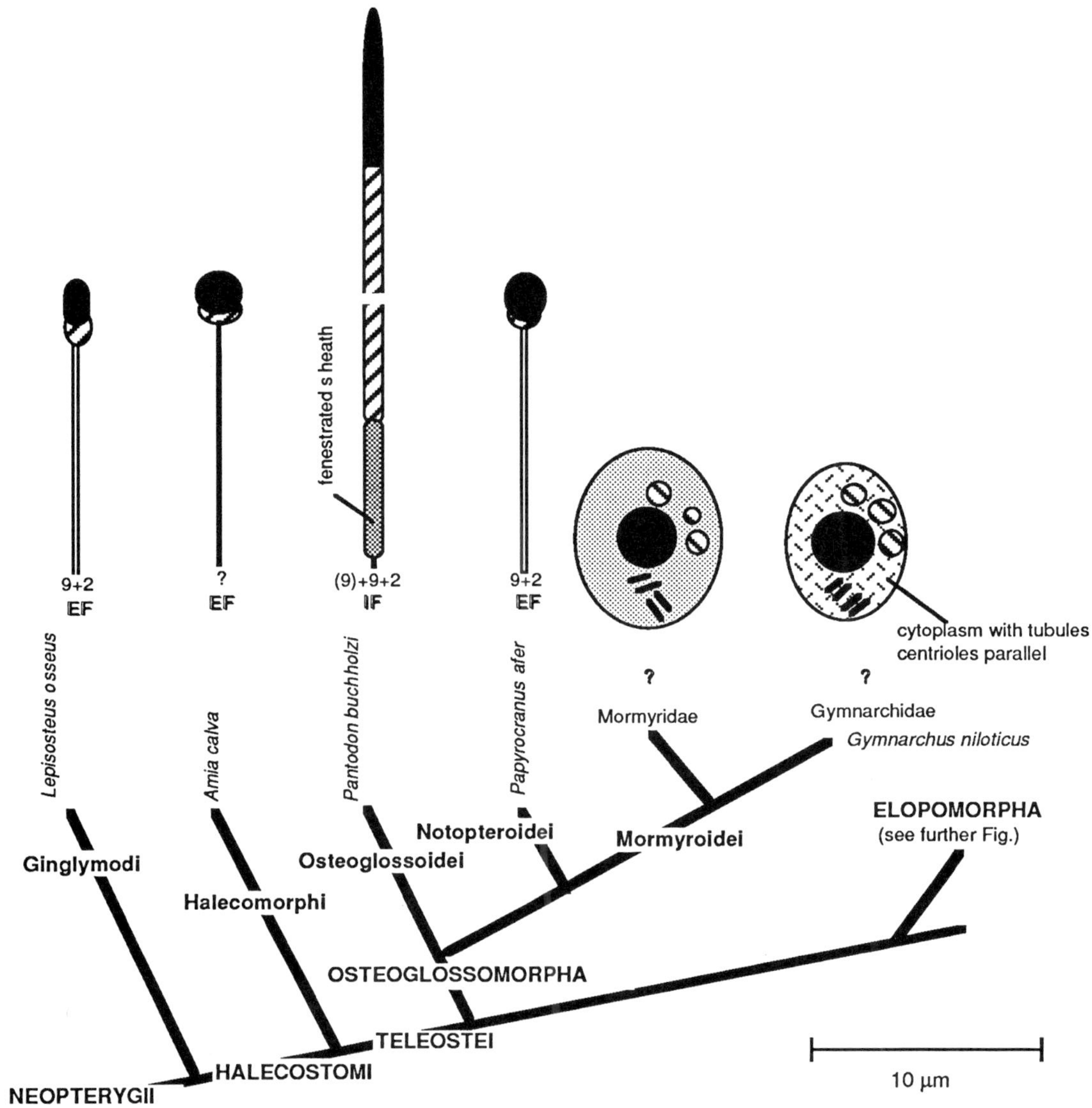

Fig. 18.2. Phylogeny of the Neopterygii, Ginglymodi through Osteoglossomorpha, showing variation in spermatozoa. Original.

a cross striated centriolar rootlet projects from the cell near the base of the nucleus. In the Muraenidae, the nucleus is rounded and the rootlet is missing.

Clupeomorpha. Clupeomorphs, with the Euteleostei, comprise the Clupeocephala. Clupeomorph sperm (Fig. 18.3), known from very few species, show a departure from the commoner teleostean aquasperm ultrastructure in penetration of the nucleus almost to its tip by the basal fossa and contained axoneme; and a ring- or C-shaped mitochondrion, a feature known in *Lepisosteus* and, in teleosts, in only five families: the clupeiform families, Clupeidae (*Ethmalosa fimbriata*) and Engraulidae (*Anchoa guineensis*), and the salmoniform families Alepocephalidae (*Xenodermichthys* sp.), Searsidae (*Searsia* sp.), Salmonidae (*Salmo gairdneri*) and Galaxiidae (*Galaxias olidus*).

Ostariophysi. The euteleost superorder Ostario-

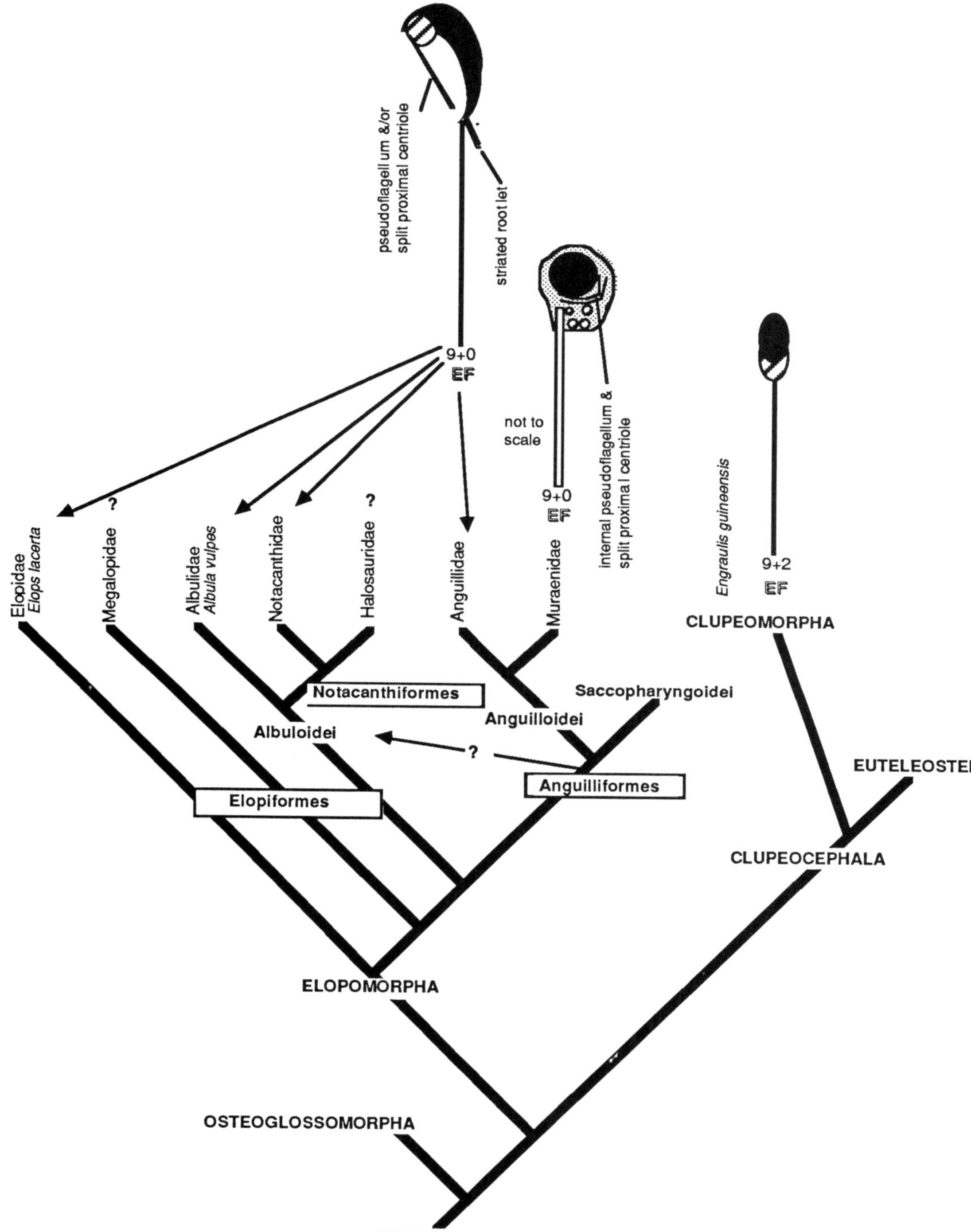

Fig. 18.3. Phylogeny of the Teleostei, Elopomorpha through Clupeomorpha, showing variation in spermatozoa. Original.

physi (Fig. 18.4) is unified by a clearly synapomorphic loss of flagellar fins in all three constituent orders (Cypriniformes, Siluriformes (Siluroidei and Gymnotoidei), and Characiformes). The mutual presence of simple anacrosomal aquasperm is presumably a symplesiomorphy but some siluriforms (ictalurids) have biflagellate anacrosomal aquasperm. No phylogenetic relationship can be detected in the possession of biflagellate sperm in other fish noted above.

Salmoniformes. In the Order Salmoniformes (of the now obsolete Superorder Protacanthopterygii), excepting lepidogalaxiids, the head of the anacrosomal aquasperm is somewhat [but only slightly] elongated compared with the subspherical head of sperm of the Cyprinidae and Esocidae. There is a narrow and moderately deep basal nuclear fossa. Relationship of the Salmonoidei and Argentinoidei is endorsed by the synapomorphic ring-shaped single mitochondrion and slight elongation of the nucleus. The presence of flagellar fins is a symplesiomorphy but contrasts with the apomorphic loss of these in the related Ostariophysi. Occlusion of the A tubules of doublets 1 2 5 6 and 7 may be a synapomorphy of argentinoids and *Galaxias*. A primitive position for the salmoniforms is possibly supported by presence of a transient acrosome in *Salmo* spermatids and persistence of an acrosome in the mature introsperm of *Lepidogalaxias*. In esocids, normally placed in the salmoniforms but here placed in an order Esociformes below the Ostariophysi (Fig. 18.4), the midpiece overlaps the head, in the pike, as it does in cyprinids and unlike salmonids. *Esox* differs from cyprinids in having a fin on the flagellum

Scopelomorpha. In the Neoteleostei (Fig. 18.5), the few sperm known for the Superorder Scopelomorpha have provided little phylogenetic information. The Aulopiformes have a uniflagellate anacrosomal aquasperm with a (single?) large asymmetrically located mitochondrion. In the

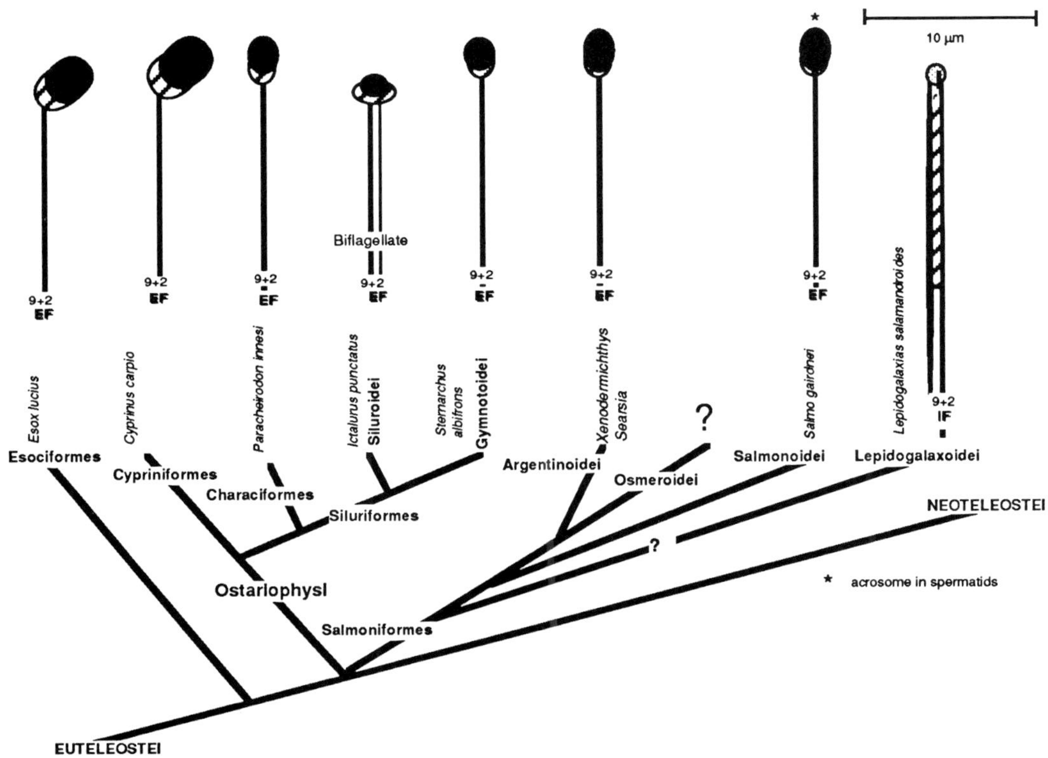

Fig. 18.4. Phylogeny of the Euteleostei, Esociformes, Ostariophysi through Salmoniformes, showing variation in spermatozoa. Original.

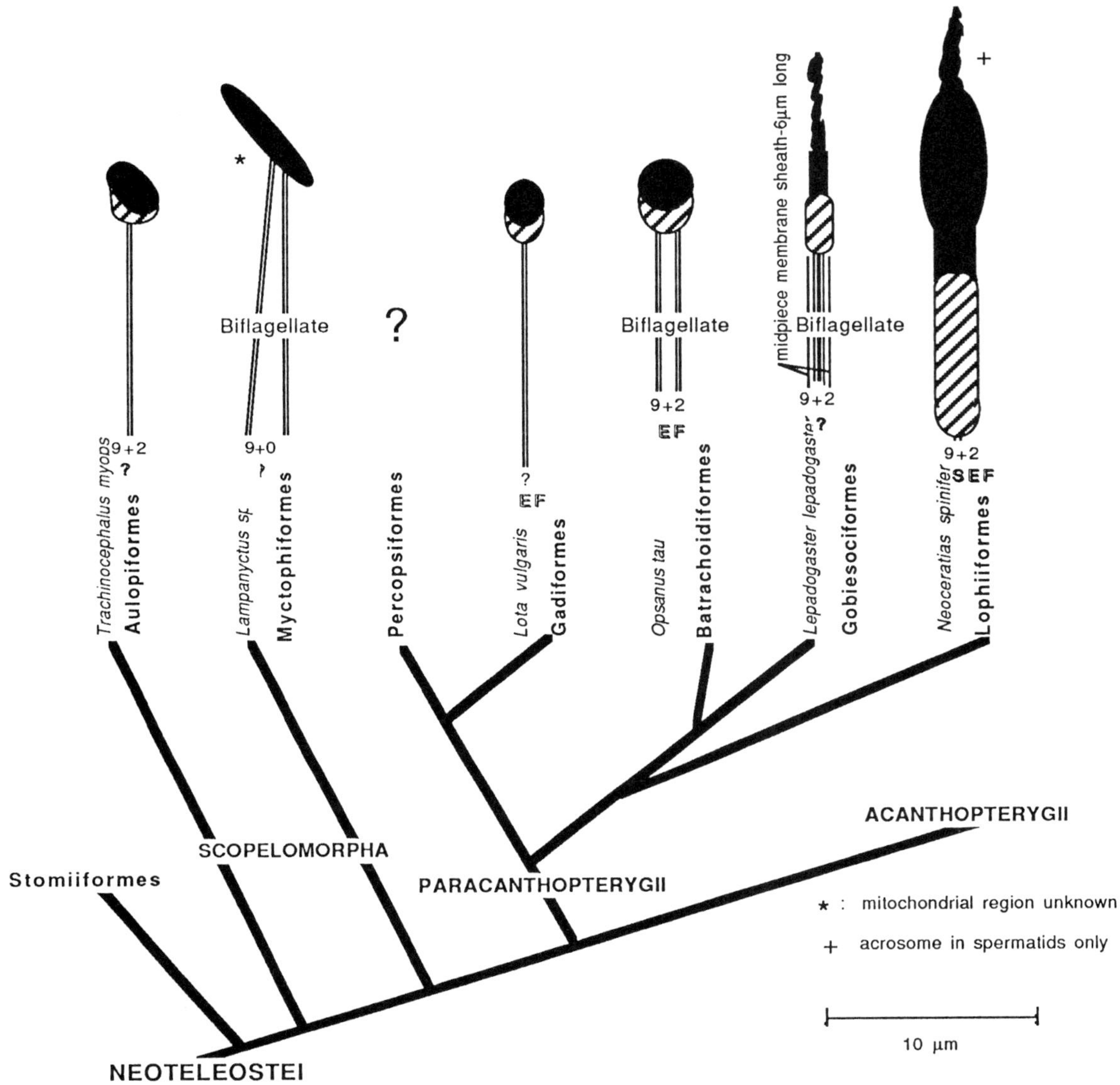

Fig. 18.5. Phylogeny of the Euteleostei, Scopelomorpha through Paracanthopterygii, showing variation in spermatozoa. Original.

Myctophiformes *Lampanyctus* has a biflagellate sperm with 9+0 axonemes; the flagella appear to be inserted considerably proximal to the base of the S-shaped nucleus. A report of aflagellate sperm in *Lampanyctodes* and *Diaphus* awaits confirmation.

Paracanthopterygii. In the Superorder Paracanthopterygii, the gadiform, *Lota vulgaris*, has the simple uniflagellate anacrosomal aquasperm. A sister-group relationships of the Batrachoidiformes and Gobiesociformes is supported spermatologically. The two investigated species of the Batrachoidiformes, *Opsanus tau* and *Porichthys notatus*, both spawners with parental care, have biflagellate aquasperm, simple in the former, more complex in the latter. The sole investigated Gobiesociform, *Lepadogaster lepadogaster*, also has biflagellate sperm which we consider to resemble the complex sperm of *Porichthys*. More species need to be studied to confirm if biflagellarity is general for these sister-groups. The sperm of the

Lophiiformes reveal no special relationship with these other paracanthopterygians, though the spiral nucleus, a rare condition, may indicate relationship with *Lepadogaster* rather than that here postulated with the batrachoids. Both investigated lophiiform species have uniflagellate sperm. Of these *Neoceratias spinifer*, with a parasitic male, has filiform, modified sperm adapted to the very intimate external fertilization while *Antennarius senegalensis*, presumably a free spawner, has a basic teleostean ect-aquasperm.

Acanthopterygii. Percomorpha. In the Superorder Acanthopterygii (Percomorpha + Atherinomorpha) (Fig. 18.6) most investigated percomorphs have simple anacrosomal aquasperm, with or without some modification (such as invagination of the nucleus, and/or asymmetrical location of mitochondria and the axoneme) and have external fertilization. They are exemplified by the Zeiformes; Syngnathiformes; Dactylopteriformes; some Scorpaeniformes; Pleuronectiformes; Tetraodontiformes; and many Perciformes - familes Mullidae, Haemulidae, Centropomidae, Percichthyidae, Serranidae, Carangidae, Centracanthidae, Sparidae, Cichlidae and Pomacentridae. These sperm are, however, referable to two types, though intermediates sometimes occur (Mattei, 1970, 1988): type I, with the centrioles in the nuclear fossa, and type II, with the centrioles external to any fossa and the flagellum asymmetrically located.

Some percomorph species with unknown but probably external fertilization (*Dactylopterus volitans*, *Ophioblennius atlanticus*) have grossly asymmetrical sperm, albeit referable to type I. In *Blennius pholis*, the mitochondria have become apical on the nucleus. The asymmetrical sperm of the eleotrid gobioid *Hypseleotris galii* (Perciformes), a spawner, resemble those of *Dactylopterus volitans* (Dactylopteriformes), in extreme attenutation of the chromatin in part of the nucleus as a very thin layer so that apical and basal nuclear envelopes are virtually in apposition. This suggests close relationship of the two orders. Furthermore, similarity of the sperm of Zeiformes, recently placed in the Tetraodontiformes (Rosen, 1984), in deep invagination of the nucleus, though not unique, supports a special relationship of the two orders.

Internal fertilization has evolved independently in a few percomorph groups (e.g. the scorpaeniforms *Oligocottus maculosus* and *Sebastiscus marmoratus*, and the perciform embiotocid *Cymatogaster aggregata*). In these the spermatozoa are elongated, though again derivable from the anacrosomal aquasperm, but sperm structure and details of fertilization differ between groups and offer little evidence for recognition of phylogenetic affinities. It is, nevertheless, probable that examination of additional species will support interspecific relationships. *Cymatogaster* has true spermatophores containing parallel spermatozoa, a notable difference from the spermatozeugmata of poeciliids (in the Atherinomorpha), in which internal fertilization is clearly an independent development. Poeciliids have the sperm heads located peripherally to a core of flagella. Examined mugiloid and blennioid perciforms, like the two pleuronectiforms studied, have anacrosomal aquasperm with, in blennioids, a marked tendency to asymmetry.

Acanthopterygii. Atherinomorpha. The Atherinomorpha (Atheriniformes, Cyprinodontiformes and Beloniformes) (Fig. 18.6) stand apart from other fish, exemplified by the Salmoniformes, Perciformes, and Cypriniformes in which spermatogonia are distributed along the entire length of the tubules, in having the spermatogonia restricted to the distal end of the testicular tubules. This telogonic condition is a clear autapomorphy for the Atherinomorpha. A tendency to an anterior acrosome-like vesicle is seen in *Melanotaenia* sperm (Atheriniformes) and in the spermatid of *Gambusia* (Cyprinodontiformes).

Atheriniformes. The sperm of atheriniforms are simple anacrosomal aquasperm but, despite their highly evolved position, some have a satellite apparatus, around the distal centriole, which is unusually well developed for fish sperm.

Cyprinodontiformes. Reproduction in the Cyprinodontiformes is exceptionally varied, with oviparity (with external or internal fertilization), ovoviparity, viviparity and functional hermaphroditism. Viviparity may have evolved at least four times in the order. Sperm structure is correspondingly variable though often family specific. The basic sperm type is presumably that seen in the Aplocheiloidei (rivulines) as these have external fertilization. It is a simple anacrosomal aquasperm with one to three fins on the tail, the number varying intraspecifically or even

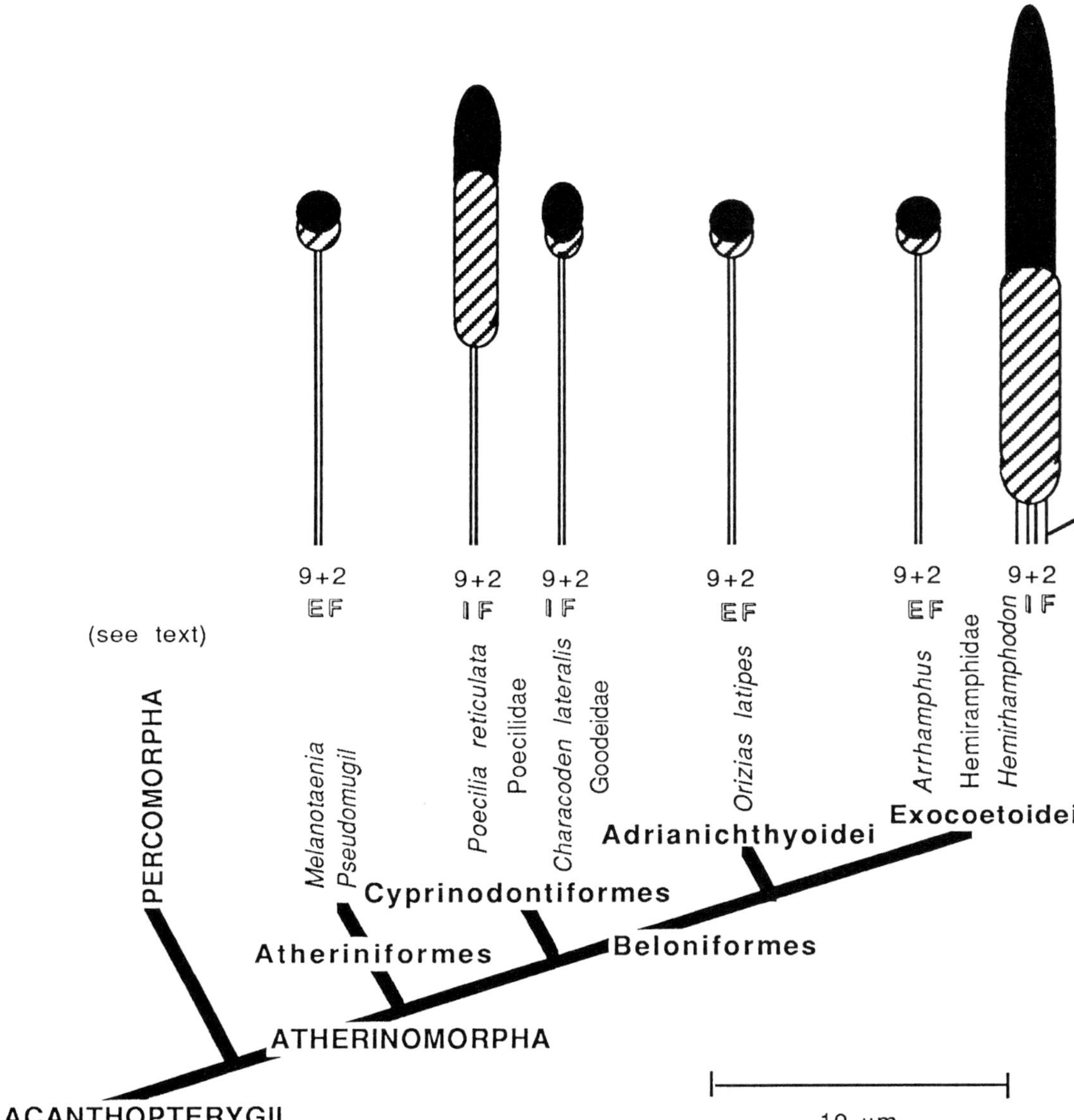

Fig. 18.6. Phylogeny of the Acanthopterygii, Atherinomorpha, showing variation in spermatozoa. Original.

along the length of a single tail. Within the Cyprinodontoidei, the Cyprinodontidae, exemplified by *Fundulus*, also have a simple anacrosomal aquasperm but it resembles that of the internally fertilizing, live-bearing Goodeidae in having the flagellum implanted in the "ventral" fossa of a slightly elongate nucleus. At least in goodeids, and poeciliids, a pair of lateral flagellar fins is present.

Viviparous poeciliids and the sole jenynsiid genus differ from goodeids and cyprinodontids in having complex sperm with elongation of both the nucleus and the midpiece. Relationship to cyprinodontids and goodeids is seen in the "ventral" nuclear fossa. Origin from externally fertilizing ancestors is seen in persistence of a pair of albeit small lateral fins. The phylogenetic value of sperm is exemplified by the uniformity of the complex sperm of poeciliids and differences from the jenynsiid

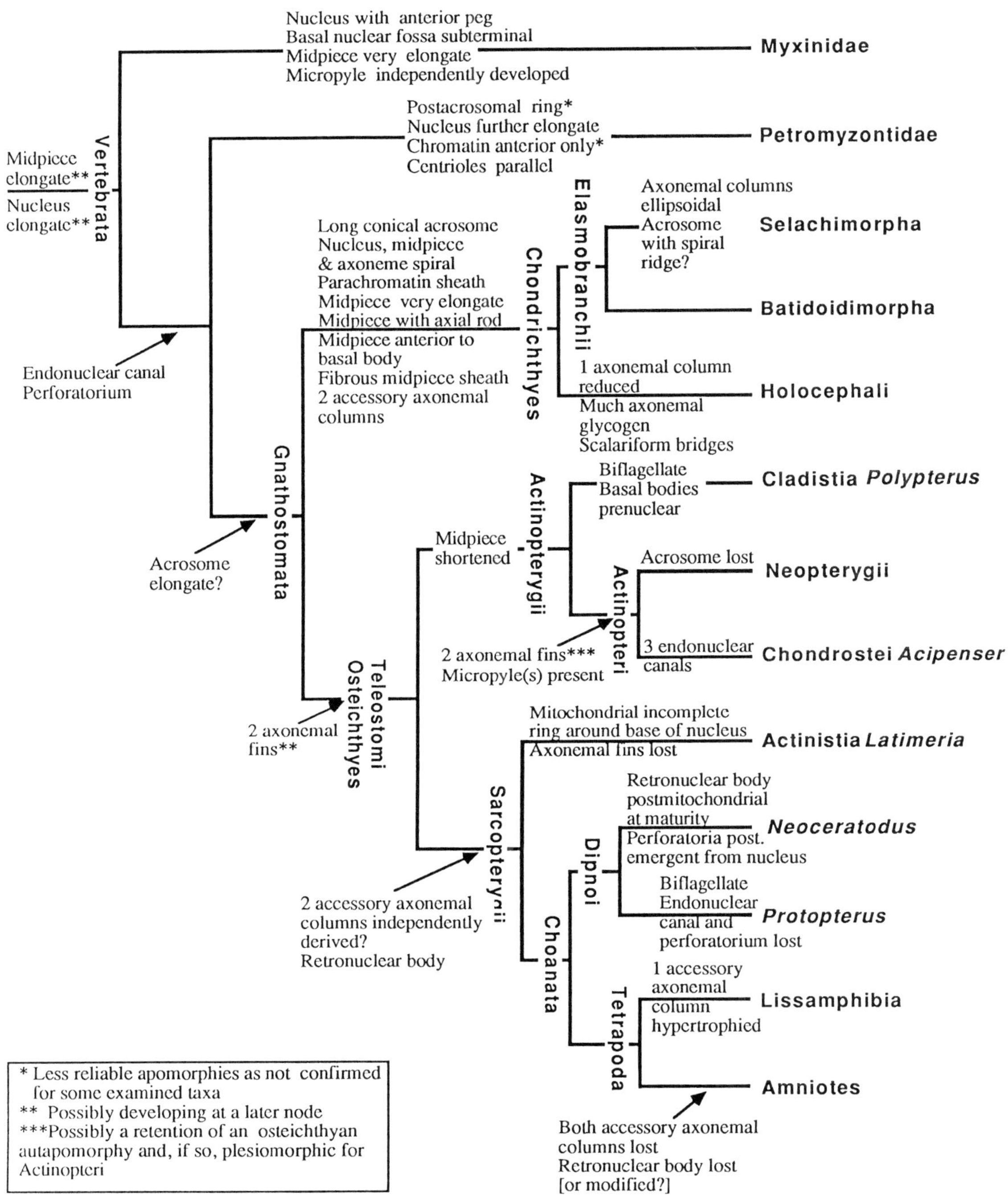

Fig. 18.7. Chief phylogenetic trends in vertebrate sperm, Myxini through Tetrapoda. Jamieson, unpublished.

sperm, including the presence of a single mitochondrial derivative in the latter.

Beloniformes. The rounded nucleus and basal location of the fossa in the sperm of the externally fertilizing *Oryzias latipes* support its exclusion from the Cyprinodontidae in which it is sometimes placed. Sperm structure does not at present allow a decision between placement of *Oryzias* in the beloniform suborder Adrianichthyoidei, as here accepted, or in the Cyprinodontiformes. In the suborder Exocoetoidei, the spermatozoon of *Fodiator* is a simple teleostean anacrosomal aquasperm while that of the internally fertilizing halfbeak *Hemirhamphodon pogonognathus* shows modifications, notably elongation of the nucleus, extension of the mitochondria of the midpiece as an elongate sheath and unique radial rodlets around the axoneme. In view of the teleostean aquasperm in the externally fertilizing halfbeak *Arrhamphus*, these similarities between *Hemirhamphodon* and poeciliid and jenynsiid sperm are clearly homoplasic. Fins, present in *Arrhamphus* are lost in *Hemirhamphodon*.

Chapter 19

PRINCIPLES OF BIOLOGICAL CRYOPRESERVATION

by L.K.-P. Leung

I. Introduction

The freezing and thawing of biological material involves a series of complex and dynamic physicochemical processes of heat and water transport between cells and their surrounding medium. Understanding of these fundamental aspects of cryobiology is essential to the application of cryopreservation, and will be briefly outlined in section II below. The basic principles of cryopreservation are discussed in sections III to V. In the last section (VI), some practical aspects of cryopreservation will be examined.

II. Relevant Physicochemical Phenomena

Knowledge of physicochemical phenomena relevant to cryobiology is essential for a thorough understanding of the basic principles of biological cryopreservation and some treatment of these phenomena is therefore included here. Much of the material of this section can be found in fundamental physics textbooks and has been reviewed by Luyet (1966; 1970), Luyet and Gehenio (1940), and Meryman (1966), to which the reader is referred. More recent information is added.

A. Freezing of Water

It is generally believed that a crystal nucleus must be present to initiate thermal crystallization. Nucleus production is a result of either random aggregations of molecules (homogeneous nucleation) or catalyzed aggregations of molecules (heterogeneous nucleation). The latter usually occurs in bulk water at, or slightly below, its freezing point. In the absence of heterogeneous nuclei, water can be supercooled by heat removal at a uniform rate down to approximately -40° C, at which temperature homogeneous nucleation takes place spontaneously, followed by ice propagation (see Fig. 19.1). At this point, the mixture of water and ice is warmed to freezing point by the release of the latent heat of fusion. The temperature of the mixture remains at 0° C until all of the water has frozen. An approximate 10% increase in volume results from the change from water to ice because the structure of ice is more open than that of water.

B. Freezing of Solutions

A schematic cooling curve of a salt solution is shown in Fig. 19.1. This cooling curve has lower freezing and homogeneous nucleation points compared with that of pure water, owing to the presence of a solute. Depression of the freezing point is approximately proportionate to the molal concentration of the solution. At equilibrium, osmolality of the partially frozen solution approximates to:

$$Me = dT/1.86,$$

where Me is the equilibrium osmolality, dT is the number of degrees below 0° C, and 1.86 is the molal freezing point depression constant for water. The relationship between solute concentrations and these two temperature points is illustrated in Fig. 19.2.

As water is frozen out during cooling, the solution becomes more concentrated and its freezing point is accordingly lowered. However, the linear relationship between molal concentration and freezing no longer holds true for concentrated solutions at low temperature (see Section II.D). The freezing point of the solution is thus along the curve "d-e" in Fig. 19.1. It is possible to predict how these various factors influence the rate of nucleation in biological systems. Knight (1986) has discussed this matter in some

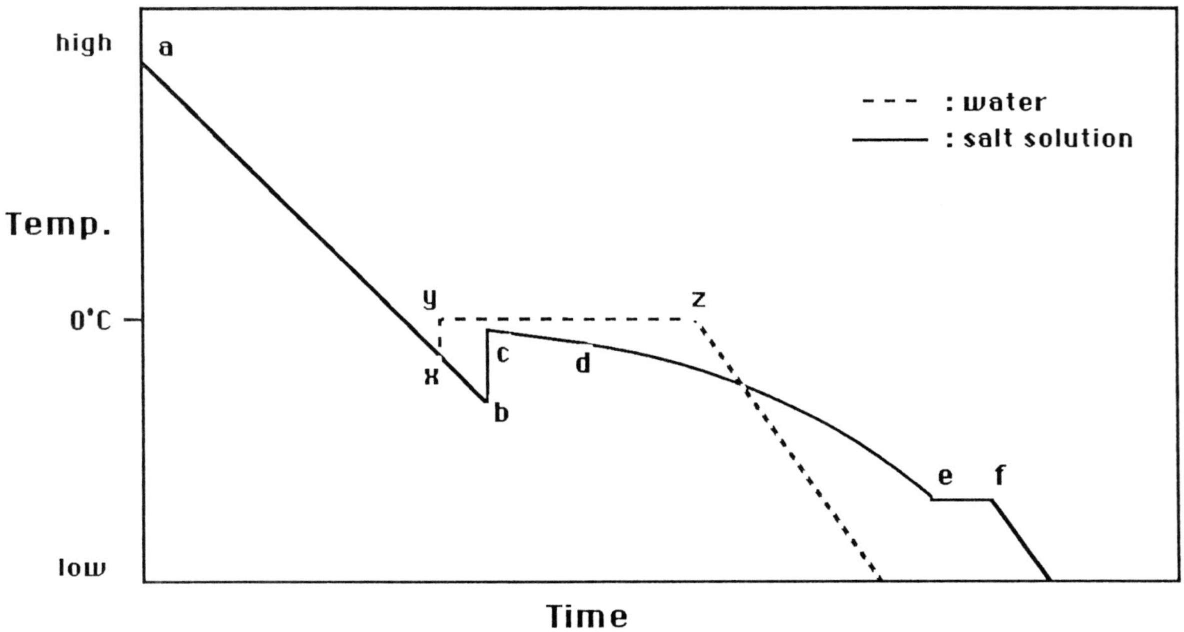

Fig. 19.1. Schematic cooling curves of pure water (broken line) and salt solution (solid line) with constant rate of heat removal.
For water: a-x, initial cooling plus supercooling. x-y, warming by release of heat of fusion. y-z, ice propagation.
For solution: a-b, initial cooling plus supercooling. b-c, warming by release of heat of fusion. c-d, linear depression of freezing point and ice formation. d-e, non-linear depression of freezing point and ice formation; e-f, ice and salt crystal propagation (eutectic freezing). f onward, cooling of the frozen solution.

detail.

C. Eutectic Freezing

Ultimately, during cooling, all the water that can be crystallized as ice is frozen out, leaving only the solute and its water of hydration. At this temperature (eutectic point), the concentration of the solution reaches its maximum. Further cooling results in the solidification of the remaining solution, a process termed eutectic freezing (see Fig. 19.1, curve "e-f"). There are two types of eutectic freezing: (1) formation of crystalline hydrate e.g. $NaCl.H_2O$ with, or without ice formation; (2) formation of a mixture of ice and crystalline salt. Table 19.1 lists the eutectic temperature and the type of eutectic freezing of some commonly used electrolytes. However, observed eutectic points may be 10-20° C lower than the theoretical values, a phenomenon eqivalent to supercooling.

Table 19.1. Eutectic freezing characteristics of some electrolyte solutions.

Electrolyte	Eutectic Point °C	Type of Solidification
$NaHCO_3$	-2.3	hydrate
KNO_3	-2.9	hydrate + ice
KCl	-11.1	salt + ice
NaCl	-21.8	hydrate + ice
CaCl	-54.9	hydrate

Observed eutectic point may be 10-20° C lower than the theoretical value.

D. Vitrification

When water molecules are deprived of the chance to crystallize during cooling, they solidify as a noncrystalline structure, a process termed vitrification. The temperature at which vitrification begins is termed the glass transformation temperature (e.g. -13° C for water). At a sufficiently low temperature, highly concentrated solutions become too viscous for ice growth, significant crystallization can be bypassed and the solution vitrifies when cooled down to the glass transformation temperature. Another cause of vitrification is an extremely high cooling rate which allows insufficient time for water molecules to crystallize.

The effects of increasing solute concentration and hydrostatic pressure is illustrated in Fig. 19.2 (from Fahy *et al.*, 1984).

presses the homogeneous nucleation temperature but in a more complex manner than with the melting point depression. In contrast, the glass transformation temperature rises with solute concentration. Hydrostatic pressure affects these temperature points in approximately the same manner as solute concentration; increasing the hydrostatic pressure lowers the homogeneous nucleation point and raises the glass transformation temperature.

E. Devitrification and Recrystallization

Homogeneous nucleation and ice growth take place in a vitrified system when it is warmed above

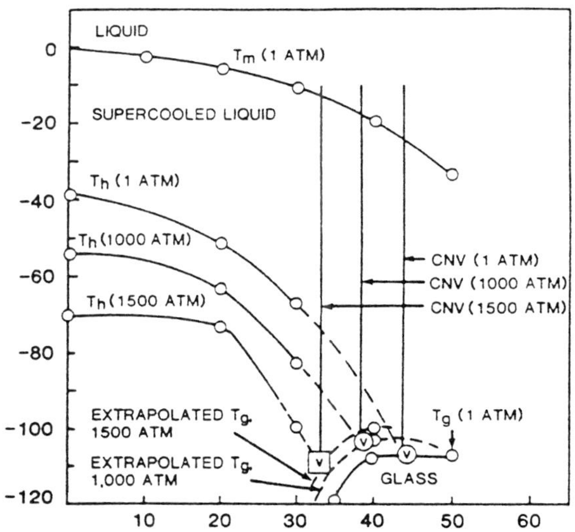

Fig. 19.2. Phase diagram of 1,2 propanediol-water showing both the effects of increasing solute concentration and of hydrostatic pressure on ice formation. Increasing solute concentration depresses the melting point (Tm) and the homogeneous nucleation temperature (Th), and raises the glass transformation temperature (Tg). Increasing the hydrostatic pressure also depresses the Th and raises the Tg. The lines labelled CNV indicate the circumstances necessary for vitrification of solutions at three different hydrostatic pressures. From Fahy *et al.*, 1984. *Cryobiology* **21**, 407-426. Fig. 4.

For an aqueous solution of 1,2-propanediol, the melting point is depressed by increasing the solute concentration. The melting point is not necessarily identical with the freezing point since the solution can supercool in the absence of heterogeneous nuclei down to the homogeneous nucleation temperature, at which point spontaneous freezing takes place. Freezing, therefore, can be initiated at or between the melting point and the homogeneous nucleation point. Increasing the solute concentration also de-

its glass transformation temperature, a process termed devitrification. When a frozen solution containing only small crystals is warmed slowly above its glass transformation temperature, many small crystals are converted into a few large ones, a process termed recrystallization. Smaller crystals have a higher effective vapour pressure due to topological factors, and, therefore, have a greater tendency to lose surface molecules than large ones. The large crystals will tend to grow at the expense of the

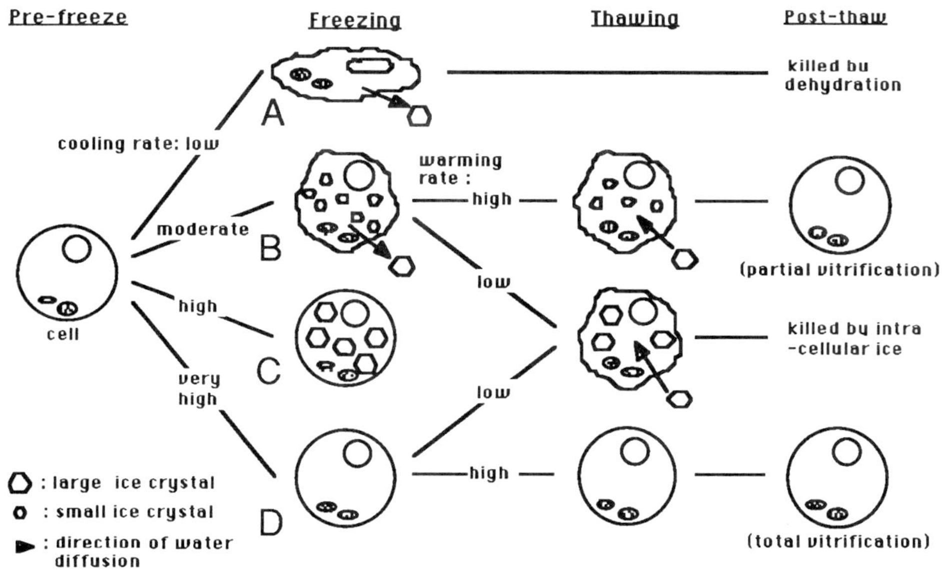

Fig. 19.3. Diagrammatic illustration of a cryopreservation model.
A. When the cooling rate is low, osmotic equilibrium is maintained. Freezable water leaves the cell. Cell death is caused by excessive dehydration (some organisms survive desiccation).
B. Partial vitrification. When the cooling rate is moderate (optimum), osmotic equilibrium is not maintained. Only part of the freezable water leaves the cell. The remaining water vitrifies or forms small ice crystals which are tolerable if thawing is fast enough to avoid recrystallization. The use of cryoprotectants will increase the rate of vitrification, thus optimizing post-thaw survival.
C. When freezing is rapid, little or no freezable water leaves the cell. Large intracellular ice crystals form; when thawing is slow, devitrification and recrystallization occur to form large ice, which is lethal.
D. Total vitrification: when the cooling rate if very high, water vitrifies without forming ice. Such a cooling rate is unfeasible. Vitrification solutions suppress ice formation but allow glass formation. (Original, based on Ashwood-Smith, 1986, and Meryman *et al.*, 1977).

smaller ones.

III. Principles of Cryopreservation

The basic principles of cryopreservation are illustrated in Fig. 19.3 and are discussed at different stages of a generalized cryopreservation procedure as follows.

A. Freezing

When cells are cooled in an aqueous solution, both cells and solution supercool to some extent; then heterogeneous nucleation takes place, usually in the extracellular solution. If it should occur intracellularly, the resultant nuclei will be isolated by plasma membranes from the other unfrozen cells. As water is frozen out, the extracellular solution becomes progressively more concentrated. If the cooling rate is low, there is sufficient time for the cells to lose enough water to remain in osmotic equilibrium with the concentrating extracellular solution. However, prolonged exposure to concentrated solution is generally lethal (see section IVE).

If the cooling rate is high, there is insufficient time for water to diffuse out of the cells to the ice crystal. The cells will equilibrate by intracellular freezing initiated either by homogeneous or heterogeneous nucleation. Intracellular freezing is considered fatal (see section IVC). If the cell is sufficiently small, not exceeding a few micra, as in microorganisms, so much water can be withdrawn during freezing that they are effectively desiccated (Luyet, 1938).

A balanced situation may exist which allows survival when the cooling rate is high enough to minimize the time of exposure to concentrated solution and yet is low enough to minimize the amount of intracellular ice below a damaging level. Certain chemicals can increase the dimension of this balance between the effects of intracellular ice and concentrated solution, thus improving survival. These chemicals, termed cryoprotectants, are discussed in section VA.

It is possible to predict such optimum cooling rates using mathematical models (Mazur, 1963; Mazur *et al.*, 1976; Ashwood-Smith, 1980; Toner and Cravalho, 1986) provided that values of factors required for the models are known. These factors are: (1) the viscosity of the intracellular solution; (2) the surface to volume ratio of the cell; (3) the rate of diffusion of water through the cell membrane; (4) the distance between the cell and the nearest ice crystal; (5) the viscosity of the extracellular solution; and (6) the permeability of the added cryoprotectant.

An extremely high cooling rate theoretically allows survival if the thawing rate is equally high. Under these conditions, intracellular nuclei may form but have insufficient time to grow or glass transformation may take place with little nucleus formation. However, these extreme rates of cooling and warming are not feasible for biological material. High freezing and thawing rates (510-760° C/min) are necessary for survival of the membranes of rat liver mitochondria (Tsvetkov, 1986).

B. Thawing

During thawing, the same physicochemical processes take place in reverse order. Theoretically, the thawing rate should be the same as the corresponding cooling rate. However, there is usually a minute amount of small intracellular ice present even when an optimum cooling rate is used. Recrystallization invariably occurs during thawing, forming lethal intracellular ice. A high warming rate is usually employed to minimize the degree of recrystallization. When thawing is rapid, there is insufficient time for the dehydrated cells to absorb the amount of water lost during freezing. Most cells and tissues appear to be tolerant to rapid thawing, mammalian embryos constituting a notable exception. Schneider (1986) suggests that rapidly changing solute gradients cause membrane damage to mouse embryos during rapid thawing.

C. Storage

Storage temperature should be -130° C or below for reasons given in section VB. A commonly used storage temperature is -196° C (the temperature of liquid nitrogen). This is simply because liquid nitrogen is a convenient storage medium. At these low temperatures, all biological molecules become motionless and, therefore, cannot participate in any biochemical reactions. Theoretically biological material could be kept indefinitely in this frozen state.

However, reactions at atomic level can still take place at the temperature of liquid nitrogen. Of these reactions, decomposition of cell nuclei caused by background radiation is the major concern in cryopreservation because it is potentially mutagenic. DNA damage caused by background radiation is cumulative since no DNA repair would take place during storage at -196° C (Ashwood-Smith, 1986). However, mouse embryos exposed to the equivalent of about 2,000 years of background radiation did not show observable deterioration, excepting those with mutant genes or chromosome anomalies (Glenister and Lyon, 1986). Under most circumstances, background radiation is of no importance in cryopreservation and the period of storage under these conditions is almost indefinite.

IV. Cryoinjuries

Fig. 19.4 lists the major known injuries associated with freezing and thawing (cryoinjuries) in relation to their temperature ranges during a generalized cryopreservation procedure. The temperature range 0° C - 40° C is most critical as most cryoinjuries take place over this range.

The causal relationships for cryoinjuries are illustrated in Fig. 19.5. There are two global causes:

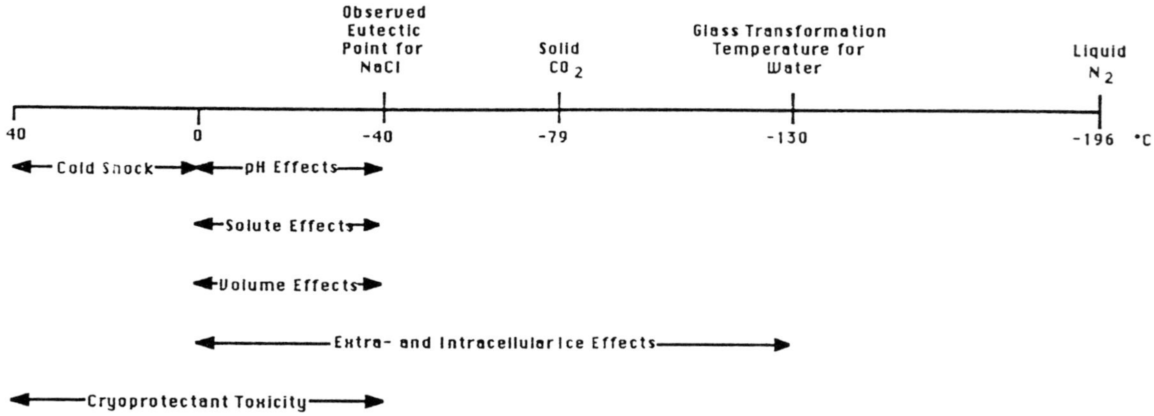

Fig. 19.4. Schematic diagram illustrating different cryoinjuries in relation to their temperature ranges.

heat removal and the addition of cryoprotectants. Individual causes of cryoinjuries are discussed below.

A. Cold Shock

This phenomenon has been recently reviewed, chiefly as a comprehensive bibliography, by Morris and Watson (1984). Cold shock is mainly caused by the change of membrane lipids from the liquid to the solid phase during initial freezing, particularly when this is done rapidly, at temperatures ranging from 10 to 16° C. Low temperature (down to -18° C without freezing) alone may not be sufficient to cause phase changes in certain membrane lipids; at that temperature ice crystals spontaneously form and effectively dehydrate the lipid of the bilayer; as a result of this solvent-induced effect the lipid undergoes a lamellar liquid crystal to gel phase transition (Caffrey, 1986). Cold shock haemolysis of human erythrocytes is supposedly caused by increase in membrane tension resulting from temperature reduction (Mcgrath and Thomas, 1986). Most biological material, so far as investigations have proceeded, appears not to be affected by cold shock excepting embryos of the pig (Polge *et al*., 1974) and most mammalian spermatozoa (Blackshaw, 1954). For mammalian spermatozoa, species differences in susceptibility to cold shock are correlated to differences in calcium uptake at temperature below 16° C, which in turn may be correlated with differences in cell morphology (Watson, 1986).

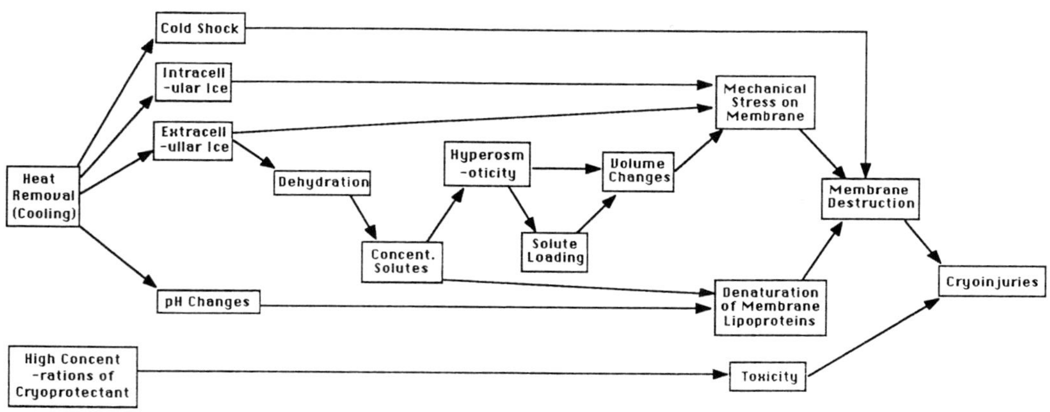

Fig. 19.5. Path diagram illustrating interrelationships of cryoinjuries during cooling and on addition of cryoprotectants.

B. pH Effect

The eutectic points of most biological salts are different and range approximately from 0 to -55° C (see Table 15.1). Freezing and thawing within this temperature invariably destroys the buffering capacity of these salts and, therefore, markly changes the pH of the biological solution. This pH fluctation can be caused by freezing (van den Berg and Rose, 1959)

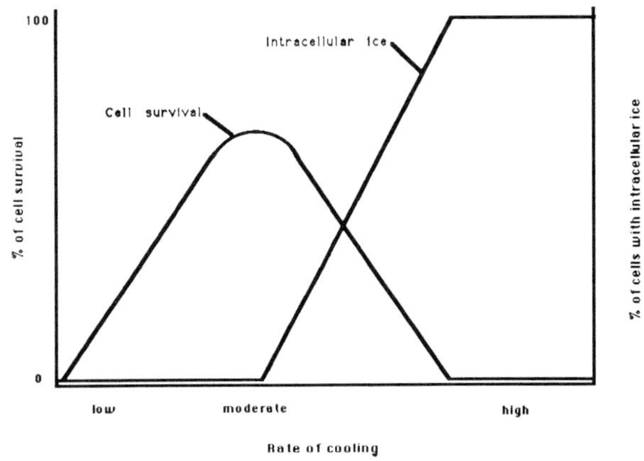

Fig. 19.6. Theoretical relationship between intracellular ice formation, cell survival and cooling rate. (From Ashwood-Smith, M.J. (1986). *Human Reproduction* **1**, 319-332. Fig. 5.

or, in some circumstances, such as presence of potassium phosphate, by addition of cryoprotectants (van den Berg and Soliman, 1969). Denaturation of proteins invariably takes place when their pH limits are exceeded.

C. Intracellular Ice Effects

The likelihood of the formation of intracellular ice increases with the cooling rate (see Fig. 19.3) and the effects of intracellular ice usually appear during thawing rather than freezing. The degree of injury is in proportion to the size of the ice crystals. Small ice crystals produced during rapid freezing may not be detrimental, but the recrystallization of these small crystals, inside and outside the cell, results in large damaging ice which causes mechanical destruction of membrane structure (Fujikawa, 1978). The degree of membrane lesions has three recognizable levels (Tsvetkow *et al.*, 1986). The theoretical relationship betweem intracellular ice formation, cell survival and cooling rate is illustrated in Fig. 19.6 (after Ashwood-Smith, 1986).

D. Extracellular Ice Effect

Membrane injury (or fungal hyphae) caused by mechanical stress resulting from extracellular ice formation has been demonstrated by Fujikawa and Miura (1986) using a very low cooling rate [about 0.67° C/min]. It is possible that cell injury is caused by physical forces which arise when the expanding ice field places contraints on the shapes that can be assumed by the shrinking cells. Extracellular ice is usually considered not to be detrimental to most biological material frozen at commonly used cooling rates. Meryman *et al.* (1977) proposed that injury from slow freezing is due to membrane stress produced by osmotic dehydration resulting from extracellular ice formation (see next section).

E. Solute Effect

As water is frozen out during freezing, both extra- and intracellular solutions become progressively more concentrated. Both slow and rapid freezing will produce the same final concentration. However, exposure to high solute concentrations prolonged by slow freezing is detrimental. Membrane injuries associated with concentrated solution are either the result of the denaturing effects of the salts on the membrane lipoproteins (Lovelock, 1957) or result from direct osmotic effects of the concentrated solution on the cell membrane (Meryman, 1968; Meryman *et al.*, 1977).

F. Volume Effects

Occurrence of cryoinjury through volume effects is not yet fully established. In the proposal by Meryman *et al.* (1977) of "solution effects", membrane stress is caused either by forces induced by the volume reduction or by some direct osmotic effects of the concentrated solution (Meryman, 1968). The "solute effect" proposed in this review, as implied by its name, includes only effects which are immediately caused by concentrated solute, hence different from the "solution effects" of Meryman *et al.* (1977).

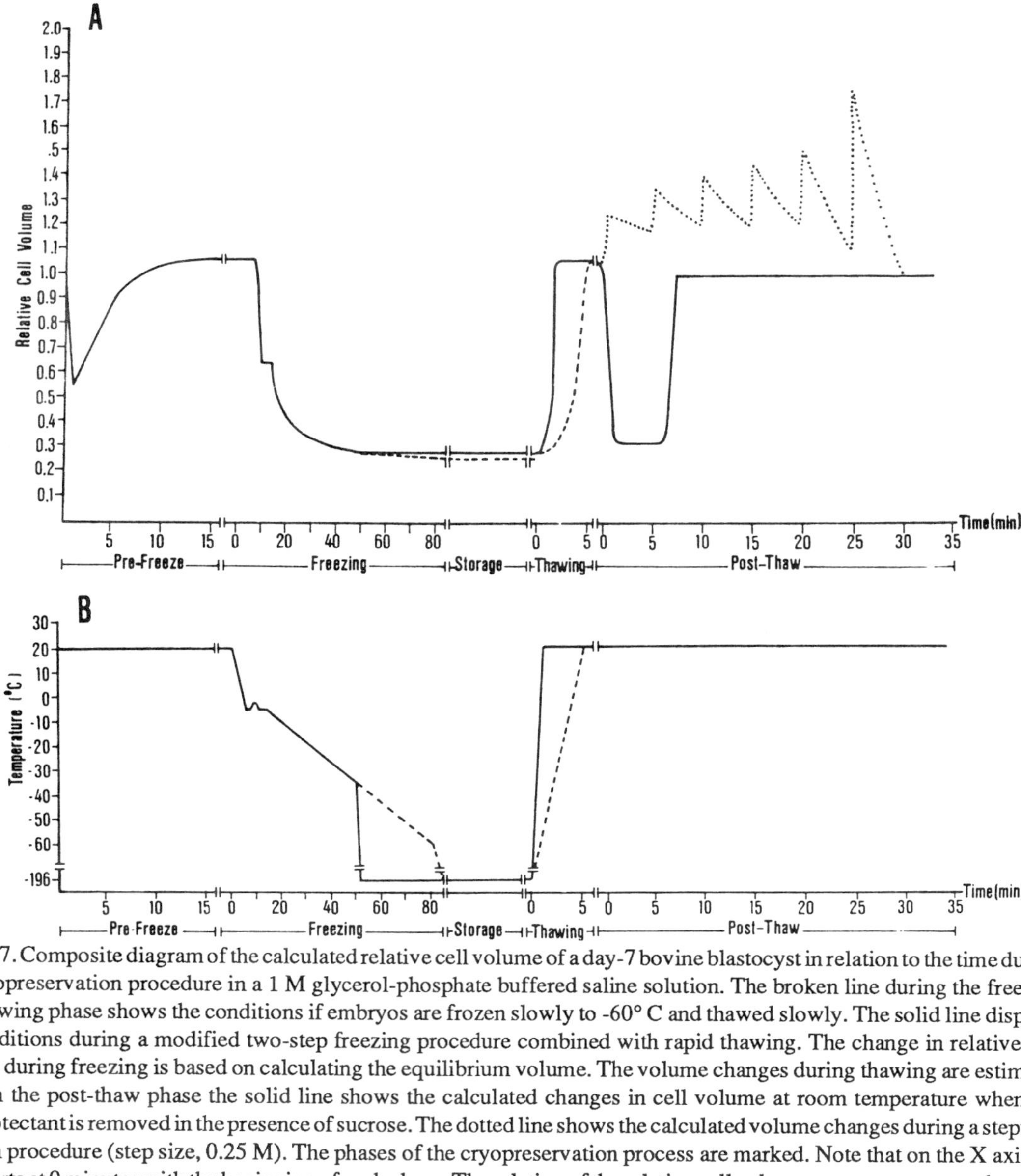

Fig. 19.7. Composite diagram of the calculated relative cell volume of a day-7 bovine blastocyst in relation to the time during the cryopreservation procedure in a 1 M glycerol-phosphate buffered saline solution. The broken line during the freezing and thawing phase shows the conditions if embryos are frozen slowly to -60° C and thawed slowly. The solid line displays the conditions during a modified two-step freezing procedure combined with rapid thawing. The change in relative cell volume during freezing is based on calculating the equilibrium volume. The volume changes during thawing are estimates only. In the post-thaw phase the solid line shows the calculated changes in cell volume at room temperature when the cryoprotectant is removed in the presence of sucrose. The dotted line shows the calculated volume changes during a stepwise dilution procedure (step size, 0.25 M). The phases of the cryopreservation process are marked. Note that on the X axis the time starts at 0 minutes with the beginning of each phase. The relation of the relative cell volume to temperature can be made by comparison with (B) which shows temperature changes in relation to time during cryopreservation of mammalian embryos. From Schneider, U. (1986). *Journal of in Vitro Fertilization and Embryo Transfer* **3**, 3-9. Fig. 1.

It is appropriate to define the volumetric cause of their "solution effects" separately as supporting evidence has become availabe recently. The term "volume effect" is proposed here to explain cryoinjury caused by cellular volume changes (see Fig. 19.5).

Cell volume fluctuation during freezing has been demonstrated for bovine embryos (Schneider, 1986) and in plant cells (Williams, 1974). The relative cell volume of a day-7 bovine blastocyst in relation to the time and the temperature during a cryopreservation procedure is shown in Fig. 19.7. There are limits of tolerance to cell volume changes. Cell volume reduction to 40-50% have been observed to be lethal in plant cells (Williams, 1974). Mammalian eggs and early embryos can tolerate a 50% decrease or a 200% increase in cell volume for 30 minutes without losing their developmental capacity, but their tolerance is markedly reduced just after thawing (Renard, 1986).

Shrinkage of the cell during freezing is caused by hyperosmoticity resulting from extracellular ice formation. The membrane of the shrunken cell is compressed at the sites of convexity and stretched at the sites of concavity. In addition to such osmotic stress, there are three major factors in freezing injury: chilling, cold shock and dilution shock, as demonstrated for human polymorphonucleocytes (Takahashi *et al.*, 1985). During rapid thawing, the loaded solutes cannot leave the cells fast enough and cause water influx and swelling. Cell volume fluctuation can also be caused by addition and removal of cryoprotectants but it can be minimized when this is done in a stepwise manner (Schneider and Mazur, 1984; see also Fig. 19.7).

Swelling invariably stretches the cell membrane more intensively than shrinkage. Stretching of the cell membrane exerts tension on membrane pores and increases the diameter of the pores, allowing leakage of macromolecules (Pushkar *et al.*, 1980). Increase in permeability, loss of membrane material and irreversibe membrane disintegration all result from excesssive cell volume reduction (Williams, 1974).

G. Cryoprotectant Toxicity

Cryoprotectants can supress most cryoinjuries (see Section V) but, when used at higher (more effective) concentrations, most of them become toxic to biological materials. Fig. 19.8 illustrates how the use of higher concentrations of cryoprotectant results in increased injury of fish spermatozoa rather than in increased cryoprotection. Toxicity of most commonly used cryoprotectants has been reported.

Dimethylsuphoxide (DMSO) inhibits catalase and peroxide activity (Rammler, 1967); ethylene glycol decreases the polarity of the aqueous phase and changes the partition of hydrophobic molecules between the cell membrane and the external phase; the resultant dehydration of the phospholipid bilayer

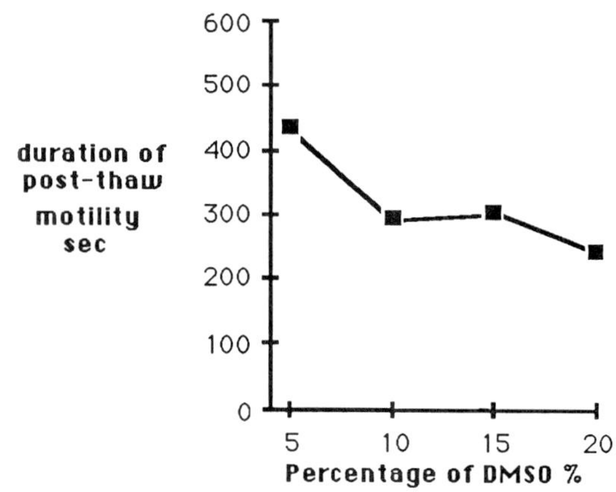

Fig. 19.8. Toxicity from high concentrations of Dimethyl sulphoxide (DMSO). Data from Leung, 1987a.

causes membrane damage (Arnold *et al.*, 1983); DMSO and polyethylene glycol, in combination, are exceedingly dangerous, inducing membrane fusion (Ashwood-Smith, 1986); toxicity of methanol causes failure of hearts to recover after thawing if frozen to below 30° C (Rapatz 1973). Toxicity from high concentrations of DMSO, however, can be suppressed by addition of amides. Comparison of toxicity between cryoprotectants is discussed in section V.A.1 (seealso Fahy, 1982).

V. Cryoprotection

The addition of some chemicals (cryoprotectants) can minimize cell damage associated with ice formation or, when used at high concentrations, will suppress any ice formation. Such a process is termed cryoprotection.

A. Cryoprotectants

Some commonly used cryoprotectants are listed in Table 19.2. The chemical and physical properties of most cryoprotectants are listed by Nash (1966). Of these properties, water solubility and low toxicity are the most important. Cryoprotectants must be highly water soluble in order to alter the physicochemical properties of water during freezing. Their toxicity, if any, must be less than their protective ability, or they would offer damage rather than protection (see section IVG). Cryoprotectants can be divided into two

Table 19.2. Commonly used cryoprotectants.

Chemical	Molecular Formula	Molecular Weight	Density
Dimethylsulphoxide	$CH_3SO\ CH_3$	78.13	1.10
Glycerol	$CH_2OH\ CHOH\ CH_2OH$	92.10	1.47
Methanol	CHOH	32.04	0.79
1,2-Propanediol	$CH_3CHOH\ CH_2OH$	76.09	1.04

groups: those permeable to the cell membrane (strictly, to which the membrane is permeable) , and those not.

1. Permeating Cryoprotectants

Permeating cryoprotectants serve (1) to reduce the rate of diffusion of water from the cell to the ice crystal; (2) to reduce the amount of cell volume change / salt concentration colligatively; (3) to lower the homogeneous nucleation temperature; (4) to reduce the rate of ice crystal growth; and (5) to raise the glass transformation temperature. Choice of the cryoprotectant is normally on the basis of toxicity and permeability to the cells, and solubility in water during freezing.

Commonly used permeating cryoprotectants are DMSO, glycerol, methanol and 1,2-propanediol (Table 19.2). Of these compounds, glycerol is the least toxic to most biological materials but also the least permeable to the cell membrane and hence takes longer to equilibrate with glycerol osmolality. Nevertheless it was found to give equal or superior protection to the deleterious effects of freezing and thawing on mouse embryos when compared with DMSO (Rall *et al*, 1984). The large difference between the permeabilities of water and glycerol causes volume effects during introduction and removal of this cryoprotectant (Schneider and Mazur, 1984). Methanol, on the other hand, is highly permeable to cell membranes but is generally considered the most toxic, excepting a few cases of remarkably low toxicity (e.g. mouse embryos, Rall *et al.*, 1984; fish spermatozoa, Harvey, 1983; isolated mammalian cells, Ashwood-Smith and Lough, 1975; erythrocytes, Meryman, 1968). DMSO is fairly permeable to membranes but its toxicity, intermediate between glycerol and methanol, can be minimized by reducing the temperature; it is also, like glycerol, a radioprotective agent (Ashwood-Smith, 1967). Experiments using radiolabeled cryoprotectants show the degree of permeation of methanol, dimethyl sulphoxide [DMSO] and glycerol is inversely proportional to the molecular weight of the compounds; glycerol does not penetrate the egg, while methanol, which penetrates with the greatest rapidity, achieves no more than 23% of the expected equilibrium concentration after 2 h exposure at 0°C (Harvey and Ashwood-Smith, 1982).

Unlike glycerol, the permeability of DMSO, is not markedly affected by low temperature (Whittingham, 1977). Probably for these reasons, DMSO is the most widely used permeating cryoprotectant. A relatively recent addition to the list of cryoprotectants, 1,2-propanediol (propylene glycol), is possibly better than DMSO (Boutron and Kaufman, 1979; Renard and Babinet, 1984; Renard *et al.*, 1984). Its success is due, at least partly, to the fact that it has a higher glass formation tendency than glycerol or DMSO as indicated by scanning differentiation calorimetry studies (Renard and Babinet, 1984). The mechanism of action of permeating cryoptotectants is discussed in Section VB.

The use of vacuum equilibration to facilitate entry of cryprotectants into eggs and sperm (Jamieson and Marshall, 1990) is discussed in Chapter 20.

2. Nonpermeating Cryoprotectants

Non-permeating cryoprotectants include sugars (e.g. sucrose, glucose), polymers (e.g. dextran, hydroxyethyl starch, polyvinylpyrrolidone, PVP) and proteins, (e.g. egg yolk, serum, skim milk, and the antifreeze proteins found in polar fish and in freeze-resistant and freeze-tolerant insects) (Ashwood-Smith *et al.*, 1972; Connor and Ashwood-Smith, 1973; Ashwood-Smith, 1975; Knight and Duman, 1986). These compounds, since they do not penetrate into cells, should not be able to promote any colliga-

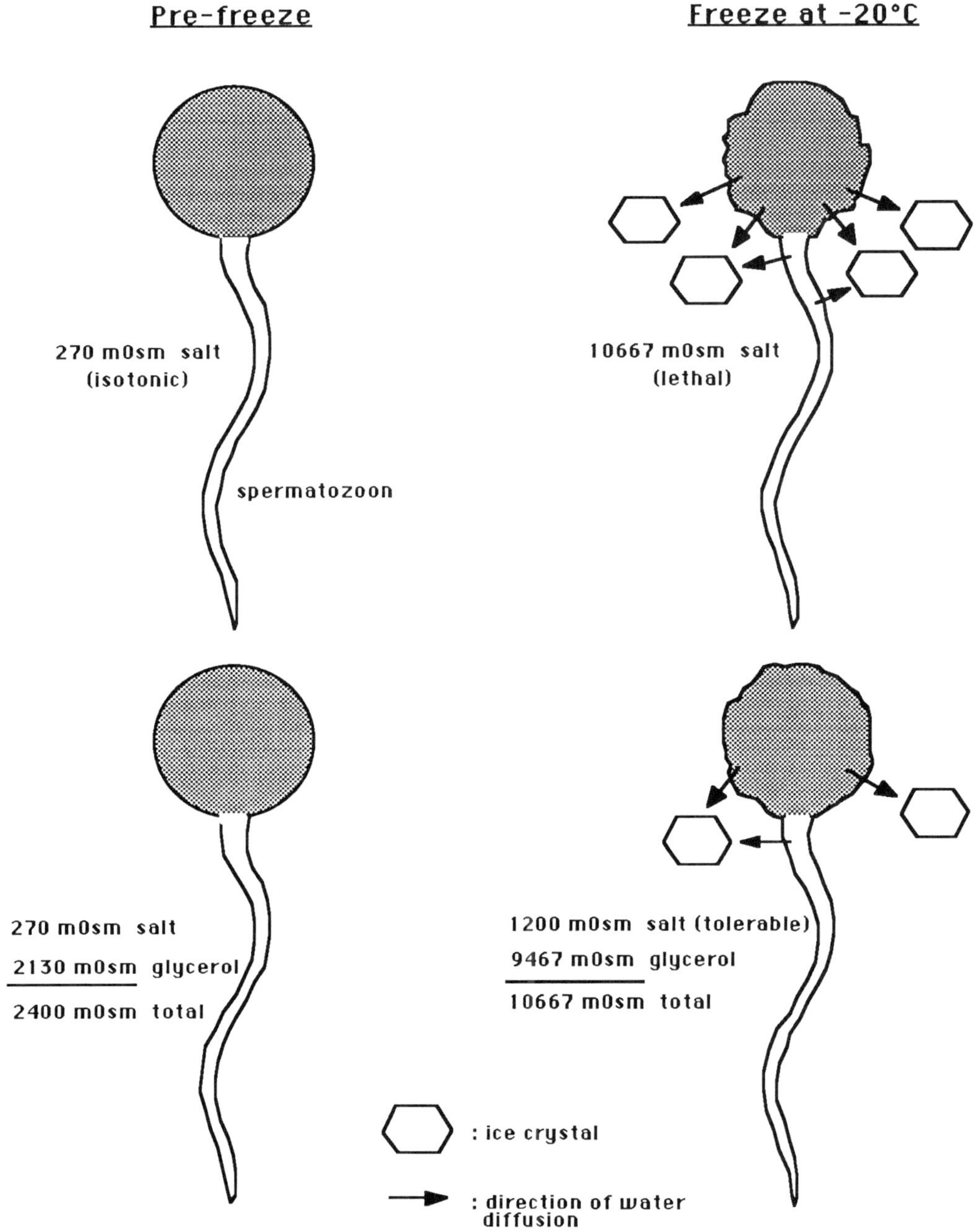

Fig. 19.9. Colligative cryoprotection. This schematic diagram illustrates how a permeating cryoprotectant reduces the salt concentration attained following freezing. The equilibrium osmolality at -20° C is 10667 mOsm. In the absence of a cryoprotectant, salt concentrations which reach this equilibrium level after freezing at -20° C are fatal to spermatozoa. With glycerol added, the salt concentration at -20° C is tolerable. Data from Leung, 1987a; based on Lovelock, J.E. (1953). *Biochimica et Biophysica Acta* **11**, 28-36.

tive protection. Their cryoprotective ability, usually in conjunction with a permeating cryoprotectant, is related to their ability to depress the freezing point and to raise the glass transformation temperature of the extracellular solution.

Specific lipids, with potential for lowering membrane phase transition temperatures, are used to minimize membrane damage during initial cooling (i.e. cold shock) and during freezing (Graham and Foote, 1987).

B. Partial vitrification

The basic mechanism of permeating cryoprotectants for partial vitrification (see below) is illustrated in Fig. 19.9 (based on Lovelock, 1953, with an example from Leung, 1987a).

The limit of tolerance to salt concentration of the spermatozoa of *Lates calcarifer* is estimated to be about four times isotonic, i.e. 1200 mOsm, which is approximately equivalent to that of seawater which the spermatozoa normally encounter during spawing or artificial insemination (M. Mackinnon, personal communication, 1986). In other words, the spermatozoa can tolerate a dehydration leading to a reduction of approximately three quarters of the cell water. The optimum concentration of glycerol for cryopreserving viable *Lates* spermatozoa using an initial cooling rate of 31° C/min (Leung, 1987b and unpublished) is 2130 mOsm. Both the extra- and intracellular solutions are presumably supplemented by this amount of glycerol. Using the equation from Section IIB (Me = dT/1.86), the equlibrium osmolality at the temperature -20° C is estimated to be 10,667 mOsm. If the extracellular solution were the presumably isotonic fish ringer (270 mOsm), the salt concentration produced at -20° C would be forty times isotonic, many times more than the limit of tolerance. However, because of the elevation in initial osmolality contributed by the glycerol, only approximately three quarters of the water needs to be frozen out to acquire the equilibrium osmolality, resulting in a degree of cell dehydration that is easily tolerated.

A slow cooling rate must be used to allow sufficient time to reach equilibrium osmolality. Further decrease in temperature below -20° C will increase the equilibrium osmolality beyond the tolerated level, resulting in cryoinjuries. By more rapid freezing (e.g. 31° C/min or the "moderate cooling rate" as defined in Fig. 19.3), a much lower, stablizing temperature can be reached before the spermatozoa lose three quarters of the cell water. The reasons for this are: (1) rapid freezing allows insufficient time for water to leave the cells; and (2) the rate of diffusion of water out of the cells is reduced by increased viscosity of the added cryoprotectant. There exists the risk that the spermatozoa may equilibrate by forming intracellular ice. However, the increasing intracellular cryoprotectant concentration and viscosity resulting from progressive dehydration will minimize the amount and the size of intracellular ice below a damaging level through (1) lowering the homogeneous nucleation temperature; (2) raising the glass transformation temperature; and (3) reducing the rate of ice growth.

When the glass transformation temperature of the remaining intracellular solution is reached, the remaining solution vitrifies, a process termed partial vitrification. It is self-evident that this noncrystalline water fraction determines the extend of dehydration and, therefore, the survival of the cells. Below this glass transformation temperature, the metastable glass phase is stabilized and ice formation no longer takes place. The storage temperatures, therefore, should be at or below the corresponding glass transformation temperature, taken conservatively as -130° C for pure water. Under these conditions, rapid thawing should be used to avoid recrystallization of intracellular ice and additional cell dehydration by extracellular ice.

C. Total Vitrification

When the cryoprotectant concentration is high enough to bring the homogeneous nucleation temperature and the glass transformation temperature into coincidence (see Fig. 19.2), the solution will transform from a supercooled to a glass state without any formationof ice, a process termed total vitrification. Total vitrification of biological material can certainly avoid damage associated with ice and solutes, and is particularly important in cryopreserving tissues and organs as they are easily damaged even by extracellular ice (Fahy *et al.*, 1984).

Total vitrification is a new approach to cryopreservation though its conceptual basis was laid by

Luyet and Gehenio in 1940. It has been used successfully as a means of cryopreserving mouse embryos (Rall and Fahy, 1985). The major problem of total vitrification of biological material is the toxicity associated with the highly concentrated cryoprotectants (Fahy *et al.*, 1984). The addition of some chemicals can minimize this toxicity (Fahy, 1982). For the same glass-forming tendency, some mixtures of cryoprotectants may have a lower total concentration and, therefore, a lower toxicity than other mixtures. Under these considerations, three vitrification solutions, based on DMSO, glycerol, and propylene glycol, have been developed (Reid and Rall, 1986). The molecular and biological aspects of these vitrification solutions may be consulted in Fahy *et al.* (1986).

VI. Cryopreservation in Practice

The guidelines for designing a procedure for cryopreserving cells with a partial or total vitrification approach will now be briefly discussed.

A. Partial Vitrification

1. Choice of Cryoprotectant

For most applications, DMSO is preferred because of its rapid penetration into cells and relatively low and somewhat controllable toxicity at low temperatures. When DMSO toxicity remains a problem, other permeating cryoprotectants should be tried. The biological material may be less susceptible to 1,2-propanediol or to methanol toxicity. If toxicity still remains a problem, glycerol will be the preferred cryoprotectant. However, special care must be taken to ensure sufficient equilibration time for glycerol. Stepwise addition and removal of glycerol at high concentrations is also recommended.

A non-permeating cryoprotectant is used to supplement extracellular viscosity and, therefore, to reduce (1) the optimum cooling rate to a manageable level or (2) the concentration of permeating cryoprotectant used to a nontoxic level.

The concentrations of cryoprotectants to be used will depend on their toxicity and the hyperosmotic-tolerance of the material, and can be determined by experimenting with a variety of concentrations. Caution must be taken with regard to the highly interactive nature of the factors discussed in section VB.

2. Selection of Cooling Rate

Theoretically, an optimum cooling rate can be estimated using the mathematical models from section III.A. However, it is difficult to determine the values of all the parameters requried for the models. The optimum cooling rate can be easily determined empirically by experimenting with various cooling rates. A variety of cooling rates can be easily achieved by varying the temperature of the cooling medium and/or the type of sample container (e.g. straws, ministraws, plastic or glass vials). Two step cooling, slow to -40°C and rapid to -196° C, has been found advantageous for cryopreservation of mouse embryos (Wood and Farrant, 1980).

3. Selection of Warming Rate

For most applications, rapid thawing (e.g. warming at room temperature or in a water bath at 35-40° C) is prefered as recrystallization is minimized. In cases in which the material cannot tolerate rapid thawing (e.g. mouse embryos, Whittingham *et al.*, 1972), reducing the warming rate is the only compromise.

4. Storage Procedure

Storage in liquid nitrogen (-196° C) or its vapour (-120° C) is the general practice for reasons given in section VB. It is essential to maintain the storage temperature by the use of alarm systems or scheduled inspections on the level of the liquid nitrogen.

B. Total Vitrification

1. Choice of Vitrification Solution

These are generally balanced salt solutions containing various permeating cryoprotectants at high concentrations. A variety of these solutions is now being developed to improve "non-toxicity". VS1, VS2 and VS3 are based on DMSO, glycerol and propylene glycol (Reid and Rall, 1986). Polyalcohols (e.g. 1,2-propanediol, 1,3-butanediol) have been recently recommended as components of vitrification solutions (Boutron and Mehl, 1986). The choice of vitrification solution depends mainly on its toxicity to the material.

2. Selection of Cooling and Warming Rates

Cooling and warming rates are not critical in total vitrification but are usually high so as to minimize the time of exposure to the vitrification solution and, if any, the chance of ice formation.

3. Storage Procedure

The storage procedure is the same as for partial vitrification (see section VIA.4).

Chapter 20

LIVE PRESERVATION OF FISH GAMETES

by L.K.-P. Leung and B.G.M. Jamieson

I. Introduction

Successful cryopreservation of fish gametes is well established for sperm cells from many species, but a technique for true cryopreservation of fish ova has proved elusive (see reviews by Harvey and Ashwood-Smith, 1982; Schmehl and Graham, 1986). Cryopreservation or long term storage of fish eggs or embryos would be beneficial to the fish aquaculture industry. It would provide a method of retaining specific genetic lines of fish without the expense of maintaining broodstock populations and would provide a secondary source of a genetic line in case of broodstock loss (Schmehl and Graham, 1986) or would allow the preservation of endangered genetic lines in wild populations.

Reports of successful freezing of viable Rainbow Trout eggs to -55° C (Zell, 1978) or to -20° C (Erdahl and Graham (1980) have subsequently been interpreted as supercooling (Harvey and Ashwood-Smith, 1982). However, a procedure for viable freezing of fish eggs to -196° C has been introduced, pending further trials, by Jamieson and Marshall (in preparation) and will be discussed below.

Most of the fish species which have been the subject of gamete preservation are commercially important species reared in hatcheries (see list below and review by Scott and Baynes, 1980). For wild fish, gamete cryopreservation has been recommended to "save" endangered species (Stoss and Donaldson, 1983). The major benefits which devolve from fish gamete cryopreservation include 1, economy in maintenance of stocks; 2, potential increased protection of stocks from disease; 3, continuous supply of gametes for optimum ultilization of hatchery facilities or for experimentation; 4, ease of transport of stocks between hatcheries; 5, potentially greater efficiency of selective breeding; and 6, a potential greater range of stocks.

The following species have been the subject of cryopreservation studies which have contributed to the present account. It will be seen that most of the investigations have been directed to cryopreservation of salmonid sperm, with the Rainbow Trout, *Salmo gairdneri*, as a major species.

CLUPEIFORMES

Clupeidae

Clupea harengus (Atlantic Herring), Blaxter, 1953 (Dry ice); Rosenthal *et al.*, 1978.

Clupea pallasii (Pacific Herring), Rosenthal *et al.*, 1978.

ESOCIFORMES

Esocidae

Esox lucius (Pike), de Montalembert *et al.*, 1978; Stein and Bayrle, 1978; Koldras and Moczarski, 1983.

SALMONIFORMES

Salmonidae

Various salmonids, in four useful reviews: Horton and Ott, 1976; Fredrich, 1984; Munkittrick and Moccia, 1984; Stein and Bayrle, 1985; also Terner, 1986.

Salmo gairdneri (Rainbow Trout), Graybill and Horton, 1969; Horton and Ott, 1976; Billard, 1978b, 1983a, b; Büyükhatipoglu and Holtz, 1978; Stein and Bayrle, 1978; Stoss, Büyükhatipoglu and Holtz, 1978; Erdahl and Graham, 1980; Kurokura and Hirano, 1980; Legendre and Billard, 1980; Van der Horst *et al.*, 1980; Craig *et al.*, 1983 (4° C); Stoss and Donaldson, 1983; Stoss and Holtz, 1981a,b, 1983 a,b,c; Baynes and Scott, 1987. (Further references in Scott and Baynes, 1980).

Salmo salar (Atlantic salmon), Hoyle and Idler, 1968; Truscott and Idler, 1969; Horton and Ott, 1976; Mounib, 1978; Stoss and Refstie, 1983, (-196° C); Yoo *et al.*, 1987 (-80° C); Alderson and MacNeil, 1984.
Salmo trutta fario (Brown Trout), Stein and Bayrle, 1978; Billard, 1983a; Erdahl *et al.*, 1984.
Salmo trutta (Sea Trout), Stein, 1979; Stoss and Refstie, 1983. (Further references in Scott and Baynes, 1980).

Salvelinus fontinalis (Brook Trout), Stein and Bayrle, 1978; Erdahl and Graham, 1980; Erdahl *et al.*, 1984. (Further references in Scott and Baynes, 1980).

Hucho hucho (Danube Salmon), Stein and Bayrle, 1978.

Thymallus thymallus (Grayling), Stein and Bayrle, 1978.

Onchorhynchus gorbuscha (Pink Salmon), Ott, 1975; Horton and Ott, 1976. (Further references in Scott and Baynes, 1980).
Oncorhynchus keta (Chum Salmon), Horton and Ott, 1976; Jensen and Alderdice, 1984 (3-15° C only).
Onchorhynchus nerka (Sockeye Salmon), Horton and Ott, 1976
Onchorhynchus kisutch (Coho Salmon), Horton and Ott, 1976; Stoss and Donaldson, 1983.
Onchorhynchus tschawytscha (Chinook Salmon), Horton and Ott, 1976; Erdahl *et al.*, 1984. (Further references in Scott and Baynes, 1980).

Coregonidae
Coregonus muksun (Whitefish), Piironen and Hyvärinen, 1983a.

SILURIIFORMES

Clariidae
Clarias gariepinus (Sharptooth Catfish), Steyn *et al.*, 1985; Steyn and Van Vuren, 1987.

Ictaluridae
Ictalurus puncatus (Channel Catfish), Guest *et al.*, 1976.

Pangasiidae
Pangasius sutchi (Catfish), Withler, 1980, 1982.

CYPRINIFORMES

Cyprinidae
Cyprinus carpio (Carp), Pavlovici and Vlad, 1976; Stein and Bayrle, 1978; Withler, 1980, 1982; Kurokura *et al.*, 1984; Koldras and Bieniarz, 1987.
Aristichthys nobilis (Bighead Carp), Withler, 1980, 1982.
Brachydanio rerio (Zebra Fish), Harvey *et al.*, 1982.
Ctenophangodon idella (Grass Carp), Hulata and Rothbard, 1979 (chilling to 5° only); Withler, 1980, 1982; Durbin *et al.*, 1982; Drokin *et al.*, 1985.
Labeo rohita, *Cirrhinus mrigala* and *Catla catla* (all carp species), Kumar, 1989.
Labeo rohita (Indian Carp), Withler, 1980, 1982.
Puntius gonionotus (Thawes Carp), Withler, 1980, 1982.

Characidae
Leporinus silvestrii (Piau fish), Cóser *et al.*, 1987.

PERCIFORMES

Percidae
Stizostedion vitreum vitreum (Walleye), Moore,1987.

Centropomidae
Lates calcarifer (Barramundi), Leung, 1987b.

Cichlidae
Tilapia spp. including *Oreochromis aureus*, *O. mossambicus*, *O. niloticus*, *Tilapia zillii*, *O. niloticus* x *O. aureus* hybrid and *Oreochromis* sp. (Red Tilapia), Chao *et al.*, 1987; *Sarotherodon mossambicus*, Harvey, 1983. *O. niloticus*, *O. aureus* and *O. mossambicus*, Rana and McAndrew, 1989.

Gadidae
Gadus morhua (Cod), Mounib *et al.*, 1968; Mounib, 1978.

Eleotridae

Hypseleotris galii (Fire-tail Gudgeon), Marshall and Jamieson, unpublished.

Mugilidae
Mugil cephalus (Grey Mullet), Chao *et al.*, 1975.

Percicthyidae
Dicentrarchus labrax (Sea Bass), Billard, 1978b).
Macquaria novemaculeata (Australian Bass), Jamieson and Marshall, 1989.

Scombridae
Thunnus thynnus (Bluefin Tuna), Doi *et al.*, 1982.

Serranidae
Epinephelus tauvina (Grouper), Withler and Lim, 1982).
Dicentrarchus labrax (Sea Bass), Billard, 1984.
Morone saxatilis (Striped Bass), Kerby, 1983; Kerby *et al.*, 1985.

Sparidae
Sparus auratus (Sea Bream), Billard, 1978b; 1984.
Acanthopagrus schlegeli (Black Porgy), Sin and Hirano, 1972; Chao *et al.*, 1986.

Suborder Chanoidei

Chanidae
Chanos chanos (Milkfish), Hara *et al.*, 1982.

PLEURONECTIFORMES

Pleuronectidae
Pleuronectes platessa (Plaice), Pullin, 1972.
Hippoglossus hippoglossus (Atlantic Halibut), Bolla, 1987.

The principles of biological cryopreservation have been discussed in Chapter 19. In the present chapter, current knowledge of fish gamete preservation is briefly reviewed. Some biological aspects of fish gametes relevant to cryopreservation are discussed in the next section, followed by a review of short and long term preservation of fish gametes. A useful, though not exhaustive, review of articles on cryopreservation of fish, poultry and mammalian sperm from 1979-1984 is given by Graham *et al.* (1984).

II. Relevant Aspects of Fish Gamete Biology

A knowledge of the biology, including ultrastructure, of fish gametes, is important for the understanding and design of preservation procedures for the gametes. The morphology, metabolism and motility of spermatozoa in relation to cryopreservation are discussed by Hoar *et al.* (1983).

A. Morphology of Spermatozoa

The ultrastructure of the spermatozoa of fish, including teleosts, has been reviewed in detail in Chapters 5-18 and does not require further treatment here. It will be seen from the list (above) of fish species which have been subjects for sperm cryopreservation that these are all externally fertilizing teleosts with anacrosomal aquasperm.

B. Metabolism of Spermatozoa

Spermatozoa of fish with internal fertilization, exemplified by the Shiner Surfperch, *Cymatogaster aggregata* (Embiotocidae), and the Guppy, *Poecilia reticulata* (Poeciliidae), are capable of glycolysis, an anaerobic process. It is likely that sperm metabolism is supported by ovarian sugars (Gardiner, 1978c). In contrast, sperm from externally fertilizing, oviparous fish show limited glycolytic activity and depend mainly on oxidative metabolism, oxidizing endogenous and exogenous substrates. They are not specifically adapted for utilization of exogenous energy sources, although they appear to be capable, to a limited extent, of doing so when provided with external nourishment (Mounib, 1967; Harvey and Kelley, 1984a). Sperm in 1 ml of *Oncorhynchus keta* semen at 10° C consume 21 mm^3 of oxygen per hour (Okada *et al.*, 1956).

After release into the spawning environment, spermatozoa cease swimming when the cellular reserves of energy have been exhausted. In viviparous fish, spermatozoa can convert exogenous sugars to lactic acid and use the energy to prolong motility (Gardiner, 1978c).

C. Motility of Spermatozoa

The motile apparatus is the axoneme, usually of

the 9+2 microtubular pattern, of the flagellum or tail (see Gibbons, 1981). In external spawners there is generally a positive relationship between motility and fertility, and motility is often used as an indicator of fertility in artificial insemination because it is easy to assess microscopically. However, spermatozoa of the Herring, *Clupea pallasii*, are almost immotile in seawater and move actively only if they come in contact with activating substances found near the micropyle of the egg (Yanagimachi, 1957). Motility is thus not obligatory for fertility.

D. Induction of Sperm Motility

For a review of factors influencing sperm motility see Morisawa (1985). Fish spermatozoa are immotile in the testis, and often also in the seminal plasma. Induction of motility is related to the particular environmental conditions during spawning. In general, spermatozoa of feshwater and seawater spawners are activated by the hypotonicity and hypertonicity, respectively, of their surrounding media (Morisawa and Susuki, 1980; Morisawa *et al.*, 1983). Potassium ions, present in the seminal plasma, inhibit motility in salmonid sperm (Schlenk and Kahmann, 1938; Benau and Terner, 1980; Baynes *et al.*, 1981; Morisawa, 1985) and motility is therefore induced when the sperm are spawned into fresh water or 100 mmol/kg NaCl solution (Morisawa, 1985). The effect is due to reduction in K^+ ions rather than reduction in osmolality as motility can experimentally be induced at the osmolality of seminal fluid (300 mosmol/kg) if Na^+ replaces K^+ (Morisawa *et al.*, 1983).

Normal ovarian fluid induces spermatozoal motility but presence of broken ova progressively suppresses this ability and reduces the success of fertilization, as shown for the Coho Salmon, *Oncorhynchus kisutch*, by Wilcox *et al.* (1984). Loss of motility is attributed to decreasing ratio of Na^+ relative to K^+ in the ovarian fluid because of liberation of K^+ rich egg contents into the fluid. Removal of contaminated ovarian fluid by rinsing ova in isotonic $NaHCO_3$ completely restored fertility and is recommended as a routine procedure where presence of ruptured eggs is suspected.

However, it has been shown for cyprinids that although the high osmolality, contributed particularly by potassium, in the seminal plasma suppresses motility, potassium increases the viability and speed of sperm movement when at concentrations below that in the seminal plasma. Osmolality other than that produced by K^+ is the factor for initiation of motility in freshwater cyprinids (Morisawa *et al.*, 1983; Morisawa, 1985). Similarly motility of the sperm of some marine teleosts such as the puffer, flounder and cod is not specially influenced by K^+, though here it is induced by hypertonicity relative to the seminal fluid (Morisawa and Suzuki, 1980; Morisawa, 1985).

In the freshwater catfish, *Rhamdia sapo*, Cussac and Maggese (1988) observed no effect of potassium levels on sperm motility, even at concentrations found to be inhibitory for salmonids. The effect of increased K^+ concentration in depressing sperm motility in salmonids is further discussed under gamete quality, below.

A change in potassium concentration or osmolality induces intraflagellar cyclic AMP flux through

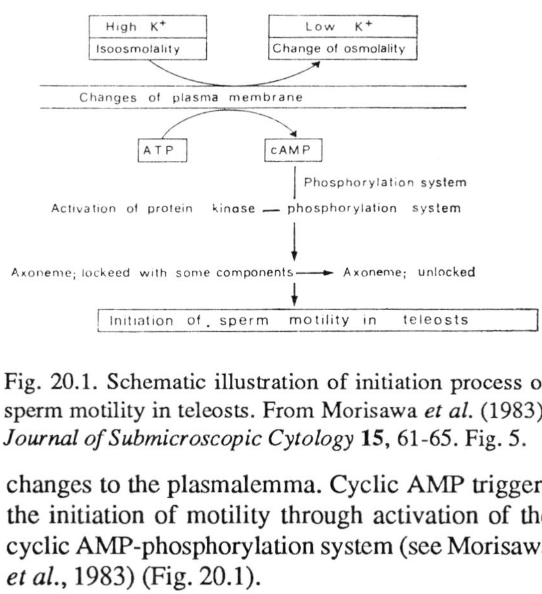

Fig. 20.1. Schematic illustration of initiation process of sperm motility in teleosts. From Morisawa *et al.* (1983). *Journal of Submicroscopic Cytology* **15**, 61-65. Fig. 5.

changes to the plasmalemma. Cyclic AMP triggers the initiation of motility through activation of the cyclic AMP-phosphorylation system (see Morisawa *et al.*, 1983) (Fig. 20.1).

The pH of the medium also affects motility. Thus the motility of salmonid sperm is suppressed when the pH value is below 7.8 (Baynes *et al.*, 1981) and enhanced by higher pH values (Scott and Baynes, 1980). The semen of the Rainbow Trout and the Atlantic Salmon has a pH of 8.3 (Stoss and Holtz, 1981b).

E. Duration of Sperm Motility

Basic methodology for evaluating motility of salmonid sperm is briefly presented by Terner (1986). Motility of fish spermatozoa is confined to a brief period of seconds or minutes in freshwater spawners because of osmotic injury (Billard, 1978a). Thus the very short duration of vigorous movement (1.5-7 min) in fresh water and physiological solutions make Rainbow Trout (*Salmo gairdneri*) spermatozoa difficult subjects for cryopreservation studies (van der Horst *et al.*, 1980). Trout sperm (species not given) may be motile for as little as 30 s and Atlantic Salmon (*Salmo salar*) sperm for 1-2 min (Terner, 1986). Motility is considerably longer in saltwater spawners, for example, 7 minutes in the Tuna, *Thunnus thynnus* (Doi *et al.*, 1982), 9 minutes in the Barramundi, *Lates calcarifer* (Leung, 1987b); 15 minutes in the Atlantic Cod, *Gadus morhua* (Davenport *et al.*, 1981). They are capable of motility at any time during several days in Herring, but only for the short period that they are near the egg micropyle (Yanagimachi, 1957). Sperm of the Black Porgy, *Acanthopagrus schlegeli*, showed good motility for up to 10 days in vials hanging in a water bath at 4° C (Chao *et al.*, 1986).

Osmotic injury has been observed in spermatozoa after release into their natural spawning medium. This is intensive and irreparable, and often is fatal within seconds for most freshwater species (Billard, 1978a; Maggese *et al.*, 1984). In cryopreservation of sperm of Rainbow Trout, *Salmo gairdneri*, sperm thawed in solutions of different ionic strength had the highest fertility (85%) at the lowest osmolality tested (249 mOsm). Fertility dropped significantly with increasing osmolality to a lowest level of 61% at 365 mOsm (Stoss and Holtz, 1983c).

The actual osmolality of fish semen is rarely stated. That of the cichlid *Oreochromis mossambicus* is 321 mosmol/kg. This species is euryhaline and the response to increasing environmental salinity is a shift to greater motility at higher osmolality (Harvey and Kelley, 1984c). The osmolality of the semen of six species of tilapia ranged from 240 to 380 mOsmol/kg (Chao *et al.*, 1987); that of Black Porgy, *Acanthopagrus schlegeli*, milt is 385 mOsm/kg (Chao *et al.*, 1986); it is 300 mosmol/kg in Rainbow Trout (Morisawa *et al.*, 1983).

F. Composition and quality of milt and ova

Milt. The high variability in fertilization achieved using cryopreserved milt and the corresponding low reliability of the milt is probably due as much to biological variation of the milt and eggs as to the various well established methodologies for cryopreservation. Indeed, it has been questioned whether changes in existing methods of cryopreservation will significantly improve performance (Baynes and Scott, 1987).

The constitution of milt in terms of performance and numbers of spermatozoa, chemical composition and osmolality varies interspecifically and even within the same individual with time. The time of collection of milt is therefore of relevance to successful cryopreservation. Where donor fish are collected by gill-net it has been shown that, if death of the fish ensues, sperm remain viable for not more than a few hours (Billard, 1978b).

Interspecific variation. The milt composition of 6 freshwater teleosts was studied by Piironen and Hyvärinen (1983b). The measured parameters showed clear specific differences between species. The highest spermatocrits and sperm densities were observed in Perch (*Perca fluviatilis*) and Burbot (*Lota lota*) and the lowest in Rainbow Trout (*Salmo gairdneri*) and the Whitefish (*Coregonus lavaretus*). Also studied were *S. salar sebago* and *S. trutta lacustris*. Fructose concentrations in the seminal plasma were small compared to mammalian values. The glucose concentrations in the seminal plasma were 5 times higher than those of fructose, and higher in Landlocked Salmon and Rainbow Trout than in the other species (but see *Salmo trutta*, below). The citric acid concentration of all species except Whitefish showed a significantly positive correlation to either spermatocrit or sperm density. The role of citric acid in the seminal plasma of fishes was assumed to be important. The glycerol concentration in the seminal plasma was comparatively high, and highest in Whitefish (Piironen and Hyvärinen, 1983b). Stein and Bayrle (1985) found the highest glucose content in the seminal plasma of *Salmo trutta fario*, at 12.2 mg/100 ml, compared with 3.7, 1.8 and

Table 20.1
Sperm densities in teleosts

Species	Mean no. of million sperm/ml	Reference
Mugil cephalus	53 000	Chao *et al.*, 1975
Tilapia zillii	770	Chao *et al.*, 1987
Oreochromis mossambicus	27 400	Chao *et al.*, 1987
Gadus morhua	13.1	Mounib, 1978 [Erroneous?]
Acanthopagrus schlegeli	2 870	Chao *et al.*, 1986
Thunnus thynnus	49 600-64 800	Doi *et al.*, 1982
Brachydanio rerio	40-80	Harvey *et al.*, 1982
Esox lucius	2-86 (mean 27.8)	Koldras and Moczarski, 1983
Salmo gairdneri	10 000	Schlenk and Kahmann, 1938
	9-16 000	Bratanov and Dikov, 1961
	26 000	Clemens and Grant, 1965
	12 000	Hamor, 1966
	10-25 000	Billard *et al.*, 1971
	10 700	Piironen and Hyvärinen, 1983
S. trutta	9-26 000	Ginsburg, 1968
S. trutta lacustris	14 100	Piironen and Hyvärinen, 1983b
S. salar	12-30 000	Truscott and Idler, 1969
	12.2	Mounib, 1978 [Erroneous?]
S. salar sebago	9 200	Piironen and Hyvärinen, 1983b
	4 000-12 000	Piironen, 1985
Coregonus lavaretus	6 800	Piironen and Hyvärinen, 1983b
Lota lota	37 500	Piironen and Hyvärinen, 1983b

8.8 mg/100 ml for *S. gairdneri*, *S. trutta lacustris* and *Coregonus* sp., respectively.

The specific differences between the testis, milt and sperm of six species of tilapia including *Oreochromis aureus*, *O. mossambicus*, *O. niloticus*, *Tilapia zillii*, *O. niloticus* x *O. aureus* hybrid and red tilapia, *Oreochromis* sp., were studied by Chao *et al.* (1987). The pH values of individual milts ranged from 6.2 to 8.2 and the osmolality from 240 to 380 mOsmol per kg. The quantity of milt obtained by stripping averaged only about 0.3 ml, and only in the *O. niloticus* x *O. aureus* hybrid did it exceed 3 ml. Sperm motility graded from weak to moderate was determined for the stripped tilapia milt. Tilapia sperm was active in various salinity ranges such as 0-5°/oo for *O. niloticus*, and 0-15°/oo for *O. mossambicus* and *T. zillii*.

Some specific sperm counts are shown in Table 20.1. The highest estimate is for the Bluefin Tuna, *Thunnus thynnus*, at between 50 and 65 thousand million per ml (Doi *et al.*, 1982). Neverthless, densities of the same order, varying from 1-53 000 x 10^6, are seen throughout most species for which counts are available. The lowest figures are 2 million, mean 27.8, per ml for the Pike, *Esox lucius*, given by Koldras and Moczarski (1983) and 12.2 million per ml, for the Atlantic Salmon, *S. salar*, given by Mounib (1978). The latter, at least, is presumably an error and is, indeed, cited as 12 x 10^9 by Scott and Baynes (1980), also rendering Mounib's figure of 13 million for *Gadus morhua* suspect. Otherwise the lowest figure is 40 to 80 million per ml for *Brachydanio rerio* (Harvey *et al.*, 1982), a range which exemplifies the great variation between individual fish of the same species noted by Scott and Baynes (1980).

Intraindividual variation. Legendre and Billard (1980) noted that at the end of spermiation in the Rainbow Trout, sperm fitness for cryopreservation and fertilization decreased, perhaps due to sperm senescence. Pooling the sperm of several males partially compensated for this decline.

In contrast, in the Finnish Landlocked Salmon, *Salmo salar sebago*, sperm density was about 4×10^9 in the first third of the spawning period but rose to a maximum of 12×10^9 at the end of spawning; the volume of milt obtained per stripping rose from 15-20 ml at the beginning to 50 ml at the end of the season (Piironen, 1985). Properties of milt and the chemical composition of the seminal plasma in this fish were studied in the course of a spawning season. Milt samples were stripped six times in succession and both inorganic and organic components of seminal plasma were analyzed. Milt properties were comparatively stable during the first half of the season. However, as indicated, milt production and sperm density as well as the concentrations of seminal sodium, potassium and chloride, and the osmolality increased continuously towards the end of the season. The organic components of seminal plasma, on the other hand, showed specific changes throughout the spawning season. Osmolality of the seminal plasma clearly exceeded the calculated ionic sum at the beginning of spawning but the reverse situation was found at the two last samplings, suggesting the presence of some other osmolality active organic substances in the seminal plasma. Highly significant correlations were found between sperm density and seminal citric acid ($r = 0.958$), sodium ($r = 0.823$), postassium ($r = 0.849$), chloride ($r = 0.792$) and osmolality ($r = 0.855$) during spawning. The clear differences observed in the composition of milt during spawning were considered of significance for development of cryopreservation methods for the sperm (Piironen, 1985).

Three individuals of the Atlantic Halibut, *Hippoglossus hippoglossus*, showed significant differences in fertilization performance of their sperm, whether fresh or cryopreserved (Bolla *et al.*, 1987).

In cryopreservation of the sperm of the Zebra Fish, *Brachydanio rerio*, variability in motility and hatching was not correlated with sperm volume or age of the fish and was attributed to difference in sperm quality between individuals as well as technical constraints imposed by the short duration of motility in the thawed spermatozoa (Harvey *et al.*, 1982).

Ova. Fertility of ova relative to the same batch of sperm varies greatly and is an obstacle to the practical application of short term preservation of Rainbow Trout sperm. For some eggs stored at 4° C, high levels of fertilization were recorded after 14 days whereas the capacity for fertilization of eggs of other females dropped in one day (Billard, 1981).

G. Morphology of Ova

A detailed consideration of the morphology of fish ova is beyond the scope of this volume. Mature teleost ova generally contain yolk in the centre and ooplasm peripherally. Oil droplets may be present in some species. Cortical alveoli are found near the microvillous plasma membrane. They are often spherical, dimensionally much larger than spermatozoa, and are always enclosed by a multilayered membrane which forms an effective barrier to spermatozoa. In Neopterygii, penetration by spermatozoa takes place only through a micropylar apparatus (rarely absent), which plays an important part in prevention of polyspermy (see Chapter 5 and Ginzburg, 1968; Kudo,1980; Kobayashi and Yamamoto, 1981).

H. Activation of Ova

Syngamy is regularly accompanied by "activation" of the egg, which includes water hardening of the egg membrane (Zotin, 1958), closure of the micropyle (Kobayashi and Yamamoto, 1981), calcium (Gilkey *et al.*, 1978) and electrical waves (Nuccitelli, 1980), and expulsion of the contents of the cortical alveoli (Brummett and Dumont, 1979). In salmonids, fertilization and activation of the egg can be easily separated experimentally, and salmonid eggs can be activated by immersion in water for 40 seconds (Ginzburg, 1968). In carp, ova autoactivate after natural or artificial spawning (Yamamoto, 1961). Spontaneous activation resulting from handling was reported for ova of Thai Carps and Catfishes (Withler, 1980).

The fertilizability of fish ova is limited to a period of a few seconds or minutes after spawning, and becomes impossible after the ova are activated. Ova of Trout and Pike lose fertilizability after one minute in fresh water (Billard and Jalabert, 1974), and for ova of Sea Bass, Sea Bream and Turbot, fertilizability decreases following immersion in sea water for more than a few minutes (Billard, 1978a). As in the case of sperm, the fertilization rates of eggs fertilized with frozen-thawed sperm differ significantly between individual females (Stoss and Holtz, 1981b).

I. Initial Gamete Union

Before fertilization, the spermatozoon must swim through the micropylar canal. Brummett and Dumont (1979) observed the occurrence of plasma membrane fusion during fertilization of *Fundulus* eggs. The first spermatozoon reaching the ooplasmic outgrowth at the end of the micropylar canal is caught by numerous microvilli or protrusions of the ooplasmic surface. The egg plasma membrane fuses with various regions of the plasma membrane of the enclosed spermatozoon. The spermatozoon continues penetration into the ovum and its plasma membrane is left on the ooplasmic surface.

Because the plasma membrane of the spermatozoon has no affinity with that of other spermatozoa, this incorporation into the surface of the micropylar outgrowth appears to prevent additional sperm penetration before the breakdown of cortical vesicles (Kobayashi and Yamamoto, 1981; Kudo, 1980). The micropylar apparatus is, therefore, the major means of preventing polyspermy. For a further discussion of micropylar events, see Chapter 5.

III. Storage at or near 0° C

Most work on short-term storage has been done on milt of salmonid species. Recent success in refining techniques has increased storage time to more than a month for milt from Rainbow Trout, *Salmo gairdneri* (Stoss and Holtz, 1983a).

Walleye (*Stizostedion vitreum vitreum*) semen was frozen at 0° C with four different modified extenders containing DMSO and other additives and thawed at either 32.1 or 21.1° C. Maximum egg fertilization rates were 83.2% for thawed semen and 96.6% for fresh semen (Moore, 1987). The extender used is given in Table 20.2.

For long term (-196° C) storage, addition to this extender of DMSO, mannitol, bovine serum albumin and soy protein as cryoprotectants is advocated.

The most important factors determining the success of short term storage are discussed below:

Table 20.2
Walleye extender of Moore, 1987

$CaCl_2.2H_2O$	0.234 g
$MgCl_2.6H_20$	0.267 g
Na_2HPO_4	0.472 g
KCl	3.744 g
NaCl	13.155 g
Glucose	20.000 g
Citric acid monohydrate	
$HOCCOOH[CH_2COOH]_2.2H_2O$	0.200 g
40 ml NaOH (1.27 g NaOH/100 ml H2O)	
40 ml Bicine (5.3 g bicine/100mlH2O)	

A. Storage Temperature

The storage temperature plays an important part in reduction of cell metabolism for preservation. For short term storage at temperatures from 3-15° C, Jensen and Alderdice (1984) found an exponential decay relationship ($y = ae^{bx}$) between estimates of time to 90 % and 50% fertilization success (y) and temperature (x) for Chum Salmon, *Oncorhynchus keta*, gametes. The ova, though affected by storage sooner than sperm, lost fertility at a slower rate than sperm.

Low temperatures just above freezing point have not been reported to have detrimental effects on gametes of coldwater fish species, at least within short time periods. Accordingly, storage temperatures close to 0° C are generally used for short term storage. As an example unfertilized salmonid ova have been stored at -1° C for up to 20 days in artificial media and in ovarian fluid, with longer retention of fertility in the latter medium (Harvey, Stoss and Butchart,1983). Fertility of Rainbow Trout sperm kept under O_2 at 4° C was similar after 5 days to that of fresh sperm and, for some females, high levels of fertilization were recorded after 14 days (Billard, 1981). Fertility of Atlantic salmon, *Salmo salar*, sperm was maintained for up to 10 days when 2 mm thick samples were stored at 0° C under an O_2 atmosphere in the presence of antibiotics (125 IU penicillin and 125 µg streptomycin per ml sperm). Fertility was completely lost after 24 days. Sperm stored without antibiotics fertilized 100% of eggs after 6 days (Stoss and Refstie, 1983).

There appears to be an optimum storage temperature above 0° C for gametes from warm water species. In Tilapia, *Sarotherodon mossambicus*, sperm stored at 0° C had a more abrupt decline in motility than those stored at at 5° C. However, dilution (1:1) of milt with a simple egg yolk-citrate diluent prolonged post-activation motility for up to 18 days (Harvey and Kelley, 1984a). The optimum storage temperature of tilapian ova was 20° C (Harvey and Kelley, 1984a, 1984b). Abnormal development was observed in embryos from ova stored below or above this optimum temperature. Thus, it is essential to determine the optimum storage temperatures for warmwater species.

B. Gaseous Exchange

Because of dependence on aerobic metabolism, gamete fertility can be prolonged by ensuring optimal oxygen availability for respiration during storage. Rainbow Trout (*Salmo gairdneri*) sperm maintained high fertility when stored under oxygen for five days compared with storage under air (Billard, 1981). Regular flushing of the storage containers with pure oxygen further extended the storage life up to 34 days at 0° C in the presence of penicillin and streptomycin (Stoss and Holtz, 1983a). The post-storage fertility of tilapian ova was increased from 35% to 55% by oxygenation of the sample container (Harvey and Kelley, 1984b).

In general, oxygen demand is greater for sperm than for ova and for higher rather than lower storage temperatures.

C. Diluent and Prevention of Gamete Activation

The prevention of gamete activation is most critical in short-term storage, as fertility declines soon after activation (see Section IID and E). As gametes remain arrested in seminal plasma or ovarian fluid, they are usually stored undiluted. Composition of diluents for fish gamete storage has been reviewed by Scott and Baynes (1980). The effects of medium volume and osmolality are discussed by Harvey *et al.* (1983).

Sperm. Dilution of milt appears essential for some warmwater species. In these species, sperm motility is induced by handling or contamination during stripping and is not suppressed by the storage temperature or addition of potassium ions (Harvey, 1982; Harvey and Kelley, 1984a). Dilution of nutrient is therefore required to supply energy needs during storage.

Storage of Milkfish (*Chanos chanos*) sperm diluted with the fish serum was superior to undiluted storage (Hara *et al.*, 1982). Milt from the Grouper, *Epinephelus tauvina*, was diluted with balanced salt solution containing 10% DMSO for improved storage at 5-10° C (Withler and Lim, 1982). Dilution of milt from Tilapia with egg yolk-citrate diluent containing glucose, glycine and sucrose extended the storage life from 5 to 18 days (Harvey and Kelley, 1984a).

Ova. Unwanted mechanical activation is the major problem in storage of unfertilized ova, for example in Zebra-fish (*Brachydanio rerio*) (Harvey, 1982); Thai Carps and Catfishes (Withler 1980). However, activation poses no such problem in salmonids as their ova can be held unactivated by immersion in isotonic solution (Ginzburg 1968). Diluent consisting of 12% Ficoll was used to delay activation in zebra fish ova and activation was completely inhibited by immersing the ova in silicone oil (Harvey, 1982).

D. Antibiotics

The use of antibiotics is essential for prolonged but nevertheless short-term storage, as bacterial and fungal contamination of gametes from excreta cannot be avoided (Stoss *et al.*, 1978; Stoss and Holtz, 1983a). Fertility of Atlantic Salmon (*Salmo salar*) sperm, stored with and without added penicillin and streptomycin, reached zero after 24 and 6 days respectively (Stoss and Refstie, 1983). Although antibiotics inhibit bacterial growth, they have a detrimental effect on spermatozoa of the Rainbow Trout at levels of 9 000 i.u. penicillin or 9 000 μg streptomycin per ml (Stoss *et al.*, 1978).

Gram-negative bacteria were found after 19 hours storage of Tilapian ova and growth of these could be inhibited by Kanamycin sulfate (Harvey and Kelley, 1984b). Walleye (*Stizostedion vitreum vitreum*) semen has been successfully stored using ampicillin. The semen was collected by aspiration, mixed with extenders containing the ampicillin and refrigerated for up to 14 days at 1° C. Egg fertilization trials resulted in mean fertilization values of 90.8% for extender containing 180 mu.g ampicillin/ml, 86.4% for extender with 3,325 mu.g ampicillin/ml, and 93.2% for fresh semen (Moore, 1987).

IV. Storage by Supercooling

In this mode of storage, gametes are kept supercooled to several degrees below 0° C. The freezing point of the suspending medium and cells can be depressed by the addition of cryoprotectants (see Chapter 19, Section VB). This mode of storage is suitable for coldwater species and usually obtains longer storage life of gametes than storage above 0° C. The success of storage by supercooling depends on prevention of intracellular ice formation and the tolerance of cells to low temperature and cryoprotectants. These factors are discussed in Section V, below.

Sperm. In salmonids, sperm fertility was maintained at a satisfactory level after five weeks of supercooling (Sanchez-Rodriguez and Billard, 1978). Storage of milt from warmwater species by supercooling has not been reported.

Ova. Unfertilized ova of coldwater fishes tolerate cold storage and supercooling. Harvey (1982) obtained fertile salmonid eggs after one month's storage of the eggs at -2° C. Harvey and Ashwood-Smith (1982) obtained supercooling. They showed that the freezing point of isolated, unfertilized eggs of Rainbow Trout is -1.7° C. For this species, using radiolabelled cryoprotectants, they demonstrated that the degree of permeation of methanol, DMSO and glycerol was inversely proportional to the molecular weights of the compounds: glycerol did not penetrate (contrast the *Brachydanio rerio* experiment, below), while methanol, which penetrates with the greatest rapidity, achieved no more than 23% of the epected equilibrium concentration after 2 hr exposure at 0° C. Eggs in Fish Ringer treated with 10% DMSO and 10% sucrose resisted intracellular freezing longer than those in Ringer; reversible supercooling to -18° C occurred if seeding by ice crystals was avoided. Reports of viable freezing are attributed to supercooling (Harvey and Ashwood-Smith, 1982). The medium for supercooling is shown in Table 20.3.

In later work salmonid ova were stored at -1° C for up to 20 days and the effects of medium volume, osmolality, antibiotics and cryoprotectants were investigated (Harvey, Stoss and Butchart, 1983).

Table 20.3
Supercooling medium for salmonid eggs (Harvey and Ashwood-Smith, 1982)

a. Fish Ringer Solution	
NaCl	6.50g
KCl	0.25g
$NaHCO_3$	0.20g
$CaCl_2$	0.30g
Distilled water 1 litre	
b. 10% sucrose w/v in Fish Ringer	
c. 10% DMSO v/v in Fish Ringer	
About 150 eggs to 50 ml of this medium.	

By contrast, the ova of warmwater fish appear to be unable to withstand supercooling and no success has been reported. Even short term refrigeration has proved unsuccessful. Thus stripped ova stored without media at 2-9° C lost all capacity for fertilization in 1 day (Indian Carp), or in 2 to a few hours (Common Carp and *Pangasius*) (Withler, 1980). Again, unfertilized ova stripped from *Sarotherodon mossambicus* showed declining post-storage fertility after only 1.5 hr below 18-20° C. The optimal temperature for storage for 19 hours was 20° C and produced 35% fertility, compared with high fertility for fresh controls (Harvey and Kelley, 1984b). It is the inability to tolerate supercooling which distinguishes warmwater eggs. However, even salmonid eggs if untreated survive only a few days at 0° C (Billard,1980)

V. Cryopreservation

Although freezing of fish eggs has hitherto been lethal (with the exception of the preliminary technique of Jamieson and Marshall, below), spermatozoa preserved in this mode can have an almost indefinite storage life. It might be questioned whether a decrease in fertilizing ability of sperm which has been observed with time is real or is due to a difference in the quality of eggs used. Thus, using an insemination ratio of 0.3 ml milt/20 ml (.±.7600) ova, Steyn and Van Vuren (1987) found that cryopreserved and fresh milt were equally effective in hatching trials. Fertilization with both fresh milt and milt

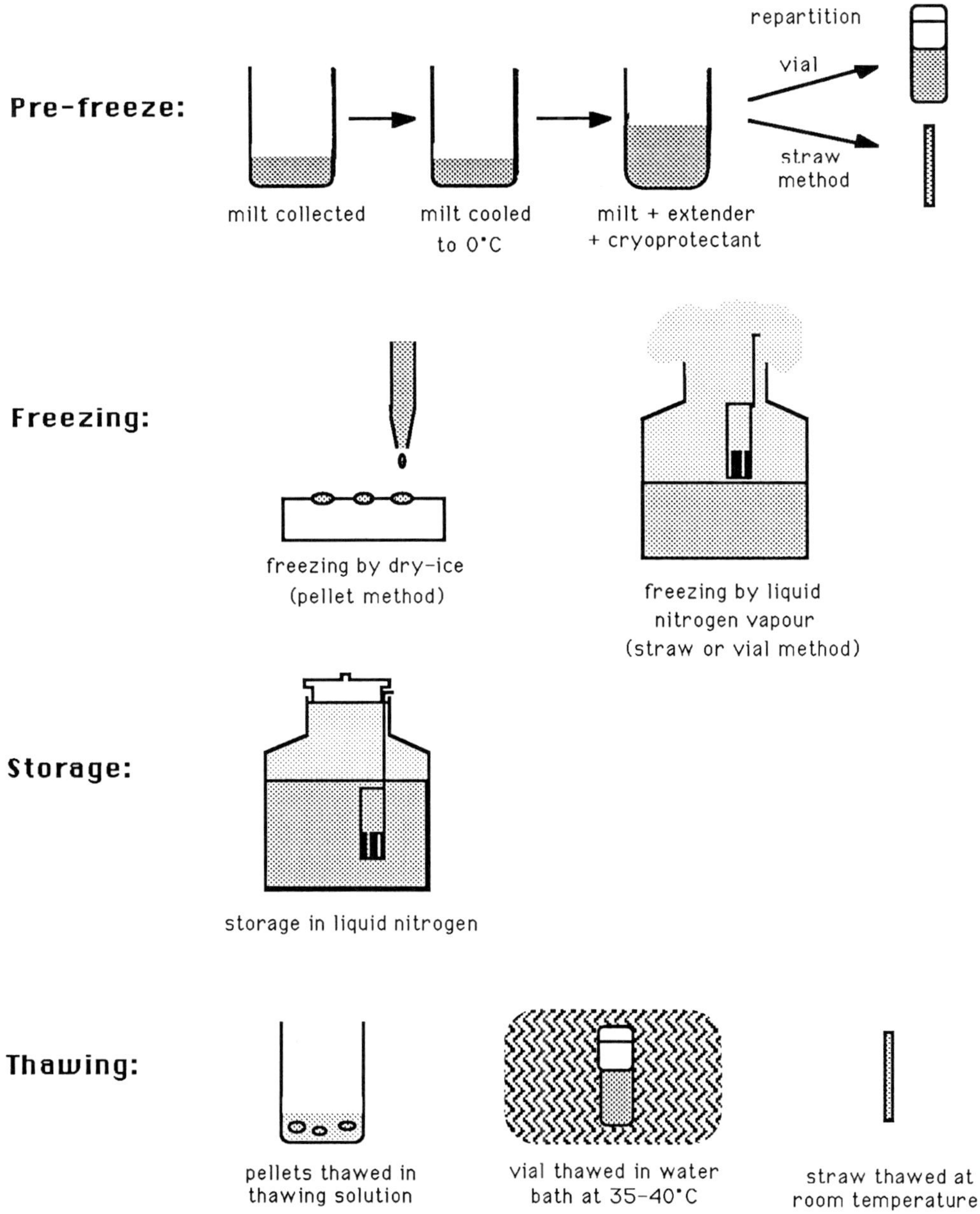

Fig. 20.2. Cryopreservation procedures for fish sperm: pellet, straw and vial methods. From Leung, unpublished.

cryopreserved for 14 days resulted in a maximum of 51% hatched ova. However, milt which was cryopreserved for 16 months resulted in a lower hatching rate of 41%.

Similarly, Kerby *et al.* (1985) found that the number and total weight of juveniles of Striped Bass (*Morone saxatilis*) harvested did not differ significantly ($P < 0.05$) between those produced by fertilization with cryopreserved as opposed to fresh sperm. Nearly all fish were healthy.

There is, however, some evidence for a real decrease in fertilizing ability of cryopreserved sperm with time. Black porgy (*Acanthopagrus schlegeli*) sperm, after being cryopreserved for 1, 7, 7 and 342 days, showed decreasing fertilities of 99.0, 93.2, 91.9 and 91.5% respectively (Chao *et al.*, 1986).

The success of cryopreservation mainly depends on the prevention of cryoinjury during freezing and thawing. Most work has been done on salmonid species and success has been reported in spermatozoa, (e.g. Mounib, 1978; Stoss and Holtz, 1981a,b; Stoss and Refstie, 1983; Harvey, 1983), but not in ova (Harvey and Ashwood-Smith, 1982; Schmehl and Graham, 1986; Stoss and Donaldson, 1983).

Major factors determining post-thaw survival are discussed below at each phase of a generalized cryopreservation procedure, as illustrated in Fig. 20.2. Many of these factors interact and are, therefore, difficult to analyse separately.

A. Pre-Freeze Phase

This phase includes collection and dilution of gametes, addition and penetration of cryoprotectant and, usually, initial cooling down to 0° C.

1. Gamete Quality

Differences in post-thaw fertility between different individual fishes (both sperm and eggs) have been observed (Billard, 1978b; Harvey and Kelley, 1984a; Stoss and Holtz, 1981b, 1983a). These are probably due to the use of aged or immature gametes. Attention should therefore be paid to the parent's state of readiness. In Rainbow Trout, maximum egg and fry survivals are achieved if the eggs are stripped 4 - 6 days post-ovulation. The success of subsequent development is closely correlated with the initial level of fertilization which can therefore be used as a predictor of later success (Springate *et al.*, 1984). See also IIF, Quality of Milt, above.

2. Interval between Gamete Collection and Freezing

Gametes should be cryopreserved immediately after collection, as a decrease in post-thaw fertility has been observed with increased storage times prior to freezing (e.g. for Rainbow Trout sperm, Stoss and Holtz, 1981a). In their tests a solution which activated sperm motility was used in the thawing process and the decline in fertilization capacity was explained by a rapid decline of sperm motility after activation. However, in tests on the Carp (Kurokura *et al.*, 1984) no thawing solution was used and the spermatozoa were not activated on thawing; decline in fertilization was reasonably attributed to accumulating injury from the freeze-thawing process. Reduction of fertility because of loss of motility by exhaustion may additionally be postulated.

The susceptibilty of the ovum membrane of salmonids to osmotic shock was increased by such storage effects (Stoss and Donaldson, 1983).

3. Extenders

Undiluted gametes are not suitable for freezing, and they must be diluted with a suitable extender, which is a solution of balanced salts and, sometimes, organic compounds. This usually serves also as the cryoprotectant. One of the functions of the salts or organic compounds is to inhibit osmotically the activation of ova and motility of sperm. For composition of typical extenders, see Scott and Baynes (1980).

Sperm. Although there have been many extender media reported in the literature, it has been claimed that there is little difference in post-thaw fertility between them (Stoss and Holtz, 1981b) while some workers claim superiority for certain extenders.

One of the better known and most successful extenders is Mounib's medium, pH 7.57 (Mounib, 1978) which has been modified by Legendre and Billard (1980) (Table 20.4). This extender was originally developed because of its ability to preserve the activity of four carbon dioxide fixing enzymes in sperm when frozen. This was found much superior to Ménézo's medium (see Egg yolk, below).

An extender consisting of 300 mM glucose has been used with success in salmonids (Stoss and Refstie, 1983). This was found to be more effective for Atlantic salmon sperm than Mounib's extender but no difference between treatments was found for sea trout sperm. Extender composition seems relatively unimportant to spermatozoa because sperm are usually exposed to the extender for only a few

Table20.4
Mounib's medium

sucrose	125 mM
reduced glutathione	6.50 mM
potassium bicarbonate	100 mM
7 parts of the medium:1 part DMSO	
The whole mixed 3:1 with semen	
5000 i.u./ml penicillin then added	

Modification of Legendre and Billard

Add to Mounib's medium
10 mg/ml bovine serum albumin
10 p. 100 DMSO
egg yolk 10 p. 100 of the final diluent

seconds prior to freezing and after thawing.

Stoss and Holtz (1981a) obtained good fertility for Rainbow Trout sperm with an extender, modified from Büyükhatipoglu and Holtz (1978), consisting (in m mol), of 101 NaCl, 23 KCl, 5.4 $CaCl_2$, 1.3 $MgSO_4$, 200 *Tris* (hydroxymethylaminomethane) and citric acid to pH 7.25, with addition of 4 mg bovine serum albumin, 7.5 mg Promine D (Soybean protein) and 0.12 ml DMSO; dry ice pellet freezing was followed by liquid nitrogen. However, the same authors (Stoss and Holtz, 1981b) later found that superior fertility (mean 87.1% eyed eggs) was obtained for this species if buffer was eliminated from the diluent. This held for eight buffers tested, including *Tris*. They concluded that, at least for the short time of about one minute for which the semen is exposed to the medium before freezing, the buffering capacity of seminal plasma was presumably sufficient to avoid damage due to the low pH of the unbuffered diluent. They confirmed the findings of other workers that for cryopreservation of salmonid semen the medium should be slightly alkaline prior to addition of semen and DMSO.

Stoss and Holtz (1983c) investigated the suitability of sucrose (0, 100, 200 or 300 mmol) and KCl (0, 6.7, 13.4, 20.2, 26.9 or 33.6 mmol) as extender components in cryopreservation of Rainbow Trout sperm, using DMSO (see Cryoprotectants below). Sucrose did not influence the fertility of cryopreserved sperm in the presence of KCl. The use of 6.7 mmol KCl showed a slight advantage over KCl-free extenders. The absence of sucrose and KCl resulted in reduced sperm survival (Stoss and Holtz, 1983c).

Kurokura and Hirano (1980) developed an effective extender, to be used in cryopreservation with DMSO, for Rainbow Trout sperm which gave 40% eyed eggs. The extender consisted of 0.035 M NaCl, 0.018 M KCl, 0.06 M aminoethylpropanediol-HCl buffer (pH 8.3) and 1.8 g/100 ml serum albumin.

For spermatozoa of the Striped Bass (*M. saxatilis*) no less than 14 extenders were tested in combination with three concentrations of four cryoprotectants at various sperm:medium ratios. Best results were obtained with extender OH-189 (Kerby, 1983). This consisted of 7.30 g NaCl, 0.38 g KCl, 0.23 g $CaCl_2.2H_2O$, 5.00 g $NaHCO_3$, 0.41 g $NaH_2PO_4.H_2O$, 0.23 g $MgSO_4.7H_2O$, 5.00 g fructose, 7.50 g lecithin and 5.00 g mannitol in 968.95 g distilled H_2O. It was optimally combined with 5.0% DMSO and mixed in a 1:4 sperm:medium volume ratio. The highest fertilization obtained with cryopreserved semen was 87.7%. No fertilization was obtained with glycerol as the cryoprotectant.

Achievement of 70% eyed eggs of *Cyprinus carpio* after sperm cryopreservation with DMSO indicates the sucess of two extenders containing NaCl, KCl, $CaCl_2$ and $NaHCO_3$, with or without $MgCl_2$, though results without extender are not given (Kurokura *et al.*, 1984). One of these extenders adopted by Leung (1987b) for cryopreservation of *Lates calcarifer* sperm, giving high post-thaw motility with DMSO (fertility untested), consisted of 0.75 g NaCl, 0.02 g KCl, 0.02g $CaCl_2$ and 0.02g $NaHCO_3$.

Various extenders, containing KCl, NaCl, glucose, sodium citrate, Ringer's solution, cow serum and milkfish serum were used by Hara *et al.* (1982) to preserve milkfish (*Chanos chanos*) sperm at near-zero temperatures (0-4° C) and in liquid N (-196° C). Milkfish serum was a superior extender in both cases. After 5 days, comparatively good motility (> 30%) and fertilizing capacity (6.7-18.9%) were observed in the near-zero liquid samples, while in other extenders, sperm ceased to show motility after 2 days. The fertilization success of 4-5 days cryopreserved sperm averaged 67.5% (n = 2) with milkfish serum, 60.5% (n = 2) with 400 mM glucose, 58.0% (n = 2) with 150 mM NaCl, 41.2% (n = 1) with Ringer's solution and 31.9% (n = 2) with cow serum.

A saline extender used in the cryopreservation of Atlantic Halibut (*Hippoglossus hippoglossus*) sperm, diluted 1:3, as pellets or in straws, and thawed in marine fish Ringer gave fertilization results similar to fresh sperm and independent of whether the sperm pellets were thawed at 10 or 40° C (Bolla *et al.*, 1987).

For cichlids, Chao *et al.* (1987) used an extender containing 15% milk and 5% methanol to prepare a milt mixture before cooling rapidly to -35° C and then at 5° C per min to -75° C for storage in liquid nitrogen (-196° C). Fertility tests on frozen tilapia milt resulted in a fertilization rate of 72.7% (vs. control 85.7%) for the 22-day frozen milt of the *O. niloticus* x *O. aureus* hybrid used to fertilize the eggs of *O. honorum*, and 93.4% (vs. control 90%) for the 304-day frozen sperm of red tilapia used to fertilize conspecific eggs.

Kumar (1988) investigated a range of extenders for spermatozoa of three carp species, *Labeo rohita*, *Cirrhinus mrigala* and *Catla catla*, preserved under cryogenic conditions (-196° C). Among the seven extenders (blood sera, Alsever's solution, Ringer's solution, extender Ma and Mb, urea-egg-yolk and egg-yolk citrate) tried, egg-yolk citrate gave the best results in terms of post-thawing motility, closely followed by urea-egg-yolk and extender Ma. Although a large number of chemicals or their combinations were used as extenders, the simplest formulation was most successful. The major problem faced during the investigation was the enormous variability in the quality of semen (Kumar, 1989).

A successful extender for short term storage or, with modification, for storage at -196° C. of Walleye sperm (Moore, 1987), is given above in Table 20.2.

Table 20.5
Extender 189M of Horton
(Withler, 1980)

NaCl	730 mg
$NaHCO_3$	500 mg
Fructose	500 mg
Vegetable lecithin	750 mg
Mannitol	500 mg
Distilled water	100 ml

Nine parts of this extender were mixed with 1 part of DMSO as cryoprotectant. Four parts of this diluent were added to a part milt (Withler, 1980).

For sperm of the Indian Carp, *Labeo rohita*, fertility (reaching 79% live larvae) superior to that of fresh sperm was obtained after liquid nitrogen storage in a diluent containing an extender (termed 189M), and attributed to Horton, indicated in Table 20.5. This medium also preserved the motility of sperm of the Grouper, *Epinephelus tauvina* (Withler and Lim, 1982).

Ova. No specific extenders has been developed for fish ova.

4. Cryoprotectants

Cryoprotectants for fish gametes have been reviewed by Scott and Baynes (1980). The most commonly used one is DMSO (see below).

Permeating cryoprotectants

Sperm. In early work, on the Grey Mullet, Chao *et al.* (1975) found no significant difference in fertility of sperm cryopreserved for 1 year in liquid N with 5-10% glycerol, or DMSO diluted 1:1, 1:5 or 1:10, as the cryoprotectant. The rates, <3%, were very low, a notable contrast with results for Black Porgy sperm by Chao *et al.* (1986) in which commercially valuable results were obtained when glucose-glycerol cryopreservation of sperm gave fertility superior to that of fresh sperm.

Glycerol has been used for marine fish sperm (e.g. Atlantic cod, *Gadus morhua*, Mounib *et al.*, 1968), and can be replaced by DMSO (Mounib, 1978). For Tilapia sperm cryoprotective ability of methanol was higher than that of the DMSO or glycerol (Harvey, 1983). In some cases well known cryoprotectants have failed to achieve any fertility of frozen-thawed sperm (see Hara *et al.*, 1982; Kerby, 1983, below).

DMSO

Dimethyl sulphoxide is a better cryoprotectant for most cells than glycerol; probably because it enters and leaves the cell much faster than glycerol, a process which is far less dependent on temperature for DMSO than for glycerol (see review by Ashwood-Smith, 1986; contrast, however, faster glycerol penetration of Zebra Fish embryos, below). The gene activation properties of DMSO have suggested caution in use of this cryoprotectant for human embryos (Ashwood-Smith, 1985) and the possibility of genetic effects in fish cryopreservation should be investigated.

DMSO at concentrations of 5-20 % (v:v) has been used successfully for cryopreservation of the sperm of Pike, *Esox lucius* , de Montalembert *et al.*, 1978; Herring, *Clupea harengus* and *C.pallasii*, Rosenthal *et al.*, 1978; salmonids Kurokura and Hirano, 1980; Legendre and Billard, 1980; Scott and Baynes, 1980; Stoss and Refstie, 1983; Stoss and Holtz, 1983c; Erdahl *et al.*, 1984; Milkfish, *Chanos chanos*, Hara *et al.*, 1982; Sea Bream, *Sparus auratus* and Sea Bass, *Dicentrarchus labrax*, Billard, 1978b , 1984; Common Carp, *Cyprinus carpio*, Withler, 1980, 1982; Kurokura *et al.*, 1984; the catfish, *Pangasius sutchi*, Thawes Carp, *Puntius gonionotus* and Indian Carp, *Labeo rohita*, Withler, 1980, 1982; Striped Bass, *Morone saxatilis*, Kerby, 1983; Grey Mullet, *Mugil cephalus*, Chao *et al.*, 1975; Grouper, *Epinephelus tauvina*, Withler and Lim, 1982; Cod, *Gadus morhua*, Mounib, 1978; and Barramundi, *Lates calcarifer*, Leung, 1987b.

Toxic effects have been observed when using DMSO at high concentrations or when equilibration time is extended (e.g. Stoss and Holtz, 1983c). The post-thaw motility of sperm from the Barramundi (*Lates calcarifer*) was reduced when the DMSO concentration was higher than 5 % (v:v) (Leung, 1987b). DMSO at more than 20 % (v:v) reduced the post-thaw fertility of Rainbow Trout sperm even when equilibration time was kept to a minimum. Detrimental effects of DMSO were reduced when it was added in a stepwise manner (Stoss and Holtz, 1983c).

Stoss and Refstie (1983) found a very simple extender consisting of 0.3 M glucose and 10% DMSO to be the most successful of a number of extenders in cryopreservation of the sperm of Atlantic salmon and sea trout (*Salmo trutta*). The extenders tested were: an aqueous solution of 10% DMSO; the DMSO-glucose mixture; that of Stoss and Holtz, 1981a); and that of Mounib (1978). Semen was mixed with extender and frozen on dry ice (pellets) with subsequent storage in liquid N. Sperm pellets were thawed in a 0.12M $NaHCO_3$ solution at 10° C before insemination. Insemination with cryopreserved Atlantic salmon sperm resulted in 36-91.3% eyed eggs (control = 100). The differences were attributed to the type of extender and the batch of gametes employed. Results with cryopreserved Sea Trout sperm ranged between 38.6-54.8% eyed eggs, but unlike Atlantic Salmon showed no difference between treatments (Stoss and Refstie, 1983).

Instead of glucose, van der Horst *et al.* (1980) added sucrose to DMSO for cryopreservation of Rainbow Trout spermatozoa, as did Stoss and Holtz (1983c, see 3.Extenders, above). Solutions consisting of 250-280 mmol sucrose and 5-12% DMSO (4 parts) did not activate trout spermatozoa (1 part), but after dilution with fresh water vigorous motility could be fully restored. These sucrose-DMSO solutions were employed in cryopreservation studies. Using straws and a fast freezing - fast thawing procedure, post-thaw dilution with fresh water resulted in motility of 25-60% of the spermatozoa (van der Horst *et al.*, 1980).

Erdahl and Graham (1980) used an extender consisting of 7% DMSO mixed 1:1 with Brown Trout, *Salmo trutta fario*, and Rainbow Trout, *Salmo gairdneri*, semen collected by aspiration. After pellet or straw freezing and storage at -79° C and -196° C, thawed spermatozoa of both the species resulted in fertilization rates in excess of 80% when introduced to fresh ova. For Brown Trout and Brook Trout, *Salvelinus fontinalis* dilution 1:2 with 7% DMSO gave an average fertility of 54% (range 3-98%). Glycerol as cryoprotectant gave negligible fertility while ethylene glycol gave intermediate but acceptable results (Erdahl *et al.*, 1984).

Stoss and Holtz (1983c) demonstrated the suitability of DMSO for cryopreservation of salmonid (Rainbow Trout) sperm while showing that increasing concentration and equilibration time were detrimental. The effects of various extenders, of added ions, and osmolality were also shown. Effects of varying sucrose and KCl are noted above under 3. Extenders. The concentration of DMSO added to the extender was varied from 5-20 ml/100 ml extender and simultaneously the sperm were equilibrated in these extenders for less than 1, 2, 4 and 60 min at 0° C. DMSO concentrations between 10 and 20% provided high post-thaw fertility (77%) when sperm equilibration time was kept at a minimum. With increasing DMSO concentration, the detrimental effect of any equilibration time was more pronounced. Sperm thawed in solutions of different ionic strength, obtained by varying the concentration of the thawing solution ($NaHCO_3$), had the highest fertility (85%) at the lowest osmolality tested (249 mOsm). Fertility dropped significantly with increasing osmolality to a lowest level of 61% at 365 mOsm.

For Sea Bass (*Dicentrarchus labrax*) and Sea Bream (*Sparus auratus*) Billard (1984) used a diluent consisting of NaCl (19.5 g/l), $MgSO_4$ $7H_20$ (0.25 g/l), $CaCl_2$ $2H_2O$ (0.25 g/l), glycine (6.25 g/l), Tris (2.4 g/l), at pH 8.5, with 10% DMSO; the optimal rate of dilution was 1 volume of sperm to 2 volumes of diluent. The diluent did not have to be equilibrated. At a cooling rate of 10° C per min, the straws could be plunged into liquid nitrogen as soon as the temperature reached -80° C. The diluent had the disadvantage of mobilizing the sperm, with the result that freezing had to be performed very soon after dilution. A diluent in which the sperm remained immotile would be preferable.

Koldras and Bieniarz (1987), investigating cryopreservation of carp (*Cyprinus carpio* L.) sperm at -196° C, using 18 diluents and 3 cryoprotectants, found that the best percentages of fertilization and number of small fry were obtained when sperm was frozen in the presence of Stein's diluent with an addition of 10% of DMSO. However, if carp blood plasma was used 10% glycerol gave similar results. Cryopreservation was not, however, successful as in both cases up to 98% mortality of embryos occurred after 20 hours of normal development. Similarly, cryopreservation in liquid N of piau fish (*Leporinus silvestrii*) semen using DMSO at concentrations of 5 and 10% with two extenders resulted after thawing in only 0-10% sperm motility. Semen and extender were mixed at the ratio of 1:4.5 (volume:volume), equilibrated for 2 or 4 minutes and frozen in plastic flasks (capacity of 1.0 ml); thawing was in a waterbath (50-60° C) (Cóser *et al.*, 1987). It seems possible that these poor results were due to low quality milt.

DMSO at a concentration of 5% with either 15% milk powder or 20% egg yolk gave optimal results in cryopreservation of *Lates calcarifer* (post-thaw motility of spermatozoa: 70-100% for 7 min) compared with glycerol and very inferior results for methanol (q.v.) (Leung, 1987b).

DSMSO was found inferior to methanol (with skim milk) in cryopreservation of spermatozoa of Tilapia (*Sarotherodon mossambicus*) (Harvey, 1983) (see Methanol, below).

Glycerol

At equimolar concentrations cryoprotection by glycerol is almost identical to that of DMSO but because glycerol with its slower permeation of cells takes longer to reach equilibrium than DMSO, its removal after freezing and thawing can cause severe osmotic stress (see review by Ashwood-Smith, 1986). Indeed, glycerol could not be removed without resort to special techniques from *Brachydanio rerio* embryos exposed to a 1 M concentration for 1 hr and caused disruption of the cell layers, both serious impediments to use of glycerol in cryopreservation (Harvey, Kelley and Ashwood-Smith, 1983). Nevertheless, these workers state that although glycerol has a larger molecule than DMSO it penetrates embryos more rapidly.

Glycerol is usually considered inferior to DMSO for cryopreservation of sperm and, correspondingly, gave zero fertility for Striped Bass sperm (as was also the case for ethylene glycol or propylene glycol) whereas DMSO was highly successful (Kerby, 1983). The same was true for the Grouper, *Epinephelus tauvina* (Withler and Lim, 1982). Glycerol is nevertheless superior to DMSO in reducing loss of proteins of Atlantic salmon (*Salmo salar*) spermatozoa into the seminal fluid during storage at -80°. Glycerol more effectively fostered retention of sperm proteins, especially those of higher molecular weight. The massive loss and extensive labelling of sperm proteins with 125I was taken as an indication of damage to salmon sperm membranes during freezing, storage, and thawing (Yoo *et al.*, 1987). We will see that the effectiveness of glycerols vs. DMSO is highly dependent on the length of the equilibration time (see Doi *et al.*, 1982, below).

Earlier, fertility of 36 ± 12% had been obtained for sperm of the Cod, *Gadus morhua*, frozen at -79° C (dry ice-acetone) or -196° C (liquid N) in a freezing mixture containing 1 part semen and 3 parts of a medium which consisted of 10 parts glycerol: 40 parts 0.4 M NaCl-0.1 M glycine: 8 parts 1.3% sodium bicarbonate; a freezing rate of 5° C/min was more successful than 1° C/min (Mounib *et al.*, 1968).

Using glycerol only as a cryoprotectant for spermatozoa of the Whitefish, *Coregonus muksun*, but six different extenders, with -40° C, and -196° C as storage temperatures, the highest rate of fertility was obtained using an extender consisting of 0.3 M glucose with 20% glycerol (Piironen and Hyvärinen, 1983a). The high value of glycerol as a cryoprotective agent for the Whitefish, *Coregonus lavaretus*, sperm has been correlated with the fact that in six teleost species (Perch, *Perca fluviatilis*; Burbot, *Lota lota*; the salmonids, *Salmo gairdneri*; *S. salar sebago*; and *S. trutta lacustris*; and the Whitefish) the glycerol concentration in the seminal plasma was highest in the Whitefish. The high glycerol concentration in all six species (lowest in the Rainbow Trout, *Salmo gairdneri*) was also assumed to be related to the lipolytic capacity of the testis (Piironen and Hyvärinen, 1983b).

Very high fertility was obtained for black porgy (*Acanthopagrus schlegeli*) sperm using an extender containing 5% glucose mixed with glycerol, as the cryoprotective agent, at a 4:1 ratio. The milt was diluted with the extender at a 1:1 ratio (Chao *et al.*, 1986).

Cryopreserved semen of the Bluefin Tuna, *Thunnus thynnus*, in liquid N, with glycerol or DMSO, as cryoprotective agent, and Mounib's medium or 10% glucose, basic medium, successfully maintained its spermatozoal motility. Glycerol showed its cryoprotective effectiveness at equilibration times longer than 20 min, while DMSO was effective at equilibration times shorter than 20 min (Doi *et al.*, 1982).

We have seen above that carp sperm cryopreserved at -196° C, in 10% glycerol in carp blood plasma gave similar (<98%) fertilization to DMSO in Stein's solution, though in both cases development did not proceed (Koldras and Bieniarz, 1987).

For the Sharptooth Catfish, *Clarias gariepinus*, sperm, extender with 9% glycerol gave a slightly higher percentage of hatched ova than with the same strength of DMSO, 44.5% compared with 40.5%, with a 1:1 mixture of semen and diluent (Steyn and Van Vuren, 1987).

Glycerol gave acceptable results when used at high concentration (20%) with 15% milk powder for liquid N cryopreservation of *Lates calcarifer* sperm although a DMSO regime (q.v.) was preferable (Leung, 1987b).

Ova. By comparing uptake of labelled (^{14}C-methyl) DMSO and (2-3^{H}) Glycerol, both 1M in fish ringer, into normal and dechorionated Zebra Fish, *Brachydanio rerio*, eggs Harvey, Kelley and Ashwood-Smith (1983) showed that glycerol enters the normal embryo more easily (despite its larger molecule) than does DMSO, although reaching only about 8% of the expected equilibrium level after 2 hr at 23° C. DMSO reaches only about 2.5% of the expected level. On removal of the chorions, permeation increased several fold. Because of the resistance of the chorion to permeation, embryos are unaffected by exposure to 1 M DMSO for 1 hr but there is significant decrease in viability at 1.5 and 2 M. Embryos exposed to 1 M glycerol for 1 hr showed disruption.

The relative penetration of the chief three cryoprotectants is illustrated in Fig. 20.4.

Methanol

Sperm. The excellent cryopreserving qualities of methyl alcohol for living cells are discussed under ova below. For *Brachydanio rerio* (Zebra Fish), 51±35.6% hatching, positively correlated with sperm motility, was obtained using milt mixed with 5 volumes of a diluent consisting of 9ml fish ringer, 1 ml methanol and 1.5 g skim milk powder (Harvey *et al.*, 1982). Fertility was as high in some cases as 103% of that obtained with fresh sperm, the highest fertilility we are aware of in all fish sperm cryopreservation studies, confirming that methanol should not be rejected as a cryoprotectant. These authors note that inclusion of powdered (skim) milk counteracts a tendency to aggregation of sperm tails. Both fresh and powdered egg yolk, as well as lecithin, were ineffective.

15% methanol with 15% milk powder also gave successful cryopreservation of Tilapia, *Sarotherodon mossambicus*, sperm. Cryopreserved sperm yielded 64.3 ± 34.2% of embryos developing to the stage of blastopore closure, compared with 57.5 ± 10.5% for control, unfrozen milt. Post-thaw motility was roughly double that obtained by substituting 5 or 10% DMSO for methanol (Harvey, 1983). Rana and McAndrew (1989) also found that 10% methanol, of a number of cryoprotectants tested, gave maximum cell protection. However, methanol provided little protection against freezing damage to the sperm of the Barramundi, *Lates calcarifer* (Leung, 1987b).

For the Sharptooth Catfish, *Clarias gariepinus*, sperm, extender with 9% methanol gave inferior hatching of ova, 18.9%, compared with the same strength of DMSO or glycerol (see above) with a 1:1 mixture of semen and diluent (Steyn and Van Vuren, 1987).

Ova. Methanol penetrates into cells faster than DMSO or glycerol (Ashwood-Smith, 1986; from Harvey and Ashwood-Smith, 1982) (Fig 20.4). Its entry into mouse blastomeres is so rapid that no volume changes associated with osmotic imbalance is observable. Fears that it is toxic may be allayed by the observation that cryopreservation in 10% methanol results in a cellular concentration some 15 000 times less than the lethal concentration for gastrointestinal absorption in man. However, it ceases to act as a cryoprotectant and becomes toxic, as high concentrations build up, if cooling below -45° C is slow (see review by Ashwood-Smith, 1986).

Ethylene glycol

In cryopreservation of the sperm of Chinook Salmon, *Oncorhynchus tschawytscha*, and Brook Trout, *Salvelinus fontinalis*, dilution 1:2 with 7% ethylene glycol gave acceptable results (~53% fertility), though inferior to DMSO (Erdahl *et al.*, 1984). However, it gave zero fertility for Striped Bass, *Morone saxatilis*, sperm (as was also the case for glycerol or propylene glycol) whereas DMSO was highly successful (Kerby, 1983).

Non-permeating cryoprotectants

A non-permeating cryoprotectant is often used in conjunction with a permeating cryoprotectant.

Egg yolk

The most commonly used non-permeating cryoprotectant is egg yolk. For salmonid sperm, 20% egg yolk is recommended (Billard, 1983c).

The presence of 5-10% hen's egg yolk in a sucrose-based extender has been shown to significantly improve post-thaw fertility of cryopreserved rainbow trout (*Salmo gairdneri*) spermatozoa compared to use of the extender without egg yolk. Despite variation in the performance of different batches of eggs, the extender containing 10% yolk consistently gave high post-thaw fertility in samples of cryopreserved milt (67.3 ± 3.0% standard error) in thirty replicated trials. This medium was reliable for cryopreserving Rainbow Trout milt and fertilizing small quantities of eggs (Baynes and Scott, 1987). Again, Legendre and Billard (1980) found that Mounib's medium gave better results than Ménézo's medium for cryopreservation of Rainbow Trout sperm, especially after 10% of tellurite egg yolk was added. The optimal deep-freeze conditions were : 1/3 dilution, no equilibration after dilution but immediate deep-freezing at a rate of 10-40° C per min. Thawing had to be carried out rapidly in 10 sec. The spermatozoa apparently underwent degenerative changes during the freezing-thawing process, and therefore, during insemination, more frozen spermatozoa were used to equal the fertilization rate obtained with non-frozen sperm. The fertile spermatozoa gave normal embryogenesis and no abnormal development was seen up to the vesicle resorption stage. At the end of spermiation, sperm fitness for deep-freezing decreased, perhaps due to sperm senescence. Pooling the sperm of several males partially compensated for the fertilizing ability loss seen at the end of the reproductive period.

Milk

Despite the above claimed superiority of egg yolk, skimmed milk powder was more cryoprotective than egg yolk for Barramumdi sperm where DMSO at a concentration of 5% with either 15% milk powder or 20% egg yolk gave better results (post-thaw motility of spermatozoa: 70-100% for 7 min) than were obtained with glycerol or methanol (Leung, 1987b).

A diluent consisting of 1.5 g skim milk powder (Carnation brand), 9ml fish ringer and 1 ml methanol gave successful cryopreservation for *Brachydanio rerio* (Zebra Fish). The milk counteracts a tendency to aggregation of sperm tails (Harvey *et al.*, 1982).

Other proteins

Stoss and Holtz (1983b) conducted three experiments to improve the technique of pellet-freezing of rainbow trout semen. When comparing different sources of protein in the diluent, a combination of the soybean protein Promine D and bovine serum albumin gave the best result (77.5% eyed eggs vs. 91.5% for fresh semen controls). Considerable individual variation was observed when comparing postthaw fertility of semen from different males. Mixing of semen from two males did not have a notable effect on fertilization rate. This is in contradiction to a previous finding (Stoss and Holtz, 1981b) that mixing semen of different males gave higher fertility.

Ova. Prior to recent work in this laboratory (Jamieson, unpublished; Jamieson and Marshall in preparation) no success was obtained in attempts to cryopreserve fish ova. This failure has reasonably been attributed to the very low permeability of eggs to most cryoprotectants (Harvey, Kelley and Ashwood-Smith, 1983; Schmehl and Graham, 1986). It occurred to the senior author that exposure of eggs to cryoptrotectant (DMSO) in a vacuum might achieve sucessful penetration of the cryoprotectant and subsequent cryopreservation. It was first shown that gametes (sperm ejaculate) of the fire-tailed gudgeon, *Hypseleotris galii*, survived 10 minutes exposure to a vacuum (Jamieson, unpublished). Subsequently, in preliminary trials, successful cryopreservation of sperm and of ova of this species and of the Australian Bass, with DMSO as the cryoprotectant, was obtained after what we have termed vacuum equilibration (Jamieson and Marshall, in preparation). Details of this method are set out below after a brief review of previous procedures in attempting to cryopreserve fish eggs.

Previous procedures

The freezing point of isolated, unfertilized eggs of rainbow trout (*Salmo gairdneri*) is -1.7° C. Eggs treated with 10% DMSO and 10% sucrose in fish Ringer resist intracellular freezing longer when cooled at 0.01° C per min than do those in fish Ringer alone; intracellular freezing is inevitable with both cryoprotectants, although eggs that remained unfrozen for several hours appeared viable upon slow rewarming (Harvey and Ashwood-Smith, 1982). These workers consider that earlier studies reporting survival of cryopreserved eggs were probably dealing with supercooled eggs. Scanning electron micrographs of cooled and frozen DMSO-treated eggs show a progressive deterioration of the outermost layer of the zona radiata following cooling and freezing, and suggest intracellular freezing is occurring following nucleation by way of pore canals in the zona that are exposed as cooling progresses (Harvey and Ashwood-Smith, 1982).

Toxicity from high concentrations of cryoprotectants on salmonid ova has been observed by Stoss and Donaldson (1983) who investigated the action of DMSO (0, 1 and 2 mol) and the tolerance of Coho eggs to temperatures between -4.6 and -30° C. DMSO at 2 and 4 mol was detrimental to Coho eggs (0-1° C). One mol DMSO had no (Coho) or reduced (Rainbow Trout) influence on egg fertility when it was added gradually. In the presence of 1 mol DMSO most eggs remained unfrozen (67-89%) when kept for 10 min in frozen artificial medium (-4.6° C) and 27.32% subsequently reached the eyed stage (control = 100%). Further cooling (0.3° C/min) to -10.° C was still tolerated (62% unfrozen, 22% eyed eggs) but not to -20° C (6% unfrozen, no development) and -30° C (no survival). Use of 2 mol DMSO did not improve the results.

Short term storage of already fertilized eggs may prove superior to that of unfertilized eggs but misses the aim of cold storage or cryopreservation of gametes of separate sexes. Haga (1982) found that for fertilized eggs of the Rainbow Trout maintained at -7° C to approximately -20° C for 3 hours, a suitable concentration of DMSO as organic cryoprotectant was 1.4-2.1 M. At -7° C, an approximately 95% hatching rate was obtained by using 10% calf serum and mineral salts solution besides DMSO as cryoprotectants, while a 14% hatching rate was obtained at -12° C. Eggs (embryos) after the eyed stage had higher tolerance for subzero temperature exposure than prior to the eyed stage.

DMSO influences the concentrations of ions in frozen eggs relative to levels in frozen but untreated eggs. X-ray microanalysis of fertilized and unfertilized eggs of the Northern Pike, *Esox lucius*, frozen in liquid N revealed that unfertilized eggs incubated without DMSO had decreased levels of Na, Cl, and K in the zona radiata and increased levels of K in the cytoplasm. Unfertilized eggs incubated with 10% DMSO showed decreased Na and Cl in the zona radiata, decreased K in the cytoplasm and increased K in the cortical alveoli. Fertilized eggs incubated in buffer with 10% DMSO showed decreased levels of Na, P, Cl, and K (zona radiata), P, Cl, and K (cytoplasm), Na (yolk), and increased Cl in the yolk (all P < .01). It was also found that concentrations of DMSO differed between treatments and in different regions of the eggs. Thus DMSO (v/v) levels in these reached 1.5-3.1% in the zona radiata, 0-3.2% in cytoplasm, 2.3-8.7% in cortical alveoli, and 0-1.6% in the yolk. Unfertilized eggs showed more Me_2SO penetration than fertilized eggs (Schmehl and Graham, 1986).

Successful supercooling, as opposed to cryopreservation of ova has been achieved (Harvey, Stoss and Butchart, 1983). Unfertilized salmonid ova were stored at -1° C for up to 20 days in artificial media and in ovarian fluid. In all cases ovarian fluid preserved fertility longer than did the synthetic media.

Vacuum equilibration

Jamieson and Marshall (in preparation) have shown that two-step freezing to -196° C in a DMSO based cryoprotectant solution, following equilibration in a vacuum, achieved modest but significant fertilization rates for eggs of the Australian Bass (*Macquaria novemaculeata*). Similar experiments on the Fire-tail Gudgeon (*Hypseleotris galii*) will not be reported here as only small numbers of eggs were obtainable from this small fish which is not exploited as a source of food.

Cryoprotective diluents were prepared using 10%, 15% and 20% DMSO (v/v) combined with 15% skim milk powder (w/v). Ova were stripped from the female and mixed with the diluents in a ratio of 1:4 (v/v) in 1.8 ml vials. 200 eggs were used in each replicate sample. The samples were immediately exposed to a vacuum (virtually total) for a few seconds, 1 min or 5 min before freezing. Controls were run in which samples were frozen after equilbration at normal, atmospheric pressure. To anticipate, no viability of ova was obtained if they were exposed to the cryoprotectant at atmospheric pressure prior to freezing.

The two-step method of freezing consisted in cooling of vials containing eggs and diluent to -70° C in a methanol-solid CO_2 (dry ice) bath for 5 minutes before plunging into liquid Nitrogen.

Two weeks after freezing, samples were thawed in 40° C water baths. They were mixed with sperm from males stripped after spawning and were incubated in aerated 2.5% seawater at 18° C. As stated above, those not vacuum-treated prior to freezing did not develop.

The eggs exposed to the cryprotectant *in vacuo* before freezing underwent embryonic development, taking 4 days to hatch at 18° C.

Fertility of frozen, vacuum equilibrated ova showed an increase from 10% to 20% DMSO (Fig. 20.3). Fertility was highest (21% fertile ova) in one of the 20% DMSO replicates. Ova were infertile in 0% DMSO. Cortical reaction was also investigated and is shown in Fig. 20.3.

Only 8% ova hatched if insemination was by sperm which had been treated to vacuum exposure in the optimal conditions empirically demonstrated for sperm of 5% DMSO-diluent with 2 step freezing and no equilibration.

Development of the vacuum treated cryopreserved eggs and their untreated controls was not followed beyond the 32 cell stage in this pilot study. The method, giving the first cryopreservation of fish eggs, is nevertheless promising and further trials should be made with DMSO and with other, less toxic, cryprotectants. In addition it seems possible that use of a vacuum to facilitate removal of cryprotectant on dilution with normal medium in the last stages of thawing might alleviate toxic or other undesirable effects of the cryprotectant.

5. Dilution of gamete sample and cryprotectant

Sperm. Most workers use a dilution ratio of one part semen to 3-4 parts extender (Scott and Baynes, 1980) but there appears to be an optimal dilution of the ejaculate with the diluent for each species. In salmonids, the optimum rate is one volume of sperm to one volume of diluent including cryoprotectant (Billard, 1978b). Further increases in dilution ratio do not improve post-thaw fertility (Büyükhatipoglu and Holtz, 1978). Harvey (1983) found little improvement when the ratio exceeded 5:1 (sperm: diluent) for tilapian sperm.

Ova. We know of no work on the dilution levels for fish ova beyond that recorded here for the vacuum

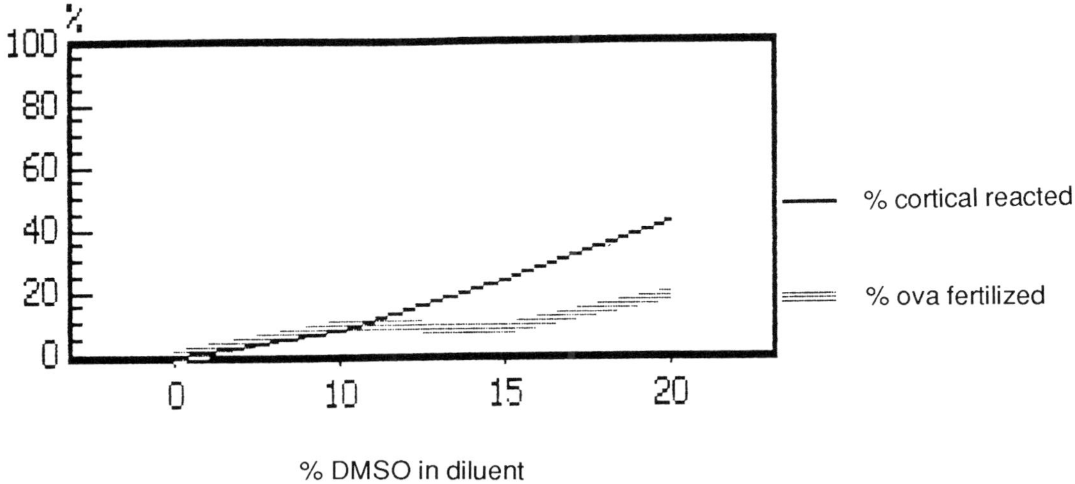

Fig. 20.3. Percentage fertility (hatching) and percentage of cortical reaction in cryopreserved Australian Bass (*Macquaria novemaculeata*) ova equilibrated for 5 minutes in vacuum in various concentrations of DMSO in diluent consisting of 15% skim milk in fish Ringer From Jamieson and Marshall, unpublished. Fig. 11.

equilibration method.

6. Equilibration Time

Sperm. Fish spermatozoa usually become motile upon mixing with diluents including the cryoprotectant. Equilibration time should, therefore, be kept to a minimum (minutes to <1 hr) to avoid exhaustion (Billard, 1978b, Chao *et al.*, 1975). This precaution will, in turn, minimize cryoprotectant toxicity, for example, in Rainbow Trout sperm (Stoss and Holtz, 1983c). As spermatozoa are sufficiently small and DMSO penetration is rapid, no lengthy equilibration period is required for this cryoprotectant (Harvey, 1983). However, penetration of glycerol is slow and a longer equilibration time is required.

Ova. For salmonid eggs methanol reaches only 23% of the expected equilibrium value after incubation for two hours (Harvey and Ashwood-Smith, 1982) (Fig. 20.4). Pike eggs reach a level of only 3.2% of DMSO in cytoplasm after two hours in 10% DMSO solution (Schmehl and Graham, 1986). Longer equilibration is clearly required. However, cryoprotectant toxicity may result from increased equilibration time (Stoss and Donaldson, 1983; also see Chapter 19, Section IVG). The vacuum equilibration technique (Jamieson and Marshall, in prepa-

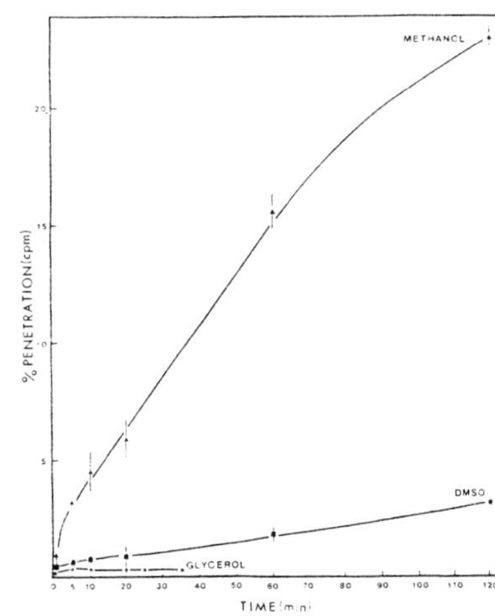

Fig. 20.4. Penetration of labelled cryoprotectants [^{14}C] glycerol, [^{14}C] methanol, and [^{3}H] DMSO into the trout egg at 0° C. Expressed as counts per minute inside/outside x 100; mean ± SD. From Harvey, B. and Ashwood-Smith, M.J. (1982). *Cryobiology* **19**, 29-40. Fig. 2.

ration), by shortening equilibration time while greatly augmenting penetration of the cryoprotectants, minimizes toxicity.

7. Equilibration Temperature

With the exception of glycerol, the permeability of most cryoprotectants is not markedly reduced by low temperatures and, therefore, equilibration is usually performed at 0° C to reduce cryoprotectant toxicity. Exposure to glycerol is preferably done at room temperature (see Chapter 19, Section VA).

B. Freezing Phase

Gametes prepared from the previous phase are cooled from 0 to -196° C during this phase.

1. Cooling Rate

Sperm. It appears that the freezing rate for fish spermatozoa has an optimal, though very wide, range. The cooling rates used successfully for salmonid and tilapian spermatozoa are, respectively, 30-160° C/min (Scott and Baynes, 1980) and 15-45° C/min (Harvey, 1983).

For Black Porgy (*Acanthopagrus schlegeli*) sperm, Chao *et al*. (1986) used an equilibration period no longer than 10 minutes; straws containing an extender consisting of 5% glucose mixed with glycerol were submerged in isopropanol at -10° C and then frozen at a rate of 2° C/min until the temperature reached -80° C or were held in liquid nitrogen vapour (-90 to -100° C) for 10 to 20 minutes. Between 50 and 90% of the post-thawed sperm were motile. After being cryopreserved for 1, 7, 7 and 342 days, sperm showed fertilities of 99.0, 93.2, 91.9 and 91.5% respectively.

For cichlids, Chao *et al*. (1987) cooled milt rapidly (with 15% milk and 5% methanol extender) to -35° C and then at 5° C min-1 to -75° C for storage in liquid nitrogen (-196° C). Fertilization rates obtained varied from 72.7% (vs. control 85.7%) to 93.4% (v. control 90%).

Ova. Because of their size and permeability to water, fish ova have very low optimum cooling rates (e.g. 0.001° C/min for salmonid eggs, Stoss and Donaldson, 1983) which are far from being feasible. Higher cooling rates (e.g. 0.01° C/min, Harvey and Ashwood-Smith, 1982; 0.3° per min, Stoss and Donaldson, 1983) resulted in intracellular ice (see Chapter 19, Section IVC). Again, the vacuum equilibration method overcomes these physical difficulties and warrants further investigation.

2. Freezing Method

Sperm. Commonly used freezing vessels for fish sperm are by vial, pellet and straw methods. These techniques are diagramatically illustrated in Fig. 20.2. The pellet method is most generally used because of its convenience. The freezing rate achieved by this technique (about 35° C per min) has been suitable for salmonid sperm (Stoss and Refstie, 1983). However, use of straws is possibly superior to pellets because recrystallization may occur when pellets are thawed (Erdahl *et al*., 1984).

Ova. No specific freezing method had been developed for fish ova prior to the introduction of the vacuum equilibration method (see above).

C. Storage Phase

Storage temperature is usually -196° C (liquid nitrogen) or 120° C (liquid nitrogen vapour), and maintanence of these storage temperature is essential. At these temperatures, gametes can be stored almost indefinitely without any deterioration (see Chapter 19, Section IIIC).

D. Thawing Phase

In this phase gametes are warmed from storage temperature to above 0° C.

1. Warming Rate

Rapid thawing is generally prefered for reasons given in Chapter 19, Section III.B. Post-thaw motility of tilapian sperm was maximum when highest warming rate (390° C per min) was used (Harvey, 1983).

2. Thawing Solution

A thawing solution is required for sperm frozen

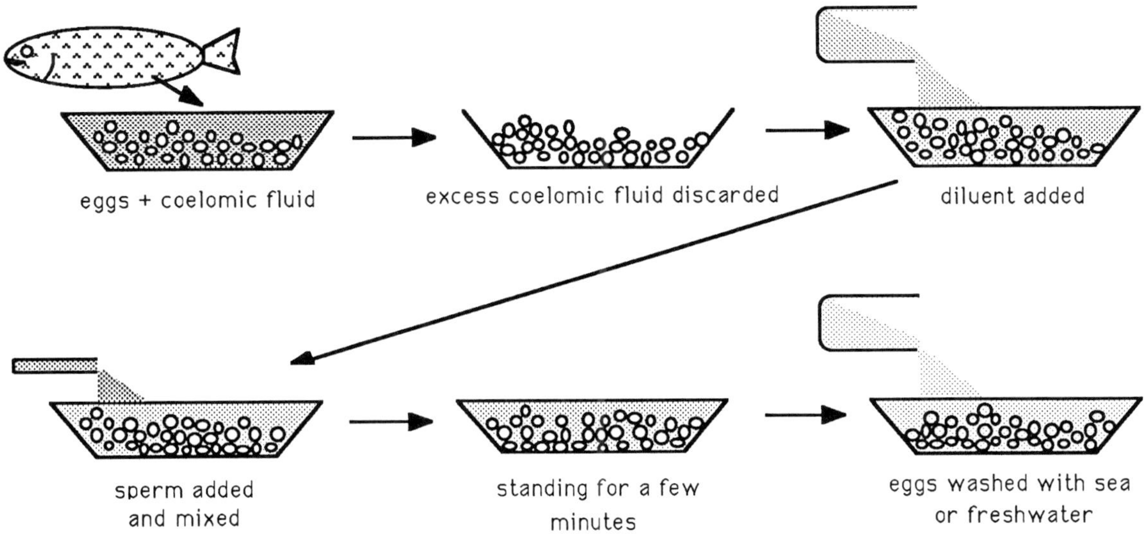

Fig. 20.5. General procedure for artificial insemination: diluent method. Based on several sources.

by the pellet method. Not only does it serve to absorb heat but it also provides a suitable medium for post-thaw survival and motility induction. The osmolality and pH of thawing solutions are important because they induce sperm motility (see Section IID). For salmonid (Rainbow Trout) sperm, $NaHCO_3$ at 249 mOsm produced the highest (85%) post-thaw fertility and the dilution effect of thawing solution was unimportant (Stoss and Holtz, 1983c). However, pellets of Rainbow Trout sperm thawed in 1% $NaHCO_3$ gave hatching results varying from 2.6-80.3% (Büyükhatipoglu and Holtz, 1978).

In a comparison undertaken by Stoss and Holtz (1981a) of several thawing solutions for cryopreserved Rainbow Trout sperm, best results were obtained with coelomic fluid, 119 mmol $NaHCO_3$, and 120 mmol NaCl (78.9, 75.3 and 69.0% fertilization, respectively, compared with 97.1% for controls). The lowest density of frozen sperm cells required to obtain satisfactory fertilization rates was determined to be 3×10^6 sperm per egg.

E. Post-Thaw Phase and Insemination

At this stage, gametes that have survived the cryopreservation procedure are ready for artificial insemination. A general procedure for this is illustrated in Fig. 20.5.

1. Estimation of motility

In evaluating the efficacy of cryopreservation procedures it is desirable to estimate post-thaw motility of sperm as a guide to their ability to successfully fertilize. It must be borne in mind, however, that only an empirical quantification of actual fertilization is a reliable indicator of success of the cryopreservation procedure. Even there initial high fertilization and early development may not result in viable embryos. Thus Koldras and Bieniarz (1987) found 98% mortality of embryos of Carp, *Cyprinus carpio*, embryos after 20 hours of ostensibly normal development following thawing and insemination of eggs by apparently successfully cryopreserved sperm. As is to be expected, the prefreezing rate of motility is not a reliable predictor of post-thaw fertility (Stoss and Holtz, 1983b).

Terner (1986) has outlined basic methodology for evaluating motility of salmonid sperm. Problems and pitfalls that are commonly encountered are identified, and suggestions are made for avoiding them. Sperm dilution in preparation for cryopreservation is discussed.

Craig *et al.* (1983) used quasi-elastic light scattering (QELS) to study the motile properties of rainbow trout spermatozoa with a view to possible extension of milt lifetime and the time available for insemination. It was possible to make motility evaluations at 10 sec intervals and to use the time course of these evaluations to study the effect of various concentrations of NaCl on motility. Concentrations of approximately 5% by weight best prolonged active swimming. The motility of undiluted samples could be initiated after storage at 4° C for at least 24 h.

2. Interval between Thawing and Insemination

After thawing, spermatozoa may show different motility characteristics from untreated sperm (Billard, 1978a; Leung, unpublished data, 1987). The duration of motility can be markedly reduced, for example, from half an hour in fresh sperm to 30 seconds in frozen-thawed sperm for the Grouper (Withler and Lim, 1982). For salmonid sperm, a delay of 30 seconds between thawing and addition of sperm to eggs reduced the fertilization rate from 72 to 56% (Stoss and Holtz, 1981a). Artificial insemination is, therefore, preferably performed immediately after thawing.

3. Insemination Temperature

Sperm.The speed of sperm movement depends on temperature. Since the energy reserve of fish spermatozoa is limited, the increased speed caused by a temperature rise shortens the lifespan (Ginzburg, 1968). The lowest limit of the temperatures for natural spawning is generally used as the temperature for artificial insemination. Commonly used insemination temperatures for salmonid sperm are between 0-10° C.

Ova. Temperature effects on fertility of ova are less understood, although Refstie *et al.* (1982) found reduced fertility in Coho eggs when insemination was carried out at 0 or 0.5° C instead of 10° C. Stoss and Donaldson (1983) showed no temperature effects between 0.5 and 9.8° C on salmonid eggs. (The ova used in both studies were stored but not cryopreserved.)

4. Insemination Medium

Besides inducing and prolonging sperm motility, the main function of the insemination medium is to give an adequate distribution of the spermatozoa in relation to the ova (Erdahl *et al.*, 1984). Medium composition is varied and has been reviewed by Scott and Baynes (1980). However, medium composition may be less important than usually thought as Billard (1983c) showed that "recommended" media, or coelomic fluid, did not differ from properly buffered diluents for insemination and in extending the life span of Rainbow Trout sperm. When the gametes were injured during the cryopreservation procedure, washing with coelomic fluid to remove seminal fluid had a more beneficial effect than washing with mineral diluents (see detailed discussion in Billard, 1983c).

Fresh water or sea water are not the best media for insemination in freshwater or marine fish, respectively, as destruction of sperm ultrastructure takes place in these media (Billard, 1978a; Maggese *et al.*, 1984). The insemination media best adapted to dilution of freshwater and marine fish sperm have a salinity of about 4-7% and 20% respectively, and a pH value about 9 for both types. In these solutions, osmotic injury is minimized (Billard, 1978b, 1983a; Maggese *et al.*, 1984). Undiluted Holtfreter medium (salinity 4%) maintained sperm morphology unchanged, and good motility, for the long-whiskered catfish, *Rhamdia sapo* (Maggese et al., 1984).

5. Dilution Ratio

Best artificial insemination results depend on an optimal sperm concentration (Erdahl *et al.*, 1984). The fertility of fish spermatozoa diluted in an insemination medium depends mainly on the dilution, and fertility declines more rapidly when the dilution exceeds 1:10 (v:v) (Billard,1983c). Not all spermatozoa survive the cryopreservation procedure, and densities of frozen-thawed sperm have to be higher than for fresh sperm to maintain the same level of fertility (Kurokura *et al.*, 1984; Legendre and Billard, 1980). For Rainbow Trout, 3 million frozen-thawed sperm per egg were required to obtain a fertility rate of ~80%. With fresh semen a minimum of 200 000 sperm per egg was necessary. Thus, with frozen semen 15 times more sperm are required (Stoss and Holtz, 1981a).

A similar ratio, of not less than 1:10, of semen to eggs has also proved desirable for Carp (*Cyprinus carpio*) fertilization following sperm cryopreservation. About 70% reached the eyed stage in fertility tests with 1 ml of cryopreserved semen added to 1 g of eggs. When 1 ml of semen was added to 100 g of eggs, few eggs reached the eyed stage. In two fertility tests with sperm preserved for 342 days, using 5 g of eggs inseminated with 1 ml of semen, percentages of eyed eggs were 31.5 and 25.5, respectively (Kurokura *et al.*, 1984).

6. Changes in semen consequent on cryopreservation

Cryopreservation of the sperm of the Grass Carp (*Ctenopharyngodon idella*) significantly decreases the content of total phospholipids, the decrease being approximately the same for all the phospholipid fractions. No relation was revealed between the degree of unsaturation of fatty acids in the sperm phospholipids and the environmental temperature. Presumably, fatty acid composition of the sperm phospholipids is affected to a higher extent by factors other than the environmental temperature of the animals (Drokin *et al.*, 1985).

Freeze-thawing has been shown to cause an increase of Na^+ and Ca^{2+} and a decrease of K^+ and Mg^{2+} in Rainbow Trout, *Salmo gairdneri*, spermatozoa. Sperm preserved for 343 days gave nearly 40% eyed eggs, a reduction of fertilizing capacity possibly related to the change in cation composition (Kurokura and Hirano, 1980).

Bibliography

Able, K.W. (1984). Cyprinodontiformes: development. In *Ontogeny and Systematics of Fishes.* Special Publication Number 1. American Society of Ichthyologists and Herpetologists, pp. 362-368.

Afzelius, B.A. (1970). Discussion in Nicander, L. 1970. Comparative studies on the fine structure of vertebrate spermatozoa. In *Comparative Spermatology* (ed. B. Baccetti), pp. 47-55. Academic Press, New York.

Afzelius, B.A. (1972). Sperm morphology and fertilization biology. In *The Genetics of the Spermatozoon* (eds. R.A. Beatty and S. Gluecksohn-Waelsch), pp. 131-143. Bogtrykkeriet Forum, Copenhagen.

Afzelius, B.A. (1977). Spermatozoa and spermatids of the crinoid *Antedon petasus*, with a note on primitive spermatozoa from deuterostome animals. *Journal of Ultrastructure Research* **59**, 272-281.

Afzelius, B.A. (1978). Fine structure of the garfish spermatozoon. *Journal of Ultrastructure Research* **64**, 309-314.

Afzelius, B.A (1979). Sperm structure in relation to phylogeny in the lower Metazoa. In *The Spermatozoon, Maturation, motility, surface properties and comparative aspects* (eds D.W. Fawcett and J.M. Bedford), pp. 243-251. Urban and Schwarzenberg, Baltimore.

Afzelius, B.A. and Ferraguti, M. (1978). Fine structure of brachiopod spermatozoa. *Journal of Ultrastructure Research* **63**, 308-315.

Afzelius, B.A. and Murray, A. (1957). The acrosomal reaction of spermatozoa during fertilization or treatment with egg water. *Experimental Cell Research* **12**, 325-337.

Afzelius, B.A., Eliasson, R. Johnsen, Ø. and Lindholmer, C. (1975). Lack of dynein arms in immotile human spermatozoa. *Journal of Cell Biology* **66**, 225-232.

Ahlberg, P.E. (1989). Paired fin skeletons and relationships of the fossil group Porolepiformes (Osteichthyes: Sarcopterygii). *Zoological Journal of the Linnean Society* **96**, 119-166.

Ahlstrom, E.H., Moser, H.G. and Cohen, D.M. (1984). Argentinoidei: development and relationships. In *Ontogeny and Systematics of Fishes.* Special Publication Number 1. American Society of Ichthyologists and Herpetologists, pp. 155-169.

Alderson, R. and MacNeil, A.J. (1984). Preliminary investigations of cryopreservation of milt of Atlantic Salmon (*Salmo salar*) and is application to commerical farming. *Aquaculture* **43**, 351-354.

Alldredge, A. (1976). Appendicularians. *Scientific American* **23**, 95-102.

Allen, G.R. (1980). A generic classification of the rainbowfishes (Family Melanotaeiniidae). *Records of the West Australian Museum* **8**, 449-490.

Allen, L.G. (1984). Gobiesociformes. In *Ontogeny and Systematics of Fishes.* Special Publication Number 1. American Society of Ichthyologists and Herpetologists, pp. 629-636.

Alvestad-Graebner, I. and Adam, H. (1977). Zur Feinstruktur der spermatogenetischen Stadien von *Myxine glutinosa* L. (Cyclostomata). *Zoologica Scripta* **6**, 113-126.

Anthony, J. (1980). Évocation des travaux français sur *Latimeria* notamment depuis 1972. *Proceedings of the Royal Society of London* B **208**, 349-367.

Anya, A.O. (1976). Physiological aspects of reproduction in nematodes. *Advances in Parasitology* **14**, 267-351.

Arnold, K., Pratsch, L., and Gawrisch, K. (1983). Effect of poly(ethylene glycol) on phospholipid hydration and polarity of the external phase. *Biochemical and Biophysical Acta* **728**, 121-128.

Asai, T. (1971). Fine structure of centriolar complex in spermiogenesis of the viviparous teleost fish *Lebistes reticulatus*. *Journal of Nara Medical Association* **22**, 371-382.

Ashwood-Smith, M.J. (1967). Radioprotective and cryoprotective properties of dimethyl sulfoxide in cellular systems. Annals of the New York Academy of Sciences **141**,45-62.

Ashwood-Smith, M.J. (1975). Current concepts concerning radioprotective and cryoprotective properties of dimethyl sulfoxide in cellular systems. *Annals of the New York Academy of Sciences* **243**, 246-256.

Ashwood-Smith, M.J. (1980). Low temperature preservation of cells, tissues and organs. In *Low temperature preservation in medicine and biology* (eds M.J. Ashwood-Smith and J. Farrant), pp. 19-44. Pitman Medical, Tunbridge Wells,

Ashwood-Smith, M.J. (1985). Genetic damage is not produced by normal cryopreservation procedures involving either glycerol or dimethyl sulfoxide: a cautionary note, however, on possible effects of dimethyl sulfoxide. *Cryobiology* **22**, 427-433.

Ashwood-Smith, M.J. (1986). The cryopreservation of human embryos. *Human Reproduction* **1**, 319-332.

Ashwood-Smith, M.J. and Lough, P. (1975). Cryoprotection of mamamlian cells in tissue culture with methanol. *Cryobiology* **12**, 517-518.

Ashwood-Smith, M.J., Warby, C., Connor, K.W., Becker, G. (1972). Low-temperature preservation of mammalian cells in tissue culture with polyvinylpyrrolidone (PVP), dextrans and hydroxyethyl starch (HES). *Cryobiology* **9**, 441-449.

Atkins, N.B. and Ohno, S. (1967). DNA values of four primitive chordates. *Chromosoma* (Berlin) **23**, 10-13.

Atwood, D.G. (1974). Fine structure of the spermatozoon of the sea cucumber *Leptosynapta clarki* (Echinodermata Holothuroidea). *Cell and Tissue Research* **149**, 223-233.

Atwood, D.G. and Chia, F.S. (1974). Fine structure of an unusual spermatozoon of a brooding sea cucumber, *Cucumaria lubrica*. *Canadian Journal of Zoology* **52**, 519-523.

Baccetti, B. (1978). L'evoluzione dello spermatozoo. In *IV Seminario sulla "Evoluzione Biologica"*. Academia Nazionale del Lincei, Rome, pp. 95-126.

Baccetti, B. (1979). The evolution of the acrosome complex. In Fawcett, D.W. and Bedford, J.M. (eds): *The Spermatozoon*. Baltimore-Munich: Urban and Schwarzenberg, pp 305-329.

Baccetti, B. (1985). Evolution of the sperm cell. In *Biology of fertilization* (eds C.B. Metz and A. Monroy), pp 3-58. Academic Press, New York.

Baccetti, B. (1986). Evolutionary trends in sperm structure. *Comparative Biochemistry and Physiology* **85**, 29-36.

Baccetti, B. and Afzelius, B.A. (1976). *The Biology of the Sperm Cell*. Monographs in Developmental Biology, Basel, Karger.

Baccetti, B., Burrini, A.G. and Dallai, R. (1972). The spermatozoon of *Branchiostoma lanceolatum* L. *Journal of Morphology* **136**, 211-226.

Baccetti, B., Burrini, A.G. and Pallini, V. (1981). Different axoneme patterns in cilia and flagella of the same animal. *Journal of Submicroscopic Cytology* **13**, 479-481.

Baccetti, B., Burrini, A.G., Dallai, R. and Pallini, V. (1979a). The dynein electrophoretic bands in axonemes naturally lacking the inner or the outer arm. *Journal of Cell Biology* **80**, 334-340.

Baccetti, B., Dallai, R., Grimaldi de Zio, S. and Marinari, A. (1983). The evolution of the nematode spermatozoon. *Gamete Research* **8**, 309-323.

Baccetti, B., Burrini, A.G., Maver, A., Pallini, V. and Renieri, T. (1979b). "9+0" immotile spermatozoa in an infertile man. *Andrologia* **11**, 437-443.

Baccetti, B., Burrini, A.G., Callaini, G., Gibertini, G., Mazzini, M. and Zerunian, S. (1984). Fish germinal cells. I. Comparative spermatology of seven cyprinid species. *Gamete Research* **10**, 373-396.

Balon, E.K., Bruton, M.N. and Fricke, H. (1988). A fiftieth anniversary reflection on the living coelacanth, *Latimeria chalumnae*: some new interpretations of its natural history and conservation status. *Environmental Biology of Fishes* **23**, 241-280.

Barnes, R. S. (1980). *Invertebrate Zoology*. 5th.

Edition. Saunders, New York.

Bartram, A.W.H. (1977). The Macrosemiidae, a Mesozoic family of holostean fishes. *Bulletin of the British Museum of Natural History (Geology)* **29**, 137-234.

Bates, W.R. (1981). Sperm-egg interactions during fertilization of Ciona intestinalis, an SEM study. *American Zoologist* **21**, 1044.

Baynes, S.M. and Scott, A.P. (1987). Cryopreservation of Rainbow Trout spermatozoa: the influence of sperm quality, egg quality and extender composition on post-thaw fertility. *Aquaculture* **66**, 53-67.

Baynes, S.M, Scott. A.P. and Dawson, A.P. (1981). Rainbow Trout, *Salmo gairdneri* Richardson, spermatozoa: effects of cations and pH on motility. *Journal of Fish Biology* **19**, 259-267.

Benau, D. and Terner, C. (1980). Initiation, prolongation, and reactivation of the motility of salmonid spermatozoa. *Gamete Research* **3**, 247-257.

Bernardini, G., Stipani, R. and Melone, G. (1986). The ultrastructure of *Xenopus* spermatozoon. *Journal of Ultrastructure and Molecular Research* **94**, 188-194.

Berrill, N.J. (1950). *The Tunicata.* Ray Society, London.

Berrill, N.J. (1955). *The Origin of the Vertebrates.* Clarendon Press, Oxford.

Bertelsen, E. (1980). Notes on Linphrynidae V: A revision of the deepsea anglefishes of the *Linophryne arborifera* - group (Pisces, Ceratioidei). *Steenstrupia* **6**, 29-70.

Billard, R. (1970a). Ultrastructure comparée de spermatozoïdes de quelques poissons Téléostéens. In *Comparative Spermatology* (ed. B. Baccetti), pp. 71-79. Academic Press, New York.

Billard, R. (1970b). La spermatogenèse de *Poecilia reticulata.* IV. La spermiogenèse. Étude ultrastructurale. *Annales de Biologie Animale Biochimie Biophysique* **10**, 493-510.

Billard, R. (1978a). Changes in structure and fertilizing ability of marine and freshwater fish spermatozoa diluted in media of various salinities. *Aquaculture* **14**, 187-198.

Billard, R. (1978b). Some data on gametes preservation and artificial insemination in teleost fish. *Actes de Colloques du Centre Nationale de l'Exploitation des Oceans (CNEXO)* **8**, 59-73.

Billard, R. (1980). Reproduction and artificial insemination in teleost fish. *International Congress on Animal Reproduction and Artificial Insemination.* 1980. Madrid. pp. 327-337.

Billard, R. (1981). Short-term preservation of sperm under oxygen atmosphere in Rainbow trout (*Salmo gairdneri*). *Aquaculture* **23**, 287-293.

Billard, R. (1983a). Ultrastructure of trout spermatozoa: changes after dilution and deep-freezing. *Cell and Tissue Research* **228**, 205-218.

Billard, R. (1983b). Spermiogenesis in rainbow trout (*Salmo gairdneri*). *Cell and Tissue Research* **233**, 265-284.

Billard, R. (1983c). Effects of coelomic and seminal fluids and various saline diluents on the fertilizability of spermatozoa in the Rainbow Trout, *Salmo gairdneri. Journal of Reproduction and Fertility* **68**, 77-84.

Billard, R. (1984). La conservation des gamétes et l'insémination artificielle chez le bar et la daurade. In G. Barnabé et R. Billard (Eds.) *L'Aquaculture du Bar et des Sparidés.* INRA Publications, Paris. pp. 95-116.

Billard, R. (1986). Spermatogenesis and spermatology of some teleost fish species. *Reproduction, Nutrition and Développement* **26**, 877-1024.

Billard, R. and Ginsburg, A.S. (1973). La spermiogenèse et le spermatozoïde d'*Anguilla anguilla* L. Étude ultrastructurale. *Annales de Biologie Animale, Biochemie, Biophysique* **13**, 523-534.

Billard, R. and Jalabert B. (1974). L'insemination artificielle de la truite (*Salmo gairdneri* Richardson). II - Comparison des effets de differents dilueurs sur la conservation de la fertilité des gametes avant et après insemination. *Annales de Biologie Animale Biochimie Biophysique* **14**, 601-610.

Billard, R., Breton, B. and Jalabert, B. (1971). La production spermatogénétique chez la Truite. *Annales de Biologie Animale, Biochimie, Biophysique* **11**, 190-212.

Bischoff, T.L.W. von (1840). Description anatomique du *Lepidosiren paradoxa. Annales de Science Naturelle* **14**, 116-159.

Bjerring, H.C. (1985). In *Facts and thoughts on piscine phylogeny* (eds R.E. Foreman, A. Gorban, J.M. Dodd. and R. Olson), pp. 31-57. Plenum

Press, New York.

Blackshaw, A.W. (1954). The prevention of temperature shock of bull and ram semen. *Australian Journal of Biological Sciences* **7**, 573-582.

Blaxter, J.H.S. (1953). Sperm storage and cross-fertilization of spring and autumn spawning herring. *Nature* **172**, 1189-1190.

Boisson, C. (1963). La spermiogenèse de *Protopterus annectens* (Dipneuste) du Sénégal étudiée au microscope optique et quelques détails au microscope électronique. *Annales de la Faculté des Sciences Université de Dakar* **10**, 43-72.

Boisson, C., Mattei, C., and Mattei, X. (1967). Troisième note sur la spermiogenèse de *Protopterus annectens* (Dipneuste) du Sénégal. *Institut Fondamental d'Afrique Noire. Bulletin* Série A. (Sciences Naturelles) **29**, 1097-1121.

Boisson, C., Mattei, X. and Mattei, C. (1968a). Le spermatozoïde de *Dactylopterus volitans*, Linné (Poisson Cephalacanthidae), étudié au microscope électronique. *Comptes Rendus des Séances de la Société de Biologie de l'ouest Africain* **162**, 820-823.

Boisson, C., Mattei, X. and Mattei, C. (1968b). La spermiogenèse de *Rhinobatus cemiculus* Geof. St-Hilaire [Sélacien *Rhinobatidae*]. Étude au microscope électronique. *Institut Fondamental d'Afrique Noire. Bulletin* Série A. (Sciences Naturelles) **30**, 659-673.

Boisson, C., Mattei, X. and Mattei, C. (1969). Mise en place et évolution du complexe centriolaire au cours de la spermiogenèse d'*Upeneus prayensis* C.V. (Poisson Mullidae). *Journal de Microscopie* (Paris) **8**, 103-112.

Bolla, S. Holmefjord, I. and Refstie, T. (1987). Cryogenic preservation of Atlantic Halibut sperm. *Aquaculture* **65**, 371-374.

Boutron, P. and Kaufmann, A. (1979). Stability of the amorphous state in the system water-1,2 propanediol. *Cryobiology* **16**, 557-568.

Boutron, P. and Mehl, P. (1986). Glass-forming tendency and stability of the wholly amporphous state in systems with water and polyalcohols with four carbons. *Cryobiology* **23**, 561-562.

Brainerd, E.L., Liem, K.F. and Samper, C.T. (1989). Air ventilation by recoil aspiration in Polypterid fishes. *Science* **246**, 1593-1595.

Bratanov, C. and Dikov, V. (1961). Sur certaines particularités du sperme chez les poissons. In *Proceedings of the IVth International Congress on Animal Reproduction* pp. 895-897. N.V. Drukkerij Trio, The Hague.

Breder, C.M. and Rosen, D.E. (1966). *Modes of Reproduction in Fishes*. The American Museum of Natural History, The Natural History Press, New York.

Briggs, J.C. (1955). A monograph of the clingfishes (Order Xenopterygii). *Stanford Icthyological Bulletin* **6**, 1-224.

Brokaw, C.J. and Omoto, C.K. (1982). The terminal piece of *Ciona* sperm flagella. *Biophysical Journal* **37**, 284a.

Brummett, A.R. and Dumont, J.N. (1979). Initial stages of sperm penetration into the egg of *Fundulus heteroclitus*. *Journal of Experimental Zoology* **210**, 417-434.

Brummett, A.R. and Dumont, J.N. (1981). Cortical vesicle breakdown in fertilized eggs of *Fundulus heteroclitus*. *Journal of Experimental Zoology* **216**, 63-79.

Brummett, A.R., Dumont, J.N., and Richter, C.S. (1985). Later stages of sperm penetration and polar body formation in the egg of *Fundulus heteroclitus*. *Journal of Experimental Zoology* **234**, 423-439.

Brusle, S. (1981). Ultrastructure of spermiogenesis in *Liza aurata* Risso, 1810 (Teleostei, Mugilidae). *Cell and Tissue Research* **217**, 415-424.

Buckland, F. (1891). *Natural History of British Fishes*. Society for the Promotion of Christian Knowledge, London.

Bullman, O.M.B. (1970). Graptolithina with sections on Enteropneusta and Pterobranchia. In *Treatise on Invertebrate Palaeonotology* pt. 5 (ed.C. Teichert), 163 pp. Geological Society of America and University of Kansas. (*Fide* Nelson, 1984).

Burighel, P., Martinucci, G.B. and Balia, G. (1982). The spermatozoa of the ascidians *Botryllus schlosseri* and *Diplosoma listerianum* (Urochordata). *Caryologia* **35**, 127-128.

Burighel, P., Martinucci, G.B. and Magri, F. (1985). Unusual structures in the spermatozoa of the ascidians *Lissoclinum perforatum* and *Diplosoma listerianum* (Didemnidae). *Cell and Tissue Research* **241**, 513-521.

Büyükhatipoglu, S. and Holtz, W. (1978). Preservation of trout sperm in liquid or frozen state. *Aquaculture* **14**, 49-56.

Caffrey, M. (1986). The combined and separate effects of temperature and freezing on membrane lipid phase behaviour. *Cryobiology* **23**, 543-580.

Carrillo, M. and Zanuy, S. (1977). Quelques observations sur le testicule chez *Spicara chryselis* C. V. *Investigación Pesquera* **41**, 121-146.

Casas, M.T., Munoz-Guerra, S. and Subirana, J.A. (1981). Preliminary report on the ultrastructure of chromatin in the histone containing spermatozoa of a teleost fish. *Biology of the Cell* **40**, 87-92.

Chao, N.-H., Chen, H.-P. and Liao, I.-C. (1975). Study on cryogenic preservation of grey mullet sperm. *Aquaculture* **5**, 389-406.

Chao, N.-H., Chao, W.-C., Liu, K.-C. and Liao, I.-C. (1986). The biological properties of Black Porgy (*Acanthopagrus schlegeli*) sperm and its cryopreservation. *Proceedings of the National Science Council of the Republic of China* Part B Life Sciences. **10**, 145-149.

Chao, N.-H., Chao, W.-C., Liu, K.-C. and Liao, I.-C. (1987). The properties of tilapia sperm and its cryopreservation. *Journal of Fish Biology* **30**, 107-118.

Cherr, G.M. and Clark, W.H. (1984a). Jelly release in the eggs of the white sturgeon, *Acipenser transmontanus*: An enzymatically mediated event. *The Journal of Experimental Zoology* **230**, 145-149.

Cherr, G.M. and Clark, W.H. (1984b). An acrosome reaction in sperm from the White Sturgeon, *Acipenser transmontanus*. *The Journal of Experimental Zoology* **232**, 129-139.

Chia, F.S. and Bickell, L.R. (1983). Echinodermata. In *Reproductive Biology of Invertebrates* (eds K.G. and R.G Adiyodi), pp. 545-620. Volume II. Spermatogenesis and Sperm Function. John Wiley and Sons, Chichester.

Chia, F.S., Atwood, D. and Crawford, B. (1975). Comparative morphology of echinoderm sperm and possible phylogenetic implications. American Zoologist, **15**, 553-565.

Chuang, S.H. (1983). Brachiopoda. In *Reproductive Biology of Invertebrates* (eds K.G. and R.G Adiyodi), pp. 517-530. Volume II. Spermatogenesis and Sperm Function. John Wiley and Sons, Chichester.

Clemens, H.P. and Grant, F.B. (1965). The seminal thinning response of carp (*Cyprinus carpio*) and rainbow trout (*Salmo gairdneri*) after injections of pituitary extracts. *Copeia* **2**, 174-177.

Cloney, R.A. and Abbott, L.C. (1980). The spermatozoa of ascidians: acrosome and nuclear envelope. *Cell and Tissue Research* **206**, 261-270.

Çolak, A. and Yamamoto, K. (1974a). An electron microscope study of spermiogenesis in the Japanese Eeel, *Anguilla japonica*. Bulletin of the Faculty of Fisheries, Hokkaido University **25**, 1-5.

Çolak, A. and Yamamoto, K. (1974b). Ultrastructure of the Japanese eel spermatozoon. *Annotationes Zoologicae Japonenses* **47**, 48-54.

Collette, B.B., McGowen, G.E., Parin, N.V. and Mito, S. (1984). Beloniformes: development and relationsips. In *Ontogeny and Systematics of Fishes*. Special Publication Number 1. American Society of Ichthyologists and Herpetologists, pp. 335-354.

Colwin, A.L. and Colwin, L.H. (1963). Role of the gamete membranes in fertilization in *Saccoglossus kowalevskii* (Enteropneusta) I. The acrosomal region and its changes in early stages of fertilization. *Journal of Cell Biology* **19**, 477-500.

Colwin, A.L. and Colwin, L.H. (1967). Behaviour of the spermatozoon during sperm-blastomere fusion and its significance for fertilization (*Saccoglossus kowalevskii*): Hemichordata. *Zeitschrift fur Zellforschung* **78**, 208-220.

Colwin, A.L., Colwin, L.H. and Philpott, D.E. (1957). Electron microscope studies of early stages of sperm penetration in *Hydroides hexagonus* (Annelids) and *Saccoglossus kowalevskii* (Enteropneusta). *Journal of Biophysical and Biochemical Cytology* **3**, 489-502.

Colwin, A.L., Colwin, L.H., and Summers, R.G. (1974). The acrosomal region and the beginning of fertilization in the holothurian, *Thione briareus*. In *The Functional Anatomy of the Spermatozoon*. (ed. B.A. Afzelius), pp. 27-38. Pergamon Press, Oxford.

Connor, W. and Ashwood-Smith, M.J. (1973). Cryoprotection of mammalian cells in tissue culture with polymers; possible mechanisms. *Cryobiology* **10**, 488-496.

Cóser, A.M.L., Godinho, H.P. and Torquato, V.C. (1987). Criopreservaçao de sêmen do peixe piau *Leporinus silvestrii* (Boulenger 1902). *Arquivo Brasileiro de Medicino Veterinaria e Zootecnia* **39**, 37-42.

Cotelli, F., De Santis, R., Rosati, F. and Monroy, A. (1980). Acrosome differentiation in the spermatogenesis of *Ciona intestinalis*. *Development, Growth and Differentiation* **22**, 561- 569.

Craig, T., Blaber, A. and Hallet, F.R. (1983). Motility of the spermatozoa of rainbow trout in solutions of various salinities as studied by Quasi-elastic light scattering. *Biology of reproduction* **29**, 1189-1193.

Cussac, V.E. and Maggese, M.C. (1988). Effects of salt solutions on the fertilizing ability and activation of gametes in the catfish, *Rhamdia sapo*, (Pisces, Pimelodidae). *Revista Brasileira de Biologia* **48**, 203-211.

Dadone, L. and Narbaitz, R. (1967). Submicroscopic structure of spermatozoa of a cyprinodontiform teleost, *Jenynsia lineata*. *Zeitschrift für Zellforschung* **80**, 214-219.

Daget, J. (1986). Position Systematique des Polypterides et Nomenclature. *Océanis* **12**, 183-188.

Dale, B., Denis-Donini, S., De Santis, R., Monroy, A., Rosati, F. and Taglietti, V. (1978). Sperm-egg interaction in the ascidians. *Biologie Cellulaire* **32**, 129-133.

Dan, J.C. (1970). Morphogenetic aspects of acrosome formation and reaction. *Advances in Morphogenesis* **8**, 1-139.

Davenport, J., Lonning, S. and Kjorsvik, E. (1981). Osmotic and structural changes during early development of eggs and larvae of the Cod, *Gadus morhua* L. *Journal of Fish Biology* **19**, 317-331.

de Montalembert, G., Bry, C. and Billard, R. (1978). Control of reproduction in northern pike. *American Fish Society Special Publications* **11**, 217-225.

De Santis, R., Jamunno G. and Rosati, F. (1980). A study of the chorion and the follicle cells in relation to the sperm-egg reaction in the ascidian, *Ciona intestinalis*. *Developmental Biology* **74**, 490-499.

de Sylva, D.P. (1984). Mugiloidei: development and relationships. In *Ontogeny and Systematics of Fishes*. Special Publication Number 1. American Society of Ichthyologists and Herpetologists, pp. 530-533.

Detlaf, T.A. and Ginzburg, A.S. (1963). Acrosome reaction in sturgeons and the role of calcium ions in the union of gametes. *Doklady Akademii Nauk SSSR* **153**, 1461-1464.

Deurs, B. van (1973). Helical, striated rootlets in the midpiece of a teleost fish spermatozoon. *Zeitschrift für Anatomie Entwicklungsgeschichte* **140**, 11-17.

Deurs, B. van (1974). The sperm cells of *Pantodon* (Teleostei) with a note on residual body formation. In *The functional anatomy of the spermatozoon* (ed. B.A. Afzelius), pp. 311-318. Pergamon Press, Oxford.

Deurs, B. van and Lastein, U. (1973). Ultrastructure of the spermatozoa of the teleost *Pantodon buchholzi* Peters, with particular reference to the midpiece. *Journal of Ultrastructure Research* **42**, 517-533.

Doi, M., Hoshino, T., Taki, Y. and Ogasawara, Y. (1982). Activity of the sperm of the bluefin tuna *Thunnus thynnus* under fresh and preserved conditions. *Bulletin of the Japanese Society of Scientific Fisheries* **4**, 495-498.

Drokin, S.I., Zabelinskii, S.A. and Kopeika, E.F. (1985). The effect of cryopreservation on phospholipids and their fatty acids from the sperm of the Grass Carp *Ctenopharyngodon idella* and Turkey Hen *Meleagris gallopavo*. *Zhurnal Evolyutsionnoi Biokhimi Fiziologii* **21**, 79-82.

Drozdov, A.L., Kolotukhina, N.K. and Maximovich, A.A. (1981). On histological structure of testes and ultrastructure of spermatozoa in the salmon *Oncorhynchus gorbuscha*. *Biologiya Morya* (Vladivostok) **1**, 49-53.

Durbin, H., Durbin, F.J. and Stott, B. (1982). A note on the cryopreservation of Grass Carp milt. *Fisheries Management* **13**, 115-117.

Ehlers, U. (1984). Phylogenetisches System der Platyhelminthes. *Verhandlungen der Naturwissenschaftlichen Vereins in Hamburg* **27**, 291-294.

Emig, C.C. (1977a). The systematics and evolution of the phylum Phoronida. *Zeitschrift für Zool Systematik und Evolutionsforschung* **12**, 128-

151.

Emig, C.C. (1977b). Embryology of the Phoronida. *American Zoologist* **17**, 21-37.

Erdahl, D.A. and Graham, E.F. (1980). Preservation of gametes of freshwater fish. *9th. International Congress on Animal Reproduction. and Artificial Insemination,* Madrid. **2**, 317-326.

Erdahl, A.W., Erdahl, D.A. and Graham, E.F. (1984). Some factors affecting the preservation of salmonid spermatozoa. *Aquaculture* **43**, 341-350.

Ezell, S.D. (1963). The lateral body of *Ciona intestinalis* spermatozoa. *Experimental Cell Research* **30**, 615-617.

Fahy, G.M. (1982). Prevention of toxicity from high concentrations cryoprotective agents. In *Organ Preservation, basic and applied Aspects* (eds D.E. Pegg, I.A. Jacobsen and N.A. Halasz), pp367-369. MTP Press, Lancaster.

Fahy, G.M., Levy, D.I. and Ali, S.E. (1986). Vitrification solutions: molecular and biological aspects. *Cryobiology* **23**, 560.

Fahy, G.M., MacFarlane, D.R., Angell, C.A. and Meryman, H.T. (1984). Vitrification as an approach to cryopreservation. *Cryobiology* **21**, 407-426.

Farmer, J.D. (1977). An adaptive model of evolution of the ectoproct life cycle. In *Biology of the Bryozoa* (eds R.M. Woollacott and R.L. Zimmer), pp. 487-517. Academic Press, New York.

Fernholm, B. (1975). Ovulation and eggs of the Hagfish *Eptatretus burgeri. Acta Zoologica* (Stockholm) **56**, 199- 204.

Field, K.G., Olsen, G.J., Lane, D.J., Giovannoni, S.J., Ghiselin, M.T., Raff, E.C., Pace, N.R. and Raff, R.A. (1988). Molecular phylogeny of the Animal Kindgom. *Science* **239**, 748-753.

Fink, W.L. (1984). Basal euteleosts; relationships. In *Ontogeny and Systematics of Fishes.* Special Publication Number 1. American Society of Ichthyologists and Herpetologists, pp. 202-206.

Fink, S.V. and Fink, W.L. (1981). Interrelationships of the ostariophysan fishes (Teleostei). *Zoological Journal of the Linnean Society of London* **72**, 297-353.

Fink, W.L. and Weitzman, S.H. (1982). Relationships of the stomiiform fishes (Teleostei), with a description of *Diplophos. Bulletin of the Museum of Comparative Zoology* Harvard **150**, 31-93.

Flood, P.R. and Afzelius, B.A. (1978). The spermatozoon of *Oikopleura dioica* Fol (Larvacea, Tunicata). *Cell and Tissue Research* **191**, 23-37.

Flower, J.W. (1967). The effect of tail shape on the propulsive efficiency of spermatozoa. *Annals of the Entomological Society of America* **60**, 639-640.

Follenius, E. (1965). Particularités de structure des spermatozoïdes de *Lampetra planeri*: etude au microscope électronique. *Journal of Ultrastructure Research* **13**, 459-468.

Fontaine, A.R. and Lambert, P. (1976). The fine structure of the sperm of a holothurian and ophiuroid. *Journal of Morphology* **148**, 209-226.

Forey, P.L. (1973). Relationships of elopomorphs. In *Interrelationships of Fishes* (eds P.H. Greenwood, R.S. Miles and C. Patterson), pp. 351-368. *Journal of the Linnean Society of London* Zoology 53 (suppl 1). Academic Press, New York.

Forey, P.L. (1980). *Latimeria*: a paradoxical fish. *Proceedings of the Royal Society of London* B **208**, 369-384.

Forey, P.L. (1986). Relationships of lungfishes. *Journal of Morphology Supplement* **1**, 75-91.

Forey, P.L. (1988). Golden jubilee for the coelacanth Latimeria chalumnae. *Nature* **336**, 727-732.

Franzén, Å. (1956). On spermiogenesis, morphology of the spermatozoon, and biology of fertilization among invertebrates. *Zoologiska Bidrag från Uppsala* **31**, 355-482.

Franzén, Å. (1958). On sperm morphology and acrosome filament formation in some annelida, echiuroidea, and tunicata. *Zoologiska Bidrag från Uppsala* **33**, 1-28.

Franzén, Å. (1970). Phylogenetic aspects of the morphology of spermatozoa and spermiogenesis. In *Comparative Spermatology* (ed. B. Baccetti), pp. 29-45. Academic Press, New York.

Franzén, Å. (1976a). On the ultrastructure of spermiogenesis of *Flustra foliacea* (L.) and *Triticella korenii* G.O. Sars (Bryozoa). *Zoon* **4**, 19-29.

Franzén, Å. (1976b). The fine structure of spermatid differentiation in a tunicate, *Corella parallelogramma* (Muller). *Zoon* **4**, 115-120.

Franzén, Å. (1977a). Sperm structure with regard to fertilization biology and phylogenetics. *Ver-*

handlungen der deutschen zoologischen Gessellschaft **1977**, 123-138.

Franzén, Å. (1977b). Gametogenesis of bryozoans. In *Biology of the Bryozoa* (eds R.M. Woollacott and R.L. Zimmer), pp. 1-22. Academic Press, New York.

Franzén, Å. (1981). Comparative ultrastructural studies of spermatids and spermatozoa in Bryozoa and Entoprocta. In *Recent and fossil Bryozoa*. (Eds G.P. Larwood, and C. Nielsen) pp. 83- 92. Olsen and Olsen, Fredensborg.

Franzén, Å. (1983a). Bryozoa Ectoprocta. In *Reproductive Biology of Invertebrates*. Volume II. Spermatogenesis and Sperm Function (eds K.G. and R.G.), pp. 491-504. John Wiley and Sons. Chichester.

Franzén, Å. (1983b). Urochordata. In *Reproductive Biology of Invertebrates*. Volume II. Spermatogenesis and Sperm Function (eds K.G. and R.G.), pp. 621-632. John Wiley and Sons, Chichester.

Franzén, Å. and Ahflors, K. (1980). Ultrastructure of spermatids and spermatozoa in *Phoronis*, phylum Phoronida. *Journal of Submicroscopic Cytology* 12, 585-597.

Franzén, Å, Woodwick, K.H. and Sensenbaugh, T. (1985). Spermiogenesis and ultrastructure of spermatozoa in *Saxipendium coronatum* (Hemichordata, Enteropneusta), with consideration of their relation to reproduction and dispersal. *Zoomorphology* **105**, 302-307.

Fredrich, F. (1984). Spermakonservierung bei Salmoniden. *Zeitschrift fuer die Binnenfischerei de DDR* **31**, 24-31.

Fribourgh, J.H. (1978). Morphology of the brook trout spermatozoon as determined by scanning and transmission electron microscopy. *The Progressive Fish-Culturist* **40**(1), 26-29.

Fribourgh, J.H. and Soloff, B.L. (1976). Scanning electron microscopy of the Rainbow Trout (*Salmo gairdneri* Richardson) spermatozoon. *Arkansas Academy of Science Proceedings* **30**, 41-43.

Fribourgh, J.H., McClendon, D.E. and Soloff, B.L. (1970). Ultrastructure of the goldfish, *Carassius auratus* (Cyprinidae), spermatozoon. *Copeia* **1970**, No.2, 274-279.

Fritzsch, B. (1987). Inner ear of the coelacanth fish *Latimeria* has tetrapod affinities. *Nature* 327(14), 153-154.

Fritzsche, R.A. (1984). Gasterosteiformes: development and relationships. In *Ontogeny and Systematics of Fishes*. Special Publication Number 1. American Society of Ichthyologists and Herpetologists, pp. 398-405.

Fuiman, L.A. (1984). Ostariophysi: development and relationships. In *Ontogeny and Systematics of Fishes*. Special Publication Number 1. American Society of Ichthyologists and Herpetologists, pp. 126-137.

Fujikawa, S. (1978). Morphology evidence of membrane damage caused by intracellular ice crystals. *Cryobiology* **15**, 707.

Fujikawa, S. and Miura, K. (1986). Plasma membrane ultrastructural changes caused by mechanical stress in the formation of extracellular ice as a primary cause of slow freezing injury in fruit-bodies of basidiomycetes (*Lyophyllum ulmarium* (Fr.) Kühner). *Cryobiology* **23**, 371-382.

Fujimura, W., Harutsugu, M., Nishiki, T. and Ito, K. (1957). Electron microscope study of sections of carp sperms. *Journal of Nara Medical Association* **7**, 122-124.

Fuke, T.M. (1983). Self and non-self recognition between gametes of the ascidian, *Halocynthia roretzi*. *Wilhelm Roux's Archives of Devel opmental Biology* **192**, 347-352.

Fukumoto, M. (1979). Tube-like structures in mitochondria of tunicate (*Pyura vittata*) spermatids. *Journal of Ultrastructure Research* **68**, 1-5.

Fukumoto, M. (1981). The spermatozoa and spermiogenesis of *Perophora formosana* (Ascidia) with special reference to the striated apical structure and the filamentous structures in the mitochondrion. *Journal of Ultrastructure Research* **77**, 37-53.

Fukumoto, M. (1983). Fine structure and differentiation of the acrosome-like structure of the solitary ascidians, *Pyura haustor* and *Styela plicata*. *Development, Growth and Differentiation* **25**, 503-515.

Fukumoto, M. (1984a). Fertilization in ascidians: acrosome fragmentation in *Ciona intestinalis* spermatozoa. *Journal of Ultrastructure Research* **87**, 252-262.

Fukumoto, M. (1984b). The apical structure in *Perophora annectens* (Tunicate) spermatozoa: fine structure, differentiation and possible role in fertilization. *Journal of Cell Science* **66**, 175-187.

Fukumoto, M. (1985). Acrosome differentiation in *Molgula manhattensis* (Ascidiacea, Tunicata). *Journal of Ultrastructure Research* **92**, 158-166.

Fukumoto, M. (1986). The acrosome in ascidians: I. Pleurogona. *International Journal of Invertebrate Reproduction and Development* **10**, 335-346.

Fukumoto, M. (1988). Fertilization in ascidians: apical processes and gamete fusion in *Ciona intestinalis* spermatozoa. *Journal of Cell Science* **89**, 189-196.

Fukumoto, M. (1990). The acrosome reaction in *Ciona intestinalis. Development, Growth and Differentiation* **32**, 51-55.

Furieri, P. (1962). Prime osservazioni al microscopio elettronico sullo spermatozoo di *Salmo trutta* L. *Bollettino della Societa Italiana Biologia Sperimentale* **38**, 1030-1032.

Gardiner, B.G. (1973). Interrelationships of teleostomes. In *Interrelationships of Fishes* (eds. P.H. Greenwood, R.S. Miles and C. Patterson) , pp. 105-136. *Journal of the Linnean Society of London* Zoology 53 (suppl 1). Academic Press, London, New York.

Gardiner, B.G. (1980). Tetrapod ancestry: a reappraisal. In *The terrestrial environment and the origin of land vertebrates. Systematics Association Special Volume* **15** (ed. A.L. Panchen), pp. 177-185. Academic Press, London and New York.

Gardiner, B.G. and Schaeffer, B. (1989). Interrelationships of lower actinopterygian fishes. *Zoological Journal of the Linnean Society* **97**, 135-187.

Gardiner, D.M. (1978a). The origin and fate of spermatophores in the viviparous teleost *Cymatogaster aggregata* (Perciformes: Embiotocidae). Journal of Morphology **155**, 157-172.

Gardiner, D.M. (1978b). Fine structure of the spermatozoon of the viviparous teleost, *Cymatogaster aggregata. Journal of Fish Biology* **13**, 435-438.

Gardiner, D.M. (1978c). Utilization of extracellular glucose by spermatozoa of two viviparous fishes. *Comparative Biochemistry and Physiology* **59**, 165-168.

Garstang, W. (1928). The mophology of the Tunicata, and its bearing on the phylogeny of the Chordata. *Quarterly Journal of Microscopical Science* **72**, 51-187.

Georges, D. (1969). Spermatogénèse et spermiogénèse de *Ciona intestinalis* L. observées au microscope electronique. *Journal de Microscopie* **8**, 391-400.

Gibbons, B.H., Gibbons, I.R and Baccetti, B. (1983). Structure and motility of the 9+0 flagellum of eel spermatozoa. *Journal of Submicroscopic Cytology* **15**, 15-20.

Gibbons, I.R. (1981). Cilia and flagella of eukaryotes. *Journal of Cell Biology* **91**, 107-124.

Gilkey, J.C., Jaffe L.F., Ridgway E.B. and Reynolds G.T. (1978). A free calcium wave traverses the activating egg of the medaka, *Oryzias latipes. Journal of Cell Biology* **76**, 448-466.

Ginsburg, see Ginzburg,

Ginzburg, A.S. (1968). *Fertilization in Fishes and the Problem of Polyspermy.* Akademiya Nauk SSSR, Institut Biologii Razvitiya. (ed. T.A. Detlaf). Translated from Russian, by Israel Program for Scientific Translations, Jerusalem 1972.

Ginzburg, A.S. and Billard, R. (1972). Ultrastructure du spermatozoide d'Anguille. *Journal de Microscopie et de Biologie Cellulaire* (Paris) **14**, 50a-51a.

Glenister, P.H. and Lyon, M.F. (1986). Long term storage of eight-cell mouse embryos at -196° C. *Journal of in Vitro Fertilization and Embryo Transfer* **3**, 20-27.

Gosline, W.A. (1960). Contributions towards a classification of modern isospondylous fishes. *Bulletin of the British Museum (Natural History)* **6**, 325-365.

Gosline, W.A. (1968). The suborders of perciform fishes. *Proceedings of the United States National Museum* **124**, 1-78.

Gosline, W.A. (1970). A reinterpretation of the teleostean fish order Gobiesociformes. *Proceedings of the Californian Academy of Sciences* **38**, 363-382.

Graham, J.K. and Foote, R.H. (1987). Effect of several lipids, fatty acyl chain length, and degree of unsaturation on the motility of bull spermatozoa after cold shock freezing. *Cryobiology* **24**, 42-52.

Graham, E.F., Schmel, M.L. and Deyo, R.C.M. (1984). Cryopreservation and fertility of fish, poultry and mammalian spermatozoa. *Proceed-*

ings of the Tenth Technical Conference on Artifical Insemination and Reproduction, Milwaukee, Apr 1984, 4-29.

Grande, L. (1985). Recent and fossil clupeomorph fishes with materials for revision of the subgroups of clupeoids. *Bulletin of the American Museum of Natural History* **181**, 231- 372.

Grassé, P. (1958). (ed). *Traité de Zoologie*. Anatomie, Systématique, Biologie. XIII Agnathes et Poissons. Masson, Paris.

Graybill, J.R. and Horton, H.F. (1969). Limited fertilization of steelhead trout eggs with cryopreserved sperm. *Journal of the Fisheries Research Board of Canada* **26**, 1400-1404.

Greenwood, P.H. (1973). Interrelationships of osteoglossomorphs. In *Interrelationships of Fishes* (eds P.H. Greenwood, R.S. Miles and C. Patterson), pp. 307-332. *Journal of the Linnean Society of London* Zoology 53 (suppl 1). Academic Press, New York.

Greenwood, P.H. (1989). Fifty years a 'living fossil' - the coelacanth fish *Latimeria chalumnae*. *Biologist* **36**, 15-19.

Greenwood, P.H., and Rosen, D.E. (1971). Notes on the structure and relationships of the alepocephaloid fishes. *American Museum Novitates* **2473**, 1-41.

Greenwood, P.H., Rosen, D.E., Weitzman, S.H. and Myers, G.S. (1966). Phyletic studies of teleostean fishes, with a provisional classification of living forms. *Bulletin of the American Museum of Natural History* **131**, 339-456.

Grier, H.J. (1973a). Aspects of germinal cyst and sperm development in *Poecilia latipinna* (Teleostei: Poeciliidae). *Journal of Morphology* **146**, 229-250.

Grier, H.J. (1973b). Ultrastructure of the testis in the teleost *Poecilia latipinna*. Spermiogenesis with reference to the intercentriolar lamellated body. *Journal of Ultrastructure Research* **45**, 82-92.

Grier, H.J. (1975). Spermiogenesis in the teleost *Gambusia affinis* with particular reference to the role played by microtubules. *Cell and Tissue Research* **165**, 89-102.

Grier, H.J. (1976). Sperm development in the teleost *Oryzias latipes*. *Cell and Tissue Research* **168**, 419-431.

Grier, H.J. (1981). Cellular organization of the testis and spermatogenesis in fishes. *American Zoologist* **21**, 345-357.

Grier, H.J. and Collette, B.B. (1987). Unique spermatozeugmata in testes of halfbeaks of the genus *Zenarchopterus* (Teleostei: Hemiramphidae). *Copeia* **1987**, 300-311.

Grier, H.J., Fitzsimons, J.M. and Linton, J.R. (1978). Structure and ultrastructure of the testis and sperm formation in goodeid teleosts. *Journal of Morphology* **156**, 419-438.

Grier, H.J., Linton, J.R., Leatherland, J.F. and De Vlaming, V.L. (1980). Structural evidence for two different testicular types in teleost fishes. *The American Journal of Anatomy* **159**, 331-345.

Grygier, M.J. (1982). Sperm morphology in Ascothoracida (Crustacea: Maxillopoda): confirmation of generalized nature and phylogenetic importance. *International Journal of Invertebrate Reproduction* **4**, 323-332.

Guest, W.C., Avault, J.W. and Roussel, J.D. (1976). Preservation of Channel Catfish sperm. *Transactions of the American Fish Society* **105**, 469-474.

Guha, T., Siddiqui, A.Q. and Prentis, P.F. (1988). Ultrastructure of testicular spermatozoon of the fish *Oreochromis niloticus*. *Proceedings of the 46th Annual Meeting of the Electron Microscopy Society of America*. San Francisco Press, San Francisco. pp. 278-279.

Gusse, M. and Chevaillier, P.H. (1978). Etude ultrastructurale et chimique de la chromatine au cours de la spermiogenèse de la roussette *Scyliorhinus caniculus* (L). *Cytobiologie* **16**, 421-443.

Haga, Y. (1982). On the subzero temperature preservation of fertilized eggs of rainbow trout. *Bulletin of the Japanese Society of Scientific Fisheries* **48**, 1569-1572.

Hámor, T. (1966). A study of the genital products of the brown trout *Salmo trutta* L. and the rainbow trout *Salmo irideus* Gibbons. *Magyar Allat.tani Kozl.ony* **43**, 63-68. (Fisheries Research Board of Canada, Translation Series No. 911, 1967). *Fide* Scott and Baynes (1980).

Hara, M. and Tanaka, S. (1986). Fine structure of spermatogenesis and mature spermatozoa in elasmobranch and chimaera fishes: A systematic consideration. *Development Growth and Differentiation* **28** (suppl), 114.

Hara, S., Canto, J.T. and Almendras, J.M.E. (1982).

A comparative study of various extenders for milkfish, *Chanos chanos* (Forsskål), sperm preservation. *Aquaculture* **28**, 339-346.

Harding, H.R., Aplin, K. and Shorey, C.D. (1986). Cladistic analysis of marsupial sperm structure: methodological perspectives. *Development, Growth and Differentiation* **28** Supplement, 62 (Abstract only).

Hardisty, M.W. and Potter, I.C. (1971) (eds.).*The Biology of Lampreys*. Academic Press, New York.

Harrington, R.W. (1961). Oviparous hermaphroditic fish with internal self-fertilization. *Science* **134**, 1749-1750.

Hart, N.H., Pietri, R. and Donovan, M. (1984). The structure of the chorion and associated surface filaments in *Oryzias* - Evidence for the presence of extracellular tubules. *Journal of Experimental Zoology* **230**, 273-296.

Harvey, B. (1982). Cryobiology and the storage of teleost gametes. In *Proceedings of the International Symposium on the Reproductive Physiology of Fish.* Wageningen, The Netherlands, August 1982. (eds H.J. Th. Goos and J.J. Richter)., pp.123-127. Pudoc, Wageningen.

Harvey, B. (1983). Cryopreservation of *Sarotherodon mossambicus* spermatozoa. *Aquaculture* **32**, 313-320.

Harvey, B. and Ashwood-Smith, M.J. (1982). Cryoprotectant penetration and supercooling in the eggs of salmonid fishes. *Cryobiology* **19**, 29-40.

Harvey B. and Kelley R.N. (1984a). Chilled storage of *Sarotheroden mossambicus* milt. *Aquaculture* **36**, 85-95.

Harvey, B. and Kelley, R.N. (1984b). Short-term storage of *Sarotherodon mossambicus* ova. *Aquaculture* **37**, 391-395.

Harvey B. and Kelley R.N. (1984c). Control of spermatozoan motility in a euryhaline teleost acclimated to different salinities. *Canadian Journal of Zoology* **62**, 2674-2677.

Harvey B., Kelley R.N. and Ashwood-Smith M.J. (1982). Cryopreservation of zebra fish spermatozoa using methanol. *Canadian Journal of Zoology* 60, 1867-1870.

Harvey, B., Kelley, R.N. and Ashwood-Smith, M.J. (1983). Permeability of intact and dechorionated zebra fish embryos to glycerol and dimethyl sulfoxide. *Cryobiology* **20**, 432-439.

Harvey, B., Stoss, J. and Butchart, W. (1983). Supercooled storage of salmonid ova. *Canadian Technical Report of Fisheries and Aquatic Sciences* **1222**, i-9.

Hawkins C.J., Kott, P., Parry, D.L. and Swinehart, H. (1983). Vanadium content and oxidation state related to ascidian phylogeny. *Comparative Biochemistry and Physiology* **76B**, 555-558.

Healy J.M. and Jamieson B.G.M. (1981). An ultrastructural examination of developing and mature paraspermatozoa in *Pyrazus ebeninus* (Mollusca, Gastropoda, Potamididae). *Zoomorphology* **98**, 101-119.

Healy, J.M., Rowe, F.W.E. and Anderson, D.T. (1988). Spermatozoa and spermiogenesis in *Xyloplax* (Class Concentricycloidea): a new type of spermatozoon in the Echinodermata. *Zoologica Scripta* **17**, 297-310.

Hearne, M.E. (1984). In *Ontogeny and Systematics of Fishes*. Special Publication Number 1. American Society of Ichthyologists and Herpetologists, pp. 153-155.

Heintz, A. (1963). Phylogenetic aspects of myxinoids. In *The Biology of Myxine*. (eds A. Brodal and R. Fänge), pp. 9-21. Universitetsforlaget, Oslo.

Hendelberg, J. (1986). The phylogenetic significance of sperm morphology in the Platyhleminthes. *Hydrobiologia* **132**, 53-58.

Hennig, W. (1966). *Phylogenetic Systematics*. University of IllinoisPress, Urbana.

Herdman, W. (1988). Report on the Tunicata collected by HMS Challenger during the years 1873-1876. *Report of the Scientific Results of the Voyage of HMS Challenger during the years 1873-1876*. **27**. British Museum, London. 157pp.

Hillis, D.M. and Dixon, M.T. (1989). Vertebrate phylogeny: evidence from 28S ribosomal DNA sequences. In *The Hierarchy of Life*. (eds B. Fernholm, K. Bremer and H. Jörnvall), pp. 355-367. Elsevier Science Publishers, Amsterdam.

Hinegardner, R. (1968). Evolution of cellular DNA content in teleost fishes. *American Naturalist* **102**, 517-523.

Hoar, W.S., Randall, D.J. and Donaldson, E.M. (1983). Fish gamete preservation and spermatozoan physiology. In *Fish Physiology* Volume 9. *Reproduction. Part B. Behaviour and Fertility*

Control. (eds W.S. Hoar, D.J. Randall, and E.M. Donaldson), pp. 305-350. Academic Press, New York.

Hoese, D.F. (1984). Gobioidei: relationships. In *Ontogeny and Systematics of Fishes*. Special Publication Number 1. American Society of Ichthyologists and Herpetologists, pp. 588-591.

Hoffman, R.A. (1963). Gonads, spermatic ducts, and spermatogenesis in the reproductive system of male toadfish, *Opsanus tau*.*Chesapeake Science* 4, 21-29.

Holland, L.Z. (1988). Spermatogenesis in the salps *Thalia democratica* and *Cyclosalpa affinis* (Tunicata: Thaliacea): an electron microscopic study. *Journal of Morphology* **198**, 189-204.

Holland, L.Z. (1989). Fine structure of spermatids and sperm of *Dolioletta gegenbauri* and *Doliolum nationalis* (Tunicata: Thaliacea): implications for tunicate phylogeny. *Marine Biology* **101**, 83-95.

Holland, L.Z. (1990). Spermatogenesis in *Pyrosoma atlanticum* (Tunicata, Thaliacea, Pyrosomatida): implications for tunicate phylogeny. *Marine Biology* (In press).

Holland, N.D. and Holland, L.Z. (1989). The fine structure of the testis of a lancelet (=Amphioxus), *Branchiostoma floridae* (Phylum Chordata: Subphylum Cephalochrodata = Acrania). *Acta Zoologica* (Stockholm) **70**, 211-219.

Holland, L.Z., Gorsky, G. and Fenaux, R. (1988). Fertilization in *Oikopleura dioica* (Tunicata: Appendicularia): acrosome reaction, cortical reaction and sperm-egg fusion. *Zoomorphology* **108**, 229-243.

Honneger, T.G. (1982). Effect on fertilization and localized binding of lectins in the ascidian, *Phallusia mammillata*. *Experimental Cell Research* **138**, 446-451.

Honneger, T.G. (1986). Fertilization is ascidians: studies on the egg envelope, sperm and gamete interactions in *Phallusia mammillata*. *Developmental Biology* **118**, 118-128.

Horton, H.F. and Ott, A.G. (1976). Cryopreservation of fish spermatozoa and ova. *Journal of Fisheries Research Board of Canada* **33**, 995-1000.

Hoshi, M. (1984). Roles of sperm glycosidases and proteases in the ascidian fertilization. In *Advances in Invertebrate Reproduction* (eds W. Engels et al.), Vol. 3, pp. 27-40. Elsevier, Amsterdam.

Hoshi, M., Numakunai, T. and Sawada, H. (1981). Evidence for participation of sperm proteinases in fertilization of the solitary ascidian, *Halocynthia roretzi*: effect of protein inhibitors. *Developmental Biology* **86**, 117-121.

Hoyle, R.J. and Idler, D.R. (1968). Preliminary results in the fertilization of eggs with frozen sperm of Atlantic salmon (*Salmo salar*). *Journal of the Fisheries Research Board of Canada* **25**, 1295-1297.

Hughes, R.L. and Potter, I.C. (1969). Studies on gametogenesis and fecundity in the lampreys *Mordacia praecox* and *M. mordax* (Petromyzonidae). *Australian Journal of Zoology* **17**, 447-464.

Hulata, G. and Rothbard, S. (1979). Cold storage of carp sperm for short periods. *Aquaculture* **16**, 267-269.

Ishikawa, H. (1977). Comparative studies on the thermal stability of animal ribosomal RNA's—V. Tentaculata (phoronids, moss-animals and lamp-shells). *Comparative Biochemistry and Physiology* **57B**, 9-14.

Iwamatsu, T. and Ohta, T. (1981). Scanning electron microscopic observations on sperm penetration in teleostean fish. *Journal of Experimental Zoology* **218**, 261-277.

Jaana, H. and Yamamoto, T.S. (1981). The ultrastructure of spermatozoa with a note on the formation of the acrosomal filament in the lamprey, *Lampetra japonica*. *Japanese Journal of Icthyology* **28**, 135-147.

Jägersten, G. (1972). *Evolution of the Metazoan Life Cycle*. Academic Press, London, New York.

Jamieson, B.G.M. (1983). Spermatozoal ultrastructure: evolution and congruence with a holomorphological phylogeny of the Oligochaeta (Annelida). *Zoologica Scripta* **12**, 107-114.

Jamieson, B.G.M. (1984). Spermatozoal ultrastructure in *Branchiostoma moretonensis* Kelly, a comparison with *B. lanceolatum* (Cephalochordata) and with other deuterostomes. *Zoologica Scripta*. **13**, 223-229.

Jamieson, B.G.M. (1985). The spermatozoa of the Holothuroidea (Echinodermata): an ultrastructural review with data on two Australian species and phylogenetic discussion. *Zoologica Scripta* **14**, 123-135.

Jamieson, B.G.M. (1986a). Onychophoran-euclitellate relationships: evidence from spermatozoal ultrastructure. *Zoologica Scripta* **15**, 141-155.

Jamieson, B.G.M. (1986b). Some recent studies on the ultrastructure and phylogeny of annelid and uniramian spermatozoa. *Development Growth and Differentiation* **28**(suppl.), 25-26.

Jamieson, B.G.M. (1987a). *The Ultrastructure and Phylogeny of Insect Spermatozoa.* Cambridge University Press.

Jamieson, B.G.M. (1987b). A biological classification of sperm types, with special reference to annelids and molluscs, and an example of spermiocladistics. In *New Horizons in Sperm Cell Research.* (ed H. Mohri), pp. 311-332. Japan Scientific Societies Press. Gordon and Breach Science Publishers, New York, London.

Jamieson, B.G.M. (1989). Complex spermatozoon of the live-bearing half-beak, *Hemirhamphodon pogonognathus* (Bleeker): ultrastructural description (Euteleostei, Atherinomorpha, Beloniformes). *Gamete Research* **24**, 247-259.

Jamieson, B.G.M. and Rouse, G.W. (1989). The spermatozoa of the Polychaeta (Annelida): an ultrastructural review. *Biological Reviews of the Cambridge Philosophical Society* **64**, 93-157.

Jamieson B.G.M., Erséus, C. and Ferraguti, M. (1987). Parsimony analysis of the phylogeny of some Oligochaeta (Annelida) using spermatozoal ultrastructure. *Cladistics* **3**, 141-151.

Jaspers, E.J., Avault, J.W. and Roussel, J.D. (1976). Spermatozoal morphology and ultrastructure of Channel Catfish, *Ictalurus punctatus. Transactions of the American Fish Society* **105**, 475-480.

Jefferies, R.P.S. (1979). The origin of chordates - a methodological essay. In *The Origin of Major Invertebrate Groups.* (ed. M.R. House), pp. 443-477. *Systematics Association Special Volume.* Academic Press, London, New York.

Jefferies, R.P.S. (1981). In defence of calcichordates. *Zoological Journal of the Linnean Society* **173**, 351-396.

Jensen, J.O.T.and Alderdice, D.F. (1984). Effect of temperature on short-term storage of eggs and sperm of Chum Salmon (*Oncorhynchus keta*). *Aquaculture* 37, 251-265.

Jespersen, Å. (1971). Fine structure of the spermatozoon of the Australian Lungfish *Neoceratodus forsteri* (Krefft). *Journal of Ultrastructure Research* **37**, 178-185.

Jespersen, Å. (1975). Fine structure of spermiogenesis in eastern Pacific species of Hagfish (Myxinidae). *Acta Zoologica(Stockholm)* **56**, 189-198.

Jespersen, Å. (1984). Spermatozoans from a parasitic dwarf male of *Neoceratias spinifer* Pappenheim, 1914. *Videnskabelige Meddelelser Dansk Naturhistorisk Forening* **145**, 37-42.

Johnson, G.D. (1975). The procurrent spur: an undescribed perciform caudal character and its phylogenetic implications. *Occasional Papers of the Californian Academy of Sciences* **121**, 1-23

Johnson, G.D. (1984). Percoidei: development and relationships. In *Ontogeny and Systematics of Fishes.* Special Publication Number 1. American Society of Ichthyologists and Herpetologists, pp. 464-498.

Jonas-Davies, J.A.C., Winfrey, V. and Olson, G.E. (1983). Plasma membrane structure in spermatogenic cells of the swordtail (Teleostei, Xiphophorus helleri). *Gamete Research* **4**, 309-324.

Jones, P.R. and Butler, R.D. (1988). Spermatozoon structure of *Platichthys flesus. Journal of Ultrastructure Research* **98**, 71-82.

Jordan, D.S. (1907). *Fishes.* Henry Holt, New York.

Kaufman, L. and Liem, K.F. (1982). Fishes of the suborder Labroidei (Pisces: Perciformes): Phylogeny, ecology, and evolutionary significance. *Breviora*, Museum of Comparative Zoology No. **472**, 1-19.

Kerby, J.H. (1983). Cryogenic preservation of sperm from Striped Bass. *Transactions of the American Fish Society* **112**, 86-94.

Kerby, J.H., Bayless, J.D. and Harrell, R.M. (1985). Growth survival and harvest of Striped Bass produced with cryopreserved spermatozoa. *Transactions of the American Fisheries Society* **114**, 761-765.

Kesteven, H.L. (1951). The origin of the tetrapods. *Proceedings of the Royal Society of Victoria* **59**, 93-138.

Kille, R.A. (1960). Fertilization of the lamprey egg. *Experimental Cell Research* **20**, 12-27.

Knight, C.A. (1986). Heterogeneous nucleating

agents. *Cryobiology* **23**, 559.

Knight, C.A. and Duman, J.G. (1986). Inhibition of recrystallization of ice by insect thermal hysteresis proteins: a possible cryoprotective role. *Cryobiology* **23**, 256-262.

Kobayashi W. and Yamamoto, T.S. (1981). Fine structure of the micropylar apparatus of the Chum Salmon egg, with a discussion of the mechanism for blocking polyspermy. *Journal of Experimental Zoology* **217**, 265-275.

Kobayashi, W. and Yamamoto, T.S. (1987). Light and electron microscopic observations of sperm entry in the Chum Salmon egg. *Journal of Experimental Zoology* **243**, 311-322.

Koch, R.A. and Lambert, C.C. (1986). Ultrastructure of mitochondrial translocation in ascidian sperm. *Journal of Cell Biology* **103**, 276a.

Koch, R.A., Lambert, G. and Lambert, C.C. (1985). Ascidian sperm mitochondrial translocation as seen by HVEM. *Journal of Cell Biology* **191**, 398a.

Koldras, M. and Bieniarz, K. (1987). Cryopreservation of Carp sperm. *Polskie Archiwum Hydrobiologii* **34**, 125-134.

Koldras, M. and Moczarski, M. (1983). Properties of Pike *Esox lucius* L. milt and its cryopreservation. *Polish Archives of Hydrobiol.*ogy **30**, 69-78.

Kott, P. (1969). Anarctic Ascidiacea. In *Antarctic Research Series* **13**, pp. 1-239. American Geophysical Union.

Kott, P. (1984). In *A Coral Reef Handbook* (eds P. Mather and I. Bennett). Second Edition, pp. 97-106. The Australian Coral Reef Society.

Kott, P. (1985). The Australian Ascidiacea. Part I, Phlebobranchia and Stolidobranchia. *Memoirs of the Queensland Museum* **23**, 1-440.

Kubo, M., Ishikawa, M. and Numakunai, T. (1978). Differentiation of apical structures during spermiogenesis and fine structures of the spermatozoon in the ascidian *Halocynthia roretzi*. *Acta Embryologiae Experimentalis* **3**, 283-295.

Kuchnow, K.P. and Scott, J.R. (1977). Ultrastructure of the chorion and its micropyle apparatus in the mature *Fundulus heteroclitus* (Walbaum) ovum. *Journal of Fish Biology* **10**, 197-201.

Kudo, S. (1980). Sperm penetration and the formation of a fertilization cone in the common carp egg. *Development, Growth and Differentiation* **22**, 403-414.

Kudo, S. (1982). Ultrastructure of a sperm entry site beneath the micropylar canal in fish eggs. *Zoological Magazine* **91**, 213-220.

Kudo, S. (1983). Response to sperm penetration of the cortex of eggs of the fish, *Plecoglossus altivelis*. *Development, Growth and Differentiation* **25**, 163-170.

Kudo, S. and Sato, A. (1985). Fertilization cone of carp eggs as revealed by scanning electron microscopy. *Development Growth and Differentiation* **27**, 121-128.

Kumar, K. (1988). A comparative study of various extenders for cryopreservation of carp spermatozoa. *Indian Journal of Animal Science* **58**, 1355-1360.

Kurokura, H. and Hirano, R. (1980). Cryopreservation of Rainbow Trout sperm. *Bulletin of the Japanese Society of Scientific Fisheries* **46**, 1493-1495.

Kurokura, H., Hirano, R., Tomita, M. and Iwahashi, M. (1984). Cryopreservation of carp sperm. *Aquaculture* **37**, 267-273.

Laale, H.W. (1980). The perivitelline space and egg membranes of bony fishes: a review. *Copeia* **1980**, 210-226.

Lagios, M.D. (1979). The coelacanth and the Chondrichthyes as sister groups: a review of shared apomorph characters and a cladistic analysis and reinterpretation. *Occasional Papers of the California Academy of Sciences* **134**, 25-44.

Laird, C.D. (1971). Chromatin structure: Relationship between DNA content and nucleotide sequence diversity. *Chromosoma* (Berlin) **32**, 378-406.

Lake, J.S. (1990). *Proceedings of the National Academy of Sciences U.S.A.* **87**, 763-766.

Lambert, C.C. (1982). The ascidian sperm reaction. *American Zoologist* **22**, 841-849.

Lambert, C.C. and Koch, R.A. (1988). Sperm binding and penetration during ascidian fertilization. *Development, Growth and Differentiation* **30**, 325-336.

Lambert, C.C. and Lambert, G. (1983). Mitochondrial movement during the ascidian sperm reaction. *Gamete Research* **8**, 295-307.

Lambert, G., Lambert, C.C. and Abbott, D.P. (1981). *Corella* species in the American Pacific North-west: distinction of *C. inflata* Huntsman, 1912 from *C. willmeriana* Herdman, 1898 (Ascidiacea, Phlebobranchia). *Canadian Journal of Zoology* **59**, 1493-1504.

Lauder, G.V and Liem, K.F. (1983). The evolution and interrelationships of the actinopterygian fishes. *Bulletin of the Museum of Comparative Zoology* **150**, 95-197.

Le, H. L. V., Perasso, R. and Billard, R. (1989). Phylogénie moléculaire préliminaire des "poissins" basée sur l'analyse de séquences d'ARN ribosomique 28 S. *Comptes Rendus Hebdomadaires des Séances de l'Académie des Sciences* III **309**, 493-498.

Legendre, M. and Billard, R. (1980). Cryopreservation of Rainbow Trout sperm by deep-freezing. *Reproduction, Nutrition et Développement* **20**, 1859-1868.

Lemire, M. and Lagios, M.D. (1979). Ultrastructure du parenchyme sécréteur de la gland postanale du coelacanthe, *Latimeria chalumnae. Acta Anatomica* **104**, 1-15.

Lester, S.M. (1988). Ultrastructure of adult gonads and development and structure of the larva of *Rhabdopleura normani* (Hemichordata: Pterobranchia). *Acta Zoologica* (Stockholm) **69**, 95-109.

Leung, L.K.-P. (1987a). *Fish Spermatology. Ultrastructure, phylogeny and cryopreservation.* Honours thesis, University of Queensland (Unpublished).

Leung, L.K.-P. (1987b). Cryopreservation of the spermatozoa of the Barramundi, *Lates calcarifer* (Teleostei: Centropomidae). *Aquaculture* **64**, 243-247.

Leung, L.K-P. (1988). The ultrastructure of the spermatozoon of *Lepidogalaxias salamandroides* and its phylogenetic significance. *Gamete Research* **19**, 41-49.

Lessman, C.A. and Huver, C.W. (1981). Quantification of fertilization induced gamete changes and sperm entry without egg activation in a teleost egg. *Developmental Biology* **84**, 218-224.

Liem, K.F. and Greenwood, P.H. (1981). A functional approach to the phylogeny of the pharyngognath teleosts. *American Zoologist* **21**, 83-101.

Lin, J., Chen, X. and Wang, D. (1987). Studies on ultrastructure in spermatogenesis of amphioxus (*Branchiostoma belcheri* Gray). *Oceanologia et Limnologia Sinica* **18**, 432-436.

Lovelock, J.E. (1953). The mechanism of the protective action of glycerol against haemolysis by freezing and thawing. *Biochimica et Biophysica Acta* **11**, 28-36.

Lovelock, J.E. (1957). The denaturation of lipid-protein complexs as a cause of damage by freezing. *Proceedings of the Royal Society* B **14**, 427-433.

Løvtrup, S. (1977). *The Phylogeny of Vertebrata.* Wiley, London.

Lowman, F.G. (1953). Electron microscope studies of silver salmon spermatozoa (Oncorhynchus kisutch [Walbaum]). *Experimental Cell Research* **5**, 335-360.

Luyet, B.J. (1938). The survival of plant cells immersed in liquid air. *Biodynamica* **33**, 284-285.

Luyet, B.J. (1966). Anatomy of the freezing process in physical systems. In *Cryobiology* (ed. H.T. Meryman), pp 115-138. Academic Press, New York.

Luyet, B.J. (1970). Physical changes occurring in frozen solutions during rewarming and melting. In *The Frozen Cell.* (Eds G.E.W. Wolstenholme and M. O'Connor), pp. 27-43. Churchill, London.

Luyet, B.J. and Gehenio, P.M. (1940). *Life and Death at Low Temperatures.* Biodynamica, Normandy.

MacDowall, R.M. (1984). Southern hemisphere freshwater salmoniforms: development and relationships. In *Ontogeny and Systematics of Fishes.* Special Publication Number 1. American Society of Ichthyologists and Herpetologists, pp. 150-153.

Maggese, M.C., Cukier, M. and Cussac, V.E. (1984). Morphological changes, fertilizing ability and motility of *Rhamdia sapo* (Pisces, Pimelodidae) sperm induced by media of different salinities. *Rev Brasil Bio* **44**, 541-546.

Marshall, C.J. (1989). *Cryopreservation and ultrastructural studies on teleost fish gametes.* Honours thesis, University of Queensland (Unpublished).

Martin, F.D. (1984). Esocoidei: Development and Relationships. In *Ontogeny and Systematics of Fishes.* Special Publication Number 1. American Society of Ichthyologists and Herpetologists, pp. 140-142.

Matarese, A.C., Watson, W. and Stevens, E.G. (1984). Blennioidei: development and relationsips. In *Ontogeny and Systematics of Fishes*. Special Publication Number 1. American Society of Ichthyologists and Herpetologists, pp. 565-573.

Matsubara, K. (1943). Studies on the scorpaenid fishes of Japan. *Transactions of the Sigenkagaku Kenyusho*. **1** and **2**. (*Fide* Washington et al., 1984a).

Mattei, C. and Mattei, X. (1973). La spermiogenèse d'*Albula vulpes* (L 1758) (Poissin Albulidae). *Zeitschrift für Zellforschung und Mikroskopische Anatomie* **142**, 171-192.

Mattei, C. and Mattei, X. (1974). Spermatogenesis and spermatozoa of the elopomorpha (teleost fish). In *The functional anatomy of the spermatozoon* (ed. B.A. Afzelius), pp 211-221. Pergamon Press, Oxford.

Mattei, C. and Mattei, X. (1976). Présence d'un système membranaire associé au noyau dans le gamète de *Pomacentrus leucostictus* (Poisson Téléostéen). *Comptes Rendus des Seances de la Société de Biologie* 170, 234-240.

Mattei, C. and Mattei, X. (1978a). La spermiogenèse d'un poisson téléostéen (*Lepadogaster lepadogaster*). I-Le spermatide. *Biologie Cellulaire* **32**, 257-266.

Mattei, C. and Mattei, X. (1978b). La spermiogenèse d'un poisson téléostéen (*Lepadogaster lepadogaster*). II-Le spermatozoide. *Biologie Cellulaire* **32**, 267-274.

Mattei, C. and Mattei, X. (1984). Spermatozoïdes biflagellés chez un poisson téléostéen de la famille Apogonidae. *Journal of Ultrastructure Research* **88**, 223-228.

Mattei, C. and Mattei, X. and Marchand, B. (1979). Réinvestigation de la structure des flagelles spermatiques: les doublets 1, 2, 5 et 6. *Journal of Ultrastructure Research* **69**, 371-377.

Mattei, C., Mattei, X., Marchand, B. and Billard, R. (1981). Réinvestigation de la structure des flagelles spermatiques: cas particulier des spermatozoïdes à mitochondrie annulaire. *Journal of Ultrastructure Research* **74**, 307-312.

Mattei, X. (1969). Contribution à l'étude de la spermiogenèse et des spermatozoïdes de poissons par les méthodes de la microscopie électronique. Thesis. Faculté des Sciences Montpellier. (*Fide* Cassas *et al.*, 1981).

Mattei, X. (1970). Spermiogenèse comparée des poissons. In *Comparative Spermatology* (ed. B. Baccetti), pp. 57-69. Academic Press, New York.

Mattei, X. (1988). The flagellar apparatus of spermatozoa in fish. Ultrastructure and evolution. *Biology of the Cell* **63**, 151-158.

Mattei, X. and Boissin, C. (1966). Le complexe centriolaire du spermatozoïde de *Lebistes reticulatus*. *Comptes Rendus Hebdomadaires des Séances de l'Académie des Sciences* D **262**, 2620-2622.

Mattei, X. and Mattei, C. (1972). L'appareil centriolaire et flagellaire du spermatozoïde *d'Albula vulpes* (Poissin, Albulidae). *Journal de Microscopie* **14**, 67a-68a.

Mattei, X. and Mattei, C. (1976a). Spermatozoïdes à deux flagelles de type 9 + 0 chez *Lampanyctus* sp. (Poisson Myctophidae). *Journal de Microscopie et de Biologie Cellulaire* **25**, 187-188.

Mattei, X. and Mattei, C. (1976b). Ultrastructure du canal cytoplasmique des spermatozoïdes de téléostéens illustrée par l'étude de la spermiogenèse de *Trichiurus lepturus*. *Journal de Microscopie et de Biologie Cellulaire* **25**, 249-258.

Mattei, X., Mattei, C. and Boissin, C. (1967b). L'extrémité flagellaire du spermatozoïde de *Lebistes reticulatus* (Poeciliidae). *Comptes Rendus des Séances de la Scociété de Biologie de l'ouest Africain* **161**, 884-887.

Mattei, X., Siau, Y. and Seret, B. (1988). Étude ultrastructurale du spermatozoïde du coelacanthe: *Latimeria chalumnae*. *Journal of Ultrastructure and Molecular Structure Research* **101**, 243-251.

Mattei, X., Boisson, C., Mattei, C. and Reizer, C. (1967a). Spermatozoïdes aflagellés chez un poisson: *Gymnarchus niloticus* (Téleostéen, Gymnarchidae). *Comptes Rendus Hebdomadaires des Séances de l'Académie des Sciences* D **265**, 2010-2012.

Mattei, X., Mattei, C., Reizer, C. and Chevalier, J-L. (1972). Ultrastructure des spermatozoïdes aflagellés des mormyres (Poissons Téléostéens). *Journal de Microscopie* (Paris) **15**, 67-78.

Mattei, X., Thiam, D., and Thiaw, O.T. (1989). Le spermatozoide de *Ophidion* sp. (Poisson, Téléostéen): particularités ultrastructurales du flagelle. *Journal of Ultrastructure and Molecular*

Structure Research **102**, 162-169.

Mazur, P. (1963). Kinetics of water loss from cells at subzero temperature and the likelihood of intracellular freezing. *Journal of General Physiology* **47**, 347-369.

Mazur, P., Kemp, J.A., Miller, R.H. (1976). Survival of fetal rat pancreases frozen to -78 and -196°. *Proceedings of the National Academy of Sciences* **73**, 4105-4109.

McGrath, J.J. and Thomas, P.C. (1986). The influence of osmotic species on human erythrocyte hypotonic cold shock hemolysis. *Cryobiology* **23**, 546.

Meryman, H.T. (1966). Review of biological Freezing. In *Cryobiology* (ed H.T. Meryman), pp 1-114. Academic Press, New York.

Meryman, H.T. (1968). Modified model for the mechanisms of freezing injury of erythrocytes. *Nature* **218**, 333-336.

Meryman, H.T., Williams, R.J. and Douglas, M.St J. (1977). Freezing injury from "solution effects" and its prevention by natural or artificial cryoprotection. *Cryobiology* **14**, 287-302.

Meves, F. (1903). Uber oligopyrene und pyrene Spermien und über ihre Enstehung nach Beobachtungen an *Paludina* und *Pygaera*. *Archives für Mikroskopische Anatomie* **61**, 1-84.

Mikodina, E.V. (1987). Surface structure of the egg membranes of teleostean fishes. *Journal of Icthyology* **27**, 19-26.

Millott, J. and Anthony, J. (1958). *Anatomie de Latimeria chalumnae. I. Squellette et muscles.* C.N.R.S., Paris. (*Fide* Forey, 1980).

Mizue, K. (1968). Studies on a scorpaenous fish *Sebastiscus marmoratus* Cuvier et Valenciennes - VI. Electron-microscopic study of spermatogenesis. *Bulletin of the Faculty of Fisheries, Nagasaki University* **25**, 9-24.

Mizue, K. (1969). Electron-microscopic study on spermiogenesis of black sailfin molley, *Mollienesia latipinna* Le Sverp. *Bulletin Faculty of Fisheries, Nagasaki University* 28, 1-17.

Mok, H.K. (1981). The posterior cardinal veins and kidneys of fishes, with notes on their phylogenetic significance. *Japanese Journal of Ichthyology* **27**, 281-290.

Moore, A.A. (1987). Short-term storage and cryopreservation of Walleye semen. *The Progessive Fish-culturist*. **49**, 40-43.

Morisawa, M. (1985). Initiation mechanism of sperm motility at spawning in teleosts. *Zoological Science*, Tokyo **2**, 605-615.

Morisawa, M. and Suzuki, K. (1980). Osmolality and potassium ion: their roles in initiation of sperm motility in teleosts. *Science* **210**, 1145-1146.

Morisawa, M., Suzuki, K. and Morisawa, S. (1983). Effects of potassium and osmolality on spermatozoan motility of salmonid fishes. *Journal of Experimental Biology* **107**, 105-113.

Morisawa, M., Okuno, M., Suzuki, K., Morisawa, S. and Ishida, K. (1983). Initiation of sperm motility in teleosts. *Journal of Submicroscopic Cytology* **15**, 61-65.

Morisawa, M., Suzuki, K., Shimizu, H., Morisawa, S. and Yasuda, K. (1983). Effects of Osmolality and potassium on motility of spermatozoa from freshwater cyprinid fishes. *Journal of Experimental Biology* **107**, 95-103.

Morris, G.J. and Watson, P.F. (1984). Cold shock injury - a comprehensive bibliography. *Cryoletters* **5**, 352-372.

Mounib, M.S. (1967). Metabolism of pyruvate, acetate and glyoxylate by fish sperm. *Comparative Biochemistry and Physiology* **20**, 987-992.

Mounib, M.S. (1978). Cryogenic preservation of fish and mammalian spermatozoa. *Journal of Reproduction and Fertility* **53**, 13-18.

Mounib, M.S., Hwang, P.C. and Idler, D.R. (1968). Cryogenic preservation of Atlantic Cod (*Gadus morhua*) sperm. *Journal of the Fisheries Research Board of Canada* **25**, 2623-2632.

Munkittrick, K.R. and Moccia, R.D. (1984). Advances in the cryopreservation of salmonid semen and suitability for a production-scale artificial fertilization program. *Theriogenology* **21**, 645-659.

Munoz-Guerra, S., Azorín, F., Casas, M.T., Marcet, X., Maristany, M.A., Roca, J. and Subirana, J.A. (1982). Structural organization of sperm chromatin from the fish *Carassius auratus*. *Experimental Cell Research* **137**, 47-53.

Nash, T. (1966). Chemical constitution and physical properties of compounds able to protect living cells against damage due to freezing and thawing. In *Cryobiology* (ed. H.T. Meryman), pp 179-211.

Academic Press, New York.

Nelson, G.J. (1969). Gill arches and the phylogeny of fishes, with notes on the classification of vertebrates. *Bulletin of the American Museum of Natural History* **141**, 475-552.

Nelson, G.J. (1972). Observations on the gut of the Osteoglossomorpha. *Copeia* **1972**, 325-329.

Nelson, G.J. (1973). Relatioships of clupeomorphs, with remarks on the structure of the lower jaw in fishes. In *Interrelationships of Fishes* (eds P.H. Greenwood, R.S. Miles and C. Patterson). *Journal of the Linnean Society of London Zoology* **53** (suppl. 1), pp. 333-350. Academic Press, New York.

Nelson, J.S. (1984). *Fishes of the World.* 2nd edition, John Wiley and Sons, New York.

Nicander, L. (1968). Gametogenesis and the ultrastructure of germ cells in vertebrates. *Proceedings of the 6th International Congress on Artificial Reproduction in Animals, Paris* **1**, 89-107.

Nicander, L. (1970). Comparative studies on the fine structure of vertebrate spermatozoa. In *Comparative Spermatology* (ed. B. Baccetti), pp. 47-55. Academic Press, New York.

Nicander, L. and Sjödén, L. (1968). The acrosomal complex and the acrosomal reaction in spermatozoa of the river lamprey. *Scandinavian Society for Electron Microscopy, Journal of Ultrastructure Research* **25**, 167-168.

Nicander, L. and Sjödén, L. (1971). An electron microscopical study of the acrosomal complex and its role in fertilization in the river lamprey, *Lampetra fluviatilis*. *Journal of Sumbmicroscopic Cytology* **3**, 309-317.

Nielsen, J.G., Jespersen, Å. and Munk, O. (1968). Spermatophores in Ophidioidea (Pisces, Percomorphi). *Galathea Report* **9**, 239-254.

Norman, J.R. (1937). *Illustrated Guide to the Fish Gallery. British Museum* (*Natural History*). Trustees of the British Museum, London.

Northcutt, R.G. (1986). Lungfish neural characters and their bearing on sarcopterygian phylogeny. *Journal of Morphology Supplement* **1**, 277-297.

Nuccitelli, R. (1980). The electrical changes accompanying fertilization and cortical vesicle secretion in the Medaka egg. *Developmental Biology* **76**, 483-498.

Nybelin, O. (1973). Comments on the caudal skeleton of actinopterygians. In *Interrelationships of Fishes* (eds P.H. Greenwood, R.S. Miles and C. Patterson), pp. 369-372. *Journal of the Linnean Society of London* Zoology 53 (suppl 1). Academic Press, New York.

Ohta, T. and Iwamatsu, T. (1983). Electron microscopic observations on sperm entry into eggs of the rose bitterling, *Rhodeus ocellatus*. *Journal of Experimental Zoology* **227**, 109-119.

Ohta, T. (1985a). Electron microscopic observations on sperm entry into eggs of the bitterling during cross fertilization.*Journal of Experimental Zoology* **233**, 291-300.

Ohta, T. (1985b). Electron microscopic observations on sperm entry and pronuclear formation in naked eggs of the rose bitterling in polyspermic fertilization*Journal of Experimental Zoology* **234**, 273-281.

Ohta, T. and Iwamatsu, T. (1983). Electron microscopic observations on sperm entry into eggs of the Rose Bitterling, *Rhodeus ocellatus*. *Journal of Experimental Zoology* **227**, 109-119.

Okada, S., Ishikawa, Y. and Kimura, G. (1956). On the viability of the sperm and the egg left in the dead body of Dog-Salmon *Oncorhynchus keta*. (Walbaum). *Scientific Reports of the Hokkaido Fish Hatchery* **11**, 7-18.

Okiyama, M. (1984). Myctophiformes: Relationships. In *Ontogeny and Systematics of Fishes.* Special Publication Number 1. American Society of Icthyologists and Herpetologists. pp. 254-259.

Omoto, C.K. and Brokaw, C.J. (1982). Structure and behaviour of the sperm terminal filament. *Journal of Cell Science* **58**, 385- 409.

Onitake, K. and Iwamatsu, T. (1986). Early response of the egg to sperm stimulation in the Medaka, *Oryzias latipes. Zoological Science* (Tokyo) **3**, 1044.

Ott, A.G. (1975). *Cryopreservation of Pacific Salmon and Steelhead Trout sperm.* Ph. D. Thesis. Oregon State University, Corvallis. (*Fide* Erdahl *et al.*, 1984).

Panchen, A.L. and Smithson, T.R. (1988). The relationships of the earliest tetrapods. In *The Phylogeny and Classification of the Tetrapods, Volume 1: Amphibians, Reptiles, Birds* (ed M.J. Benton). Systematics Association Special Volume No. 35A, pp. 1-32. Clarendon Press, Oxford.

Parenti, L.R. (1981). A phylogenetic and biogeographic analysis of cyprinodontiform fishes (Teleostei, Atherinomorpha). *Bulletin of the American Museum of Natural History* **168**, 335-557.

Paterson, H.E.H. (1982). Perspective on speciation by reinforcement. *South African Journal of Science* **78**, 53-57.

Patterson, C. (1964). A review of Mesozoic Acanthopterygian fishes, with special reference to those of the English Chalk. *Philosophical Transactions of the Royal Society London*. Ser. B **247**, 213-482.

Patterson, C. (1973). Interrelationships of holosteans. In *Interrelationships of Fishes* (eds. P.H. Greenwood, R.S. Miles and C. Patterson), pp. 233-306. *Journal of the Linnean Society of London* Zoology 53 (suppl 1). Academic Press, New York.

Patterson, C. (1977a). Cartilage bones, dermal bones and membrane bones, or the exoskeleton versus the endoskeleton. In *Problems in Vertebrate Evolution*. (eds S.M. Andrews, R.S. Miles and A.D. Walker), pp. 77-121. Academic Press, London.

Patterson, C. (1977b). The contribution of paleontology to teleostean phylogeny. In *Major Patterns in Vertebrate Evolution*. (eds M.K. Hecht, P.C. Goody and B.M. Hecht), pp. 579-643. Plenum, New York.

Patterson, C. (1980). Origin of tetrapods: historical introduction to the problem. In *The Terrestrial Environment and the Origin of Land Vertebrates. Systematics Association Special Publication* **15** (ed A.L. Panchen), pp. 159-175. Academic Press, London and New York.

Patterson, C. (1982). Morphology and interrelationships of primtive actinopterygian fishes. *American Zoologist* **22**, 241-259.

Patterson, C. and Rosen, D.E. (1977). Review of the ichthyodectiform and other Mesozoic teleost fishes and the theory and practice of classifying fossils. *Bulletin of the American Museum of Natural History* **158**, 83-172.

Pavlovici, I and Vlad, C. (1976). Some data on the preservation of carp (*Cyprinus carpio* L.) seminal material by freezing. *Rev. Cresterea Anim.* **4**, 45-48 (in Russian). *Fide* Withler, 1982.

Peterson, K.H. (1989). Egg reaction to sperm: a proposed mechanism. *Evolutionary Theory* **8**, 397-401.

Picheral, B. (1967). Structure et organization du spermatozoïde de Pleurodeles waltlii Michah. (Amphibien Urodèle. *Archives de Biologie*, Paris **78**, 193-221.

Pictet, C. (1891). Recherches sur la spermatogénèse chez quelques invertébrés de Miditerrané. *Mitteilungen aus der Zoologischen Station zu Neapel* **10**, 75-152.

Piironen, J. (1985). Variation in the properties of milt from the Finnish Landlocked Salmon (*Salmo salar* m. *sebago* Girard) during a spawning season. *Aquaculture* **48**, 337-350.

Piironen, J. and Hyvärinen, H. (1983a). Cryopreservation of spermatozoa of the Whitefish *Coregonus muksun* Pallas. *Journal of Fish Biology* **22**, 159-164.

Piironen, J. and Hyvärinen, H. (1983b). Composition of the milt of some teleost fishes. *Journal of Fish Biology* **22**, 351-361.

Pladellorens, M. and Subirana, J.A. (1975). Spermiogenesis in the sea cucumber *Holothuria tubulosa*. *Journal of Ultrastructure Research* **52**, 235-242.

Poirier, G.R. and Nicholson, N. (1982). Fine structure of the testicular spermatozoa from the Channel Catfish, *Ictalurus punctatus*. *Journal of Ultrastructure Research* **80**, 104-110.

Poirier, G.R. and Spink, G.C. (1971). The ultrastructure of testicular spermatozoa in two species of *Rana*. *Journal of Ultrastructure Research* **36**, 455-465.

Polge, C., Wilmut, I. and Rowson, L.E.A. (1974). The low temperature preservation of cow, sheep and pig embryos. *Cryobiology* **11**, 560.

Porte, A. and Follenius, E. (1960). La spermiogénèse chez *Lebistes reticulatus*. Étude au microscope électronique. *Bulletin de la Société Zoologique de France* **85**, 82-88.

Pullin, R.V.S. (1972). The storage of Plaice (*Pleuronectes platessa*) sperm at low temperatures. *Aquaculture* **1**, 279-283.

Purkerson, M.L., Jarvis, J.U.M., Luse, S.A. and Dempsey, E.W. (1974). X-ray analysis coupled with scanning and transmission electron microscopic observations of spermatozoa of the African lungfish, *Protopterus aethiopicus*. *Journal of Zoology* (London) **172**, 1-12.

Pusey, B.J. (1983). The Shannon Mud Minnow.

Journal of the Australia New Guinea Fishes Association. **1**, 9-11.

Pusey, B.J. and Stewart, T. (1989). Internal fertilization in *Lepidogalaxias salamandroides* Mees (Pisces: Lepidogalaxiidae). *Zoological Journal of the Linnean Society* **97**, 69-79.

Pushkar, N.S., Itkin, Y.A., Gordiyen, E.A. and Bronstei, V.L. (1980). Osmotic lysis as a damaging factor during low-temperature preservation of cell suspensions. *Cryobiology* **17**, 403-409.

Rall, W.F. and Fahy, G.M. (1985). Ice-free cryopreservation of mouse embryos at -196° C by vitrification. *Nature* **313**, 573-575.

Rall, W.F., Czlonkowska, M., Barton, S.C. and Polge, C. (1984). Cryopreservation of Day-4 mouse embryos by methanol. *Journal of Reproduction and Fertility* **70**, 293-300.

Rammler, D.H (1967). The effect of DMSO on several enzyme systems. *Annals of the New York Academy of Sciences* **141**, 291-299.

Rana, K. and McAndrew, J. (1989). The viability of cryopreserved Tilapia spermatozoa. *Aquaculture* **76**, 335-346.

Rapatz, G. (1973). Cryoprotective effect of methanol during cooling of frog hearts. *Cryobiology* **10**, 181-184.

Refstie, T., Stoss, J. and Donaldson, E.M. (1982). Production of all female Coho Salmon (*Oncorhynchus keta*) by diploid gynogenesis using irradiated sperm and cold shock. *Aquaculture* **29**, 67-82.

Reger, J.F. (1971). A fine structure study on spermiogenesis in the ectoproct, *Bugula* sp. *Journal of Submicroscopic Cytology* **3**, 193-200.

Reid, D.S. and Rall, W.F. (1986). DSC and cryomicroscope studies of the behaviour of vitrification solutions at low temperatures. *Cryobiology* **23**, 560-561.

Renard, J.-P. (1986). Cryopreservation of embryos. *Cryobiology* **23**, 547.

Renard, J.P. and Babinet, C. (1984). High survival of mouse embryos after rapid freezing and thawing in side plastic straws with 1,2-propanediol as cryoprotectant. *Journal of Experimental Zoology* **230**, 443-448.

Renard, J.-P., Nguyen, B.-X. and Garnier, V. (1984). Two-step freezing of two-cell rabbit embryos after partial dehydration at room temperature. *Journal of Reproduction and Fertility* **71**, 573-580.

Retzius, G. (1904). Zur kenntnis der spermien der Envertebraten. I. *Biologische Untersuchungen von G. Retzius* N.F. **11**, 1-32.

Retzius, G. (1905). Die Spermien der Leptokardier, Teleostier und Ganoiden. *Biologische Untersuchungen von G. Retzius* NF **12**, 103- 115.

Richards, W.J. and Leis, J.M. (1984). Labroidei: development and relationships. In *Ontogeny and Systematics of Fishes.* Special Publication Number 1. American Society of Ichthyologists and Herpetologists, pp. 542-547.

Rieger, R.M. (1976). Monociliated epidermal cells in Gastrotricha: Significance for concepts of early metazoan evolution. *Zeitschrift fur Zoologische Systematik und Evolutionsforschung.* **14**, 198-226.

Riehl, R. and Götting, K.J. (1974). Zur Struktur und Vorkommen der Mikropyle an Eizellen und Eiern von Knochenfischen. *Archiv für Hydrobiologie* **74**, 393-402.

Roberts, T.R. (1973). Interrelationships of ostariophysans. In *Interrelationships of Fishes* (eds. P.H. Greenwood, R.S. Miles and C. Patterson), pp. 373-395. *Journal of the Linnean Society of London* Zoology 53 (suppl. 1). Academic Press, New York.

Romer, A.S. and Parsons, T.S. (1977). *The Vertebrate Body.* W.B. Saunders Company, Philadelphia.

Rosati, F. and De Santis, R. (1978). Studies on fertilization in the ascidians. I. Self-sterility and specific recognition between gametes of *Ciona intestinalis. Experimental Cell Research* **112**, 111-119.

Rosati, F. and De Santis, R. (1980). Role of the surface carbohydrates in sperm-egg interaction in *Ciona intestinalis. Nature* **283**, 762-764.

Rosati, F., Pinto, M.R. and Casazza, G. (1985). The acrosomal region of the spermatozoon of *Ciona intestinalis*: its relationships with the binding to the vitelline coat of the egg. *Gamete Research* **11**, 379-389.

Rosen, D.E. (1964). The relationships and taxonomic position of the halfbeaks, killifishes, silversides, and their relatives. *Bulletin of the American Museum of Natural History* **127**, 219-267.

Rosen, D.E. (1973). Interrelationships of higher euteleostean fishes. In *Interrelationships of Fishes* (eds. P.H. Greenwood, R.S. Miles and C. Patterson). *Journal of the Linnean Society of London* Zoology **53** (suppl. 1), pp. 397-513. Academic Press, New York.

Rosen, D.E. (1974). Phylogeny and zoogeography of salmoniform fishes and relationships of *Lepidogalaxias salamandroides*. *Bulletin of the American Museum of Natural History* **153**, 265-326.

Rosen, D.E. (1982). Teleostean interrelationships, morphological function and evolutionary inference. *American Zoologist* **22**, 261-273.

Rosen, D.E. (1984). Zeiforms as primitive plectognath fishes. *American Museum Novitates* **2782**, 1-45.

Rosen, D.E. and Greenwood, P.H. (1970). Origin of the Weberian apparatus and the relationships of the ostariophysan and gonorynchiform fishes. *American Museum Novitates* **2428**, 1-25.

Rosen, D.E. and Parenti, L.R. (1981). Relationships of *Oryzias*, and the groups of atherinomorph fishes. *American Museum Novitates* **2719**, 1-25.

Rosen, D.E., Forey, P.L., Gardiner, B.G. and Patterson, C. (1981). Lung fishes, tetrapods, paleontology, and plesiomorphy. *Bulletin of the American Museum of Natural History* **167**(4), 159-276.

Rosenblatt, R.H. (1984). Blennioidei: Introduction. In *Ontogeny and Systematics of Fishes*. Special Publication Number 1. American Society of Ichthyologists and Herpetologists, pp. 551-552.

Rosenthal, H., Alderdice, D.F. and Velsen, F.P.J. (1978). Cross-fertilization experiments using Pacific Herring eggs and cryopreserved Baltic Herring sperm. *Fisheries and Marine Services Technical Report* **844**, i-9.

Rouse, G.W. and Jamieson, B.G.M. (1987). An ultrastructural study of the spermatozoa of the polychaetes *Eurythoe complanta* (Amphinomidae), *Clymenella* sp. and *Micromaldane* sp. (Maldanidae), with definition of sperm types in relation to reproductive biology. *Journal of Submicroscopic Cytology* **19**, 573-584.

Rowe, F.W.E., Anderson, D.T. and Healy, J.M. (1990). Concentricycloidea. In *Reproduction of Invertebrates*. (eds J.S. and V. Pearse) Academic Press, New York. (In press).

Rupple, D. (1984). Gobioidei: development. In *Ontogeny and Systematics of Fishes*. Special Publication Number 1. American Society of Ichthyologists and Herpetologists, pp. 582-587.

Russo, J. and Pisanó, A. (1973). Some ultrastructural characteristics of *Platypoecilus maculatus* spermatogenesis. *Bollettino di Zoologia* **40**, 201-207.

Sakai, Y.T. (1976). Spermiogenesis of the teleost, *Oryzias latipes*, with special reference to the formation of the flagellar membrane. *Development, Growth and Differentiation* **18**, 1-13.

Sanchez-Rodriguez M. and Billard, R. (1977). Conservation de la motilite et du pouvoir fecondant du sperme de truite are-en-ciel maintenu a des temperatures voisines de 0 C. *Bulletin Francais de Pisciculture* **265**, 143-152.

Sawada, H., Yokosawa, H., Hoshi, M. and Ishii, S. (1982). Evidence for acrosin-like enzyme in sperm extract and its involvement in fertilization of the ascidian, *Halocynthia roretzi*. *Gamete Research* **5**, 291-301.

Sawada, H., Yokosawa, H., Someno, T., Saino, T. and Ishii, S. (1984). Evidence for the participation of two sperm proteases, spermosin and acrosin in fertilization of the ascidian, *Halocynthia roretzi*: inhibitory effects of leupeptin analogs on enzyme activity and fertilization. *Developmental Biology* **105**, 246-249.

Sawada, N. (1973). Electron microscope studies on gametogenesis in *Lingula unguis*. *Zoological Magazine* **82**, 178-188.

Schabtach, E. and Ursprung, H. (1965). The fine structure of the sperm of a tunicate, *Ascidia nigra*. *Journal of Experimental Zoology* **159**, 357-366.

Schaeffer, B. (1973). Interrelationships of chondrosteans. In *Interrelationships of Fishes* (eds. P.H. Greenwood, R.S. Miles and C. Patterson). *Journal of the Linnean Society of London* Zoology 53 (suppl. 1), pp. 207-226. Academic Press, New York.

Schlenk, W. and Kahmann, H. (1938). Die chemische Zusamennsetzung des Spermaliquors und ihre physiologische Bedeutung. Untersuchung am Forellensperma. *Biochemische Zeitchrift* **295**, 283-301.

Schmehl, M.K., Graham, E.F. (1986). Changes in elemental composition of fertilized and unfertilized Northern Pike (*Esox lucius*) eggs incubated in buffer with and without dimethyl sulfoxide: An X-ray microanalysis study. *Gamete Research* **14**, 91-106.

Schmidt, H. and Zissler, D. (1979). The sperm of the Anthozoa and their phylogenetic significance. *Zoologica (Stuttgart)* **44**, 1-98.

Schneider, U. (1986). Cryobiological principles of embryo freezing. *Journal of In Vitro Fertilization and Embryo Transfer* **3**, 3-9.

Schneider, U. and Mazur, P. (1984). Osmotic consequences of cryoprotectant permeability and its relation to the survival of frozen-thawed embryos. *Theriogenology* **21**, 68-79.

Schultze, H.-P. (1987). Dipnoans as sarcopterygians. In *The Biology and Evolution of Lungfishes* (eds W.E. Bemis, W.W. Burggren and N.E. Kemp) *Journal of Morphology Supplement* **1** (1986), pp. 39-74. Alan R. Liss, New York.

Scott, A.P. and Baynes, S.M. (1980). A review of the biology, handling and storage of salmonid spermatozoa. *Journal of Fish Biology* **17**, 707-739.

Selman, K. and Wallace, R.B. (1986). Gametogenesis in *Fundulus heteroclitus*. *American Zoologist* **26**, 173-192.

Silén, L. (1954). Developmental Biology of Phoronidea of the Gullmar Fiord area (West coast of Sweden). *Acta Zoologica (Stockholm)* **35**, 215-257.

Silén, L. (1966). On the fertilization problem in the gymnolaematous Bryozoa. *Ophelia* **3**, 113-140.

Silveira, H., Rodrigues, P. and Azevedo, C. (1990). Fine structure of the spermtogenesis of *Blennius pholis* (Pisces, Blenniidae). *Journal of Submicroscopic Cytology and Pathology* **22**, 103-108.

Sin, S. and Hirano, R. (1972). Study on cryopreservation of fish sperm. *Proceedings of the Spring Meeting of the Japanese Society of Scientific Fisheries* **1972**, 115. (*Fide* Chao *et al.*, 1986).

Smith, C.L., Rand, C.S., Schaeffer, B. and Atz, J.W. (1975). *Latimeria*, the living coelacanth, is ovoviviparous. *Science*, **190**, 1105-1106.

Smith, D.G. (1984). Elopiformes, Notacanthiformes and Anguilliformes: Relationships. In *Ontogeny and Systematics of Fishes*. Special Publication Number 1. American Society of Ichthyologists and Herpetologists, p. 94-102.

Smith, M.M. (1978). Enamel in the oral teeth of *Latimeria chalumnae* (Pisces: Actinistia): a scanning electron microscopic study. *Journal of Zoology, London* **185**, 355-369.

Springate, J.R.C., Bromage, N.R., Elliott, J.A.K. and Hudson, D.L. (1984). The timing of ovulation and stripping and their effects on the rates of fertilization and survival to eying, hatch and swim-up in the Rainbow Trout (*Salmo gairdneri* R.). *Aquaculture* **43**, 313-322.

Stanley, H.P. (1964). Fine structure and development of the spermatozoan midpiece in the elasmobranch fish *Squalus suckleyi*. *Journal of Cell Biology* **23**, 88A.

Stanley, H.P. (1965a). Fine structure of the tail flagella in the spermatozoa of two chondrichthyan fishes, *Squalus suckleyi* and *Hydrolagus colliei*. *Anatomical Record* **151**, 419.

Stanley, H.P. (1965b). Electron microscopic observations on the biflagellate spermatids of the teleost fish *Porichthys notatus*. *Anatomical Record* **151**, 477.

Stanley, H.P. (1966). A fine structural study of spermiogenesis in the teleost fish Oligocottus maculosus. *Anatomical Record* **154**, 426-427.

Stanley, H.P. (1967). The fine structure of spermatozoa in the lamprey *Lampetra planeri*. *Journal of Ultrastructure Research* **19**, 84-99.

Stanley, H.P. (1969). An electron microscope study of spermiogenesis in the teleost fish *Oligocottus maculosus*. *Journal of Ultrastructure Research* **27**, 230-243.

Stanley, H.P. (1970). Differential development of homologous structures accessory to the axoneme in sperm of several vertebrate types. *Journal of Cell Biology* **47**, 201A.

Stanley, H.P. (1971). Fine structure of spermiogenesis in the Elasmobranch fish *Squalus suckleyi* II. Late stages of differentiation and structure of the mature spermatozoon. *Journal of Ultrastructure Research* **36**, 103-118.

Stanley, H.P. (1983). The fine structure of spermatozoa of *Hydrolagus colliei* (Chondrichthyes, Holocephali). *Journal of Ultrastructure Research* **83**, 184-194.

Stehr, C.M. and Hawkes, J.W. (1979). The comparative ultrastructure of the egg membrane and associated pore structure in the Starry Flounder, *Plati-*

chthys stellatus (Pallas), and Pink Salmon, *Oncorhynchus gorbuscha* (Walbaum). *Cell and Tissue Research* **202**, 347-356.

Stein, H. (1979). Cryoprotective agents for spermatozoa of Brown Trout. *Berliner und Münchener Tierarztliche Wochenschrift* **92**, 420-421.

Stein, H. (1981). Licht- und elektronenoptische Untersuchungen an den Spermatozoen verschiedener Süsswasserknochenfische (Teleostei). *Zeitschrift für Angewandte Zoologie* **68**, 183-198.

Stein, H. and Bayrle, H. (1978). Cryopreservation of the sperm of some freshwater teleosts. *Annales de Biologie Animale, Biochimie, Biophysique* **18**, 1073-1076.

Stein, H. and Bayrle, H. (1985). Gameten- und Embryokonservierung bei Salmoniden. *Bayerisches Landwirtschaftiches Jahrbuch* **62**, 236-245.

Sterba, G. (1967). *Freshwater Fishes of the World.* Studio Vista, London.

Steyn, G.J. and Van Vuren, J.H.J. (1987). The fertilizing capacity of cryopreserved Sharptooth Catfish (*Clarias gariepinus*) sperm. *Aquaculture* **63**, 187-193.

Steyn, G.J., Van Vuren, J.H.J., Schoonbee, H.J. and Chao, N-H. (1985). Preliminary investigations of the cryopreservation of *Clarias gariepinus* (Clariidae Pisces) sperm. *Water South Africa* (Pretoria) **11**, 15-18.

Stiassny, M.L.J. (1980). The anatomy and phylogeny of two genera of African cichlid fishes. Ph.D. Thesis. University of London. (*Fide* Lauder and Liem, 1983).

Storch, V. and Welsch, U. (1976). Electron Microscopical and Enzyme Histochemical Investigations on Lophophore and Tentacles of *Lingula unguis* L. (Brachiopoda). *Zoologische Jahrbucher Abteilung fur Anatomie und Ontogenie der Tiere*. **96**, 225- 237.

Stoss, J., and Donaldson, E.M. (1983). Studies on cryopreservation of eggs from Rainbow Trout (*Salmo gairdneri*) and Coho Salmon (*Oncorhynchus kisutch*). *Aquaculture* **31**, 51-65.

Stoss, J. and Holtz, W. (1981a). Cryopreservation of Rainbow Trout (*Salmo gairdneri*) sperm. I. Effect of thawing solution, sperm density and interval between thawing and insemination. *Aquaculture* **22**, 97-104.

Stoss, J. and Holtz, W. (1981b). Cryopreservation of Rainbow Trout (*Salmo gairdneri*) sperm. II. Effect of pH and presence of a buffer in the diluent. *Aquaculture* **25**, 217-222.

Stoss, J. and Holtz, W. (1983a). Successful storage of chilled Rainbow Trout (*Salmo gairdneri*) spermatozoa for up to 34 days. *Aquaculture* **31**, 269-274.

Stoss, J. and Holtz, W. (1983b). Cryopreservation of Rainbow Trout (*Salmo gairdneri*) sperm. III. Effect of proteins in the diluent, sperm from different males and interval between sperm collection and freezing. *Aquaculture* **31**, 275-282.

Stoss J, and Holtz W. (1983c). Cryopreservation of Rainbow Trout (*Salmo gairdneri*) sperm. IV. The effect of DMSO concentration and equilibration time on sperm survival, sucrose and KCl as extender components and the osmolality of the thawing solution. *Aquaculture* **32**, 321-330.

Stoss, J. and Refstie, T. (1983). Short-term storage and cryopreservation of milt from Atlantic Salmon and Sea Trout. *Aquaculture* **30**, 229-236.

Stoss, J., Büyükhatipoglu, S., and Holtz, W. (1978). Short-term and cryopreservation of Rainbow Trout (*Salmo gairdneri* Richardson) sperm. *Annales de Biologie Animale, Biochimie, Biophysique* **18**, 1077-1082.

Summers, R.G., Hylander, B.L., Colwin, L.H. and Colwin, A.L. (1975). The functional anatomy of the echinoderm spermatozoon and its interaction with the egg at fertilization. *American Zoologist* **15**, 523-551.

Swofford, D.L. (1984). *Phylogenetic analysis using parsimony.* Version 2.2. User's Manual. D.L. Swofford, Illinois Natural History Survey, 607 East Peabody Drive, Champaign, Illinois 61820.

Szarski, H. (1977). Sarcopterygii and the origin of tetrapods. In *Major Patterns in Vertebrate Evolution.* (eds M.K. Hecht, P.C. Goody and B.M. Hecht), pp. 517-643540. Plenum, New York.

Takahashi, T., Hammett, M.F., Cho, M.S. (1985). Multifaceted freezing-injury in human polymorphonuclear cells at high subfreezing temperatures. *Cryobiology* **22**, 215-236.

Tanaka, S., Hara, M. and Mizue, K. (1978). Studies on sharks. Part 13. Electron microscopic study on spermatogenesis of the squalen shark *Centrophorus atromarginatus.* *Japanese Journal of*

Icthyology **25**, 173-180.

Taverne, L. (1979). Ostéologie, phylogénèse et systématique des téléostéens fossiles et actuels du super-ordre des Ostéoglossomorphes.3. *Academie Royale de Belgique Memoirs de la Classe des Sciences Collection in Octavo* **43**, 1-168.

Terner, C. (1986). Evaluation of salmonid sperm motility for cryopreservation. *The Progressive Fish Culturist* **48**, 230-232.

Thiaw, O.T., Mattei, X., Romand, R. and Marchand, B. (1986). Reinvestigation of spermatic flagella structure: the teleostean Cyprinodontidae. *Journal of Ultrastructure and Molecular Structure Research* **97**, 109-118.

Thompson, H. (1948). *Pelagic tunicates of Australia.* C.S.I.R.O. Australia.

Thresher, R.E. (1984). *Reproduction in Reef Fishes.* TFH Publications, Brookvale-NSW.

Todd, P.R. (1976). Ultrastructure of the spermatozoa and spermiogenesis in New Zealand freshwater eels (Angullidae). *Cell and Tissue Research* **171**, 221-232.

Toner, M. and Cravalho EG (1986). Solution of water transport equation with constant rate of temperature change. *Cryobiology* **23**, 545.

Truscott, B. and Idler, D.R. (1969). An improved extender for freezing Atlantic Salmon spermatozoa. *Journal of the Fisheries Research Board of Canada* **26**, 3254-3258.

Tsvetkov, T., Tsonev, L. and Minkov, I. (1986). A quantitive evaluation of the extent of inner mitochondrial membrane destruction after freezing-thawing based on functional studies. *Cryobiology* **23**, 433-439.

Tuzet, O. and Millot, J. (1959). La spermatogenèse de *Latimeria chalumnae* Smith (Crossoptérygien coelacanthidé). *Annales des Sciences Naturelles Zoologie* 12 Ser **1**, 61- 69.

Tuzet, O., Bogoraze, D. and Lafargue, F. (1972). Recherches ultrastructurales sur la spermiogenese de *Diplosoma listerianum* (Milne-Edwards, 1841) et *Lissoclinum pseudoleptoclinum* (Von Drasche, 1883) (Ascidies Composees, Aplousobranches). *Annales des Sciences Naturelles, Zoologie*, Paris **14**, 177-190.

Tuzet, O., Bogoraze, D. and Lafargue, F. (1974). La spermatogenese de *Polysyncraton lacazei*, 1872 et *Trididemnum cereum* Giard, 1872 (Ascidies composees, Aplousobranches). *Bulletin de Biolgie de France et Belgique* **108**, 151-167.

Tyler, J.C. (1968). A monograph on plectognath fishes of the superfamily Triacanthoidea. *Monographs of the Academy of Natural Sciences Philadelphia* **16**, 1-364.

Urho, L., Hudd, R., and Hildén, M. (1984). Kalojen sittiöden liikkumisaika pH:n funktiona. *Memoranda Societatis pro Fauna et Flora Fennica* **60**, 41-42.

Ursprung, H. and Schabtach, E. (1965). Fertilization in tunicates: loss of the paternal mitochondrion prior to sperm entry. *Journal of Experimental Zoology* **159**, 379-384.

Valentine, J.W. (1973). Coelomate superphyla. *Systematic Zoology* **22**, 97-102.

Valentine, J.W. (1977). General patterns of metazoan evolution. In *Patterns of Evolution as Ilustrated by the Fossil Record* (ed A. Hallam) , pp. 27-57). Elsevier, Amsterdam.

van den Berg, L. and Rose, D. (1959). Effect of freezing on the pH and composition of sodium and potassium phosphate solutions: the reciprocal system KH_2PO_4-Na_2HPO_4-H_2O. *Archives of Biochemistry and Biophysics* **81**, 319-329.

van den Berg, L. and Soliman, F.S. (1969). Effect of glycerol and dimethyl sulfoxide on changes in composition and pH of buffer salt solutions during freezing. *Cryobiology* **6**, 93-97.

van der Horst, G. (1976). Aspects of the reproductive biology of *Liza dumerili* (Steindachner, 1869) (Teleostei: Mugilidae) with special reference to sperm. Ph. D. thesis: University of Port Elizabeth. (*Fide* van der Horst, G. and Cross, R.H.M. (1978).

van der Horst, G. and Cross, R.H.M. (1978). The structure of the spermatozoon of *Liza dumerili* (Teleostei) with notes on spermiogenesis. *Zoologia Africana* **13**, 245-258.

Van der Horst, G. Dott, H.M. and Foster, G.C. (1980). Studies on the motility and cryopreservation of rainbow trout (*Salmo gairdneri*) spermatozoa. *South African Journal of Zoology* **15**, 275-279.

Villa, L. (1975). An ultrastructural investigation of normal and irradiated spermatozoa of a tunicate, *Ascidia malaca. Bollettino di Zoologia* **42**, 95-98.

Villa, L. (1977). An ultrastructural investigation of

preliminaries in fertilization of *Ascidia malaca* (Urochordata). *Acta Embryologiae Experimentalis* **2**, 179-193.

Villa, L. (1981). An electron microscope study of spermiogenesis and spermatozoa of *Mogula impura* and *Styela plicata* (Ascidiacea, Tunicata). *Acta Embrologiae Morphologia Experimentalis* **2**, 69-85.

Villa, L. and Tripepi, S. (1981). An ultrastructural investigation on spermiogenesis and spermatozoa of *Microcosmus* sp. (Ascidiacea). Abstracts of Papers Presented at the 13th Congress of the Italian Society for Electron Microscopy, 130.

Villa, L. and Tripepi, S. (1982). An electron microscope study of spermatogenesis and spermatozoa of *Microcosmos sabatieri* (Ascidiacea, Tunicata). *Acta Embrologiae Morphologia Experimentalis* **3**, 201-215.

Villa, L. and Tripepi, S. (1983). An electron microscope study of spermatogenesis and spermatozoa of *Ascidia malaca*, *Ascidiella aspersa* and *Phallusia mamillata* (Ascidiacea, Tunicata).*Acta Embrologiae Morphologia Experimentalis* **4**, 157-168.

Walvig, F. (1963). The gonads and the formation of the sexual cells. In *The Biology of Myxine* (eds A. Brodal and R. Fänge), pp. 530-580. Universitetsforlaget, Oslo.

Washington, B.B., Eschmeyer, W.N. and Howe, K.M. (1984a). Scorpaeniformes: relationships. In *Ontogeny and Systematics of Fishes.* Special Publication Number 1. American Society of Ichthyologists and Herpetologists, pp. 438-447.

Washington, B.B., Moser, H.G., Laroche, W.A. and Richards, W.J. (1984b). Scorpaeniformes: development. In *Ontogeny and Systematics of Fishes.* Special Publication Number 1. American Society of Ichthyologists and Herpetologists, pp. 405-428.

Watson, P.F. (1986). Problems in the cryopreservation of mammalian spermatozoa. *Cryobiology* **23**, 547.

Watson, W., Matarese, A.C. and Stevens, E.G. (1984). Trachinoidea: development and relationships. In *Ontogeny and Systematics of Fishes.* Special Publication Number 1. American Society of Ichthyologists and Herpetologists, pp. 554-561.

Webb, P.W. and Brett, J.R. (1972). Respiratory adaptations of prenatal young in the ovary of two species of viviparous seaperch *Rhacochilus vacca* and *Embiotoca lateralis*. *Journal of the Fish Research Board of Canada* **29**, 1525-1542.

Weber, M. and De Beaufort, L.F. (1922). *The Fishes of the Indo-Australian Archipelago*. Brill, Leiden. Vol. 4.

Whiting, H..P. and Bone, Q. (1980). Ciliary cells in the epidermis of the larval Australian dipnoan, *Neoceratodus*. *Zoological Journal of the Linnean Society* **68**, 125-137.

Whittingham, D.G. (1977). *Some factors affecting embryo storage in laboratory animals*. In *The Freezing of Mammalian Embryos*. Ciba Foundation Symposium, Elsevier, Amsterdam, pp 97-108.

Whittingham, D.G., Leibo, S.P. and Mazur, P. (1972). Survival of mouse embryos frozen to -196° and -269° C. *Science* **178**, 411-414.

Wickstead, J.H. (1975). Chordata: Acrania (Cephalochordata). In *Reproduction of marine invertebrates. II. Ectoprocts and lesser coelomates.* (eds A.C. Giese and J.S. Pearse), pp. 283-319. Academic Press, New York.

Wilcox, K.W., Stoss, J. and Donaldson, E.M. (1984). Broken eggs as a cause of infertility in Coho Salmon gametes. *Aquaculture* **40**, 77-87.

Wiley, E.O. (1976). Phylogeny and biogeography of fossil and recent Gars. (Actinopterygii: Lepisosteidae). *University of Kansas Museum of Natural History Miscellaneous Publications* **64**, 1-111.

Wiley, E.O. (1979). Ventral gill arch muscles and the interrelationships of gnathostomes, with a new classification of the Vertebrata. *Zoological Journal of the Linnean Society* **67**, 149-179.

Williams, A. and Rowell, A.J. (1965). Brachiopod anatomy. pp. H6-57. Evolution and phylogeny. pp. H164-199. In *Treatise on Invertebrate Paleontology* (ed. R.C. Moore). Part H. Brachiopoda. Volume 1. The Geological Society of America, Inc. and the University of Kansas Press.

Williams, R.J. (1974). A unified model of plant hardiness based on osmotic stress. *Cryobiology* **11**, 555 .

Winterbottom, R. (1974). The familial phylogeny of the Tetraodontiformes (Acanthopterygii: Pisces) evidenced by their comparative myology. *Smithsonian Contributions to Zoology* **155**, 1-201.

Withler, F.C. (1980). Chilled and cryogenic storage of gametes of Thai Carps and Cat Fishes. *Canadian Technical Report of Fisheries and Aquatic Sciences* **948**, i-15.

Withler, F.C. (1982). Cryopreservation of spermatozoa of some freshwater fishes cultured in South and Southeast Asia. *Aquaculture* **26**, 395-398.

Withler, F.C. and Lim, L.C. (1982). Preliminary observations of chilled and deep-frozen storage of Grouper (*Epinephelus tauvina*) sperm. *Aquaculture* **27**, 389-392.

Wolenski, J.S. and Hart, N.H. (1987). Scanning electron microscope studies of sperm incorporation into the Zebrafish (*Brachydanio*) egg. *Journal of Experimental Zoology* **243**, 259-273.

Wolenski, J.S. and Hart, N.H. (1988). Sperm incorporation independent of fertilization cone formation in the Danio egg. *Development, Growth and Differentiation* **30**, 619-628.

Wood, M.J. and Farrant, J. (1980). Preservation of embryos by two-step freezing. Cryobiology **17**, 178-180.

Woollacott, R.M. (1977a). Localization of aryl sulfatase in a spermatozoan lacking an acrosome. *Journal of Cell Biology* **75**, 163a.

Woolacott, R.M. (1977b). Spermatozoa of *Ciona intestinalis* and analysis of ascidian fertilization. *Journal of Morphology* **152**, 77-88.

Woollacott, R.M. and Porter, M.E. (1975). Biochemical and ultrastructural investigations of the sperm of Ciona intestinalis. *Journal of Cell Biology* **67**, 463a.

Woollacott, R.M. and Zimmer, R. L. (1972). Origin and structure of the brrod chamber in *Bugula neritina* (Bryozoa). *Marine Biology* **16**, 165-170.

Yamamoto T. (1961). Physiology of fertilization in fish eggs. *International Reviews in Cytology* **12**, 361-405.

Yamamoto, T.S. and Kobayashi, W. (1988). Permanent blockage of polyspermy in the flounder, *Paralllichthys olivaceus*, and Alaska Pollack, *Theragra chalcogramma*, eggs. *Zoological Science* (Tokyo) **5**, 1250.

Yamashita, M., Onozato, H., Nakanishi, T. and Nagahama, Y. (1988). Necessity of breakdown of sperm nuclear envelope for male pronucleus formation. *Zoological Science* (Tokyo) **5**, 1250.

Yanagimachi, R. (1957). Studies of fertilization in *Clupea pallasii*. Parts I-III. *Zoological Magazine* (Tokyo) **66**, 218-233.

Yasuzumi, F. (1971). Electron microscope study of the fish spermiogenesis. *Journal of Nara Medical Association* **22**, 343-355.

Yoo, B.Y., Ryan, M.A. and Wiggs, A.J. (1987). Loss of protein from spermatozoa of Atlantic Salmon (*Salmo salar* L.) because of cryopreservation. *Canadian Journal of Zoology* **65**, 9-13.

Young, J.W., Blaber, S.J.M. and Rose, R. (1987). Reproductive biology of three species of midwater fishes associated with the continental slope of eastern Tasmania, Australia. *Marine Biology* **95**, 323-332.

Young, J.Z. (1981). *The Life of Vertebrates*. Third Edition. Clarendon Press, Oxford.

Zell, S.R. (1978). Cryopreservation of gametes and embryos of salmonid fishes. *Annales de Biologie Animale Biochimie Biophysique* **18**, 1089-1099.

Zimmer, R.L. (1973). Morphological and developmental affinities of the lophophorates. In *Living and Fossil Bryozoa* (ed. G.P. Larwood), pp. 593-599. Academic Press, New York.

Zimmer, R.L. and Woollacott, R.M. (1974). Morphological and biochemical modifications of the spermatozoan mitochondria of *Membranipora* (Bryozoa). *Journal of Cell Biology* **63**, 385a.

Zirkin, B.R. (1975). The ultrastructure of nuclear differentiation during spermiogenesis in the salmon. *Journal of Ultrastructure Research* **50**, 174-184.

Zotin, A.I. (1958). The mechanism of hardening of the salmonid egg membrane after fertilization or spontaneous activation. *Journal of Embryology and Experimental Morphology* **6**, 546-568.

Author Index

* Cryopreservation chapters
Bold type indicates illustrations or tables on pages not otherwise referred to

-A-

-B-

-O-

-P-

-R-

-V-

-W-

-Y-

-Z-

Subject Index

Bold type indicates illustrations where these occur on pages not otherwise referred to. Illustrations of the ultrastructure of sperm of specific taxa are, however, omitted but are given in the Taxonomic Index

* Cryopreservation chapters

-A-

-B-

-C-

-N-

-O-

-P-

-R-

-S-

Taxonomic Index

For a listing of the chief accounts of sperm for each taxon, see 'Sperm ultrastructure' in the Subject Index.
Bold type indicates illustrations on pages not otherwise referred to
* Cryopreservation chapters

-A-